A Companion to Social Geography

Wiley-Blackwell Companions to Geography

Wiley-Blackwell Companions to Geography is a blue-chip, comprehensive series covering each major subdiscipline of human geography in detail. Edited and contributed by the disciplines' leading authorities each book provides the most up to date and authoritative syntheses available in its field. The overviews provided in each *Companion* will be an indispensable introduction to the field for students of all levels, while the cutting-edge, critical direction will engage students, teachers, and practitioners alike.

Published

A Companion to Economic Geography
Edited by Eric Sheppard and Trevor J. Barnes

A Companion to Political Geography
Edited by John Agnew, Katharyne Mitchell, and Gerard Toal
(Gearoid O Tuathail)

A Companion to Cultural Geography
Edited by James S. Duncan, Nuala C. Johnson, and Richard H. Schein

A Companion to Tourism
Edited by Alan A. Lew, C. Michael Hall, and Allan M. Williams

A Companion to Feminist Geography
Edited by Lise Nelson and Joni Seager

A Companion to Environmental Geography
Edited by Noel Castree, David Demeritt, Diana Liverman, and Bruce Rhoads

A Companion to Health and Medical Geography
Edited by Tim Brown, Sara McLafferty, and Graham Moon

A Companion to Social Geography
Edited by Vincent J. Del Casino Jr., Mary E. Thomas, Ruth Panelli, and Paul Cloke

Also available:

The New Blackwell Companion to the City
Edited by Gary Bridge and Sophie Watson

The Blackwell Companion to Globalization
Edited by George Ritzer

The Handbook of Geographic Information Science
Edited by John Wilson and Stewart Fotheringham

A Companion to Social Geography

Edited by

Vincent J. Del Casino Jr., Mary E. Thomas,
Paul Cloke, and Ruth Panelli

A John Wiley & Sons, Ltd., Publication

Library of Congress Cataloging-in-Publication Data

A companion to social geography / edited by Vincent J. Del Casino, Jr. ... [et al.].
    p. cm. – (Wiley-Blackwell Companions to Geography)
  Companion v. to: Social geography / Vincent J. Del Casino, Jr. 2009"
  Includes bibliographical references and index.
  ISBN 978-1-4051-8977-4 (hardcover: alk. paper)
  1. Human geography. I. Del Casino, Vincent J. II. Del Casino, Vincent J., Social Geography.
GF41.C5735 2011
304.2–dc22
                          2010041343

A catalogue record for this book is available from the British Library.

ePDF: 978-1-4443-9519-8
epub: 978-1-4443-9520-4
oBook: 978-1-4443-9521-1

Set in 10/12.5 pt Sabon by Toppan Best-set Premedia Limited

01   2011

Contents

Illustrations

Figures

Table

Boxes

Contributors

Gavin Brown is a Lecturer in Human Geography at the University of Leicester, United Kingdom. He is an urban social and cultural geographer specializing in the study of geographies of sexualities; social movement activism; education and youth policy; and urban responses to climate change and peak oil. He has published his work in *Environment & Planning A*; *Emotion, Space and Society*, and *Social and Cultural Geography* amongst other journals. He is co-editor of *Geographies of Sexualities: Theory, Practices and Politics* (2007).

Kath Browne is a Senior Lecturer in the University of Brighton, United Kingdom. Her work spans and intersects the areas of geographies, sexualities, genders, methodologies, and social engagements. In 2007 Kath was awarded the Gill Memorial Award from the Royal Geographical Society/Institute of British Geographers recognizing young researchers who have shown great potential. Kath has written over 40 publications across a range of disciplines using diverse formats, and is the co-author of *Queer Spiritual Spaces: Sexuality and Sacred Spaces* and co-editor of *Geographies of Sexualities: Theory, Practices and Politics*, and *Queer Methods and Methodologies: Queer Theory and Social Science Research*. She continues to work on the Community–University Research project, *Count Me In Too*.

Emilie Cameron is an Assistant Professor in the Department of Geography at Carleton University. Her doctoral research examined the materiality of stories and the ways in which stories order geographies of race, nature, political mobilization, and resource extraction in the Central Canadian Arctic. She is currently investigating the cultural-historical and political-economic dimensions of mineral exploration and mine development in northern Canada.

Paul Chatterton is a writer, researcher, and campaigner. He is currently Reader in Cities and Social Change in the School of Geography at the University of Leeds where he heads up the Cities and Social Justice Research Cluster and is co-founder of the MA in Activism and Social Change. He has written extensively on urban regeneration, youth cultures, self-managed politics, and movements for social and ecological justice. He is a co-editor of the journal *Antipode* and senior editor of the journal *City*. He is a founder member of the Trapese popular education collective who wrote a handbook for *Do It Yourself Politics* with Pluto Press. Paul helped to establish the Common Place Social Centre in Leeds city center, is founder member of a pioneering low impact housing cooperative called Lilac, and is a co-director of a community interest research company called Love it Share it. All his work can be found at www.paulchatterton.com.

Paul Cloke is Professor of Human Geography at the University of Exeter. He has co-produced a series of books on the ontological and epistemological bases of human geography, including *Introducing Human Geographies, Approaching Human Geography, Practising Human Geography and Envisioning Human Geography*. His substantive research interests are in rural social geographies, geographies of homelessness, and geographies of ethical participation, and recent research is being published in two forthcoming books for the Blackwell RGS-IBG Series, *Swept Up Lives* (with Sarah Johnsen and John May) and *Consuming Ethics* (with Clive Barnett, Nick Clarke, and Alice Malpass). His current focus is on how new geographies of ethics can help develop understandings of how people shape their self-identities through participatory devices of care, charity, and consumption. This work has been published in journals such as *Environment and Planning D: Society and Space, Progress in Human Geography, Geoforum, Antipode, Environment and Planning A, Social and Cultural Geographies, International Journal for Urban and Regional Research, Journal of Rural Studies*, and *Cultural Studies*.

David Conradson teaches Human Geography at the University of Canterbury, New Zealand. He is interested in the complex intertwining of emotion, subjectivity, and space, and has recently explored this with respect to places of retreat in southern England. He is curious about ideas of well-being and flourishing and what these might offer social geography. His work been published in journals such as *Environment and Planning A, Social and Cultural Geography, Health and Place, Urban Studies*, and *Mobilities*.

Brad Coombes is a Senior Lecturer in Geography and Environmental Management at the University of Auckland, New Zealand. His research interests focus on Indigenous participation in or resistance to natural resource management, and he has long contributed to the Treaty of Waitangi settlement process in New Zealand. Brad is currently Chair of the Indigenous Peoples, Knowledges, and Rights Commission of the IGU and Joint Director of Te Whare Kura, a thematic research initiative of the University of Auckland on Indigenous culture and development.

Mike Crang is a Reader in Geography at Durham University. He works in cultural geography broadly defined. He has researched the transformations of space and time through electronic technologies, with specific work based around Singapore's "Wired City" initiative and the "digital divide" in UK cities. He completed an ESRC project on "Multi-Speed Cities and the Logistics of Daily Life" and is now working on the notion of a "sentient city" and the politics of locative media. He has published on these issues in journals such as *Urban Studies, Environment and Planning A, Information Communication & Society*, and edited *Virtual Geographies* (1999, with Phil Crang and Jon May).

Joyce Davidson is Associate Professor of Geography at Queen's University, Canada. Her doctoral research on first-hand accounts of agoraphobia at Edinburgh University formed the basis of *Phobic Geographies: The Phenomenology and Spatiality of Identity* (2003). She has since developed a research and teaching program focused around health, embodiment, and different or "disordered" emotions. Organizer of the First and Second Interdisciplinary Conference on Emotional Geographies (Lancaster 2002, Queen's 2006), she is lead and founding editor of the new journal, *Emotion, Space and Society*, and has co-edited *Emotional Geographies* (2005), and *Emotion, Place and Culture* (2009).

Gail Davies is Senior Lecturer in Geography at University College London, United Kingdom. She is interested in the contemporary intersection of the social, the spatial, the biological and the ethical, particularly in the constitution of biological related-ness and difference, and in the relations between different ways of knowing biology. She has published in journals such as *Economy and Society*, *Environment and Planning*, *Geoforum*, *Health and Place*, *Progress in Human Geography*, *Public Understanding of Science*, and *Transactions of the Institute of British Geographers*. She is currently tracing the unfolding spaces of postgenomics and the shifting location of global science through the changing use of mice as model organisms.

Vincent J. Del Casino Jr. is Professor and Chair of Geography at California State University, Long Beach. He has published articles and book chapters related to his interest in social geography, health geography, sexuality and sexual politics, homelessness, HIV/AIDS, and critical social theory. He has also co-published work on the intersections between critical cartography and tourism studies with Stephen P. Hanna, Mary Washington College. This includes pieces that have been published in *Progress in Human Geography*, *Social and Cultural Geography*, and *ACME: An International E-Journal for Critical Geographies* as well as the edited volume *Mapping Tourism*, which was published by the University of Minnesota Press in 2003. In addition, Dr. Del Casino has published work related to his ongoing research on HIV/AIDS related health care and prevention research in both Thailand and Long Beach, California. This includes articles published in the journals *Health and Place* and *The Professional Geographer* as well as book chapters in the recently edited *Population Dynamics and Infectious Diseases in Asia*. He has recently published *Social Geography: A Critical Introduction* for Wiley-Blackwell (2009).

Sarah de Leeuw, a social-historical geographer and creative writer, is an Assistant Professor with the Northern Medical Program at UNBC, the Faculty of Medicine at UBC. Her work, both as a geographer in a faculty of medicine and as a creative writer, engages questions of power, place, and landscape, colonialism, social justice, and marginalization. She is the author of *Unmarked: Landscapes along Highway 16* (2004) and her book, *The Geographies of a Lover*, is a collection of poetry forthcoming (2012) with NeWest Press. Her essays "Quick-quick. Slow. Slow" (2009) and "Columbus burning" (2008) both won CBC Literary Awards for creative non-fiction. Her poetry has appeared in a number of Canadian literary journals, including *Fiddlehead*, *Wascana*, and *The Claremont Review*. Her academic writing, which is broadly concerned with (post)colonial geographies, Indigenous peoples, and the social determinants of health, appears in venues ranging from *The Canadian Family Physician* and *Children's Geographies*, to *The Journal of Native Education* and *The Canadian Geographer*. In 2007/8, prior to moving back to northern British Columbia, she was a Fulbright Fellow with the University of Arizona.

Lorraine Dowler is an Associate Professor of Geography and Women's Studies at Penn State University. Her interests focus in the intersections of gender, nationalism, and war. She is the author of several publications focusing on issues of gender and war in Northern Ireland. Her current research project is a feminist examination of the critical geopolitics of the Cold War, the War on Terror and the New Military. This research project examines how individual women and men were/are viewed as ethical or deviant as their actions were interpreted by way of adaptation or transgression of the national moral landscape. As part of this analysis Dr. Dowler has conducted extensive interviews with women who trained to be astronauts during the NASA Mercury program, contemporary women firefighters and female soldiers returning from Iraq and Afghanistan.

Isabel Dyck is Professor of Geography, Queen Mary, University of London, United Kingdom. Her theoretical interests concern the interconnections between place, gender, and health, with a focus on issues of identity and embodiment. Research projects include investigation of: the home and work experiences of women with chronic illness; international migrant women's health and illness management; the reconstitution of home and family for international migrants; and the home as a site for long-term care. She has published widely from these studies and on qualitative methodology. Books include the co-edited volume, *Geographies of Women's Health* (2001), and *Women, Body, Illness*, with Pamela Moss (2002).

Marcia England is an Assistant Professor in the Department of Geography at Miami University. She is an urban, cultural, and feminist geographer, whose research interests focus in three areas: access to public space; the politics of representations; and the socio-spatial regulation of marginalized persons. Current research centers on the regulation of public space and how it relates to violence against homeless persons. Other interests include geographies of media and popular culture, including horror films and internet pornography, using a feminist lens to understand geographies of the body. Her publications on these topics can be found in journals such as *Gender, Place and Culture*, *Environment and Planning A*, *Space and Polity*, and *Social and Cultural Geography*.

Lieba Faier is Assistant Professor of Geography at the University of California, Los Angeles. Her work brings ethnographic and feminist approaches to understanding the spatial and cultural dynamics of contemporary transnational processes, particularly as these involve people's lives in Japan, the Philippines, and the United States. Her first book, *Intimate Encounters: Filipina Women and the Remaking of Rural Japan* (2009) is an ethnography of cultural encounters that explores how Filipina migrants and Japanese residents in rural Nagano remake meanings of Japanese and Filipino culture and identity through their shared daily lives. She is also involved in a collaborative research project about *matsutake* (pine mushroom) commodity chains across the Pacific Rim.

Carolyn Gallaher is an Associate Professor in the School of International Service. She is a broadly trained human geographer with interests in political and cultural geography. Her research has focused on two main substantive areas: right wing politics in the United States and right wing paramilitarism. She has written two books. The first, *On the Fault Line: Race, Class, and the American Patriot Movement*, traces the rise and fall of the militia movement in the state of Kentucky. Her second book, *After the Peace: Loyalist Paramilitaries in Post-accord Northern Ireland*, examines the internal divisions within the Ulster Volunteer Force that inhibited timely demilitarization. She has also co-published a textbook entitled *Key Concepts in Political Geography*. Her work has appeared in numerous geography journals including *Society and Space*, *Social and Cultural Geography*, and *Space and Polity*.

Nicole Gombay is a Lecturer in Geography at the University of Canterbury, New Zealand. Her research focuses on the experience of Indigenous peoples living in the context of settler societies with a particular emphasis on the impacts of the inclusion of Indigenous populations in the political and economic institutions associated with the state.

Nik Heynen is an Associate Professor in the Department of Geography and the Associate Director of Center for Integrative Conservation Research at the University of Georgia. His research interests include urban political ecology, social theory, and social movement theory with specific interests in environmental and anti-hunger politics. His main research foci relate to the analysis of how social power relationships, including class, race, and gender are inscribed in the transformation of nature/space, and how in turn these processes contribute to understanding, and ensuring, the geographies of survival. He is an editor for *Antipode* and book series editor for the *Geographies of Justice and Social Transformation* series at the University of Georgia Press. Heynen has published in *Antipode*, *The Annals of the Association of American Geographers*, *The Professional Geographer*, *Urban Geography*, *ACME: An International E-Journal for Critical Geographies*, *Urban Affairs Review*, *Environment and Planning A*, and *Capitalism Nature Socialism*, among other journals.

Nancy Hiemstra is a Doctoral Candidate in the Department of Geography at Syracuse University. Her research interests are grounded in political and feminist geography, and focus on human mobility, migration policy-making, and the role of the state in shaping daily life. Previous research examined Latino immigration to small-town Colorado. Current research focuses on impacts in Ecuador of migration to the United States, as well as migrant detention and deportation policies and practices in the United States. She has published articles in *Antipode* and *Social and Cultural Geography*.

Julian Holloway is Senior Lecturer in Human Geography at the Department of Environmental and Geographical Sciences, Manchester Metropolitan University, United Kingdom. His research focuses upon the geographies of religion, spirituality, and the supernatural, with particular attention given to affect, embodiment, and practice. He has published numerous articles related to these interests in journals such as *Environment and Planning A*, *Annals of the Association of American*

Geographers, and *Cultural Geographies*. He is also the co-editor (with Binnie, Millington, and Young) of *Cosmopolitan Urbanism* (2005).

Richie Howitt is Professor of Human Geography, Department of Environment and Geography, Macquarie University, Sydney. He has undertaken applied social research as a geographer working in Indigenous Australia since the late 1970s. He has contributed to major social impact studies in Queensland and the Northern Territory and leads Macquarie's postgraduate SIA program. His research deals with the social impacts of mining on Indigenous peoples and local communities, and is generally concerned with the interplay across scales of social and environmental justice, particularly in relation to Indigenous rights. His book *Rethinking Resource Management* (2001) advocated deep integration of social, environmental and economic dimensions of justice into natural resource management systems. His teaching and applied research focuses on social impact assessment, corporate strategy, Indigenous rights, regional planning, social theory, human rights, and resource and environmental policy. He received the Australian Award for University Teaching (Social Science) in 1999 and became Distinguished Fellow of the Institute of Australian Geographers in 2004.

Sarah Johnsen is a Senior Research Fellow in the School of the Built Environment at Heriot-Watt University, Edinburgh. A social geographer by background, most of her current work is policy-oriented – focusing particularly on homelessness, street culture (begging/panhandling, street drinking, etc.), and other forms of social exclusion. She has written a number of government reports on these issues, published in several geography journals, and was co-author (with Paul Cloke and Jon May) of *Swept Up Lives?: Re-envisioning the Homeless City* (Wiley-Blackwell, 2010).

Jay T. Johnson is an Assistant Professor of Geography and Global Indigenous Nations Studies at the University of Kansas, United States. His research interests concern the broad area of Indigenous peoples' cultural survival with specific regard to the areas of resource management, political activism at the national and international levels and the philosophies and politics of place that underpin the drive for cultural survival. Much of his work is comparative in nature but has focused predominately on New Zealand and North America. He is currently co-editing a volume with Soren C. Larsen entitled *A Deeper Sense of Place: Stories and Journeys of Indigenous-Academic Collaboration* for the First Peoples: New Directions in Indigenous Studies publishing initiative.

mrs c. kinpaisby-hill (sometimes mrs kinpaisby) is an (imperfect) anagrammic pseudonym for an ongoing collective writing project/friendship between Caitlin Cahill, Rachel Pain, Sara Kindon, and Mike Kesby ("authorial order by height," Marston, Jones, and Woodward 2005, "Human geography without scale," *Transactions of the Institute of British Geographers* 30: 428). Etymologically, her disarmingly quaint *nom de guerre* (pseudonym used in war) began with a dare to use humor seriously: a merging of personas usefully reflects the participatory principles and transformations she espouses; the claiming of a socially produced gender identity speaks to the feminist politics that inform our perspective; the use of a married (as opposed to academic) title incites questions about scholars' weddedness to the academy, their divorcement from other social relations and spheres, and highlights the absence of ordinary voices in most academic literature; and the name as a whole is a provocation and an irreverent gesture to citation obsessed processes of research audit. The sites of her work include the City University of New York and the Universities of Durham, Victoria in Wellington, and St. Andrews, infrequent international conference sessions and the time–space convergent digital arena of email and document attachments. She has written a number of book chapters and papers on the subjects of participatory approaches, collective writing, and community/ university interactions.

Audrey Kobayashi is a Professor of Geography and Queen's Research Chair at Queen's University. She has published widely in social, cultural, and political geography on topics such as anti-racism, citizenship, immigration policy, employment equity, and disability studies, and has published in such journals as the *Annals of the AAG, The Canadian Geographer, Antipode, Professional Geographer, Gender, Place, and Culture*, and others. She participates in a number of community anti-racism and accessibility projects and is currently the co-investigator in two Community University Research Alliances, one on immigration policies in Ontario cities, and the other on disability policy in Canada. Since 2002, she has been the Editor, People, Place, and Region for the *Annals of the Association of American Geographers*. In 2009, she was the recipient of the annual Lifetime Achievement Award of the Association of American Geographers.

Jason Lim is a Teaching Associate in the Department of Geography at Royal Holloway, University of London, United Kingdom. His research interests include embodiment, affect, everyday practices, race, ethnicity, sexualities, and gender. He is co-editor of *Geographies of Sexualities: Theory, Practices and Politics*.

Geoff Mann teaches political economy and economic geography in the Geography Department at Simon Fraser University. He has written one book, *Our Daily Bread: Wages, Workers, and the Political Economy of the American West* (2007), and articles for a range of journals, from *Ethnic and Racial Studies* to *New Left Review*. Current research focuses on money, democracy, and climate change.

Katharine McKinnon is a Lecturer in Human Geography at Macquarie University, Sydney. She is a cultural geographer working on the geopolitics of development with a particular interested in how dominant social and political discourses come in to being and how they are challenged and altered through the everyday actions of ordinary people. Much of her research is based in the northern borderlands of Thailand exploring development professionalism, the practice of community development, and the Indigenous rights movement. In 2005 her doctoral research was awarded J.G. Crawford Prize for academic excellence and the Asian Studies Association of Australia Presidents' Prize. Her work has been published in journals such as *Social and Cultural Geography, The Annals of the Association of Geographers*, and *Pacific Viewpoint*. She is currently involved in a collaborative project on gendered community economies in the Pacific.

Chris McMorran is Visiting Fellow in the Department of Japanese Studies at the National University of Singapore, where he teaches courses on cultural and environmental aspects of contemporary Japan. His research centers on gendered labor, theories of mobility, exclusionary landscapes, and the commodification of the notion of home, particularly in tourist destinations.

Katharine Meehan is Assistant Professor of Geography at the University of Oregon, United States. Her research brings together urban political ecology and feminist political economy to understand how urban dwellers develop "off-grid" forms of water supply in the global South, particularly Mexico. Her work has been supported by grants and fellowships from the National Science Foundation, the National Oceanic and Atmospheric Administration, Fulbright-Hays, and the Social Science Research Council.

Louise Meijering is Assistant Professor at the University of Groningen, The Netherlands. Her research interests include aging and well-being, geographies of exclusion, intentional communities, migration, rurality, and qualitative methodologies. In her teaching, she addresses these themes in the context of courses on research methodology in an undergraduate program on Human Geography and Planning, a masters program in Population Studies, and a training for PhD students. She has published in journals such as *Area*, *Journal of Rural Studies*, and *Tijdschrift voor Economische en Sociale Geografie*, and contributed to edited volumes on sense of place and masculinities.

Alison Mountz is Associate Professor of Geography at Syracuse University and author of *Seeking Asylum: Human Smuggling and Bureaucracy at the Border*. Her research and teaching span the fields of urban, feminist, and political geography. Her work on transnational migration is driven by a desire to understand how migrants, states, and activists negotiate immigration and refugee policies. Recent projects examine border enforcement, detention practices, and the shrinking of space of asylum.

Caroline Nagel is an Assistant Professor in the Department of Geography at the University of South Carolina, Columbia, United States. She specializes in the politics of identity, citizenship, and integration in immigrant-receiving contexts, especially in Britain and the United States. Her current research examines the role of faith communities in shaping immigrant identities and citizenship practices in the American South. She also has a long-standing interest in public space issues in Beirut, Lebanon, and she was awarded a Fulbright Scholarship in 2010 to pursue research on this topic. Her work has appeared in several edited volumes and in journals such as *Antipode, The Professional Geographer*, and *Social and Cultural Geography*. She was co-editor, with Ghazi-Walid Falah, of *Geographies of Muslim Women: Gender, Space, and Religion* (2005).

Ruth Panelli is a former Reader in Human Geography at University College London, now participating in community work and research in rural New Zealand. She is the author of *Social Geographies: From Difference to Action* (2004) and the Social Geography entries for both the *International Encyclopedia of Human Geography* (2009) and *Encyclopedia of Human Geography* (2006). She completed also the 2007-9 Social Geography reports for *Progress in Human Geography*. In addition to commentary on social geography, her research interests are diverse as they explore the navigation of social differences (gender, youth, sexuality, and ethnicity) and the individual and collective responses people make to such differences (in personal, community, industrial, political, and environmental contexts). This work has been published in journals such as *Area, Childhood, Ecohealth, Environment and Planning A, Gender Place and Culture, Geoforum, Health and Place, Journal of Rural Studies, Policy and Politics, Sociologia Ruralis,* and *Transactions of the Institute of British Geographers.*

Hester Parr is Reader in Geography at the Department of Geographical and Earth Sciences at the University of Glasgow, Scotland. She works on the social geographies of mental health and has recently published a monograph: *Mental Health and Social Space: Geographies of Inclusion?* (RGS-IBG series, Wiley-Blackwell, 2008). She is interested in finding creative ways to facilitate the voices of people with mental health problems into conceptually challenging research, and has worked through a range of visual, textual, and ethnographic methods to achieve this aim.

Chris Philo is Professor of Geography at the University of Glasgow. Chris's research interests are diverse: ranging from geographies of mental ill-health, including asylum and post-asylum geographies; to various strains of social, cultural and animal geographies; and to the history and theory of human geography. Chris has fairly recently published a book out of his asylums research: *A Geographical History of Institutional Provision for the Insane from Medieval Times to the 1860s in England and Wales: The Space Reserved for Insanity* (2004) – as well as a co-authored methodological text: Paul Cloke et al., *Practising Human Geography* (2004).

Jennifer L. Rice is a Visiting Scholar in the Department of Geography at the Ohio State University, United States. She engages in interdisciplinary studies at the intersection of urban political geography, nature-society theory, and science-policy studies. Her current research examines the role of US cities in global carbon governance, with an emphasis on the relationship between the state, society, and science in urban climate mitigation efforts. She also explores the production and circulation of scientific knowledge in the creation and implementation of climate adaptation programs among urban water providers in the western United States. Both of these research areas consider the possibilities and limitations for social change via everyday urban environmental governance. Recent publications of this work have appeared in the *Annals of the Association of American Geographers* and the *Journal of the American Water Resources Association.*

Amy Ross is Associate Professor of Geography at the University of Georgia, and affiliate faculty for the Institute of Women's Studies and the Latin American Studies Program. Her main interests are on genocide, crimes against humanity, and war crimes. Her research focuses on transformations in power and space through the struggles to achieve justice and accountability in the wake of mass atrocity. She has researched truth commissions and international courts in Latin America, Africa, and Europe.

Joanne Sharp is a Senior Lecturer in Geography at the University of Glasgow, Scotland. Her research and teaching interests span feminist, cultural, political, and postcolonial geographies, specifically around issues of identity and geopolitics, the role of community involvement in arts-based urban regeneration, and issues of voice and agency in development and political geographies. To address these themes Jo has undertaken research in the United Kingdom, United States, Egypt, and Tanzania. She has written a number of books including *Condensing the Cold War: Reader's Digest and American Identity* (2000, University of Minnesota) and *Geographies of Postcolonialism: Spaces of Power and Representation* (2009, Sage), and has published in journals such as *Third World Quarterly*, *Society and Space*, *Political Geography*, and *Cultural Geographies*.

Wendy S. Shaw is Senior Lecturer at the University of New South Wales, Sydney, Australia. Her research interests includes the contemporary field of whiteness in postcolonial Australia, theoretical debates about identity around urbanism and urbanity, cosmopolitanism, and the complex realities of (post)colonialism. Recent research also includes projects in Papua New Guinea (on the coffee industry), Indigenous and rural Australia – on homelessness, sea-change, and the cotton industry. Wendy S. Shaw publishes in international and Australian journals, and is the author of *Cities of Whiteness* (Wiley-Blackwell 2007).

Deborah Thien is Assistant Professor of Geography at California State University, Long Beach, United States. A feminist geographer, she specializes in the study of emotion, health, and well-being; the effects and affects of gender; and rural, remote and northern places. To understand these concerns, she has conducted fieldwork in Canada, Scotland, New Zealand, and California's Sierra Nevada. She has published her research in journals including *Area, Health and Place, Gender, Place and Culture; The Canadian Journal of Public Health*; and *The Journal of Geography in Higher Education*. She has also authored several book chapters, for example, in *Emotion, Place & Culture* (2009). She is the book review editor for *Emotion, Society and Space*.

Mary E. Thomas holds a joint appointment in Geography and Women's Studies at the Ohio State University, United States. Her work explores how subjects learn about and reproduce social difference through individual identities like gender, sexuality, race, and class, and she asks how sexuality, racism, and economic privilege structure identity formation. She looks to the lives of teenage girls in the United States to understand how race comes to be a primary separating force in and through high school and urban social spaces. This work can be found in the journals *Environment and Planning A, Gender Place and Culture, Social and Cultural Geography, Environment and Planning D: Society and Space*, and *The Professional Geographer*. Her book on teen girls, racial segregation, and urban education in Los Angeles, California, is forthcoming.

Dan Trudeau is an Assistant Professor of Geography and Director of the Urban Studies Program at Macalester College, Minnesota, United States, where he teaches courses in political geography, research methods, and urban geography of the United States. His research interests include processes of racial segregation; the creation of exclusionary landscapes; the cultural politics of belonging; and social geographies of the shadow state.

Bettina van Hoven is Assistant Professor of Cultural Geography at the University of Groningen, The Netherlands. Her interest is in geographies of belonging and identities, and qualitative methodologies. Her work has been published in *Area*,

Journal of Rural Studies, Environment and Planning D: Society and Space. She has also recently completed a video production on "making place" in the Great Bear Rainforest in British Columbia, Canada (a forest for the future).

Keith Woodward is Assistant Professor of Geography at the University of Wisconsin-Madison, United States. His research explores intersections of affect, politics, and ontology as means for developing new understandings of social movements, political change, direct action, and autonomous organization. He is also the co-author, with Sallie A. Marston and John Paul Jones III, of a series of papers engaged in the development of "site ontology" for human geography, a project that has raised several challenges to scalar theory within Human Geography.

Chapter 1

Introduction

Vincent J. Del Casino Jr., Mary E. Thomas,
Paul Cloke, and Ruth Panelli

Questioning the Normative "We"

> To the people of poor nations, we pledge to work alongside you to make your farms
> flourish and let clean waters flow; to nourish starved bodies and feed hungry minds.
> And to those nations like ours that enjoy relative plenty, we say we can no longer
> afford indifference to the suffering outside our borders, nor can we consume the world's
> resources without regard to effect. For the world has changed, and *we must change*
> *with it*. (United States President Barack Obama, Inaugural Address, January 20, 2009,
> our emphasis)

Less than two years after the first African-American won the US Presidency, the US
government's commitment to the agenda of "free trade" has not wavered (disaffect-
ing millions of poor farmers worldwide), economic recovery programs seek to get
people living in the United States to earn and thus spend money to stimulate the
capitalism that eagerly consumes the world's resources, millions of gallons of oil
have spewed into the Gulf of Mexico just after the President signed legislation
allowing for more offshore oil exploration, and anti-immigration sentiments entrench
through local and state laws like Arizona's, which gives police the right to search
anyone suspected of being an undocumented migrant. (In fact, the number of depor-
tations of undocumented migrants in the United States has *increased* during Obama's
administration.) The intense global attention given to the assent of a black man to
the "most powerful position in the world" presented for some a sense of post-racial
politics, of hope for those living in racist and disempowered America, and of a
possible progressive turn in American national politics. But the representational
power of Obama as President has thus far largely failed to connect with a wider
progressive messaging. The office is not primed for dramatic change, even as the
President had hoped it could be. This is even more clear now that the 2010 midterm

A Companion to Social Geography, First Edition. Edited by
Vincent J. Del Casino Jr., Mary E. Thomas, Paul Cloke, and Ruth Panelli.
© 2011 Blackwell Publishing Ltd. Published 2011 by Blackwell Publishing Ltd.

elections have remade Washington, DC, once again. Obama's "we" therefore remains an imagined community. The polarities of rich and poor, of young and old, of white and not, of party politics, mark the inability of a national geography to align toward a global "good." And let's be clear here: Obama is addressing a certain middle-class constituency. He beseeches the polarity of national politics with a "hope" that there is a middle ground of diplomatic agreement. American economic and military hegemony go unquestioned in an assumption that its intervention is what the world needs to cure itself of illness, poverty, environmental degradation, and hunger (not to mention terrorism). President Obama's words might shape a sentiment for goodwill, but in the practice of the US government under his watch it is largely business as usual.

While it is tidy to identify the United States for its many behaviors condemning the world's poor, let's not forget that the 2010 Conservative victory in the United Kingdom – which put the Tories and new Prime Minister David Cameron in power for the first time in 13 years – also illustrates the effectivity of anti-immigration sentiments, free market ideology, and the sentiment that post-national unity (i.e., a European Union) is more trouble than it is worth. Indeed, the UK elections represent more than simply the fact that the Labour Party has moved so far away from labor politics that it might as well be the US Democratic Party, it also represents a clear set of fissures in societies throughout the world, where conflict and tension make socio-spatial differences more distinct. Examples of these fissures can be witnessed as well in countries such as Thailand, where street protests by various political factions have met, to differing degrees, with outright state repression by national leaders, such as Prime Minister Abhisit Vejjajiva. Resting at the heart of these tensions is the very question of what it means to be "Thai" in Thailand. On one side are those "loyal" to the monarchy, who threaten both national and expatriot with imprisonment for maligning the King, while others line up on the side of open democratic debate, while others still support a constitutional monarchy, which is both "democratic" and "monarchical."

President Obama's call for change confronted George W. Bush's policies of direct and egoistic disregard for life beyond his simplistic worldview. We thus started the book's introduction with Obama's quote to draw attention to the ways that even so-called progressive messaging can quickly be mired in status quo hierarchies of social difference. Changing the inequities of the world means fundamental redefinition of how life itself is defined and valued. This book therefore addresses the assumptions underpinning dominant ideals and norms about what is "good" for the world and for individual people. These kinds of ideals and norms are deeply problematic for their erasure of radical particularity, of space and its differences, and of an ignorance and disregard for unequal sociality, privilege, and positionality. Powerful assumptions about the "good" life are mired in normative racism, capitalist consumption, nationalism, colonialism, and heteronormativities about family and nation, whether they come from men with state power like Obama or Abhisit, or whether they result from the banalities of everyday life that reproduce these assumptions. Critique of these normative underpinnings is absolutely necessary, and authors in this volume address issues at the heart of why and how such messaging work on and through all subjects, and how resistance to the social norms of differentiation can be possible. This book is about space and social difference, the spatial contingencies that frame real and possible social life, and scholarship that attempts not only to make sense of these social-spatial relations, but that advocates for more just social geographies.

Social geographers in this text thus help to disentangle how any politics of difference must carefully contemplate the underlying categories that propel and constrain resistance. Power relations frame the differentiation of subjects and spaces, despite all too-easy calls to ignore the foundational categorization of the world into "we" and "you." It is therefore incumbent on social geographers to continually work against the grain of these messages to help produce new social geographies that can meet the goals of the larger progressive agenda established by the words of President Obama, and others. With this in mind, we now briefly move into a discussion of social geography, the subdiscipline, to ask what frames this field of inquiry and what challenges it faces as its practitioners seek to present new social geographic possibilities, confront inequalities, and promote social justice more broadly.

Social Geographical Turns

Social geography is a broad field that attends to the socio-spatial differences, power relations, and inequalities that shape every person's life. Social geography is also a way of going about the intellectual work that focuses in on these very political questions and issues. How exactly social and spatial differences are embodied and reproduced through communities, individuals' identities and subjectivities, and indeed societies, are of course issues of debate. Not all geographers agree on how to theorize social-spatial life, or how difference exists and gets reproduced in the world, what matters when and where, and the mechanisms for the reproduction of social categories and power relations. It is thus important to recognize that social geographies have developed over time using contrasting theoretical traditions that have different ideas of the world. This means that social geographers do not produce knowledge about social geography in the same ways. The question of how social geographers know what they know – or what we would call epistemology – is very much tied to their theories of what the world is or could be – what we call ontology. What social geographers know is connected to how they examine and explore social geographies – that is, their methodologies. Our quest for this volume has been to illuminate the different ontologies, epistemologies, and methodologies that make up today's social geographies. We revel in the differences found throughout these chapters, which means we do not necessarily advocate for a synthesis of all social geographic thought and practice. Illustrating the tensions that propel social geographers' work as scholars, theorists, and sometimes activists, can energize debate and research. New thinking about the social and spatial differences that constrain and enable life can grapple with questions that are crucial to a range of different everyday lives in different places. It is this sort of energy that animates social geography today.

Social geography has occupied an important position in the wider canon of human geography scholarship for many years now, and we maintain that this role for geographies of the social remains a vital one in the contemporary world. Since the late 1980s, however, there has been some disquiet within the debates surrounding social geography. In much of Anglophone geography, for example, there has been an ongoing tendency to subsume the social within the cultural, by emphasizing the use of cultural texts, a heightened reflexivity towards the role of language and representation in the constitution of "reality," and a particular interpretation of poststructural epistemologies that points to the close relationship among language, power, and knowledge and toward a set of emergent and hopeful possibilities for new social geographies. There are many benefits from this co-immersion with the cultural, but there is also some concern that social geography has risked losing sight of some of its core foci – e.g., the study of inequality and difference – during this period. We address very briefly the four most significant concerns surrounding the shifts in social geography since the 1980s. At the same time, we want to be mindful of the fact that these challenges have also enabled a whole new set of social geographic possibilities.

The first main concern has been to examine the effect of what is often called "the cultural turn" in human geography. This "turn" is primarily defined as "cultural," which detracts from the ways that its debates proceeded along very social lines. Thus, while the turn toward "culture" might indicate a process that has *desocialized* human geography, thereby rendering social geography moot, in fact we can evidence a healthy flourishing of social geographies throughout the 1990s and 2000s. We do caution, however, against a tendency to withdraw from the everyday processes – the social practices, relations, and struggles – that constitute the stuff of everyday life when relying heavily on a largely "cultural" social geography. Nicky Gregson (2003) has argued, in fact, that although the social has not been replaced by the cultural, it is nevertheless the case that the social has been increasingly refracted through the cultural. Thus, a new set of concerns for cultural difference and resultant identity politics might encourage attention away from the structures and spaces of inequality. That said, because of the engagements with cultural theory, many social geographers theorize subjectivity and identity through the ongoing importance of gender, sexuality, race-ethnicity, nationality, age, health, and class. Simply put, social categories still matter.

Second, because social categories still frame the ways that subjects and spaces become delineated, the material ramifications of difference also must be examined. There is a price to pay for social difference in spaces of inequality. Chris Philo (2000) argues that human geography has become "less attentive to the more thingy, bump-into-able, stubbornly there-in the-world kinds of matter" (p. 13, see also Chapter 21, this volume) and his diagnosis is to insist on a re-emphasis on reclaiming the materiality of the everyday world. A *dematerialized* human geography is preoccupied with immaterial processes, the intersubjective nature of meaning and the working out of identity politics through texts, signs, and symbols, without an accompanying exploration of how these result in winners and losers under capitalism, heterosexist patriarchy, colonialism, nationalism, and globalization. Social geographies offer a vital imprint on the material outcomes of social differentiation.

Third, political quiescence draws attention away from contemporary forms of globalization and neoliberalism, which sustain and transport the cut-throat politics of the market throughout the world, with violent results on everyday lives and environments. A *depoliticized* human geography re-routes research away from the analysis of and intervention in social struggles (cf., Mitchell 2000). Social geographies, particularly feminist, queer, and anti-racist geographies, on the other hand, insist that processes and spaces of social differentiation involve elitist power relations, normative social meanings and identification, and uneven material distribution (e.g., Jeffrey and Dyson 2008; Gilmore 2007; Oswin 2008; Thomas 2008; Wright 2006). These arguments demand a political stance against injustice, disenfranchisement, and hierarchies of subjectivity.

Fourth, it has been claimed that human geography has been insufficiently *deconstructionist*, by adopting a rather lukewarm and over-conservative approach to non-representational approaches, approaches that could open up new and exciting possibilities relating to the performative and the immanent as models for the emerging social (see Thrift 2000). In this way, it is feasible to imagine social geographies that are more fully engaged with the mundane everyday acts of social violence or hope that mark out spaces of both domination and resistance, both planned and spontaneous. At the same time, calls for a more robust ethics based in deconstruction have challenged the concern that we must always operate in the realm of "affect" as a way to engage with the performance of just and moral social geographies (Popke 2003, 2007, 2009; see also Chapter 19 on affective life in this volume). Indeed, social geographers continue to engage the relational (inter)subjective nature of ethics and justice in ways that provide possibilities for more fully deconstructionist social geographies.

Thus, social geographers offer important responses to these broader human geographic concerns, and in fact, over time they have deflated attempts to unproductively distinguish between "culture" and "society." The chapters in this book illustrate this point well. We also want to remember that debates in human geography always have different levels of intellectual purchase in different parts of the world, and some of the resultant perceived excesses and limitations of debate apply to a relatively small part of the global and heterogeneous "we" of social geography. As the chapter authors in this volume survey current scholarship and research in social geography through this *Companion,* we see a strong awareness of, and response to, the risks outlined above. We see a social geography that is in different ways highly socialized, with a capacity to address longstanding social differences relating to gender, race-ethnicity, sexuality, nationality, and class. The chapters that follow explore in exciting ways the importance of other social groupings and new theorizations of materiality and agency. There is considerable diversity in the ways in which geographers delimit and address the human-social, and we see that diversity as richly productive. It is clear that social geographers are maintaining and expanding their politically active edge, resisting by word, thought, and deed the reactionary forces that would seek to govern these spaces to the detriment of the marginalized, and being attentive to new political lines of flight, including the integral deployment of both ethics and aesthetics in emerging practices of protests and resistance.

We want to suggest, then, that there is currently a healthy array of diverse awareness about the risks of desocializing, dematerializing, depoliticizing, and underwhelming deconstructing geographies of difference (e.g., Pain 2003, 2004, 2006). Part of our excitement as editors of this *Companion* is to see in the chapters that follow a willingness to consider different ontological and epistemological avenues, without any seeming need to produce a synthetic "we must" approach to social geography. We appreciate this diversity, this openness to different inquiry, and this determination to contextualize knowledge and to engage actively with its various politics and ethics. Social geography involves wrestling with contradictions and tensions, and we salute the energy that these practices and performances engender in the hope that social geographies of the future can continue to participate in the positive assertion of social difference and challenge the broader inequalities that distinguish our world today.

Mapping *A Companion to Social Geography*

While social geographers have worked to identify (and often critique) the differences and inequalities shaping societies and everyday lives, the intellectual mechanisms for building this knowledge have varied enormously. This has resulted in a (sometimes) confusing array of approaches and styles for students and scholars to understand (see e.g., Cloke et al. 1991; Del Casino 2009; Panelli 2004, 2009). At the same time, in a *Companion* of this type, it is important to recognize that social geographies (and social geographers for that matter) are intellectually and politically diverse. It is also essential to acknowledge the reasons for these differences in *how* social geographers understand – and create – social geographies, reasons tied as much to the history of discipline as it is to the situatedness of various research programs and perspectives in particular institutional contexts. In the following chapters, many authors convey important contemporary foci and debates as well as institutional tensions surrounding those debates, while also drawing upon the diversity of approaches from and through which social geography has flourished. This is important for them to do if readers are to understand how to generate social geographies that are both intellectually robust and politically relevant.

Consequently, this volume sketches out – in differing ways – how various historical practices have shaped geographers' engagement with the social and the spatial, sociality and spatiality. It does so, however, without chapters outlining "traditional" theoretical approaches – spatial scientific models, marxism(s), feminism(s), poststructuralism(s), etc. – but rather through a series of discussions that draw from these approaches differentially across four main parts: (1) ontological tensions in/of society and space; (2) thinking and doing social geographies; (3) matters and meanings; and (4) power and politics. In each of these parts, authors were asked to consider some of the larger concerns of social geography – e.g., how do we conceptualize difference or belonging?; how do we frame "the field" or write about our social geographies?; how do we theorize life and its contours?; and how do we study and engage with questions related to marginalization or care and caring? These and many other questions animate the chapters that follow, which attempt to both trace social geography as a subdiscpline and offer a trajectory for how to "think" and "do" social geography moving forward.

This *Companion* is thus framed as an intervention into the major debates in and practices of social geography as a subdiscipline, while also engaging in the conceptual tensions that inform how social geographers think about, on the one hand, the meanings of nature or economy, and on the other hand, the sociality of geopolitics.

While each editor outlines their part's chapters in more depth (see the individual introductions preceding each part), it is important to briefly situate each part in terms of the larger project. In Part I ("Ontological Tensions in/of Society and Space"), the authors in six chapters map out some of the core ontological tensions with which social geographers most often have to struggle – difference, identification, social natures, economies, community, and belonging. In each of these chapters, authors trace not only the "current" theorizations of these concepts but the larger historical trajectories of how these concepts have been "taken up" in social geography. This provides both context and nuance to what are sometimes represented as easily delineated and defined concepts. That difference cannot be easily grounded in one ontological definition engages with the wider argument made with this volume – social geography is a dynamic, diverse, and contested (in the best sense of the word) intellectual and political terrain.

Part II ("Thinking and Doing Social Geographies") steps back (or better yet repositions) readers to think about how social geographers go about "doing" social geography. This part does not eschew theoretical questions but stays at the level of methodology – the theory of method. In so doing, authors in this part ask readers to think about their own positionality as researchers: why we ask the questions we do, what we do with the answers once we have them, and how we go about "doing" something politically effective with all those data and answers. Moving back and forth between the broader methodological concerns in social geography and some of the on-the-ground methods and approaches that social geographers employ, this part provides partial insight into the complex and contested spaces of "the field" and the tensions that are brought to life in and through the processes of analysis and writing.

Part III ("Matters and Meanings") offers perhaps this volume's most original entry point into the debates and discussions animating social geography today. Social geography is after all a subdiscipline intimately concerned with questions of life, and not just human life but life writ large. The authors in this part were thus asked to consider how life is not only understood but also socially and spatially made "real" through bodies, feelings, affects, and practices. Taken together, this part provides some provocative insights into both how social geographers theorize socio-spatial relations and how social geographers take on the tensions and debates that inform social geographic theory and practice today.

The final part of this volume, titled "Power and Politics," stays true to the rest of this *Companion* by engaging in a conversation about social geography's larger role in challenging dominant relations of oppression, inequality, and difference. In many ways, readers should be able to work across this part and back through chapters from the other parts they think about, for example, how theories of difference (Chapter 2) inform the study of marginalization (Chapter 25) or transnationalism (Chapter 28). In fact, readers should begin to see that areas of scholarship historically situated in "other" subdisciplines, such as political geography (see Chapter 24 on Geopolitics), are primed for social geographic engagement and inquiry. Moreover, as readers take on the chapters in this volume, they will likely find themselves challenging its very organizational principles – reading chapters on embodiment (Chapter 20) or emotion (Chapter 18) alongside chapters on participatory research (Chapter 13) and social action (Chapter 29). Indeed, the editors and authors of this text encourage readers to do just that – disrupting the linearity that remains so entrenched in academic publishing.

We believe that this volume does more than catalogue the field of social geography; it challenges readers to reflect critically on the diversity and differences found within the broadly organized field of social geography. We did not ask the contributors to resolve these tensions for readers. Rather, we asked the authors to be both engaging and engaged. The overall text demands that readers think about their own theoretical and political positions and consider how their intellectual choices compel them to think reflexively about the way that they epistemologically and methodologically frame their social geographies. While this volume is intended to set out an agenda for key topics and concerns that inform the histories of the subdiscipline and drive the current field of social geography, it is also a volume riddled with theoretical, methodological, and political differences that continue to drive social geographic research, practice, and politics. Readers are asked to engage this *Companion* in the spirit in which the editors and authors engaged it – as another important moment for social geographers to reflect on what we mean by social geography in all its complexities.

References

Amin, A. and Thrift, N. (2002) *Cities: Re-imagining the Urban.* Cambridge: Polity Press.

Cloke, P., Philo, C., and Sadler, D. (1991) *Approaching Human Geography: An Introduction to Contemporary Theoretical Debates.* London: Guilford Press.

Del Casino, V.J., Jr. (2009) *Social Geography: A Critical Introduction.* Oxford: Wiley-Blackwell.

Del Casino, V.J., Jr., and Marston. S.P. (2006) Social geography in the United States: everywhere and nowhere. *Social and Cultural Geography* 7 (6): 995–1009.

Gilmore, R. (2007) *Golden Gulag: Prisons, Surplus, Crisis, and Opposition in Globalizing California.* Berkeley: University of California Press.

Gregson, N. (2003) Reclaiming "the social" in social and cultural geography. In K. Anderson, M. Domosh, S. Pile, and N. Thrift (eds.) *Handbook of Cultural Geography.* London: Sage, pp. 43–58.

Jeffrey, C. and Dyson, J. (eds.) (2008) *Telling Young Lives: Portraits of Global Youth,* Philadelphia: Temple University Press.

Mitchell D. (2000) *Cultural Geography: A Critical Introduction.* Oxford: Blackwell.

Oswin, N. (2008) Critical geographies and the uses of sexuality: deconstructing queer space. *Progress in Human Geography* 32 (1): 89–103.

Pain, R. (2003) Social geography: on action-orientated research. *Progress in Human Geography* 27 (5): 649–57.

Pain, R. (2004) Social geography: participatory research. *Progress in Human Geography* 28 (5): 652–63.

Pain, R. (2006) Social geography: seven deadly myths in policy research. *Progress in Human Geography* 30 (2): 250–9.

Panelli, R. (2004) *Social Geographies.* London: Sage.

Panelli, R. (2009) More-than-human social geographies: posthuman and other possibilities. *Progress in Human Geography* 34 (1): 74–87.

Philo C. (2000) More words, more worlds: reflections on the "cultural turn" and human geography. In I. Cook, D. Crouch, S. Naylor, and J. Ryan (eds.), *Cultural Turns/Geographical Turns.* Harlow: Prentice-Hall, pp. 26–54.

Popke, J. (2003) Poststructuralist ethics: subjectivity, responsibility and the space of community. *Progress in Human Geography* 27: 298–316.

Popke, J. (2007) Geography and ethics: spaces of cosmopolitan responsibility. *Progress in Human Geography* 31 (4): 509–18.

Popke, J. (2009) The spaces of being-in-common: ethics and social geography. In S. Smith, S. Marston, R. Pain, and J.P. Jones, III (eds.), *Handbook of Social Geography*. London: Sage.

Thomas, M. (2008) The paradoxes of personhood: banal multiculturalism and racial-ethnic identification among Latina and Armenian girls at a Los Angeles high school. *Environment and Planning A* 40 (12): 2864–78.

Thrift N. (2000) Introduction: dead or alive? In I. Cook, D. Crouch, S. Naylor, and J. Ryan (eds.), *Cultural Turns/Geographical Turns*. Harlow: Prentice-Hall, pp. 1–6.

Wright, M. (2006) *Disposable Women and Other Myths of Global Capitalism*. New York and London: Routledge.

Part I   Ontological Tensions in/of Society and Space

Introduction

Paul Cloke

The history and development of social geography has often viewed as a series of *isms*, reflecting a development of study that is characterized by particular theoretical and philosophical periods in which dominant ways of thinking tended to emphasize particular ideas over others. Thus neo-Marxism's concerns for the power relations of capital restructuring and attendant social recomposition have been set against humanism's emphasis on the experiences, potential, and needs of particular people.

A Companion to Social Geography, First Edition. Edited by
Vincent J. Del Casino Jr., Mary E. Thomas, Paul Cloke, and Ruth Panelli.
© 2011 Blackwell Publishing Ltd. Published 2011 by Blackwell Publishing Ltd.

Postmodernism's deconstruction of metanarrative and celebration of relative rather than normative "truths" has led on to poststructuralism's exploration of difference, interrelations and hybrids. That these thumbnail sketches of vast swathes of scholarship represent unacceptably crude caricatures underscores the risk that *ismic* social geographies will lead us into somewhat self-contained understandings, and misunderstandings, of how society and space are organized, performed, and simply become. Theory in this manner can build walls around ideas. The assumption sometimes unfolds that unless you are fully wired into a particular ontological framework you can neither understand its power nor grasp how it overrules other potential sets of ideas. The result can be groups of geographers sitting within their chosen walled academic gardens engaging in conversation only with those of a similar ontological persuasion. The result *has* in my view been a certain lack of conversation – and respect – across these ontological boundaries.

Talking to my students in Exeter about these issues I have picked up at least three kinds of broad approach to theory in social geography. First, there are those for whom theory represents a kind of mysterious necessary evil; something that they are told is important, but that just simply does not fire the academic or intellectual imagination. Theory in this case is often just a matter of dropping key ideas into the conversation because somehow they ought to be there, rather than because they help to explain or explore the subject concerned. The ideas are often presented as easily defined and readily encapsulated into what are otherwise atheoretical narratives. Secondly, there are those for whom theory becomes significant and understood in and through events, circumstances and relations happening *on the ground*. Ontology comes into play either when these grounded phenomena suggest the significance of particular ideas, or when such ideas unlock understandings of society in space. Either way, the emphasis is not on theory for theory's sake, but on the deployment of ontological ways of thinking in order to enliven everyday social geographies. Thirdly, there are those who wholeheartedly embrace philosophy and theory, but often from a rather narrow platform. The tendency here is to ask, "what would Foucault or Deleuze or Badiou (and here substitute your own favorite thinker …) have to say about this?" Practical social geographical circumstances are sometimes only evoked in this manner as loose illustrations of the ideas concerned.

I find myself, perhaps unexpectedly, drawing on each of these three broad sets of attitudes to negotiate a role for the ontological both in teaching and in research. From the "theory as necessary and mysterious evil" approach comes an often healthy cynicism about the desire to present social geography in ontological terms. At its most lazy this approach makes unthinking use of key ideas, but in more critical form it provokes a valuable debate about presuppositions in human geography that theory is more important than political normativity. Grounded theory can result in undue pragmatism, but it emphasizes the need to work through ideas in different contexts and thereby challenges any presumed uniformity or universalism in ontological thinking. Detailed readings of the philosophical frameworks of theory preempt superficiality and over-easy theorization. They teach us to struggle with ideas so as to go beyond what may seem obvious.

It is this sense of struggle that we want to pursue in this opening part of the *Companion*. We want to recognize that key ideas in social geography are sometimes presented as easily defined and delineated, and that their utility can become polarized; for example the argument that community is dead and is a mere historical relic in postmodern times, can be contrasted with the argument that community is a key aspect of social cohesion that needs to be rebuilt. Such arguments have very significant political ramifications, but they depend on some common understanding of what community is – an understanding that is often simply assumed to exist rather than being struggled over. We are not arguing, however, that key ideas are so complex that they are impossible to use. On the contrary, we want to insist that the criticality of social geography depends on our ability to recognize it as a dynamic, diverse and contested intellectual and political terrain. It follows that key ideas need to be grappled with, struggled over, re-enchanted, re-politicized. This opening part includes six contributions that trace particular ontological tensions that need to be grappled with in this manner.

In Chapter 2, Sarah de Leeuw, Audrey Kobayashi, and Emilie Cameron explore the idea of *difference*. They discuss how difference has been theorized, and how social geographers have approached different aspects of difference and its material effects. There is a long history in social geography of seeking to understand difference in terms of a shopping list of otherness, involving for example, gender, ethnicity, disability, and alternativeness, each of which have multiple social and spatial manifestations, and involve the production of particular subjectivities through space, discourse, and performance. The authors argue, however, that social geographers have done less to break down the same-other dualisms on which these understandings of difference are often founded. Too often, social subjects are recognized as "different" in relation to particular social norms. Difference in these terms involves being opposite and other to those norms. In grappling with this idea of difference, Sarah, Audrey and Emilie urge us to pay more attention to the relationships between subjects, considering how multiply-othered subjects interrelate both with each other and with the processes by which social norms are constructed. Such a re-orientation re-envisages space and place as potentially liberating agents of social change – a ready partner in the politics of social justice.

In Chapter 3, Katharine McKinnon helps us to grapple with the idea of *identity*, exploring how we make ourselves, give ourselves descriptive indicators, and often become the labels by which we explain ourselves to others. She explains how identification works in multiple and complex ways to bring the self in line with discursively constituted subjectivities and subject-positions, involving both less-than-deliberate concatenations of events and circumstances, and more deliberate and conscious acts to claim identities and to reshape the self and the world. Identification is thus a powerful political tool, both for maintaining existing systems of power and control, and for bringing new kinds of subjects into being, and thereby provoking new possibilities of reshaping our ways of being in the world.

In Chapter 4, Katharine Meehan and Jennifer Rice trace how the idea of *nature* has been used to inform social geography. Typically, nature has been assumed until recently to be produced by society, in a one-sided nature-culture binary in which human beings, their economic demands and their cultural proclivities, have colonized the natural other for their own needs and ends. By this token, nothing that remains is truly natural – nature has been totally reshaped by society. However, as the authors explain, this unwieldy view of the social has been rigorously unsettled in human geography, particularly in attempts to recognize the complex and emergent associations between humans and non-humans that make up the world. Nature–society relations, then, have been re-envisaged as a series of networks of associational agency involving humans, animals, plants, technology, discourses, and so on. Katharine and Jennifer grapple with ideas that suggest that social and physical landscapes are always in dynamic co-constitution. In this, place-contexts matter in the process of socio-ecological formation, suggesting that site-ontologies of social natures are crucial in rendering visible emergent associations between the human and the non-human.

In Chapter 5, Geoff Mann considers the relationship between social geography and the *economy*. So often the social and the cultural have been presented as a "natural" intellectual pairing, for example in the examination of ideas about difference and identity, but Geoff argues through the example of money that economic theoretical and categorical concerns are significant in the work that social geographers do. In particular he points to the dynamics that constitute seemingly macro-scale economic phenomena such as globalization and neoliberalism, and argues that an increasingly blurred set of boundaries between the economic and the social is something to be grappled with and celebrated. Thinking critically about the category "economic," demands that we think critically about the category "social," and vice versa. To do so, we need to continue to explore deeply embodied and emplaced human lives *at the same time as* exploring the macro relations that they co-construct and interact with. We need to make social geographies of the economy sensitive to the myriad and sometimes contradictory forces that shape the economic as a category of production and performance, rendering the economic as a tangible realm of social experience that is sometimes distinct from the social according to the social's own perceived norms.

In Chapter 6, Marcia England examines *community*, an idea that contributes meaning to social geography in multiple ways. She wrestles with the ways in which community affects both the organization of social life, and the spaces that emerge from those social interactions. Once again we are shown that the idea of community is often given a false sense of coherence through the incorporation of a set of key binaries – inclusion/exclusion, unity/division, heterogeneity/homogeneity – in its definition and assumed inherent meaningfulness. Marcia argues that only by unsettling these binaries can we place the idea of community into everyday contexts, both imagined and material, and in so doing explore the potential for some kind of innate need for social cohesion amongst humans. Having spent so long deconstructing the social construct that is community, social geographers may now need to revisit the question of why people form communities and others seek to join them. Far from being an essentially exclusionary space, communities may yet be a platform from which to offer exciting new ways of being in the world.

Finally, in Chapter 7 Caroline Nagel discusses *belonging*, an idea that can impart both a very positive sense of security and comfort, and more disciplinary senses of having to fit in with boundaries and norms established by actors more powerful than ourselves. Belonging is therefore an idea that encompasses a necessary ambivalence, around being part of, and being made to fit in, that which is structured and experienced in particular spaces. Using the issue of migration as an example, Caroline explores emotional, formal, discursive and normative and negotiated senses of belonging. She argues for a multi-layered approach in order to sensitize our studies of social geography to the ambiguities of belonging in the spaces of everyday life; to the multiple frameworks of formal and informal regulations that establish who belongs; and how they belong; and to the creation of new jurisdictions of power alongside the fragmentations of older ones.

This opening part, then, is designed to present a series of explorations of key ideas about social geography and of the ontological tensions they harbor. These ideas are fundamental in and of themselves, but these opening chapters also perform an important pedagogic purpose. They help us to wrestle with complex and shifting ideas, to ask critical questions about theoretical suppositions, to appreciate the significance of grounded context, and to get beyond over-easy and simplistic theorization. It is this sense of grappling with theoretical ideas that we believe is key to a re-enchantment, and a re-politicization of social geography. Struggle, then, is an essential component of the struggle. Enjoy ...

Chapter 2

Difference

Sarah de Leeuw, Audrey Kobayashi,
and Emilie Cameron

Courtenay is a tiny town on Canada's southwest coast. It has a population of less than 25,000 and an economy reliant on fishing, logging, tourism, and the military base located a few kilometers up the local highway. In early July 2009, on a bright, sunshiny afternoon in a downtown parking lot, three white men violently assaulted Jay Phillips, a black man, who had moved to town six years earlier. In what police call an unprovoked "apparent hate crime" (CBC News 2009), the three men approached Phillips, began hurling racial slurs at him, and then attacked him. According to Phillips, and corroborated by video recording of the incident, after throwing Phillips onto the pavement, and as they delivered kick after kick, punch after punch, the three men screamed that Courtenay was "a white town" and Phillips should "get the fuck out" (Hui and Lavioe 2009).

Courtenay is one of many places around the world that few geographers have heard of, and one of many places where conceptions of "difference" have under-pinned violent actions. Violence was meted out to Jay Phillips based upon an (il) logic that some people have rights to, and power or privilege within, certain geographies, while others have neither the right to be there nor the most basic human right of safety. The vicious brutality of this event, and the ongoing scars that it will leave upon Phillips, his family, and the diverse communities of people who feel the direct or vicarious impact of race-based violence, makes one thing abundantly clear: difference matters. It matters because for the most part difference is linked with power and powerful categorizations that result in social hierarchies with the poten-tial to materialize in just the kind of violence that unfolded in Courtenay's downtown.

With the deep hope that in some not-too-distant time and place the idea of difference will no longer spark hatred in the minds of some, we explore in this chapter why and how difference matters. We discuss how scholars across the

A Companion to Social Geography, First Edition. Edited by
Vincent J. Del Casino Jr., Mary E. Thomas, Paul Cloke, and Ruth Panelli.
© 2011 Blackwell Publishing Ltd. Published 2011 by Blackwell Publishing Ltd.

humanities and social sciences have theorized difference, how social geographers in particular have engaged with difference and its material effects, and how social geographers might think *differently* about difference. Social geographers have mainly considered difference as it relates to class, sexuality, race, ethnicity, gender, ability, and questions of power as expressed spatially. Geographers have made important contributions to a spatial understanding of how particular subjects are produced as different. They have paid relatively little attention, though, to the relationships *between* subjects who are conceptualized as different in reference to powerful ideas about what constitutes social norms. In other words, until recently, geographers have tended to work with a dualistic and antagonistic understanding of difference wherein subjects produced as "different" are understood as opposite and other than a social norm, rather than consider the ways in which multiply-othered subjects engage with one another and in relation to such norms. Recently, however, geographers have begun to explore alternative ways of conceptualizing difference. We welcome this shift, and suggest that geography – with its realization that spatiality and place are dynamic and thus potentially libratory forces of social change – is well positioned to reconceptualize difference as part of efforts to move toward social justice.

Theorizing Difference

To understand the construction of difference in Western and particularly geographic thought requires an adjustable lens that can take a wide angle on the historical processes that have systematically sorted the world's people according to differential categories that fit with imperial, colonial, and capitalist expansion at a global scale, as well as zoom in on the everyday practices through which difference is constituted among people in direct contact with one another. We can find many precedents in early European history, from the "Crusades," which set Europe apart particularly from the "oriental" world, through the period of European exploration and expansion from the fifteenth to the eighteenth centuries, during which a Manichean order of things was firmly established in colonial discourse.

Theories of difference began to crystallize during the eighteenth-century "Enlightenment" when scientific methods and practices were directed toward the classification of all manner of living entities, including human beings, according to their most obvious physical properties: sexuality and the phenotypical characteristics that became known as "race." The "geometer of race," to use Gould's (1994) phrase, was legitimized by Linneaus as a geographical division of the world's people into four broad categories. His student, Hans Blumenbach, established the five categories – Caucasian, Mongolian, Ethiopian, American, and (the additional category) Malay – that became the basis of both scientific and popular discourse. It was not until well into the twentieth century that this classification, or some variant of it, received significant challenge in either realm (see UNESCO 1956).

Immanuel Kant, the first recognized university chair holder in geography, ascribed to the four-category Linnean model, expanding on his theories of natural division, hereditary difference, and environmental determinism as the engine of evolution. Kant's geography of human difference, propounded in *Physical Geography* (Kant 1997), links everything from phenotypical charactistics such as hair and skin color to morality, intelligence, civility, and the capacity to appreciate beauty, to the effects of climate on human adaptability. His public debate with Herder over climatic versus cultural explanations of human difference helped to elevate the search for theories of difference to the highest ranks of scientific endeavor, creating the most significant platform for geographical debate of the eighteenth and nineteenth centuries (Livingstone 2002).

If earlier scientists had fixated upon classifications, the overwhelming preoccupation of the twentieth century was variability. The first half of the twentieth entury saw a major shift as geographers began to shed notions of environmental determinism, and the idea that "race" is a product of climate. In the period following World War II, at a time when the United Nations was putting in place the international human rights documents that sought to end violence and oppression based on global definitions of differences, geographers were seeking ways to systematize their science so that variables could explain geographical distributions on the basis of spatial pattern. Although such thinking rejected environmental determinism, geographers nonetheless continued to assume racial difference for some time. The debates that occurred during the rise of the so-called "quantitative revolution" were over how to explain such variables. It was not until the so-called "relevance debates," which began during the early 1970s that for the first time geographers began in large numbers to pay attention to the ways in which ascriptions of difference, including those of race, class, and gender, result in disadvantage and oppression (Kobayashi and Peake 1994).

In the twenty-first century, it is widely accepted among geographers that "difference" is a social construction, by which we mean that the differences between people are neither inherent nor inevitable but historically and spatially produced through human actions, which include social, cultural, and economic practices and relationships, in place. The turn to poststructuralist explanations is particularly important because it suggests that if differences are humanly produced, rather than determined by climate or any other external structure, then humans have the capacity to create new historical circumstances. To do so, however, would require that we overcome powerful social constructions that are the result of violent and oppressive actions, and particularly a history of colonialism.

Difference, like other slippery but important words in the humanities and social sciences, is a protean term denoting concepts that are equally contested and mutable. Across disciplines, and in various literatures including those in geography, definitions of difference tend to include words like dissimilarity, distinction, or unlikeness, that emphasize a comparison, but more often a contrast, between subjects, conditions, or qualities (OED Online 2nd edition 1989; see also Mayhew 2009). Sameness and occasionally equality are often offered as antonyms of difference. Most theories of difference ask questions about power as the ability to exercise control or authority over other persons or objects.

Many theories about differences focus on the way that difference is produced and then deployed as a means to categorize "other" subjects and entities in order to exert or maintain social, cultural, economic, or spatial control over them. By grouping entities and subjects together based on constructed ideas of sameness, categories of what constitutes normal are also produced. These categories, in turn, function to exclude subjects who do not adhere to the characteristics of sameness or normalcy. Difference in turn becomes abnormal and deviant. As a number of scholars have documented, the production of difference as deviance has made people and practices categorized as "different" vulnerable to powers interested in enforcing normalcy and/or fixing, altering, subjugating, or erasing what is constructed as deviant (see Fabian 1983 and Foucault 1972, 1977).

The interconnectedness of power, difference, and deviance has especially interested scholars working in poststructuralist, postmodern, and feminist traditions. These scholars pay close attention to language and to cultural "texts" in an effort to tease out the ideological and cultural dimensions of representation. Jacques Derrida, for example, in efforts to destabilize and expose the structures of power embedded in text and speech, argued that many of society's inequities can be traced to the structure of languages. Concepts and words, he argued, are understood in relation to what they are not – or that which they are different from. In his work on "différance," he observed that understandings of the world are problematically premised – at least from an emancipatory perspective – on seeing and speaking of one thing in relation to its opposite. Those things which are the opposite, or the "not," tend to be subjugated or deferred. For example, yellow is understood in relation to not-yellow, with not-yellow deferential to the primacy of yellow; house is understood by that which it is not, such as a cabin or a hut. When conceptualizing yellow or house, all that is not house or not yellow is deferred or subjugated (Derrida 1982; see also Spivak 1994). Given that we understand ourselves, each other, and our world in great part through language that has developed over centuries of use, hierarchies of power premised on differences are maintained in the very way we speak or think about subjects.

Indeed, as Derrida and other theorists have recognized, difference is not a neutral category; it is socially meaningful and legible in relation to ideas of normalcy, deviance, and hierarchy. Michel Foucault (1972, 1977, 1991) argued that knowledge and associated belief systems or discourses (which form the subjects to which they refer) turn upon people accepting a common or universal knowledge against which something different is a deviation. Deviations come to be conceptualized as punishable, or to be actually punished, through violence or the threat of violence. Categories like sanity and heterosexuality become normalized by authorities (for example, by government figures or church and medical officials) and then taken up as normal, and thus acceptable, by those upon whom authorities act. Once normalized, non-authorities then take up the multiple practices of maintaining striations and hierarchies of power, in part by circulating them as truth. Foucault was particularly interested in how we come to monitor ourselves and each other for signs of deviations from the normal we have internalized. We mete out violence or the threat of violence accordingly. For Foucault, power rests on the production of difference, on a differentiation between right and wrong, moral and immoral, normal and abnormal.

Although for Foucault there is an intentionality of power, power's trick is that it need not be consciously (or even always violently) reproduced nor actively made material. Instead, it can operate opaquely, as that which is assumed to be natural. Power also operates and maintains traction through that which it is not, namely, resistance to power. In this relationship, power is strengthened and reproduced through the resistance mounted against it. In other words, power exists only because it is able to differentiate itself from something else; thus the inextricability of power from difference or from the production of something different. There is a circularity to Foucault's logic that has been critiqued as preemptively refusing the possibly of disrupting power, or of envisioning subjects as ever fully having agency to combat or refute power. It is important, however, to remember that power and difference are deeply interconnected, an idea not lost on postcolonial theorists upon whom geographers also draw (see Crampton and Elden 2007).

Orientalism, Edward Said's concept of a persistent and pervasive production of an eastern other for western domination and consumption, also rests on ideas of difference (Said 1978, 1994). Saidian postcolonial theory, including later work on imperialism as opposed to orientalism, turns upon the idea that cultural products – and consequent social systems and structures – have for thousands of years subtly produced the west/Europe as superior to the east/Asia. The west requires a homogenous subjugated other in order to understand itself. Other scholars have taken up the concept of orientalism to show that popular media and cultural products (movies, literatures, theatre, music) work to produce an oriental other category that is different from a similarly produced category of western normalcy, or non-otherness. The production of an other rests on reducing diverse times, spaces, and societies into an undifferentiated mass imbued with characteristics like barbarism and savagery against which the west understands itself as modern, cultured, and civilized. This dualistic relationship positions the west as entitled, if not required from a morally righteous perspective, to intervene into the lives and lands of those categorized as different or othered. Cultural production of difference, from a Saidian perspective, forms a legitimating foundation upon which rests social and geographic imperialism, including the dispossession of people from their ways of life, their homelands, and their resources. Said's greatest contribution has been to inspire geographers to think about how the world is spoken and written (and thus conceptualized) through differentiation and deferral of the different, and about how power is produced, enacted, and reliant upon categories of difference (see e.g., Gregory 1995, 2004). Difference matters. It is simultaneously constructed and eminently material. It is also the focus of many who want to destabilize the structures of power that rely upon it.

Categories of otherness have been consistently populated, historically and into the present day, with certain people and places. These include, to name some but by no means all: women; the poor; racialized peoples; Indigenous peoples; displaced peoples; nations and citizens of the global south; nations and citizens colonized by European countries or, more recently, invaded by the United States of America; gay, lesbian, bisexual, or transgendered peoples, and those with restricted mobility or those who live with illness. Those who are (re)produced as different, as other, have always worked to destabilize power, or to resist the strictures of power that attempt to confine or direct them based on difference. How such resistance occurs, and the people who enact it, is the focus of many critical theorists, including feminists, anti-racist scholars, queer theorists, and anti-colonial theorists (e.g., Butler 1993; Fanon 2004 [1963]; Haraway 1989; Spivak 1988, 1999). At the core this scholarship is a problematization and unsettlement of the categories of difference, particularly as those categories work to (re)produce unequal relations of power. These theories do not support an erasure or a disavowal of unique characteristics or the possibilities of variations and diversities: rather, and like this chapter, such approaches take issue with the way that social powers are (re)produced through the fixing of categories of difference. As we explore, places and spaces of difference, subjects marked as different, and the relationship of difference to power are of significant and growing interest to social geographers.

Putting Difference in Its Place: How Geographers Deal with Difference

Social geographers have been concerned with "difference" for decades, but the theoretical and empirical focus of their research has shifted over time. Perhaps the most notable shift characterizing social geography over the past several years has been its (sometimes tense) alignment with cultural geography, and particularly the use of theories of power, representation, and knowledge to understand processes of social differentiation and inequality. As Smith (2000: 25) notes, whereas in past years social geographic inquiry into questions of race and gender would have been "viewed through political economic lenses," beginning in the 1990s such questions became "increasingly subject to cultural deconstruction." Today, the intersection of social and cultural geographic inquiry appears to be on the move again, as social geographers seek to investigate social differentiation and inequality in specific contexts and to overcome some of the limitations of cultural geographic inquiry.

The impact of the so-called "cultural turn" on social geographic scholarship has been extensively documented and debated since the late 1980s (e.g., Gregson 2003; Pain and Bailey 2004; Smith 2000). Most commentators observe that prior to the 1990s social geographers working in a Marxian political-economic tradition primarily understood "difference" in relation to class (Jackson 2000). Although feminist and antiracist scholars have long insisted on the constitutive importance of race and gender in structuring geographies of difference (see Katz 1992; Pratt and Hanson 1994; Rose 1992), race and gender were often understood as subsets of political-economic differentiation rather than as axes of difference in their own right. With the advent of cultural theories, "difference," including class, came increasingly to be conceptualized as a social and cultural construction with material effects, rather than wholly as a result of political-economic stratification (see for instance Domosh 2005). Geographers became increasingly interested in the discursive production of difference, in difference as a form of identity, and in a multiplicity of differences, while interest in the manifestation of social inequalities in housing, education, health, or labor occupied less attention.

The impact of cultural geographic understandings of difference on social geographic scholarship is perhaps most marked in studies of the colonial and postcolonial, where the theoretical insights of Foucault, Said, Bhabha, and Spivak, among others, have exerted significant influence. The ways in which colonial processes produce, inscribe, and maintain "difference" between Indigenous and non-Indigenous peoples has been a rich focus of study in geography, including considering the production of colonizing discourses and the specific interventions made into Indigenous peoples' bodies, lands, and lives (e.g., Anderson 2007, 2001; Blunt and Rose 1994;

Braun 2002; Cameron 2008; Clayton 2000, 2003; de Leeuw 2007; Gregory 2004; Harris 1997, 2002; Lester 2003; McKittrick 2006; Nash 2002). While this line of scholarship has enabled critical engagement with the geographical dimensions of colonial knowledge, representation, and power, a number of scholars have recently identified a worrisome lack of engagement with the colonial *present* (Gregory 2004) in these studies, and a tendency to attend to the representational at the expense of the more pressing material dimensions of colonial injustice (Cameron et al. 2009; Gilmartin and Berg 2007; Noxolo et al. 2008).

Indeed, while exploration of the cultural dimensions of difference has yielded important new insights into difference and its production, cultural approaches to difference are also marked by distinct limitations. These include a reinforcement of dualistic understandings of difference and power, a relative over-emphasis on the textual and representational (as opposed to the lived, specific, and material), and a tendency to focus on the distant and the past rather than the proximal and the present. Social geographers have aimed to redress some of these limitations over the past several years, and to reinvigorate social geographic understandings of difference by attending to the relational, material, specific, and contemporary geographies of difference.

Although recent work in social geography can be partly understood as an effort to redress some of the limitations of more explicitly "cultural" approaches to difference, it is important not to overstate the supposed differences between the social and the cultural, or to shore up a dangerous and untenable division between the "real world" (understood as the domain of social geographic inquiry) and the imagined or constructed world (understood as a the focus of cultural geographic scholarship). Some have indeed cast research into the representational dimensions of social and cultural processes as an abandonment of "the social," understood not just as class but by attention to the lived materiality of difference (see Smith 2000). Others insist that what was at work was an expansion of the "social" to include a broader and more complex set of social experiences and a much broader and more complex understanding of the ways in which social life is constructed. Many of these debates have rested on a false dichotomization of the "real" or material and the imagined or constructed. Geographers are just beginning to redress this dichotomization, particularly as they seek to recuperate social geographic inquiry as a site of engaged, political scholarship. Cultural geographic inquiry has indeed led to an interest in social and cultural construction, but many geographers have mistakenly understood such a focus to involve something that is not material. A more helpful approach is to break down the distinctions, or the perceived differences, between the social and the cultural and to recognize that it is not "rematerialization" that is needed but rather a critical geography that attends to a more comprehensive materiality (Castree 2000, 2004).

The concept of spatiality (as opposed to "space") has been promoted by some social geographers as a useful tool for accommodating this more comprehensive materiality, and for avoiding the tendency to differentiate cleanly between that which is socially constructed and produced from that which is "real," ontological, or material. Particularly in reference to categories of difference that separate people by citizenship, race, or sexuality, we agree with the wide consensus that typologies of difference are deeply imaginary, a product of social construction. We suggest too, and strongly, that these imaginaries bear down upon and materialize in violent and terrifying ways, affecting the ways knowledge is produced and altering people's bodies, lives, and the places in which we live. Thus, as we have written elsewhere, our discussions about difference are anchored in ongoing attempts to disrupt even basic differentiations between the "material" and the "immaterial" (Boyle and Kobayashi forthcoming; Cameron 2009, forthcoming) and to insist that discourses over difference, no matter how many degrees of separation they may involve between particular bodies and their representations, always occur in material land-scapes and always occur through concrete spatialities. Indeed, one of the most significant and fruitful areas of geographic investigation is to understand how the construction of difference involves a complex set of spatial relations created through varying degrees of spatial separation from the site of the colonized body to the global system of capitalist reproduction.

Some interpret a more vigorously "material" social geography to include explicit focus on lived, human experiences of differentiation. For instance, the study and documentation of immigration laws and policies in Canada, Australia, and the United States demonstrate that the histories of each nation relied upon – and to some degree continue to rely upon – the idea that some people are more deserving than others of entry into the countries (Kobayashi 1995). Generally, those thought to be less deserving are neither white, of European decent, nor Christian. The relega-tion of "different" citizens to marginal places – whether urban "ghettos," rural internment camps, or, more recently, specially modified prisons constructed in remote Canadian and Australian landscapes for "illegal" migrants – has been well documented as a geographically specific materialization of difference. Furthermore, and of growing importance since 11 September 2001, migrants to Europe or North American who are seen – to the exclusion of other characteristics – as Muslim, or as coming from places conceptualized as Muslim landscapes, face heightened immi-gration scrutiny, are categorized as deviant, and are understood as having the potential to do harm to normalized (i.e., white, Anglo-Saxon Christian) social structures. These migrants face threats of being deported, tortured, imprisoned without reason, or – in more opaque fashions – simply living under a constant social gaze that views and constructs them as violent extremists who must be feared. In some cases, and as consequence of being cast as different, these citizens do not just face an idea of violence and deportation: for instance, as a direct outcome of state-endorsed ideas about Muslim men being dangerous deviants, Canadian citizen Maher Arar was illegally and secretly flown to Syria after traveling abroad on busi-ness. In Syria he was detained and tortured for over a year, all with the knowledge, sanction, and even – some suggest – the support of the Canadian government (Pither 2008; see also Hyndman and Mountz 2007). Ideas of difference, coupled with the power of entities like states and law-enforcement agencies, have material and embodied outcomes that play out across various global spaces. Indeed, geographic interest in examining the connections between discursive productions of difference and the lived dimensions of difference have been particularly rich in studies of citi-zenship, territoriality, and war (e.g., Cowen and Gilbert 2008; Gregory and Pred 2007).

Other compelling examples of the spatialized and embodied outcomes of differ-ence have also been documented by geographers, including those studying the lives of Indigenous peoples. Around the world, from Canada, New Zealand, Australia and the United States to countries like Bolivia and Guatemala in Central and South America, or in Russia or countries with significant circumpolar Indigenous populations, the social differentiation of Indigenous from settler colonial peoples was the historical impetus for territorial divisions that have extant embodied outcomes (Anderson 2007; Harris 2002, 2004; Gregory 2004). Premised upon typologies that cast them as different, dying, deviant, or undeserving, Indigenous peoples were forcibly relocated to segregated territories that undermined their ways of life and any opportunities for healthy sovereign existences (Alfred and Corntassel 2005; Raibmon 2005). The result is that, globally and when compared with non-Indigenous peoples, Indigenous peoples experience disproportionately higher levels of unemployment, loss of language and culture, lack of education, and disease. These collude and result in high rates of poverty, community violence, and overall ill health (Adelson 2005; de Leeuw et al. 2009). Differentiation, then, has had spatial ramifications and continuing social consequences and individually embodied outcomes.

While these studies of differentiation and categorization add much to discussions about difference in social geography, they also highlight the concept's deep complexity, namely, the impossibility of understanding it within or through dualisms. Thus, in studies about neo- or postcolonial landscapes, it would be a vast oversimplification to theorize social relationships as solely between Indigenous and non-Indigenous peoples. Such a conceptualization risks homogenizing both categories that, when looked at carefully, are actually comprised of vastly varied peoples. To think in nuanced and complex ways about the social geographies of difference in these landscapes means conceptualizing differences and power relationships between differently othered peoples – for instance between racialized (im)migrant peoples and Indigenous peoples, or between Indigenous men and women, or between all the people who hail from different geographies or territories (Kobayashi and de Leeuw 2010).

In response to the limitations of dualistic understandings of difference, some social geographers have turned to "new materialist" theoretical resources in an effort to attend to the constitutive importance of both human and nonhuman "things" in social life and, crucially, to rethink dualistic understandings of race, gender, class, and other locations of difference (Panelli 2010). Arun Saldanha, for example, has been exploring the ontological and material dimensions of race in an effort to move beyond some of the limits of past work emphasizing race as a social product (e.g., Saldanha 2006). Catherine Nash's investigations into the material dimensions of national belonging and relatedness cover similar ground, insofar as they aim to revisit traditional geographic interest in social differentiation through investigations of the embodied, fleshy aspects of human relations. Her study of blood (Nash 2005), for example, investigates traditional social geographic themes such as citizenship and national belonging through a study of the materiality and relational effects of blood. Joyce Davidson's interest in the relational and experiential dimensions of autism spectrum disorder also offers important perspectives on issues of social exclusion (Davidson 2010). These works and others have in common their insistence on understanding difference as a fundamentally spatial relationship, one that is inadequately accounted for through antagonistic and dualistic understandings of power and difference.

Social geographers have not only been concerned with the material effects of difference upon the lives of others, but also with the material ramifications of the ways in which we produce knowledge, the ways we (re)produce the geographies we have chosen for study, and the types of scholarship and research we have produced about those geographies (Kwan 2004). Although significant shifts have occurred in the discipline, including the infusion of critical anti-hegemonic philosophies such as Marxism and feminism, the cultural turn and more recently a plethora of other "turns" such as the ontological and emotional turns, there remains a realization that research foci and methodologies run the risk of (re)entrenching ideas of normal versus different, acceptable versus unacceptable, or interesting versus uninteresting. These categories, in the realm of producing knowledge and writing or researching geography, are infused with questions of power that hold the potential of rendering some subjects legitimate and others not. In more recent decades, these questions of subjects' legitimacy has resulted in questions about the pervasive whiteness of geographical inquiry, the preponderance of studies about public as opposed to private spheres, or the enduring Anglicized nature of the discipline (e.g., Delaney 2002; Desbiens and Ruddick 2006). These, among other focuses that produce certain categories of legibility and legitimacy, are situated within a growing body of inquiry arguing, first, that bounded or categorized ways of conceiving subjects, spaces, and societies are themselves social productions; and, second, that the categories that rely on difference are proving virtually impossible to justify or maintain.

If knowledge has material effects, some social geographers argue that we should engage in research and writing that is more "participatory," "public," "relevant," and "political" (for various interventions into this issue, see DeLyser and Pawson 2005; Dempsey and Rowe 2004; Fuller 2008; Massey 2004; Mitchell 2004; Pain 2004; 2006; Smith 2009). While many argue that, like the division between the "real" and the "imagined," appeals to be more public and political rest on dubious and untenable dichotomies, there is nevertheless a longstanding interest in social geography to *participate* in processes of differentiation and facilitate social transformation. For some, this has involved efforts to study the alternatives we wish to see in the world (see Gibson-Graham 1996, 2008), rather than build on existing work emphasizing marginalization, exclusion, and violence. Indeed, Valentine (2008) documents a shift away from emphasizing social exclusion and conflict, particularly in urban studies, and toward interest in geographies of conviviality, cosmopolitanism, and new forms of belonging. This shift in focus has yielded an important re-examination of previous assumptions about social life, but as Valentine observes, seems to have been "too quick to celebrate everyday encounters and their power to achieve cultural destabilization and social transformation" (p. 334). Moving away from studies of how difference manifests itself as social inequality and exclusion, in other words, runs the risk of abandoning attention to the persistence of difference as a locus of social power.

In sum, if social geographers experienced a lag in the subdicipline when preoccupation with the cultural gained significant disciplinary traction, today they are theorizing difference in new and intriguing ways as part of a process of reinventing social geographic inquiry. If, as Stuart Hall (1993) observed people's capacity to live with difference is the prescient question of the twenty-first century, then recent interest among social geographers in understanding the theoretical, political, and ethical dimensions of difference represents an important development. While we welcome these new developments, we suggest that there might be still other ways of conceptualizing difference in geography, and in the following section we consider these alternatives.

Toward a Different Difference

As discussed, geographers are increasingly destabilizing divisions between subjects, dualisms, and divisions that are the basis of and result in categories of difference (Pile and Keith 1997; Rose 2002; Sharpe et al. 2000; Sparke 2008). These efforts demonstrate the insufficiency of thinking about difference simply in terms of dichotomies or dualisms, particularly in relation to the categories that so much scholarship on otherness has addressed (see also Li 2007; Mawani 2009). To date, however, much geographic research has focused precisely upon such relationships, on the differences of those with power or without power and between those classically conceived as different from each other, including men and women, the rich and the poor, European and other, hetero- and homosexual, white and not-white. Focusing on the relationships between subjects defined appositionally (against that which they are not) has three risks. First, it can miss the complexities of difference *within* (as opposed to *between*) various categories or variously othered subjects. Second, it risks overlooking the ways that subjects might embody various types of characteristics, all of which mark them as different in various times and spaces. Third, there is risk that by remaining with theoretical frameworks reliant on more straightforward categories of difference, or in continuing with modes of thinking that are reliant on dualisms and oppositions, we entrench the very categories and social powers our critiques are trying to destabilize.

Social dynamics tend not to unfold in clean or clear polarities. Subjects change, in relation to others or even in the ways they self-identify, depending on their social and spatial contexts. For example, a first generation Indo-American woman, fighting for her right to marry her same-sex partner in the state of California, might in the context of an interview with Fox Television News choose to highlight her sexual orientation as opposed to her family lineage and connections with an Indo-American community. Furthermore, the same woman may reside on contested land claimed by the Cahuilla Indian Tribe. Add to this complexity, for instance, opposition to her efforts for same-sex marriage by members of the Korean Pentecostal community who are simultaneously, and within the same landscape, fighting a battle against racism that has resulted in members of their community having businesses vandalized. A picture emerges of the complex geographies and relationships between differently positioned subjects who are each uniquely and individually marked as different within a broader white and heteronormative society. If difference simultaneously relies upon and engenders categories that perpetuate social power imbalances, it follows that to undo or destabilize those imbalances requires the dismantling of ideas about difference and of the categories upon which the concept and its outcomes rest and rely. Given the traction that difference has in the ways that societies and places are conceptualized and subsequently organized, and given the depth and longevity of difference as a means of understanding ourselves and each other, it would be overly optimistic to suggest any quick ways of simply doing away with difference. There may be, however, some new ways of thinking about difference, particularly as it unfolds spatially, which trouble its fixity or its taken-for-granted-ness as a necessary means of thinking about and organizing the world in which we live.

How might social geographers think differently about difference? Here we consider a series of artistic interventions into the question of difference, and ponder what they might offer social geographers as they seek to carry out politically engaged and theoretically rich investigations into the geographies of difference. It may seem that in so doing we are flying in the face of recent criticisms that "re-materialized" geographies need to rely on "more than representation." To the contrary, however, we argue, following Foucault, that representation is a profoundly material practice, and insofar as representations assist in demarcating the discursive frames within which social and political change might be possible, representation continues to demand our attention.

In 1966, in London's Indica Gallery, a still relatively unknown artist named Yoko Ono mounted her *Ceiling Painting (YES Painting)*, an installation comprised of paper, glass, a metal frame, metal chain, and a painting ladder. Viewers were invited to climb the ladder and, through a magnifying glass hanging from the ceiling, read the tiny text printed on the gallery's ceiling. Now famous, it was the word "YES" stenciled on the ceiling that purportedly moved John Lennon to fall almost instantly in love with Yoko Ono. "YES" unhinged a series a sociospatial divides. It was printed on the ceiling, eschewing expectations that letters be spatially confined to texts. It hovered above readers, demanding that they climb, and then lean backwards to read it. "YES" thus subverted the generally differentiated relationships between reader/onlooker and word/represented subject wherein the former (a viewer) maintains some power over the latter (a word) by looming over it. And, perhaps somewhat in the realm of cultural myth, "YES" also quickly and effectively obliterated the divides between Yoko Ono and John Lennon, spurring a set of relations and creative collaborations that throughout the 1960s and 1970s worked toward social justices and the dismantling of categories infused with power. As an art object and creative expression, *Ceiling Painting (YES Painting)* is not unique in subverting or undermining generally accepted categories of difference as means of orienting to the world. Unsettling divides and shaking categories that differentiate subjects from one another is the work of many artists and the works they produce.

Two other artists, both from Canada, provide more interesting ways to think about and see the world beyond established or well-honed categories of difference. Shelley Niro, an Aboriginal artist, produces works aimed expressly at destabilizing viewers' assumptions and taken-for-granted categories about Aboriginal identity. Her photograph series *Mohawks in Beehives* features close-up portraits of her mother and aunties, staring frankly and amusedly at a more than likely not-Mohawk viewer, fresh from a beauty salon and festooned with carefully set "up-do" hair, glittery accessories, and intricately applied makeup. Niro's purpose is to disrupt normative assumptions about Indigenous women's identity and to establish the intractability of categories that differentiate Indigenous from non-Indigenous.

Viewers of Niro's work are shown modern, urban, happy resilient Aboriginal women who clearly enjoy beauty spas and shopping at the mall. The images refute categorizations of Indigenous peoples as pre-modern, suffering, anti-capitalist or intrinsically and spiritually close to nature, tropes that, from Niro's perspective, have served falsely to differentiate, and thus other, Indigenous people from non-Indigenous Canadians. In her efforts to destabilize categorical assumptions about Indigenous women, Niro works toward new and more socially just geographies in Canada. Her photographs self-consciously return the gaze of a colonial country invested in maintaining the power structures that accompany categorical differentiations between various citizens and, from Niro's perspective, serve to perpetuate the longstanding marginalization of Indigenous women. About her work, Niro observes that "some people think that to be Indian, you have to do certain things, but I'm just saying that you're Indian no matter what you do ... I'm always thinking about that" (Shelly Niro quoted in Abbot 1994/6). In other words, what Niro offers is a lesson in refuting differences or binaries as the means to construct identities about self or others, by depicting them as "out of place" according to established norms. She offers a lesson in side-stepping dichotomous categories that perpetuate difference, and in doing so undermines techniques of power that rely on people and subjects being conceptualized as different and divided. She does not in the process erase difference, but challenges the history of making it and the necessity of inhabiting it on others' terms.

In a very different way, but still destabilizing binaries and dualisms as means of categorizing the world, sound and installation artists Janet Cardiff and George Bures Miller work with the concepts of antiphonality and binaurality. In their installation *The Forty Part Motet*, in which forty speakers are arranged in a large circle with each speaker projecting a single unique voice, listeners/viewers can either stand close by one speaker to hear a single voice or they can choose to stand in the center of the circle and listen to the voices sing collectively. The project reveals the falsity of categorizations, separations, and oppositionalities: we perceive sound with two ears, but not by categorizing sounds as distinct and different per se. Instead, and even if they are transmitted from two sources, which could suggest or engender dualisms and privileging of one element over another, we actually understand sounds as behaving stereophonically together. As Cardiff and Miller's installation renders tangible, sound and the act of listening to sound are only possible because two (or multiple) elements work together, responsively, alternating in turns, resulting in an antiphon. Difference, then, is unhinged from privileging and, by extension, from categorizations of power.

Each of the three artists provides an example of conceptualizing the world beyond categories of difference, beyond divisions that fix one thing as over, or under, or as different from something else, the result of which are relationships infused with power imbalances. What might these works offer to geographers as we re-examine our understandings of difference? Recent interest in theories of intersectionality offers some intriguing possibilities. As alluded to earlier in the chapter, geographers concerned with questions about social operations and techniques of power have most often turned their attention toward "othered" subjects. What has been missing from some of these studies of difference and power, however, is a way to understand power and difference as variegated and multidimensional forces that ebb and flow depending on who is enacting or embodying them. It is no longer sufficient, as is increasingly being documented particularly by multiply othered subjects, to straightforwardly suggest that one subject has power while another does not, or that power (or lack of power) is one-dimensional or fixed. Put another way, one subject's power does not always come at the expense of another's and differences differ depending on a great number of factors. One way to capture and address these conceptual conundrums is to understand social geographies through a lens of intersectionality (Valentine 2007). An intersectionality-type approach to theorizing social geographies encourages power to be theorized as reliant on systems and matrices, and difference to be conceptualized as an unfixed, relational, and complex categorization. Intersectionality is a relatively recent theory that has yet to be fully explored either across geography writ large or within social geography. Intersectionality is a theory that wrestles with, and attempts to explain, how socioculturally constructed categories (predominantly but not exclusively categories such as gender, ethnicity, and sexual orientation) interact with and impact upon one another to produce differentially lived social inequalities among peoples (Buitelaar 2006; Cole 2008; Collins 2000; Davis 2008; Hancock 2007; see also Anthias 1998, 2001). It makes effort to destabilize fixed categories of difference and to understand instead that one mode of difference does not necessarily hold sway over another. Instead, identities and spaces are comprised of multiple, interlocking axes of power that are best understood by their overlaps as opposed to their divides.

The theory of intersectionality arose as a means to examine, in nuanced and complex ways, how varied forms of otherness interact to produce differently striated realities of social marginalization and social exclusion (Collins 2000). The theory arose in response to enduring understandings of power relationships and inequalities that focused predominantly on gendered otherness as opposed to other categories that equally result in social marginalization. This focus, some critics leveled, led to gender and sex as more broadly understood to be mutable categories, subject to change, whereas typologies of race, ethnicity, or states of wellness and illness remained biological and thus not subject to critiques about power (see also Moore et al. 2003). Intersectionality, then, might best be broadly understood as a theory that accounts for the relationships between categories of otherness and the ensuing marginalization resulting from those relationships (McCall 2005). Theories of intersectionality offer the potential to understand otherness, including the othered social geographies of so many people, as complex aggregates comprised of shifting and always relational parts that are always being constituted in particular times and places.

There are a number of reasons why intersectionality-type research fits well with geographic inquiry. In its conceptualization of power not as static or fixed with reference to difference, but as a malleable and shifting process of differentiation, as a force that operates in and through specific sites, scales, historical contexts, and times, an intersectionality approach to social geographies can assist in illuminating the spatial specificities of social hierarchies or power imbalances. Although it might seem easier to understand social geographies through categorizations or cohorts – based for instance on population census or distinct neighborhoods or specific typologies such as race or sexual orientation – what such an analysis elides is the complexities of peoples' spatial relationships, with each other in different locations, and with the variable ways that people shape and are shaped by the places in which they live. Intersectionality embraces complexities and complexifies difference, all the while realizing that geography is an integral component of questions about categories, power, and difference. As a theoretical lens, this approach to power and categories of difference has potential for topics increasingly being taken up by social geographers, including questions about (im)migration, nation-states, political violence, relationships in settler-colonial landscapes, or life in global cities.[1] In the case of settler-colonial countries, for instance, there must be recognition about the arrays of differently racialized peoples, many of whose lineages can be traced to forced migration, who now co-exist on territories claimed by Indigenous peoples. Moreover, with both (im)migrant and Indigenous peoples, there are ranges of education and income levels, assortments of sexual orientations, and multiple other states of being. Landscapes of overlapping differences and intersecting categories of otherness are a reality in globalized cities around the world. So are the distances that difference creates and sustains. Movement and intersection are hallmarks of these spatialities, and identities and relationships between multiple subjects are in constant states of flux, their power dynamics never fully fixed. Intersectionality recognizes and validates the vast power imbalances that exist across various geographies and between multiple subjects. What the theory avoids, though, is approaching such gaps and disparities in simple or categorical ways, and this is what makes intersectionality a "different" approach to difference.

Concluding Differences

We began this chapter with the supposition that difference matters. Difference is the historical condition of human relationship. How difference is understood, how it is produced, and how it manifests also matter. The power to differentiate runs the gamut from personal relations to colonizing powers, and includes the power of the scholar, from Linneaus with his compulsion to categorize to contemporary thinkers who wish to study the process of differentiation itself. One of the most compelling reasons difference has been studied is its collusive relationship with power. Categorizations, and the differences that categories engender, produce power relations that ultimately result in social hierarchies across vast times and distances, at the expense of a great many others. By theorizing ways to destabilize categories of difference and by examining new ways of understanding difference that are not reliant on firm or fixed categories, we make effort to sort through possible ways of examining social relationships that do not perpetuate power imbalances. We cannot erase difference. We suggest that social geography, specifically a social geography concerned with social justice, holds the potential of theorizing spatialities and human relations in ways that might move us away from difference as a dominant means of organizing and understanding the world in which we live.

Note

1   Intersectionality is not the only theoretical lens through which social geographers might re-examine the social geographies of difference. We note, for example, the growing

interest in "care" as a framework for making sense of social relations (e.g., Conradson 2003; Lawson 2007; Pratt 2009; Raghuram et al. 2009).

References

Abbot, L. (1994/6). Interview with Shelley Niro. *A Time of Visions*. Accessed October 2008, www.britesites.com/native_artist_interviews/sniro.htm.

Adelson, N. (2005) The embodiment of inequalities: health disparities in Aboriginal Canada. *Canadian Journal of Public Health* 96: 45–61.

Alfred, T. and Corntassel, J. (2005) Being indigenous: resurgences against contemporary colonialism. *Government and Oppression* 40: 597–614.

Anderson, K. (2001) The nature of race. In N. Castree and B. Braun (eds.), *Social Nature: Theory, Practice and Politics*. Oxford: Blackwell, pp. 64–83.

Anderson, K. (2007) *Race and the Crisis of Humanism*. London and New York: Routledge.

Anthias, F. (1998) Rethinking social divisions: some notes towards a theoretical framework. *Sociological Review* 46 (3): 557–80.

Anthias, F. (2001) Beyond feminism and multiculturalism: locating difference and the politics of location. *Women's Studies International Forum* 25 (3): 275–86.

Boyle, M. and Kobayashi, A. (forthcoming) A critical interrogation of metro-centrism in Sartre's theory of colonialism: towards a theory of situated universals.

Buitelaar, M. (2006) "I am the ultimate challenge": accounts of intersectionality in the life-story of a well-known daughter of Moroccan migrant workers in the Netherlands. *European Journal of Women's Studies* 13 (3): 259–76.

Blunt, A., and Rose, G. (eds.) (1994) *Writing Women and Space: Colonial and Postcolonial Geographies*. New York: Guilford Press.

Braun, B. (2002) *The Intemperate Rainforest: Nature, Culture, and Power on Canada's West Coast*. Minneapolis: University of Minnesota Press.

Braun, B. (2003) "On the raggedy edge of risk": articulations of race and nature after biology. In D. S. Moore, J. Kosek, J., and A. Pandian (eds.), *Race, Nature, and the Politics of Difference*. Durham, NC: Duke University Press, pp. 175–203.

Butler, J. (1993). *Bodies That Matter*. London: Routledge.

Cameron, E. (2008). Indigenous spectrality and the politics of postcolonial ghost stories. *Cultural Geographies* 15 (3): 383–93.

Cameron, E. (2009). The ordering of things: narrative geographies of Bloody Falls and the central Canadian Arctic. PhD thesis, Department of Geography, Queen's University, Kingston, ON.

Cameron, E. (forthcoming) Copper stories: imaginative geographies and material orderings of the central Canadian Arctic. In A. Baldwin, L. Cameron, and A. Kobayashi (eds.), *Rethinking the Great White North*. Vancouver: UBC Press.

Cameron, E., de Leeuw, S., and Greenwood, M. (2009) Indigeneity. In N. Thrift and R. Kitchen (eds.), *International Encyclopedia of Human Geography*. London: Elsevier, pp. 352–7.

CBC News (July 10, 2009) Apparent B.C. hate crime leads to 3 arrests. Vancouver: Canadian Broadcasting Corporation. Accessed March 13, 2010, www.cbc.ca/canada/british-columbia/story/2009/07/07/bc-courtenay-hate-crime-reaction.html#ixzz0i6qBYT2r.

Castree, N. (2000) What kind of critical geography for what kind of politics? *Environment and Planning A* 32: 2091–5.

Castree, N. (2004) Differential geographies: place, indigenous rights, and "local" resources. *Political Geography* 23: 133–67.

Cole, E.R. (2008) Coalitions as a model of intersectionality: from practice to theory. *Sex Roles* 59: 443–53.

Collins, P.H. (2000) *Black Feminist Thought: Knowledge, Consciousness, and the Politics of Empowerment*. New York: Routledge.

Conradson, D. (2003) Geographies of care: spaces, practices, experiences. *Social and Cultural Geography* 4 (4): 451–4.

Cowen, D. and Gilbert, E. (eds.) (2008) *War, Citizenship, Territory*. New York: Routledge.

Clayton, D. (2000) *Islands of Truth: The Imperial Fashioning of Vancouver Island* Vancouver: UBC Press.

Clayton, D. (2003) Critical imperial and colonial geographies. In K. Anderson, M. Domosh, N. Thrift, and S. Pile (eds.), *The Handbook of Cultural Geography*. London: Sage, pp. 354–68.

Crampton, J.W. and Elden, S. (eds.) (2007) *Space, Knowledge and Power: Foucault and Geography*. Aldershot: Ashgate.

Davidson, J. (2010) "It cuts both ways": a relational approach to access and accommodation for autism. *Social Science and Medicine* 70 (2): 305–12.

Davis, K. (2008) Intersectionality as buzzword: a sociology of science perspective on what makes a feminist theory successful. *Feminist Theory* 9 (1): 67–85.

Delaney, D. (2002) The space that race makes. *Professional Geographer* 54 (1): 6–14.

de Leeuw, S. (2007) Intimate colonialisms: the material and experienced places of British Columbia's residential schools. *Canadian Geographer* 51 (3): 339–59.

de Leeuw, S., Greenwood, M., and Cameron, E. (2009) Deviant constructions: how governments preserve colonial narratives of addictions and poor mental health to intervene into the lives of indigenous children and families in Canada. *International Journal of Mental Health and Addictions* 81 (2): 282–95.

DeLyser, D. and Pawson, E. (2005) From personal to public: communicating qualitative research for public consumption. In I. Hay (ed.), *Qualitative Research Methods in Human Geography*. Oxford: Oxford University Press, pp. 266–74.

Dempsey, J., and Rowe, J.K. (2004) *Why poststructuralism is a live wire for the left*. In D. Fuller and R. Kitchin (eds.), *Radical Theory/Critical Praxis: Making a Difference Beyond the Academy?* Vernon and Victoria: Praxis (e)Press, pp. 32–51.

Derrida, J. (1982). Differance. In J. Derrida (ed.), *Margins of Philosophy*. Chicago: University of Chicago Press, pp. 3–27.

Desbiens, C., and S. Ruddick. (2006) Speaking of geography: language, power, and the spaces of Anglo-Saxon "hegemony." *Environment and Planning D: Society and Space* 24 (1): 1–8.

Domosh, M. (2005) An uneasy alliance? Tracing the relationships between feminist and cultural geographies. *Social Geography* 1: 37–41.

Fabian, J. (1983) *Time and the Other: How Anthropology Makes Its Object*. New York: Columbia University Press.

Fanon, F. (2004) [1963] *The Wretched of the Earth*. New York: Grove Press.

Foucault, M. (1972) *The Archaeology of Knowledge*. Tr. A. M. Sheridan. New York: Pantheon Books.

Foucault, M. (1977) *Discipline and Punish: The Birth of the Prison*. London: Penguin.

Foucault, M. (1991) Questions of method. In G. Burchell, C. Gordon, and P. Miller (eds.), *The Foucault Effect: Studies in Governmentality with Two Lectures and an Interview with Michel Foucault*. Toronto: Harvester Weatsheaf, pp. 73–86.

Fuller, D. (2008) Public geographies: taking stock. *Progress in Human Geography* 32 (6): 834–44.

Gibson-Graham, J.K. (1996) *The End of Capitalism (As We Knew It)*. Oxford: Blackwell.

Gibson-Graham, J.K. (2008) Diverse economies: performative practices for "other worlds." *Progress in Human Geography* 32 (5): 613–32.

Gilmartin, M. and Berg, L. (2007) Locating postcolonialism. *Area* 39 (1): 120–4.

Gould, S.J. (1994) The geometer of race. *Discover* November: 65–9.

Gregory, D. (1995) Imaginative geographies. *Progress in Human Geography* 19: 447–85.

Gregory, D. (2004) *The Colonial Present: Afghanistan, Palestine, Iraq.* Oxford: Blackwell.

Gregory, D. and Pred, A. (eds.) (2007) *Violent Geographies.* New York: Routledge.

Gregson, N. (2003) Reclaiming the "social" in social and cultural geography. In K. Anderson, M. Domosh, N. Thrift, and S. Pile (eds.), *The Handbook of Cultural Geography.* London: Sage, pp. 43–57.

Hall, S. (1993) Culture, community, nation. *Cultural Studies* 7: 249–63.

Hancock, A.-M. (2007) When multiplication doesn't equal quick addition: examining intersectionality as a research paradigm. *Perspectives on Politics* 5 (1): 63–79.

Haraway, D. (1989) *Primate Visions: Gender, Race, and Nature in the World of Modern Science.* New York: Routledge.

Harris, C. (1997) *The Resettlement of British Columbia: Essays on Colonialism and Geographical Change.* Vancouver: UBC Press.

Harris, C. (2002) *Making Native Space: Colonialism, Resistance, and Reserves in British Columbia.* Vancouver: UBC Press.

Harris, C. (2004) How did colonialism dispossess? Comments from an edge of empire. *Annals of the Association of American Geographers* 94 (1): 165–82.

Hui, A and Lavoie, J. (July 2009) Two more arrested after racist attack caught on video. *The Vancouver Sun.* March 13, 2010, www.vancouversun.com/news/more+arrested+after+racist+attack+caught+video/1768710/story.html?id=1768710

Hyndman, J. and Mountz, A. (2007) Refuge or refusal: geography of exclusion. In D. Gregory and A. Pred (eds.), *Violent Geographies.* New York: Routledge, pp. 77–92.

Jackson, P. (2000) Difference. In R. J. Johnston, D. Gregory, G. Pratt, and M. Watts (eds.), *The Dictionary of Human Geography.* Oxford: Blackwell, pp. 174–5.

Kant, I. (1997) [1724] This fellow was quite black … a clear proof that what he said was stupid. In E. C. Eze (ed.), *Race and the Enlightenment: A Reader.* Malden MA: Blackwell, pp. 38–64.

Katz, C. (1992) All the world is staged: intellectuals and the projects of ethnography. *Environment and Planning D: Society and Space* 10 (5): 495–510.

Kobayashi, A. (1995) Challenging the national dream: gender persecution and Canadian immigration law. In P. Fitzpatrick (ed.), *Racism, Nationalism and the Rule of Law.* London: Dartmouth, pp. 61–74

Kobayashi, A. and de Leeuw, S. (2010) Colonialism and the tensioned landscapes of indigeneity. In S. Smith, R. Pain, S. Marston, and J.P. Jones III (eds.), *The Handbook of Social Geography.* London: Sage, pp. 118–39.

Kobayashi, A. and L. Peake (1994) Un-natural discourse: "race" and gender in geography. *Gender, Place and Culture* 1 (2): 225–44.

Kwan, M.-P. (2004) Beyond difference: from canonical geography to hybrid geographies. *Annals of the Association of American Geographers* 94 (4): 756–63.

Lawson, V. (2007) Geographies of care and responsibility. *Annals of the Association of American Geographers* 97 (1): 1–10.

Lester, A. (2003) Colonial and postcolonial geographies. *Journal of Historical Geography* 29 (2): 277–88.

Li, T.M. (2007) *The Will to Improve: Governmentality, Development and the Practice of Politics.* Durham, NC, and London: Duke University Press.

Livingstone, D.N. (2002) Race, space and moral climatology: notes toward a genealogy. *Journal of Historical Geography* 28: 159–80.

Massey, D. (2004) Geographies of responsibility. *Geografiska Annaler* 86: 5–18.

Mawani, R. (2009) Colonial proximities, crossracial encounters and juridical truths in British Columbia, 1871–1921. Vancouver: UBC Press.

Mayhew, S. (2009) *Difference: A Dictionary of Geography*. Oxford University Press Oxford Reference Online. Oxford University Press. University of British Columbia. March 2010, www.oxfordreference.com/views/ENTRY.html?subview=Main&entry=t15.e885.

McCall, L. (2005) The complexity of intersectionality. *Signs: Journal of Women in Culture and Society* 30 (3): 1771–1800.

McKittrick, K. (2006) *Demonic Grounds: Black Women and the Cartographies of Struggle*. Minneapolis: University of Minnesota Press.

Mitchell, D. (2004) Radical scholarship: a polemic on making a difference outside the academy. In D. Fuller and R. Kitchin (eds.), *Radical Theory/Critical Praxis: Making a Difference Beyond the Academy?* Vernon and Victoria: Praxis (e)Press, pp. 21–31.

Moore, D.S., Kosek, J., and Pandian, A. (eds.) (2003) *Race, Nature, and the Politics of Difference*. Durham, NC: Duke University Press.

Nash, C. (2002) Cultural geography: postcolonial cultural geographies. *Progress in Human Geography* 26 (2): 219–30.

Nash, C. (2005) Geographies of relatedness. *Transactions of the Institute of British Geographers* 30 (4): 449–62.

Noxolo, P., Raghuram, P., and Madge, C. (2008) "Geography is pregnant" and "geography's milk is flowing": metaphors for a postcolonial discipline? *Environment and Planning D: Society and Space* 26: 146–68.

Oxford English Dictionary. Online (1989) 2nd edn. http://dictionary.oed.com/cgi/entry/50063748?query_type=word&queryword=difference&first=1&max_to_show=10&sort_type=alpha&result_place=1&search_id=jc6a-4oJL73–12014&hilite=50063748. Accessed September 2010.

Pain, R. (2004) Social geography: participatory research. *Progress in Human Geography* 28 (5): 652–63.

Pain, R. (2006) Social geography: seven deadly myths in policy research. *Progress in Human Geography* 30 (2): 250–9.

Pain, R. and Bailey, C. (2004) British social and cultural geography: beyond turns and dualisms? *Social and Cultural Geography* 5 (1): 319–29.

Panelli, R. (2010). More-than-human social geographies: posthuman and other possibilities. *Progress in Human Geography* 34 (1): 79–87.

Pile, S. and Keith, M. (eds.) (1997) *Geographies of Resistance*. London and New York: Routledge.

Pither, K. (2008) *Dark Days: The Story of Four Canadians Tortured in the Name of Fighting Terror*. Toronto: Penguin.

Pratt, G. (2009) Circulating sadness: witnessing Filipina mothers' stories of family separation. *Gender, Place & Culture* 16 (1): 3–22.

Pratt, G. and Hanson, S. (1994). Geography and the construction of difference. *Gender, Place and Culture* 1 (1): 5–29.

Raghuram, P., Madge, C., and Noxolo, P. (2009) Rethinking responsibility and care for a postcolonial world. *Geoforum* 40 (1): 5–13.

Raibmon, P. (2005) *Authentic Indians: Episodes of Encounter from the Late Nineteenth-Century Northwest Coast*. Durham, NC, and London: Duke University Press.

Rose, G. (1993) *Feminism and Geography: The Limits of Geographical Knowledge*. Minneapolis: University of Minnesota Press.

Rose, M. (2002) The seductions of resistance: power, politics and a performative style of systems. *Environment and Planning D: Society and Space* 20: 383–400.

Said, E. (1978) *Orientalism*. New York: Vintage Books.

Said, E. (1994) *Culture and Imperialism*. New York: Vintage Books.

Saldanha, A. (2006) Reontologising race: the machinic geography of phenotype. *Environment and Planning D: Society and Space* 24: 9–24.

Sharpe, J.P., Routledge, P., Philo. C., and Paddison, R. (eds.) (2000) *Entanglements of Power: Geographies of Domination/Resistance*. London and New York: Routledge.

Smith, N. (2000) Socializing culture, radicalizing the social. *Social and Cultural Geography* 1 (1): 25–8.

Smith, S. (2009) Everyday morality: where radical geography meets normative theory. *Antipode* 41 (1): 206–9.

Sparke, M. (2008) Political geography – political geographies of globalization III: resistance. *Progress in Human Geography* 32 (3): 423–40.

Spivak, G.C. (1988) Can the subaltern speak? In C. Nelson and L. Grossberg (eds.), *Marxism and the Interpretation of Culture*. Urbana: University of Illinois Press, pp. 271–316.

Spivak, G.C. (1994) Responsibility. *Boundary* 21 (3): 19–64.

Spivak, G.C. (1999) *A Critique of Postcolonial Reason: Toward a History of the Vanishing Present*. Cambridge, MA: Harvard University Press.

UNESCO (1956) *The Race Question in Modern Science*. Paris: UNESCO

Valentine, G. (2007) Theorizing and researching intersectionality: a challenge for feminist geography. *Professional Geographer* 59 (1): 10–21.

Valentine, G. (2008) Living with difference: reflections on geographies of encounter. *Progress in Human Geography* 32 (3): 323–37.

Chapter 3

Identification

Katharine McKinnon

Introduction

In northern Thailand highland peoples have long been subjected to exclusion and discrimination. They face the same difficulties as many Indigenous peoples across the globe: denied citizenship, scapegoated for a variety of social and environmental ills, subjected to racism, denied basic rights. But in Thailand the term Indigenous is not generally used to describe highlanders. Instead they are most often called hill tribes (*chao khao*), or ethnic minorities (*chao klum noi*). In the Thai city of Chiang Mai in mid-2007 a group of village leaders and NGO workers representing highland communities gathered at a workshop on Indigenous Futures to discuss what it means to be Indigenous. For much of the day the representatives debated whether the term "Indigenous" ought to be applied to themselves and their people, and if so what Thai language term ought to be used to translate "Indigenous" appropriately. The debate about terminology went back and forth around a handful of possible terms all of which used the term "*chon*" or "*chao*," meaning group, as a component. One prominent village leader, well known as a campaigner for highland culture and rights, made the point repeatedly and insistently that it would be better to "use "*khon*" which means people, not *chao* this or *chao* that." Referring to the autochthonous name for his people, *Pgakeryaw*, he spoke about how his grandmother brought him up to believe that "*Pgakeryaw* means we are people, not animal" (Indigenous Futures Workshop, May 29, 2007). He asked why we had to speak of being a group, instead of speaking of being simply human beings, deserving of the same respect and dignity as all human beings.

His comment pointed to the way that, as "hill tribes," highlanders have often been denigrated as primitives and denied the same rights and opportunities granted to other members of the Thai state. It was this very lack of recognition that the

A Companion to Social Geography, First Edition. Edited by
Vincent J. Del Casino Jr., Mary E. Thomas, Paul Cloke, and Ruth Panelli.
© 2011 Blackwell Publishing Ltd. Published 2011 by Blackwell Publishing Ltd.

highland representatives gathered were working to redress – as they have been for many years. Ultimately, however, the elder's call to speak of people rather than groups was ignored by his colleagues: the term *chon* was included in all of the possible translations for "Indigenous" proposed at the end of the day. This points to a vital underlying fact of contemporary life – there is no escaping being named and identified as part of a group. Furthermore, there is no escaping being identified as part of an officially recognized and sanctioned group: a gender, a race, an ethnicity, a nation. Identification is a central aspect of contemporary life and to be a fully fledged member of modern societies around the world one must be identified, possess identification papers, be a named member of a group. As others present at the workshop acknowledged, it is not enough simply to be a human being.[1] To claim a place of legitimacy and respect within the state one must be a particular kind of human being – one with the right identification.

In a way this chapter is about what it takes to be recognized as the right kind of human being by the state and thus *not* to be treated – as highlanders have been in the past – as dirty and dangerous (animals) or as primitives yet to be socialized to a civilized world. More than this however it is about the processes through which we are identified in the systems of governance and power that prevail in the contemporary world and what these processes mean both for how we are *subjected* to the machinations of power in the world and how we may act within and upon them.

This chapter considers what identification is, what it means and how it works. Identification is a way of thinking about how we make ourselves, how we name ourselves, how we place ourselves, how we become the labels by which we explain ourselves to others: I am a mother, a woman, a wife, an American, a New Zealander, a citizen, a student, a teacher, a cyclist, a dancer. What do these labels mean? How are they linked to my sense of who I am? And most importantly, what do they *do*? What possibilities come into being with particular processes of identification and what other possibilities are closed off? How do power relationships shift as certain identities are called into being?

This chapter explores some of these complex questions and introduces how theories of identification have helped geographers understand important aspects of contemporary social and political life. I begin with a brief literature review then offer my own definition of what identification is and how the concept has been used. Next, drawing on examples from my own research in Thailand I discuss how theories of identification can help to understand the emerging Indigenous rights movement and associated political wrangling around nation, ethnicity, and belonging. As with any area of social theory identification is an idea that is not only useful for how it helps us to understand the world but for how it helps us to transform it. In the last part of this chapter therefore I will explore how Indigenous activists are engaging in identity politics as a powerful tool within broader strategies for positive social change in their communities and nations.

Geographies of Identification

Social geographers have been writing about matters of identity and identification at least since the 1970s when the two very different philosophical traditions of Marxism and Humanism began to shape streams of geographic thought.

Humanistic geography was perhaps the first and most radical shift towards close consideration of the self in geography. At a time when the discipline was dominated by spatial science and positivism, humanist geographers built on the Berkeley school of cultural geography, shifting their focus to the realm of subjective human experience (see Tuan 1976; Ley and Samuels 1978). The field claimed to be establishing a "truly human" human geography focused on subjective experiences of place, giving a central position to human awareness, human agency, and the power of human creativity. Identity was for humanistic geographers a matter of "essential" human characteristics and emerged in relationship with place.

In sharp contrast, early Marxist engagements with questions of identity drew on what might be called an anti-humanist perspective which asserted that humanism had overestimated the free agency and creative power of the human subject to the exclusion of the structural conditions that shape and delimit social systems. What characterizes early Marxist engagements with identity and processes of subject formation is a concern for how they are linked with broad processes of oppression and injustice. It was the material consequences of identities that interested Marxist geographers rather than the nature of that identity itself. Class identities were understood as crucial in shaping economic opportunities and were reflected in diverse empirical settings, whether in the spatial segregations of modern cities (Harvey 1973, 1977), the structures that reproduce poverty (Peet 1977), or the cultural imperialism of Anglo-American societies (Buchanan 1977). For these scholars their primary responsibility as geographers was to understand and write about the oppressions and injustices of the world in order to contribute to a process of change (Harvey 1973). Class identities were crucial in delineating the populations whose emancipation was the focus of such change.

Working from a strict structuralist position most assumed the subordination of the individual to dominant social and economic structures. Structuration theory (Giddens 1976, 1979) drew on something of a middle ground between this foregrounding of structure and a humanist tendency to prioritize the power of individual agency. As Nelson notes, " 'the subject' within structuration theory represented a negotiation between the hopelessly determined subject of structural Marxism and the volunteerism of humanist perspectives" (1999: 334). At the same time as the subject was understood to be determined by dominant structures, those drawing on Giddens recognized the power of individuals and groups in shaping those structures.

The emergence of feminist geography challenged many of the assumptions made by geographers drawing on humanism, Marxism and structuration theory. Pointing out the gendered nature of oppression, feminists challenged the underlying representations of class struggle put forward by Marxist geographers (Massey 1994) and sought to rebalance the gender bias in geographic research (Monk and Hanson 1982). Likewise, the humanists assumptions that they could speak from a "universal human experience" were criticized by feminists who pointed out that gender (and race and class) had been made invisible by such research (Rose 1993). As Derek Gregory notes "the subject of humanism was a fiction constructed through an ideology which suppressed the multiple ways in which human beings are constructed: these erasures both promoted and privileged a white, masculine, bourgeois, heterosexual subject as the norm" (2004: 363). In questioning such norms feminist geographers started to point to the ways normative identities do not reflect innate human characteristics (as the humanists would have it) but are constructed in and through dominant discourses. This perspective developed through the 1990s as feminist critiques drew increasingly on poststructuralism to work towards greater anti-essentialism, embracing "multiplicity, difference and the 'decentred' subject" (Nelson, 1999: 334).

With the rising influence of feminism and poststructuralism in geography, the cultural turn of the 1990s saw a wider shift to anti-essentialist perspectives with the recognition of the contingent nature of knowledge and its inevitable situatedness in the machinations of social power (Benko and Strohmayer 1997; Gibson-Graham 1994; Natter and Jones 1997; Soja 1989). Drawing inspiration from the work of key poststructural thinkers such as Foucault and Derrida, geographers became increasingly attentive to texts, to discourse and discursive formations, and the politics of representation as an important pathway to understanding the social. Identity could no longer be understood as simply the reflection of a fundamental inner truth but as something constructed somewhere between the imposed norms and expectations of our social contexts and the limited agency that individuals could apply to their lives. With the influence of poststructuralist and feminist thought increasingly nuanced understandings of identity and the subject have emerged, firmly positioned within perspectives that question the existence of absolute truths and the legitimacy of meta-narratives and overarching theories employed to explain reality. Geographic research around identity began to delve more deeply into the politics of difference, unpacking how discourses of, for example, gender (Pratt 2004), race and ethnicity (Winders, Jones and Higgins 2005), class and economy (Gibson-Graham 1996, 2006), nation and democracy (Hardwick and Mansfield 2009; Kong and Yeoh, 1997), genealogy and descent (Nash 2002, 2003, 2008) intersect with intricate machinations of power to produce powerful normative subjects that seldom recognize adequately the diversity and complexity of human experience.

Although much of contemporary geography's engagements with theories of identity and the subject are significantly different to Marxist perspectives of the 1970s many geographers remain concerned with questions of injustice and social change. Whether the focus is on better understanding inequalities in the labor market (McDowell 2004), the construction of ethnic identities in the world music scene (Connell and Gibson 2004), the contingency of race (Holloway 2000), or the everyday and embodied constitutions and contestations of gender identities (Bondi 2005), there is often an underlying ethical concern to unravel the complex ways in which ideas about identity act within and upon the world. Liz Bondi, for example, explores how an ideology of separate gendered spheres is "encoded within the fabric of cities" (2005: 10) so that normative ideas about gender roles are seen to be imposed upon individuals through the way the city functions. At the same time, she argues, gender is "performed" by individuals who consciously transgress the norms of heterosexual gender roles, transcending and reshaping gender roles in everyday urban life.

Bondi's use of the idea of performance and performativity, taken from Judith Butlers work, is a good example of how contemporary social geographies explore how identification not only reveals acts of oppression, but also provide insights into how dominant paradigms may be resisted and transformed. Some geographers explicitly seek to extend the understanding of how identities can be linked to transformative acts through action research projects that work with communities towards social change. Examples include Lakshman Yapa's engagement with urban poor in Philadelphia, Gibson-Graham's work on community economies in the LaTrobe Valley and the Philippines (2005, 2006). Gibson-Graham in particular theorize that the transformative work of creating and sustaining a community economy begins with a rethinking of subjects – not as individuals subjected to the whims of global capitalism, but as capable, knowledgeable and resourceful subjects. The ways that shifts in subjectivity may be bound up with acts of social transformation is of particular relevance to the concepts and processes of identification which I explore in this chapter.

Defining Identification

The idea of identification captures a particular aspect of how subjects are brought into being. The subject as an idea, a concept, something imagined and ephemeral, comes into material being through the process of identification. In a moment of identification an identity is linked to some*body*. It is a way of thinking about subjects and subject formation that draws strongly on a Marxist tradition and is informed by feminism and poststructuralism.

Identification has become a central part of contemporary life within the nation-state. State powers require us to have a concrete identity, to be named according to gender, ethnicity, origin. It is the consequence of what Foucault identified as the extension of state power into the intimate details of people's everyday lives and the introduction of modern modes of rule which sought to govern the lives of state subjects much more closely than ever before (Foucault 1977, 1989). The modern state requires that we be identified so that we can be drawn within its "embrace" (Torpey 2000) as documented and governed citizens – and there are very few who can escape this demand (or would choose to).

The importance of identification to daily life is easiest to grasp in relation to those documents that everyone everywhere is required to have: identification papers. Whether these take the form of a birth certificate, a driving license, a national identification card, or a passport, documents of official identification such as these have both material presence and material consequences. With the right identification you are permitted to join the ranks of the legitimate citizens of your nation, to attend school, to access medical treatment, to be recorded on the state immunization register, to travel, to get a job, to pay taxes, to own property. Without the right identification these rights and freedoms can be restricted. With the wrong identification you might be expelled from the country you call home, or denied medical treatment, education or employment, or imprisoned.

The fact that it is nearly impossible to exist without an official identity, represented by the papers we are obliged to carry, indicates that processes of identification sit at the heart of contemporary social and political forms. But identification as a social and political process goes far beyond the material presence of identification papers. It is the process we engage in every time we are asked to define who we are. Questions as every day as "where are you from?," "what do you do?," or the request to identify whether you are male or female, or European/Hispanic/Asian/

Aboriginal on any official form solicits an identification from us and places us in a social, political, and territorial landscape. Such moments of identification sometimes seem so normal and natural to us that we pay no attention to them. At other times identification challenges socio-cultural norms and is thus a more troubling act, as when the neo-Nazi contests the vision of contemporary a multicultural society by declaring his allegiance to the Fatherland or the transsexual disrupts normative views of gender by declaring her newly acquired womanhood. Whether identifications seem normal, natural, strange or threatening, they are moments that position us in relation to the societies in which we live – moments in which relationships of power are enacted and revealed.

Power can be present in many different ways in and through moments of identification. Identification might be an act of subjugation by the state or compliance by the citizen, it might be an act of rebellion or transgression by the nonconformist, it might be an act of transformation that seeks to bring into being a new self, a new community, a new world. In each of these moments we are brought into being as certain kinds of subjects (cf. McKinnon 2007). In the right circumstances the act of identification makes us, for at least a moment, that thing which we have been called: woman; man; Hispanic; European; Indigenous; citizen; thief. Identifications that run counter to the norms, expectations and aspirations of those around us, however, rarely stick. To be effective (and affective) identity-claims must have wider resonance, they must, as Laclau put it, contain "a dimension of universality" (2000: 36).[2]

When effective, identification momentarily fixes us in a world that could otherwise be seen as always in a process of becoming, a world of flux and change where we and our "environment" are always in the process of being born (Ingold 2006). Each of the designations listed above carries with it complex and contested meanings. It means different things in different temporal, social, and political spaces to be, for example, a woman or a man, a Thai or a hill tribe, a Hispanic or a European. For highland peoples in Thailand, for example, there is not anything innately insulting about the term "hill tribe" or *chao khao*. Indeed, several of the representatives at the workshop on Indigenous Futures could see no reason not to continue to use that term. But it was because the term "hill tribe" has come to be associated with a particular way of understanding highland peoples as primitive, exotic, and alien, that the majority of the group agreed they needed to introduce a new term for themselves. The power a name carries, and the degree to which it is considered to reflect something *true* about the people so-called, has a great deal to do with who is doing the naming and who is listening.

Identifying someone as "*x*" ascribes a fixed identity which, from certain perspectives reflects real and essential characteristics of the person so named. The language of identity tends to speak of some interior truth, an underlying and basic essence of a person. For example, ethnic labels, such as "Hispanic" or "Mexican," are often used as if they reflect something innate about the person so described. In the right context that person named may be compelled to submit to that designation and be subjected to the abuse and prejudice it could entail. In other circumstances the label "Hispanic" may be enthusiastically claimed and celebrated in an act of ethnic pride, solidarity and resistance to dominant prejudices. In both cases the names are used in ways which give a person an identity, whether one imposed or claimed – the identity is seen to be linked to something fundamental about *who that person really is.*

In the language of subjects and subjectivity on the other hand acts of naming do not so much reflect a pre-existing identity as actually create the subjects they describe – linking people to sets of discourses that impel or encourage them to perform certain identities and shape the realm of possibilities for how they relate to and are positioned in the social world. To speak of the subject is not the same thing as speaking of the self, but the self is placed in the world through the coming into being of particular subjects. The designations of male and female are obvious examples here. Dominant norms of masculine and feminine deportment and physicality have been much challenged through second wave feminism and queer theory. Nevertheless, with the exception of societies that recognize a third sex (such as the *katoey* in Thailand or *fa'afafine* in Samoa) cases that blur the hard boundaries between male and female often cause great controversy and elicit strong emotional responses. This is so in the recent case of the South African athlete Caster Semenya who was subjected to genetic testing to "prove" she was a woman. Although she has a woman's body and she and her family believed her to have been female from birth, genetic testing "found" her to be of ambiguous gender leading the International Association of Athletics Federation to consider banning her from future international competition. Normative discourses of gender set the realm of possibility and those who challenge that norm, whether by accident of birth, a deliberate effort to cross gender boundaries using dress, deportment, or medical treatment, have no choice but to fight for a space of recognition (see Brown 1999; Ingram et al. 1997).

Identification papers and the act of identifying oneself might fix who you are in a particular space and time, but in theories of the subject that fixity is always temporary. Where identity is understood to be singular, permanent, and true, the subject is a changeable and moving thing, linked to shifting social and political discourses rather than any internal essence. The term "subject positions" is often used to indicate something of this mobility, describing how people can move between different identities or subject positions in different places and spaces – the same person might be wife and mother in the home, executive in the office, consumer in the shopping mall. But the idea of subject positions is also seen as problematic because it continues to assume a degree of fixity. The idea gives a tool for imagining the multiplicity of "selves" that individuals perform and give a basis for examining how subjects are constituted differently in different settings. Judith Butler notes, however, that the language of subject positions assumes a pre-existing position that exists external to the subject and awaits someone to step in to it like a job vacancy waiting to be filled. At the same time, it assumes a blank self who moves to occupy different positions throughout daily life. While a person may be different kinds of subjects in different spaces and places, moving from one to another is not as easy as a language of "positions" might suggest. Butler (1993) argues that this language of subject positions fails to recognize how discourse in fact materializes new subjects, rather than constructing them from pre-existing building blocks.

The language of identification extends the idea of subjects as mobile and material. Identification speaks of the moments in which subjects materialize, the moments when they come into being. It focuses attention on the process, speaking of how we are always (already) in a process of becoming. And it draws the mind to how subjects come into being in and through relationships whether they be intimate relationships with friends and family, or our relationships with the institutions, cultures, and discourses of which we are a part.

The key moments in which we come into being as subjects, the key moments in which we are identified, have been considered by a great many key theorists of modern times. Louis Althusser spoke of the moment when a policeman calls out "hey you!" and immediately the person who is being called recognizes themselves as the guilty subject (2001). The same phenomenon occurs regularly today when the sirens blare behind you when you are in your car and you suddenly think it is you who has done something wrong. Althusser called this process "interpellation," the moment when the guilty subject is hailed and suddenly comes into being as that which is named – the person that has done something wrong, broken the law, mis-behaved. In that moment the guilty subject draws on a host of societal norms and expectations to understand themselves as a transgressor, and becomes subject to the powers that decide what is right or wrong behavior (represented in this example by the policeman calling out).

Foucault took this idea further to analyze how transgressive subjects came into being in different ways as normative social and political discourses shifted through time. For Foucault, when a person is identified as criminal or insane they become subject to the normative discourses particular to the era: a criminal in 1750 was understood very differently to a criminal in 1950 (Foucault 1977). The understand-ings that arose with modern modes of power from the mid-1800s on brought into being new kinds of subjects. The "insane" subject for example only came into being in this modern era. Where before there were madmen and seers, once these same people were identified as insane they became understood as being sick and diseased people who could be "treated," and thus normalized, through the institu-tionalized interventions of the medical profession (Foucault 1989).

Judith Butler draws on Foucault's intellectual tradition to analyze an even more subtle way in which human beings are identified and brought into being as particular kinds of subjects in relation to social norms. Butler (1990) focuses on the question of gender and discusses how a girl is "girled": as soon as the family knows the sex of a baby it is immediately brought into a social world of gendered understandings. The name she is given, the color of the clothes she is dressed in, the expectations for behavior or preferences all begin to shape the child into what we expect a girl to be – far beyond what might be accounted for by biology.

The subtle and not-so-subtle processes through which people are subjected to dominant discourses *places* us in the world. In some ways what Butler, Foucault, Althusser (and others) are speaking about is not so different to the naming of a child that constitutes an often ritualized welcoming in to the social world. A name can indicate a child's heritage, the language that will be their mother tongue, the culture to which they have been born, a homeland (present or lost), a position in a family, and most of all a place in humanity. With a name, and a gender, a child can be recognized by the community, the gods, spirits and ancestors, and, of course, the state. It is this latter recognition that is most relevant to the question of identifica-tion. Along with the naming of a child and the gendering of a child through informal secular rituals or religious ceremonies comes the official, state sanctioned recogni-tion with the granting of the birth certificate.

Identification and the State

The birth certificate is the first official identification document we are granted and one that we must carry through life, and produce again and again to prove who we are and where we belong. It the first of a series of formal recognitions and certifications that are a confirmation of our legitimate being in the world. The system of formal identification that requires us to have birth certificates, identification cards, drivers licenses, passports etc., is a core component of being a member of a modern nation-state. Within this system we must be formally placed in relation to our family, our sex, our ethnicity, and our nation in a reified way that is distinctly different from family and community based processes of naming.

Identifications granted by "the authorities" are vastly overdetermined, meaning that they carry a heavy load of additional significance and meaning that is deeply embedded in the constitution of the global political systems in which we live. The system is based on the assumption that every single person on the planet belongs (or ought to belong) to a given territory, a particular nation-state. These identifications are so deeply embedded in our sense of social norms that it is usually taken for granted that such official documents are normal and necessary. It is only when the process is somehow disrupted or competing systems of identification clash – moments in which these identifications are suddenly made strange to us – that we become aware of these formal recognitions as fabrications.

An example of when practices of formal identification become strange and problematic is in cases of people who are not automatically recognized as belonging to the country of their birth. Highland Indigenous peoples in northern Thailand, whose circumstances I introduced at the beginning of this chapter, are one such case. Many highlanders born in Thailand do not have Thai citizenship, even if their families have been in the country since the first citizenship legislation was passed in 1956. In the decades since, the majority of highlanders have been issued with registration papers of some kind, but a significant proportion remain without citizenship (see McKinnon 2005). Without citizenship papers highlanders rights within their home state are limited: the ability to move freely within the country is curtailed, opportunities to access education, healthcare, or employment are limited, and non-citizens are under the constant threat of "repatriation" across the border into Burma or Laos.

There have been moves since the 1990s to grant citizenship to those highlanders who are eligible. Under Thai legislation eligibility depends upon presenting proof that one's parents were born on Thai soil. If you are able to present household registration papers and birth certificates that can prove this (which is not always an easy task for communities in which, until very recently, mothers rarely gave birth in hospitals) then you have the right to apply for citizenship.

The application process itself can be extremely intimidating – while (if accepted) it constitutes a confirmation of your legitimacy and your belonging within the Thai nation, that confirmation can also seek to displace the community and family based belonging that came before. One example of this is the story of one Akha[3] woman who applied for citizenship in her late teens. She was told by the government official processing her application that she must choose a new name, a Thai name. That Thai name, written on her new Thai ID card, grants her the right to stay in the land where she was born and where her family live, to attend its hospitals and schools, to travel freely within its borders and to get a job. Even with the card the process of recognition is not complete: every time she is required to show the card she is again in a moment of becoming the Thai citizen, again asking the officials to recognize her legitimacy. But by demanding that she take a new name the state here sought to displace her Akha name and diminish, at least on paper, her membership of a community considered fundamentally not-Thai.

The position of the Akha in Thailand is not so dissimilar to that of many peoples around the world who find themselves somehow outside of a system of national identification: the stateless children of eastern Malaysia whose parents fled conflict in the Philippines and who have grown up without access to any of the education or healthcare that Malaysian children are entitled to (Kassim 2009); the Serbs who suddenly found themselves declared non-citizens when independent Croatia came into being in 1941 (Denich 1994); Palestinians who no longer have a state (Akram 2002; Mavroudi 2008); the Roma whose settlements cross multiple European states (Ringold et al. 2005). The Akha, the Palestinians, and the Roma are examples of people who do not fit with efforts to identify belonging in relation to the nation-state. Examples like this make strange the assumption that everyone belongs somewhere, to some territory and to some state authority. They reveal the constructedness of this idea, the way in which it is a fabrication – but a fabrication with very concrete, material implications.

To briefly recap, identification is first a process: that of the coming into being of subjects as they are named and therefore given a place in the world. While the act of naming is part of every human culture, the kind of identification I am referring to is particularly relevant to the operation of the modern nation-state system. In this system everyone, regardless of whether they live in Europe, Africa, America, or Asia, must have official identification papers that give them a place within the state. In practice many people lack such documents and suffer for it as there is simply no space for people to exist outside this system. This is why the plea made by the village elder to simply be recognized as being human is not enough – there is no room to *just* speak of common humanity in a contemporary state-based system of identification. This is also why the highlanders gathered at the meeting in Thailand were so interested to debate what name they should use for themselves – one's name, one's identification is vitally important in this contemporary political system. What you are called by those who represent state authority has very real, very material implications. While identifying or naming oneself or others is a rhetorical act it is also tied very closely to the concrete world. Particular identifications in particular places can determine whether you are considered to belong inside or outside the borders of a nation or the boundaries of community and family. They can enable you to move freely, to seek a livelihood, to obtain an education, or they can bar you from all of these things.

Through identification the self becomes intimately connected to the larger machinations of the political, the social and the cultural. When we are identified and placed, the singular embodied self comes to mean something against these broader structures and processes. Thus it is in identification that one can see the inseparability of the discursive (the act of naming or being named) and the material. Because of this identification is a potent site of social and political change and transformation.

Transformative Possibilities

It is well recognized that identification and subject making cannot only be processes of being made subject to dominant discourses but also processes of intentional and unintentional change. Butler speaks of how each time the performance of normative identities is repeated there is room to shift and alter (1993). Laclau discusses how the subject is a political agent in which ideology and identity merge to form the foundation of hegemonic political struggles, articulated as the "interpenetration between universality and particularity" (Laclau 2000: 305; see also Devenney 2004; Howarth 2004). Amongst geographers, Gibson-Graham (1996, 2006) posit that affective transformations of subjectivity are the starting point for local community based efforts to step outside capitalocentrism – the discursive dominance of global capitalism. None of these authors accept the volunteerism of a humanistic perspective, which tends to take "an expansive view of what the human person is and can do" (Tuan 1976: 266), emphasizing the creative power of humanity and playing down the role of dominant systems and structures in society. While they do not accept that human beings are entirely free agents, there is also a recognition that within dominant structures there is still room to maneuver and to reshape subjects. The room for reshaping emerges due to the inevitable incompleteness of any identity-claim. The fixity of identities is never complete, never unopposed or whole, and "'identity' itself is never fully constituted" (Butler et al. 2000: 1). Because identities are intimately linked with larger machinations of the political, the social and the cultural, the reshaping of subjects also holds the potential to reshape wider communities linked by social, political, or economic ideologies and practice.

Processes of identification are one such conduit to reimagining subjects and thus acting upon the world. As well as being about moments in which we are subjected to dominant discourses, identification is also about the moments in which we identify *ourselves*. These can be well thought out and clearly articulated political movements but can also be much more everyday and less deliberate. Foucault speaks of the daily self-discipline we impose upon ourselves, shaping ourselves to fit the norms of society through appropriate dress and deportment. This daily discipline can take many forms and be tied up with aspirations to become a certain kind of person. Contemporary Lacanians (such as Slavoj Zizek 1992, 1997, or geographer Paul Kingsbury 2005, 2008) speak of the ways we channel a desire to become a certain person into the desire for objects which we feel will bring us closer to that goal (the *objet petit a*): the right shoes, the right watch, the "right" man, the bigger house, the better job etc. Such acquisitory desires are akin to the performances of identity through which we place ourselves as this kind of person. Flag waving and anthem singing on national holidays, choosing to buy free trade or organic, undertaking a pilgrimage to Mecca or Jerusalem or Ayodhya brings into being in oneself a certain kind of person: perhaps a patriot, a hippy, a devout member of the faith. All these actions, of acquiring the right object or companion, of performing certain rituals or duties, are in themselves a process of identifying the self – of naming, locating, materializing the self as a certain kind of subject. As we perform these acts we link ourselves to that thing with which we identify – the nation, the subculture, the faith – and thus bring that thing into being in ourselves (see Laclau 2000; Zizek 1991). Each of these acts is thus an act of becoming.

Every act of becoming is always saturated in a cultural milieu and always already overdetermined. But acts of identifying ourselves are not simply replications, what society or culture tells us to be or to desire. While no act of identification is completely new and original, neither is it mere repetition – as Butler (1993) has argued there is always room to move. Repetition never produces something completely identical. Even though thousands may celebrate their patriotism, undertake a pilgrimage, or claim an "alternative" hippy identity, each persons version is slightly different. And in some cases the performance of a particular identity is also a wilful act of reshaping that identity. Every act of becoming has this potential to also be an act of social and political change as individuals and collectives shape the ways they perform certain identities, nudging at the edges of what it can mean and shifting and changing what it is to be *this* or *that*.

The practice of identification as a transformative act is in part what the Indigenous Futures Workshop was about. It was an effort to actively and consciously shape the official language of identification used by state authorities, to identify highlanders in a language of their own choosing and potentially reshape the futures that highlanders could have within the Thai state. Rather than just an act of reshaping individual identity, however, it was an effort to begin to reshape a group identity initiated by just a small handful of leaders and activists. In the absence of a formal process for appointing community representatives this group formed an ad hoc group to represent the people and communities who could be affected, and came to the meeting with varying degrees of accountability to those communities. Some were community leaders who would be directly held accountable by members of their villages, some were activists and academics whose level of accountability relied entirely upon their personal sense of responsibility to represent their compatriots as fairly as possible.

It was the NGO staff and activists who led the push for the Indigenous designation. This group was most acutely aware that they have to act within the hegemony of the Thai state and the international nation-state system – there is no way to exist beyond it, no way to simply be human and not identified as a member of this or that group and thus be thought to legitimately belong or not. In a way they were in the role of the subaltern subject who knows that in order to speak and be heard they must speak the language of their oppressors (Spivak 1988). In contrast, the *Prageryaw* elder introduced at the beginning of this chapter perhaps hoped that he could continue to speak his own language and still be heard. The NGOs, however, know that they must engage in language of identity politics. While the elder appealed to a fundamental universal sense of shared humanity, the NGOs were enlisted in a struggle that is firmly embedded in politics, and were trying to work within this inescapable global political system. By seeking membership in and support from the international Indigenous movement they look to bypass the structures of their immediate oppressors, the Thai state. Through this they will be able to access a set of institutions established by and for Indigenous peoples as an instrument to claim certain rights and privileges from within the state – in this case simply the right to belong.

The designation of "Indigenous" is a term which expresses a set of universalizing claims about what it is to be Indigenous. These are claims which pertain to highlanders material circumstances (their marginalization in the state, their particular ways of relating to land and spirit) and which resonate at a global scale. An Indigenous identity, in other words, has a dimension of universality that gives it political clout. To be recognized as "Indigenous" by the Thai state and the international community would bring with it significant change.

Definitions of the term "Indigenous" put forward by the International Labour Organisation (ILO 2003), the International Work Group for Indigenous Affairs (IWGIA), and powerful multilateral agencies like the Asia Development Bank set the parameters for who is considered to be "Indigenous" and who is not. The definitions proposed by these organizations share two key characteristics. First, people are Indigenous if they are descendents of those who have lived in an area before colonization. The IWGIA for example states that:

> Indigenous people are the disadvantaged descendents of those peoples that inhabited a territory prior to colonization or formation of the present state (IWGIA 2009).

Second, Indigenous peoples are distinct from the dominant society and have maintained their own social, economic, cultural and political institutions since colonization and the establishment of new states. The ADB for example speaks of:

> Maintenance of cultural identities; and, social, economic, cultural and political institutions separate from the mainstream or dominant societies and cultures (ADB 2009).

Alongside these formal characteristics the term Indigenous usually reflects a special relationship with the land and the natural environment. While recognizing difference and disadvantage, the term Indigenous also recognizes the special place of Indigenous peoples and special rights and legitimacy (although these are seldom embraced wholeheartedly by the nations in which Indigenous peoples live). In addition, being Indigenous means access to international support via global Indigenous agencies, the United Nations, and the significant collection of groups and organizations who are involved in the global campaign for Indigenous rights.

For the highland NGOs present at the Indigenous Futures Workshop the ability to access these international support mechanisms was seen as an important step. Such international support could, it was hoped, provide more leverage for domestic campaigns for equal rights and would enable highlanders to come together with other marginalized groups in the South of Thailand to work for policy change at the national level.

At a more fundamental level the move to adopt the term "Indigenous" and the discussions about how best to translate that term into an acceptable Thai language phrase gave highlanders the chance to choose a pan-tribal name for themselves. As one village representative put it "now, we will not be stuck with the terms defined and called by others, but we will define ourselves, and choose the term to call ourselves" (Indigenous Futures Workshop, May 29, 2007).

In order to harness the transformative power of identification, the term chosen must first hold some recognized power. This is why the name *Pgakeryaw* has little wider resonance while the term "Indigenous" holds great weight – the latter has a globally recognized significance and respect (even though Indigenous peoples in most places still struggle for their due recognition). While there is an international platform to support Indigenous peoples, in Thailand there is as yet no space to speak of being Indigenous. As one NGO representative stated:

> We are a group of Indigenous peoples in Asia. Even though there is no space for the term in Thailand, but we are Indigenous peoples here, and we are the Indigenous peoples that do not have opportunity, and are disadvantaged.(Indigenous Futures Workshop, May 29, 2007)

In order for the term to gain power in the Thai context the group discussed strategies for opening up a space in which their Indigeneity could be recognized. Reflecting on their current involvement in the international Indigenous community a Thai academic/activist present at the workshop said:

> International participation is good because there is space [for Indigenous peoples], but at national level it is not good, because there is no space.(Indigenous Futures Workshop, May 29, 2007)

Using the international community as leverage the group seeks a new identification, creating a space to speak of themselves as Indigenous and a space in which to be recognized, to belong, and thus to be considered legitimate.

Conclusions

This chapter has discussed how identification works in multiple ways to connect discursively constituted subjects with the self: through the imposition of identities that subject individuals and groups to normative discourses; through acts that claim particular identities; and through conscious engagements that seek to reshape the self and the world. All these processes connect the autonomous embodied self to sets of ideas about who you are or ought to be. All are embedded within social, political, ideological tropes that shape and limit the types of identities it is possible to claim. But those limits themselves are malleable – even in the case of the strict demands of the state. The example of the Indigenous Futures Workshop in Thailand demonstrates one of many pathways being pursued to find new spaces in which subjects may be redefined.

The example of the Indigenous Futures Workshop also highlights the characteristics of identification that make it such a potent political tool. Identification is where the discursive becomes linked with the material. It is where selves become an embodied articulation of the ideological and the rhetorical. In the process of imposing, claiming, or contesting identities we make our *selves* meaningful in relation to the norms and ideologies to which those identities refer. This is a powerful tool for maintaining existing systems: daily reinforcements of gender norms is one example, the constant reiteration of capitalocentric discourses and the helpless worker within that is another. It is also, however, a powerful tool for acts towards social transformation. Within processes of identification there is the potential to work towards bringing new kinds of subjects into being, and with them new possibilities to reshape our ways of being in the world. The efforts of those present at the Indigenous Futures Workshop is one example. The group sought to draw on established ideas about what it is to be Indigenous and the accompanying recognition of the legitimacy of Indigenous ways of life, cultural forms and relationships with the land. By claiming an Indigenous identity the NGOs hope that highlanders can create a space for new recognition and new legitimacy within the Thai state. The creation of a new Indigenous Thai subject could, if it is successful, transform the conditions under which they are given membership of the state. It is an effort to create new subjects, but it is also an effort to create new worlds – one in which highlanders, as an Indigenous *group* – can be given the same respect and dignity as other members of the Thai state.

Notes

1   Of course, what it is to be human is itself a complex and contested question. The autochthonous names that many Indigenous groups use to name themselves often mean simply "human," distinguishing themselves from plants and animals at the same time as establishing a relationship between people and what western imaginary would label "the environment" (see Bird Rose 1991; Ingold 2006). In contrast, the United Nations Declaration of Human rights is another kind of imagining of what it is to be human, a creature born with innate rights and privileges peculiar to us as a species. However imagined humanity and human rights mean nothing if the person in question is not recognized as being "fully human" by those able to effectively dehumanize or deny a person their human rights.

2   It is important here to clarify Laclau's conceptualization of universality. Together, Judith Butler, Ernesto Laclau, and Slavoj Zizek (2000) recognize the central role of universalizing discourses in politics, but clearly reject a Habermasian sense of universality. For Habermas, the universal is real, innate, and pre-established. In contrast these three theorists "maintain that universality is not a static presumption, not an a priori given, and that it ought instead to be understood as a process or condition irreducible to any of its determinate modes of appearance" (2000: 3).

3   The Akha are one highland group among many. Other recognized groups are Hmong, Htin, Karen, Khamu, Lahu, Lisu, Lua, Mlabri, Yao.

References

ADB (2009) *Policy on Indigenous Peoples*, www.adb.org/documents/Policies/Indigenous_Peoples/ippp-002.asp.

Akram, S.M. (2002) Palestinian refugees and their legal status: rights, politics, and implications for a just solution. *Journal of Palestine Studies* 31 (3): 36–51.

Althusser, L. (2001) *Lenin and Philosophy and Other Essays*. New York: Monthly Review Press

Benko G. and Strohmayer, U. (1997) *Space and Social Theory: Interpreting Modernity and Postmodernity*. Oxford; Blackwell Publishers.

Bird Rose, D. (1991) *Dingo Makes Us Human: Life and Land in an Aboriginal Australian Culture*. Cambridge and Melbourne: Cambridge University Press.

Bird Rose, D. (forthcoming) *Wild Dog Dreaming: Love and Extinction*. Alexandria: University of Virginia Press.

Bondi, L. (2005) *Gender and the Reality of Cities: Embodied Identities, Social Relations and Performativities*, online papers archived by the Institute of Geography, School of Geosciences, University of Edinburgh.

Brown, M. (1999) Travelling through the closet. In J. Duncan and D. Gregory (eds.), *Writes of Passage, Reading Travel Writing*. London: Routledge, pp. 185–99.

Buchanan, K (1977) Economic growth and cultural liquidation: the case of the Celtic nations. In R. Peet (ed.), *Radical Geography: Alternative Viewpoints on Contemporary Issues*. Chicago: Maroufa Press, pp. 125–42.

Butler, J. (1990) *Gender Trouble: Feminism and the Subversion of Identity*. New York and London: Routledge.

Butler, J. (1993) *Bodies That Matter: On the Discursive Limits of "Sex."* New York: Routledge.

Butler, J., Laclau, E., and Zizek, S. (2000) *Contingency, Hegemony, Universality: Contemporary Dialogues on the Left*. London and New York: Verso.

Connell, J. and Gibson, C. (2004) World music: deterritorializing place and identity. *Progress in Human Geography* 28 (3): 342–61.

Denich, B. (1994) Dismembering Yugoslavia: nationalist ideologies and the symbolic revival of genocide. *American Ethnologist* 21 (2): 367–90.

Devenney, M. (2004) Ethics and politics in discourse theory. In S. Critchley and O. Marchart (eds.), *Lacalu: A Critical Reader*. London and New York: Routledge, pp. 123–39.

Foucault, M. (1977) *Discipline and Punish: The Birth of the Prison*. London: Allen Lane

Foucault, M. (1989) [1970] *The Birth of the Clinic: An Archaeology of Medical Perception*. London: Routledge.

Gibson-Graham, J.K. (1994) "Stuffed if I know": reflections on post-modern feminist social research. *Gender Place and Culture* 1 (2): 205–24.

Gibson-Graham, J.K. (1996) *The End of Capitalism (As We Knew It)* Oxford; Blackwell.

Gibson-Graham, J.K. (2005) Surplus possibililities: postdevelopment and community economies. *Singapore Journal of Tropical Geography* 26 (1): 4–26.

Gibson-Graham, J.K. (2006) *A Postcapitalist Politics*. Minneapolis: University of Minnesota Press.

Giddens, A. (1976) *New Rules of Sociological Method: A Positive Critique of Interpretative Sociologies*. London: Hutchinson.

Giddens, A. (1978) *Central Problems in Social Theory: Action, Structure and Contradiction in Social Analysis*. London: Macmillan.

Gregory, D. (2004) Humanistic Geography. In R. J. Johnston, D. Gregory, G. Pratt, and M. Watts (eds.), *The Dictionary of Human Geography*. Oxford: Blackwell, pp. 363–4.

Hardwick, S.W. and Mansfield, G. (2009) Discourse, identity and "homeland as other" at the borderland. *Annals of the Association of American Geographers* 99 (2): 383–405.

Harvey, D. (1973) *Social Justice in the City*. London: Edward Arnold.

Harvey, D. (1977) The geography of capitalist accumulation: a reconstruction of the Marxist theory. In R. Peet (ed.), *Radical Geography: Alternative Viewpoints on Contemporary Issues*. Chicago: Maroufa Press, pp. 263–92.

Holloway, S. (2000) Identity, contingency and the urban geography of "race." *Social and Cultural Geography* 1 (2): 197–208.

Holloway S. and Valentine G. (eds) (2000) *Children's Geographies*. London: Routledge.

Howarth, D. (2004) Hegemony, political subjectivity, and radical democracy. In S. Critchley and O. Marchart (eds.), *Lacalu: A Critical Reader*. London and New York: Routledge, pp. 256–76.

ILO (2003) *ILO Convention on Indigenous and Tribal Peoples 1989 (N. 169). A Manual Project to Promote ILO Policy on Indigenous and Tribal Peoples*. Geneva: ILO.

IWGIA (2009) *Indigenous Peoples: Who Are They?* www.iwgia.org/sw641.asp.

Ingold, T. (2006) Rethinking the animate, re-animating thought. *Ethnos* 71 (1): 9–20.

Ingram, G.B., Bouthillette, A.M., and Retter, Y. (eds.) (1997) *Queers in Space: Communities/ Public Places/Sites of Resistance*. Seattle: Bay Press.

Kassim, A. (2009) Filipino refugees in Sabah: state responses, public stereotypes and the dilemma over their future. *Southeast Asian Studies* 47 (1): 52–88.

Kingsbury, P (2005) Jamaican tourism and the politics of enjoyment. *Geoforum* 36: 113–32.

Kingsbury, P. (2008) Psychoanalytic approaches. In J.S. Duncan, N.C. Johnson, and R.H. Schein (eds), *A Companion to Cultural Geography*. Malden. Blackwell: 108–20.

Kong, L and Yeoh, B. (1997) The construction of national identity through the production of ritual and spectacle. *Political Geography* 16 (3): 213–39.

Laclau, E. (2000) Constructing universality. In J. Butler, E. Laclau, and S. Zizek *Contingency, Hegemony, Universality: Contemporary Dialogues on the Left*. London and New York: Verso, pp. 281–307.

Ley, D. and Samuels, M.S. (1978) *Humanistic Geography: Prospects and Problems*. Chicago: Maaroufa Press.

Massey, D (1994) *Space, Place and Gender*. Cambridge: Polity.

Mavroudi, E. (2008) Palestinians and pragmatic citizenship: negotiating relationships between citizenship and national identity in diaspora. *Geoforum* 39: 307–18.

McDowell, L. (2004) Masculinity, identity and labour market change: some reflections on the implications of thinking relationally about difference and the politics of inclusion. *Geografiska Annaler* 86 B (1): 45–56.

McKinnon, K. (2005) (Im)mobilisation and hegemony: "hill tribe" subjects and the "Thai" State. *Journal of Social and Cultural Geography* 6 (1): 31–46.

McKinnon, K. (2007) Post-development, professionalism and the political. *Annals of the Association of American Geographers* 97 (4): 772–85.

Monk, J. and Hanson, S. (1982) On not excluding half of the human in human geography. *Professional Geographer* 34 (1): 11–23.

Nash, C. (2002) Genealogical identities. *Environment and Planning D: Society and Space* 20 (1): 27–52.

Nash, C. (2003) Setting roots in motion: genealogy, geography and identity. In D. Trigger and G. Griffiths (eds.), *Disputed Territories: Land, Culture and Identity in Settler Societies*. Hong Kong: Hong Kong University Press, pp. 29–52.

Nash, C. (2008) *Of Irish Descent: Origin Stories, Genealogy, and The Politics of Belonging*. New York: Syracuse University Press.

Natter, W. and Jones, J.P (1997) Identity, space, and other uncertainties. In G. Benko and U. Strohmayer (eds.), *Space and Social Theory: Geographical Interpretations of Postmodernity*. Oxford: Blackwell, pp. 141–61.

Nelson, L. (1999) Bodies (and spaces) do matter: the limits of performativity. *Gender, Place and Culture* 6 (4): 331–53.

Peet, R. (1977) Inequality and poverty: a Marxist geographic theory. In R. Peet (ed.), *Radical Geography: Alternative Viewpoints on Contemporary Issues*. Chicago: Maroufa Press, pp. 112–24.

Pratt, G. (2004) *Working Feminism*. Philadelphia: Temple University Press.

Ringold, D., Orenstein, M.A., and Wilkens, E. (2005) *Roma in an Expanding Europe: Breaking the Poverty Cycle*. Washington: The World Bank.

Rose, G. (1993) *Feminism and Geography: The Limits of Geographic Knowledge*. Minneapolis: University of Minnesota Press.

Soja, E.W. (1989) *Postmodern Geographies: The Reassertion of Space in Critical Social Theory*. London and New York: Verso.

Spivak, G.C. (1988) Can the subaltern speak? In C. Nelson and L. Grossberg (eds.), *Marxism and the Interpretation of Culture*. Urbana and Chicago: University of Chicago Press, pp. 271–313.

Torpey, J. (2000) *The Invention of the Passport: Surveillance, Citizenship and the State*. Cambridge: Cambridge University Press.

Tuan, Y.F. (1976) Humanistic geography, *Annals of the Association of American Geographers*
    74: 353–74.
Winders, J., Jones, J.P., and Higgins, M.J. (2005) Making gueras: selling white identities on
    late-night Mexican television. *Gender, Place and Culture* 12 (1): 71–93.
Zizek, S. (1991) Beyond discourse-analysis. In E. Laclau (ed.), *New Reflections on the
    Revoution of Our Time*. London and New York: Verso, pp. 249–60.
Zizek, S. (1992) *Looking Awry: An Introduction to Jacques Lacan Through Popular Culture*.
    Cambridge, MA, and London: MIT Press.
Zizek, S. (1997) *A Plague of Fantasiesi*. London and New York: Verso.

Chapter 4

Social Natures

Katharine Meehan and Jennifer L. Rice

Introduction

In downtown Columbus, Ohio, along the banks of the Scioto River, a sign broadcasts one of the city's most stubborn environmental headaches: "WARNING: Combined Sewer Overflow. During periods of rainfall water flowing from this pipe may contain materials which may be harmful to your health." On a rainy day, the sign may not be noticeable under the haze of grey skies, but the stench of raw sewage, discharged from the pipe below, is unmistakable. In fact, more than thirty outfalls dump a potent mix of untreated sewage and stormwater into local waterways when Columbus's sewer system becomes overwhelmed with excess wastewater from heavy rains. While fish feed on the bounty of algae and invertebrates that thrive in the nutrient-rich waters, a small population of Columbus residents who cast their fishing poles to catch their day's food supply is exposed to harmful substances, such as viruses or prescriptions drugs, that can make their way into the tissues of caught fish. The US Environmental Protection Agency continues to permit combined sewage overflows in more than 770 cities with aging infrastructure (US EPA 2009), despite the well-known dangers of human exposure to raw sewage. In Columbus, debates over the legality of sewer overflows have also permeated urban growth politics, where local environmental groups have accused the city of expanding its sewer system to new, wealthier suburban areas, in lieu of repairing and replacing infrastructure in the urban core (Rice 2005).

Meanwhile, three thousand miles to the southwest, wastewater also takes center stage in the development politics of Tijuana, Mexico. A rapidly expanding desert city of nearly 2 million people and over 200 informal settlements, Tijuana suffers from limited water supply, uncontrolled urban runoff, and great wealth disparity. Public services in informal settlements are few or none, even as slum residents supply

A Companion to Social Geography, First Edition. Edited by
Vincent J. Del Casino Jr., Mary E. Thomas, Paul Cloke, and Ruth Panelli.
© 2011 Blackwell Publishing Ltd. Published 2011 by Blackwell Publishing Ltd.

critical labor for the city's industry and manufacturing sectors. These neighborhoods – poor enclaves of employment-seeking migrants from southern Mexico and Central America – pepper Tijuana's steep canyons with rickety shacks and unpaved, crumbling roads. Every winter, Pacific storms dump rain in the highly urbanized watershed, triggering "renegade flows" that pick up raw sewage, sediment, and solid waste as they race through slums. Stormwater runoff from across the city feeds into the Tijuana River, crosses the international border, and chokes aquatic habitats in the Tijuana Estuary. Oceans currents carry stormwater plumes north and southward, prompting binational beach closures, threatening public health, and eliciting protests from downstream San Diego. While many blame Tijuana's slums for the renegade flows, others point to the deep contradictions of uneven development, where local topography, aridland hydrology, state neglect, urbanization patterns, and international trade incentives (e.g., the North American Free Trade Agreement) coalesce into a sewage-spewing machine of capitalist growth and social underdevelopment.

Sewage, as these cases suggest, is simultaneously a natural and social object. In both Columbus and Tijuana, the problems of wastewater bring together a broad array of social and environmental objects: storm drains, overflowing creeks, city workers, septic tanks, leaky pipes, laborers, crop fields, industrial dump sites, slums, and thousands of toilets (Meehan 2007). Pipes and pumps, struggles over urban development, and economic disparities are animated through relations with waterways, human excrement, algae blooms, runoff, and overflows in both Columbus and Tijuana. While obvious and important contextual differences distinguish each city and its urban growth regime, key similarities rooted in the contradictions of development suggest that the production of "sewage" is not an entirely natural or politically neutral process. Indeed, wastewater problems in Columbus and Tijuana are quintessential examples of *social natures*: patchwork assemblages of people, ecologies, politics, and objects, enmeshed in a dynamic web of materialities and representations that become the moving targets of scientific explanation (Braun and Castree 2001).

Wastewater and its discontents, like many imbroglios of nature and society, clearly raise important questions for geographic research on sustainable environments, politics, and justice. How best to understand and explain the emergence of wastewater problems, in a way that does not simplify the issue to a technological question of too many or too few toilets, reveals the complexity of the puzzle at hand. Do we pinpoint blame at the point source, such as sewer outfalls or poor sanitation, or in the legacies of uneven urban development? Geographers and social scientists have shown how fundamental crises in capitalist economic valuation and exchange produce modernist categories of "waste," such as sewage and garbage (Engler 2004; Frow 2003; Laporte 2000). These politicized categories and practices of waste-making have been used as government tools to define, control, and often exclude state subjects (Gandy 1999; Melosi 2008), even as citizens use their marginalized subject positions to struggle for rights, livelihoods, and justice (Moore 2008). These scholars clearly demonstrate the twin roles of modernist ideologies and capitalist systems in transforming urban technonatures and producing inequalities; in effect, a mode of capitalism qua sewer development (e.g., Wainwright 2008).

This chapter maps these modes of explaining social natures in geography, takes stock of advances and lingering challenges, and signals some paths forward. Following this introduction, we briefly chronicle the use of "nature" in social explanation – starting with the idea of nature as "produced" and its subsequent role after the "material turn." Next, we review recent efforts to unsettle explanations of the "social," particularly within environmental subjectivities and the state–nature nexus. Building on these insights, in the fourth section, we put forward a *site ontology of social natures*, examining what becomes visible through a different way of thinking and doing (via research methods) social natures. Our discussion of the site ontology, as a way of approaching social natures, attempts to render visible the complex and emergent associations of humans and non-human that make up the world. While much scholarship in human–environment research works to show the *relations* of nature and society, a site ontology emphasis the situated *context* within which people and objects co-constitute social life, without articulating these relations through moments of "social" or "natural" influence. Furthermore, this approach demands a critical examination of moments of difference that exists within and between social sites, as possibilities for where alternative outcomes may be forged. Where possible, we ground these theoretical explorations in the problems of sewage, drawing on our own research from Columbus and Tijuana. While our appraisal of social natures is brief and inevitably limited, we hope to illustrate not only the advances in geographic thought on this topic, but also the political and practical stakes in providing situated, immanent accounts of socio-natural life in the world.

Producing Nature

Far from being straightforward sewage, Tijuana's "renegade" runoff presents a puzzle to social research. On one hand, excess urban runoff reveals the ironies and tragedies of underdevelopment in Tijuana. On the other hand, Tijuana neighborhood groups and their international collaborators use the seasonal stormwater flows as a political wedge and development opportunity (Meehan 2008). In this case, how best do we explain the presence and role of sewage in Tijuana?

The heart of this puzzle, concerning the ontological status and political stakes of "nature," has a long history and unsettled status in geographic thought. In the nineteenth and early twentieth centuries, geography was dominated by environmental determinism, a school of thought that "maintained that geographic influences determined human capabilities and cultures, with its practitioners attempting to codify that thesis into scientific practice" (Robbins 2004: 19). The pendulum eventually swung the opposite direction. Cultural ecology, as one response [among many] to determinist thought, "sought to explain the physical patterns on the land (forest cover, soil erosion, stream flows) in terms of human culture rather than the other way around" (Robbins 2004: 28–9). While these two approaches contributed a wealth of foundational knowledge about human–environment interactions, the tasks of explaining and accounting for nature remain contested and unresolved. What is this thing called "nature"? In what ways does nature matter to social explanation?

The idea that nature is *produced* through a constellation of socio-natural relations, driven by wider societal processes, threads through social geography in at least two ways: first, as the (primarily capitalist) *production of nature*; and second, as a *socially constructed* category. Both approaches share key similarities and subtle differences. Building on Marx's theory of dialectics, the production of nature thesis focuses on social relations to explain the emergence of uneven socio-natural geographies (Smith 1984; Harvey 1996). Social relations between humans and nonhumans, shaped by capitalist logics and wealth accumulation, produce uneven power geometries and mainly negative socio-natural outcomes: such as tropical deforestation in the Brazilian Amazon, soil loss and land degradation on the Ivorian savanna, or water scarcity in urban South Africa, to name a few. Even seemingly abstract natures, like global warming, are "socially produced in the fullest sense" (Smith 1996: 50).

While many scholars draw on Karl Marx to theorize environmental ills, a key distinction between the productionists and other eco-Marxists is the contrasting use of "internal" and "external" social relations (Castree 2002). For eco-Marxists, "society" and "nature" remain ontologically and analytically distinct. Nature, in this view, is an "external" lifeworld that bears the brunt of capitalism's periodic and inherent crises of over- or under-accumulation (O'Connor 1997). The productionists, in contrast, adopt a more dialectical, contingent, and embodied approach to social natures. Drawing heavily on the theoretical labor of Georg Lukács and Bertell Ollman, the productionists use an "internal relational approach" that posits nature as discursively and materially embedded within social forces (i.e. capitalism), prone to internal contradictions and transformations rather than external forces (Castree 2002; Harvey 1996). Nature, like all social objects and subjects, does not exist outside of or prior to the capitalist processes, flows, and relations that sustain or undermine it (Harvey 1996).

Moreover, the productionists see nature as a dynamic partner in the highly politicized circulation of capital. Swyngedouw (1999: 448), for example, sets up a "dialectical unity" between river flows and Spanish society that "insists on the nonneutrality of relations in terms of both their operation and their outcome, thereby politicizing both processes and fluxes." He reconstructs the historic, metabolic processes of capitalist nature – namely the growth in water demand from urbanization and export-oriented agriculture – in order to understand the material geographies of inclusion and exclusion across the Spanish waterscape (1999: 448). Not even seemingly neutral technologies, such as sewer infrastructure, can escape the grip of state and capitalist politics. For instance, the efforts in nineteenth-century Paris to provide public sewerage were responses to public health threats as much as state tactics to legitimize and deny citizen rights to the city (Gandy 1999). At root, the productionist approach is a political as well as intellectual project: drawing attention to the ontological inseparability of humans and nonhumans while underscoring the powerful force of capitalist social relations in the fate of social natures.

Social geographers also analyze "nature" as a *constructed* category of meanings – a type of production that cuts across multiple axes of difference, such as class, gender, sexuality, and race. For Hacking, an analytical philosopher, the deconstruction of "nature" entails placing "new" ideas in a contextual milieu. Unmasking the "dichotomy between appearance and reality" to reveal power geometries is a key goal because – contra many critics – "politics, ideology, and power matter more than metaphysics to most advocates of construction analyses" (Hacking 1999:

49–58). In this respect, social constructionists mirror the productionists in the thought that the world is not pre-determined or stable, that historical explanation is crucial, and that unveiling power dynamics is a fundamental project.

Moreover, some geographers hold that social construction involves a *coproduction* beyond strictly human kinship. "Scientific knowledge," argues Demeritt (1998: 181), "depends on a variety of nonhuman actors, [though] the difficulty comes in acknowledging the active role played by the objects of scientific knowledge in shaping or constraining this knowledge without falling back into some kind of epistemological realism." In other words, an "artifactual constructivist" approach sidesteps the epistemological snags of realist thought: it deconstructs the nature/ society binary and rescripts the "social" to include nonhumans, machines, and other objects. Demeritt's take differs slightly from Hacking and the productionists, in that it departs from primarily anthropocentric notions of agency, materiality, and the subject.

Tracing the construction of nature is also a powerful device for mobilizing questions of social justice. Through deconstruction of the "forest," Braun (2002) reveals how authoritative knowledge of the temperate rainforest is assembled and concealed (e.g., through scientific expertise), who frames the terms of debate (e.g., colonial forestry scientists), and who wins and loses the struggle to claim forest resources (e.g., white environmentalists over the Nuu-chah-nulth tribe). Braun unearths the "buried epistemologies" in our ideas of the forest, such as the not-so-post-colonial articulations of modern-day forest management, in order to raise the political stakes of current forest practices. Injustices of the colonial past, in this case, continue right up through our present understandings of nature.

Curiously, while productionists and constructionists are deeply concerned with nature, the old question of "how nature matters" remains unsettled. "Materiality" is often synonymous with "nature" in productionist and constructionist accounts, but the nitty-gritty intricacies of nonhuman beings and phenomena are left mostly unexplored. Of course, as Castree remarks, "some agents have far more capacity to direct the course of socio-natural relations than do others" (2002: 141). Sewage problems, for many, are explained as material *outcomes* of capitalist development and social constructions of waste – tangled in the web of production but not driving it. The ontological question of whether "nature kicks back" (Whatmore 2002) is taken up next.

The material turn

The material, as Robbins and Marks explain (2009), never really left human geography. But in recent years, geographic thought has increasingly adopted a "more-than-human" mode of explanation, drawing attention to the ontological, ethical, and epistemological implications of nonhuman subjects in social life. These efforts manifest in myriad ways. There have been calls to foreground and incorporate nature's agency via ecological and physical science (e.g., Walker 2005; Zimmerer and Bassett 2003). Scholars in agrarian political economy explore the limits to capitalism through specific ecosystem dynamics (e.g., Gidwani 2008) or with inherently "uncooperative" commodities, like water (e.g., Bakker 2004). Political ecologists and posthumanists explore flows of power through human and nonhuman networks (e.g., Robbins 2007; Whatmore 2002), rethinking the materiality of nature through forms of decentered agency (e.g., Bakker and Bridge 2006). The corporeality and performativity of nonhumans, from corncrakes to cetaceans, are argued to provide a vital force in shaping the geographies of exchange, scientific knowledge, and social power in affective and effective ways (e.g., Cloke and Perkins 2005; Lorimer 2007, 2008). The material turn clearly reaches across many geographic fields.

In spite of theoretical differences, consensus emerges on two fronts. First, the *subject* of these stories is always a more-than-human assemblage, where "nature" is a messy question that is inseparable from social production and representation. The subjectivities of people and their nonhuman kin are coproduced through situated encounters and reproduced or transformed by daily, embodied, and often mundane practices – watering a lawn, visiting the zoo – and not by transcendental, structuralist ideologies (Cloke and Perkins 2005; Robbins 2007; Whatmore 2002). Every species, then, is a multispecies crowd (Haraway 2007). The family pooch, for example, is a partner in creating "dog people," who in turn influence the landscape of urban parks, sustain an industry of pet products and dog shows, shape the pathways of genomics and science, and transform our notions of self, family, and community (Haraway 2007).

Social agency, then, is not restricted to *either* humans or nonhumans, but is a heterogeneous achievement of both. This approach, in contrast to previous accounts, clearly recognizes "the creative presence of non-humans in the fabric of social life and [registers] their part in our accounts of the world" (Whatmore 2002: 36). An "actor," be it renegade runoff or Fido the dog, is not necessarily the source of action, but "the moving target of a vast array of entities swarming toward it" (Latour 2005: 46). Bakker (2004), for example, points to the uneven spatiality and dense physicality of water as a key culprit in resource privatization failures. In this account, water is clearly a subject in political economic processes: it serves multiple functions simultaneously, it can be monopolized via spatial location, and its complicated flow regime requires large-scale capital investment to generate profit. In a different example, Lorimer (2008) narrates how scientists, aided by a range of instruments and raingear, "become-animal" in the process of listening, stalking, tracking, measuring, and cataloguing information about the corncrake (*Crex crex*), a threatened bird species that seasonally lives on the Hebrides islands. Scientific knowledge of biodiversity, in this case, is partially produced by the affective energetics of counting corncrakes, enrolling the desires, affects, emotions, and relational ethics of understanding brown little birds. This redistribution of social agency unearths the potential alliances of progressive human/nonhuman resistance in marginalized networks (Robbins 2004) and recognizes the "heterogeneous actants in multiple commodity-networks – [because] agency and power can never be strictly human" (Castree 2002: 140).

Consensus also emerges on a second front: this project is fundamentally *anti-essentialist*. While nonhumans matter to social explanation, they do not posses an essential, pre-determined meaning or existence. Mitchell (2002), for example, argues that human and nonhuman agency is, in fact, a partial and incomplete effect of historical interactions. For him, "capitalism has no singular logic, no essence" (p. 303). This thing called the capitalist economy is (in Egypt at least) built by a milieu of humans and nonhumans, mosquitoes and colonial scientists, political struggles and peasant studies, and various daily practices that animate and regulate market behavior. The ordinary mosquito, a disease vector that helped shape human settlement in the Nile Delta, is used not only to animate the "resistance of nature or materialist conditions" (p. 51), but to problematize the idea of a singular subject in the production of nature. For Mitchell, "nature was not the cause of the changes taking place. It was the outcome" (p. 35).

These materialist approaches clearly recognize nature as a heterogeneous achievement of people and plants, politics and critters, objects and discourses (Whatmore 2002). The *social* in social natures, however, remains less evident. Latour (2005), for example, observes that social scientists have long characterized and abused the "social" as an assumed domain in advance, a coherent category with exclusive explanatory force. "While it's perfectly reasonable to designate by 'social' the ubiquitous phenomenon of face-to-face relations," he explains (2005: 65), "it cannot provide any ground for defining a 'social' force that is nothing more than a tautology, a sleight of hand, a magical invocation." The question of the social we turn to next.

Unsettling the Social

Back in Columbus, city workers, civil engineers, and environmental activists are all working to "solve" the sewer overflow problem. Larger pipes, development restrictions, and water monitoring bring people, cement, microorganisms, and flows into conversation with one another. But what influence, if any, does sewage have on political practice and social life? Two theoretical investigations about how we can understand the "nature" of social life are discussed in this section, including the environmental subjectivity and the state–nature nexus.

A more-than-human subjectivity

In one of the earliest inclusions of nature in subject formation, Shildrick (1997) developed the notion of *leaky bodies* in her examination of the subjection of women in biomedical practices. Shildrick argues that new reproductive technologies (NRTs), such as surrogacy and artificial insemination, are actually challenging, perhaps even redefining, the boundaries of female bodies and natural motherhood. Anxiety surrounding these technologies, and the dominant conceptions of motherhood that they contradict, can be used in creating counter-narratives to the traditional female subject. Though Shildrick's focus was on the role of ethics in biomedical research, her work also shows that that the idea of a "natural" woman is, in fact, a highly problematic and unstable category. Shildrick (1997: 181) claims that "what makes NRTs effectively post-modern in their own right is just that refusal to accept the notion of an unchanging natural body, and their capacity to problematise the grounds of (self) identity." Critiques about the naturalness of bodily practice, such as the one from Shildrick, show how subjectivity is rooted within modernist ideologies of nature, masculinity, and motherhood, rather than innate social forms.

Unpacking the motivations for why individuals come to care for the environment is another thread of subject formation. Agrawal (2005), for example, examines the making of *environmental subjects* through his Foucault-inflected notion of "environmentality."[1] Using a study of village forest councils in North India, Agrawal shows that colonial state practices of data collection helped make forests legible and governable to post-independence bureaucracies, which also set the stage for the engagement of local communities in environmental management. The decentralization of forest administration from federal to local levels resulted in villagers becoming champions of environmental conservation, where only years before their actions and perceptions of environmental conservation were anything but compliant with state policies.

Environmentality has become foundational for unraveling how communities understand themselves and social practice in relation to a wide range of environmental contexts. Robbins (2007), for example, chronicles the production of "turfgrass subjects" in suburban, lawn-rich America. Drawing on Althusser's concept of subject interpellation, Robbins shows how a complex web of economic pressures, post-war suburbanization, chemical marketing, and suburban aesthetics have actually worked to make a kind of lawn subject – that is, a set of people who come to care for their lawns in accordance with the demanding expectations of desirable landscapes, well-behaved neighbors, and the grass itself. Taking environmentality one step further, Robbins argues that the actual materiality of the lawn is central to this process: he depicts a multi-directional relationship between the human and the turfgrass in subject formation. The grass, Robbins (2007: 134) claims, really *matters* by claiming that "The enforcement of this specific kind of political economic subject – a concerned, active, communitarian, as well as anxious, landscape producer and consumer – would be impossible without the lawn itself to enforce the daily practice, feeling, and experience of obligation and participation. The lawn interpellates the 'subject.'"

Lawn maintenance, therefore, is much more than the choice of millions of chemical-happy Americans. Seen here, it is the effect of diffuse social and economic contexts that create intense feelings of pressure and obligation to maintain one's yard free of weeds and pests, combined with the material demands of a thing that beckons its companion species (humans) as weeds grow and blades wilt. These studies have firmly linked environmental protection and the production of social natures within embodied state power and intimate processes of identity-formation.

But what has the inclusion of subject formation provided to studies of social natures? Theories of subject formation have become central for unraveling how individuals and societies understand their relationship to nature and the physical world. Perhaps the most important contribution is the idea that self, community, and society are actively coproduced in our relationship to the material objects and discursive ideas of the natural world. Furthermore, a diffuse and embodied understanding of power allows for a more nuanced understanding of subjection through a complex mixture of personal consent and state-led coercion. Social geographers have claimed that this might be used to articulate a "politics of possibility" (Gibson-Graham 1996) where understanding and acknowledging that individuals participate in alternative practices allows for the potentiality of different subject formations. In destabilizing the social world to include elements of the nonhuman, we see that the power to make us who we are lies both within the dominating and hegemonic processes of social control, as well as our everyday encounters with material objects and natures.

At the same time, however, subjectivity studies often retain a nature/society dichotomy – that is, an internal self in opposition to an external nature (Nightingale 2006). While the addition of subjectivity in social natures allows for the inclusion of environmental elements in subject formation, the focus remains of how material objects – forests, lawns, animals – make us who we are. What remains unanswered is how and why the nonhuman must be separated from the individual to then be reconnected through the processes of subject formation. Are there alternative explanations that do not pre-suppose the content of the "social" and the "natural" that can be used in our explanations?

The nature of the state

Many state theorists argue that Marx first theorized the relationship between nature and modern society, where the separation of humans from nature can only understood in relation to the historical alienation of human labor via capitalist social relations (Foster 2000). Indeed, as capitalist production continually exploits the natural environment in the process of production, the state must "mediate between capital and nature" to ward off an impending ecocrisis (Benton 1996: 104, 191). Political ecology research, offers many rich case studies on the relationships between state-led policies, capitalist development, and environmental change, demonstrating the ways local populations are often marginalized by state policies and practices in struggle surrounding natural resources and the physical environment (e.g., Zimmerer and Bassett 2003). These studies have carefully articulated how state policies and practices *do work* on the physical environment and the populations that rely on those resources for their livelihoods, but in the process, the state comes to be seen as a coherent entity *separate* from the environments over which it governs. Political ecology research frequently argues that the state "does this" or "does that," relegating its actual existence to a mysterious "'black box' in their accounts" (Robbins 2003: 641). In other words, the dialectic between nature and society is well developed in political ecology, but a similar analysis of the relationship between the state and nature is much less clear.

In an effort to better develop the state–nature nexus, Scott (1998) shows how natural and social systems deemed too messy or incoherent for the state to control are often subjected to projects of territorial simplification and legibility. Scott chronicles a brief history of technoscientific forestry, when European states replaced existing forests – which contained multiple species at various stages of growth and decline – with simplified forest monocultures in an attempt to increase timber yields. What advocates of scientific forestry did not anticipate, however, was that by simplifying the forest ecosystem, it became much more susceptible to disease outbreaks, fire devastation, or other disasters that are often buffered through more complex ecological systems (Scott 1998). While relations between the state and nature are predicated on the state's ability to make its territory (including the natural resources within) legible and recognizable to its bureaucracies, this move often results in unintended and disastrous effects.

Whitehead et al. (2007) also document the various ways that nature is ordered and framed by modern state institutions. Nature (in all its discursive, symbolic, and material diversity) is shown to be a central aspect of state formation through the centralization of knowledge about the environment via territorializing state practices. National parks, for example, are central in the making of nationalist sentiments, as they are crafted into territories of natural and national significance by state institutions. Specifically, they describe state–natures as "the various ecological emblems that are routinely used to represent national political communities ... particular places/landscapes of ecological significance ... ecological phrases, narratives, and myths ... territorial maps and landuse surveys; and even micro-biological organisms and molecules whose transformation is regulated by various national laws and restrictions" (Whitehead et al. 2007: 2). States continually work to frame and reframe environmental problems in ways that are complementary with territorial state practices, even as the material natures they wish to territorialize cross the borders and boundaries of individuals states. Stripple (2008), for example, shows how states work to "border" climate change in debates about national carbon sinks and how mitigation obligations should be shared among nations. In this way, the materiality of nature is seen to constitute state practice just as much as state institutions work to transform the physical environment over which they govern.

This reading of the state–nature nexus, via engagements of nature within critical political theory, is one where neither "the state" nor "nature" are ontological givens in society, but instead, they are mutually constructed and contested through everyday political practices and discourses, by both human and nonhuman objects (Mitchell 2002). To this point, Mitchell (1999) argues that the state is *not* an independent and coherent institution of social relations, but it is the *effect* of everyday practices of planning, information exchange, and expertise that constitute modern practices of government. Despite the reality that state practices and institutions are often inconsistent, uneven, and even contradictory, it is the distinct feature of the "modern political order" (Mitchell 1999: 77) that the state *appear* separate from society in order to maintain social and political order. This is indeed the case in the state–nature relationship, where "nature" and "the state" appear to be external to one another, but are, in fact, continually being produced and reproduced in their constant and continual engagement with one another.

A Site Ontology of Social Natures

Consider (again) sewage. Columbus and Tijuana share key similarities and differences that make straightforward explanation of their social natures difficult. Are flows of sewage in Columbus the inevitable byproduct of urban development and state practice, or have they become a key part of the urban ecology? Do wastewater flows in Tijuana signal the socioeconomic gaps produced by capitalist-driven development, or are renegade flows simply following the logics of gravity and hydrology? Can explanation include all these elements? To address this challenge, in this section we put forward a *site ontology of social natures*: an approach that treats social natures as "dynamically composed aggregate[s] whose 'map' is drawn according to its own internal 'logics,' rather than any generalizing laws" (Woodward et al. 2010). Rather than assume *a priori* that sewage problems are products of a single "social"

or "natural" force, such as nitrate ecologies or the steamroller of capitalist development, we advocate an approach that considers these cases as *sites*: distinct socio-
spatial planes where relations between actors and objects, nonhumans and humans,
are continually made and remade in situated, everyday encounters.

In this approach, the social world is seen as a context-specific, anti-essentialist
milieu of people and things, bound up in the relations of material places rather than
transcendent [and implicitly hierarchical] scales and processes. Schatzki (2003: 176)
explains: "Site ontologies contend that social life ... is inherently tied to the type of
context in which it occurs," where "social phenomena can only be analyzed by
examining the sites where human coexistence transpires." Life, according to this
view, is imminent and emergent in a "mesh of practices and arrangements" (Schatzki
2003: 191). Drawing on Schatzki's work, we too suggest that social life is the
product of human *and* nonhuman engagements, broadening the field to include
material and discursive natures and nonhuman objects. Our approach retains the
intimacy of nature–society relations established by previous scholars, while offering
an ontology that rejects persistent and underlying dichotomies of nature/society that
underlies many of these approaches. So too, it allows researchers to examine
complex and far-reaching socio-ecological events, without assuming that they play
out consistently or evenly from place to place.

What, in this approach, becomes visible that was invisible before? First, we see
that *practices* and *repetition* are key to the sustainability or transformation of social
natures. Researchers look for the ways that people and things "hang together" by
means of *socio-spatial practices*: a set of "doings and sayings" that constitute wider
social relations and event spaces (Marston et al. 2005; Schatzki 2002). These relations, however, are not haphazard or completely agent-driven, but tend toward
organized and stable social relations. At the same time, these social natures are not
"structured" in the classic sense, but are immanent and sustained (via *repetition*)
by practices and arrangements that hold the potentiality to create new arrangements
of socio-natural life (Schatzki 2003). Drawing on the work of Deleuze and Guatarri
(1987), there is a "double movement to the site: one trajectory toward continuous
variation and differentiation, the other toward repetition and moments of relative
stability" (Marston et al. 2007: 52). In the case of sewage, for example, both sites
tend toward specifically capitalist forms of urban development, where the persistent
engine of economic growth transforms hydrologic environments into industrial
dumps as it spews aside its most abject byproduct. There are, however, moments
of difference in both cases that allow for the proliferation of new socio-natural
conditions. In Tijuana, for example, stormwater flows are used by afflicted community groups as an opportunity to draw attention to the contradictions of capitalist
growth and advocate for alternative forms of development, such as pervious street
paving funded by international agencies. In Columbus, moreover, the persistent
stench of untreated sewage seeping up through man-hole covers on residential
streets can spring downtown residents into action to change the development practices of local government.

A site ontology of sewage in Columbus and Tijuana, moreover, reveals how
specific practices create and sustain certain *assemblage geographies* of wastewater,
urban development, and social injustice. Braun (2006: 644) has used the notion of
assemblages "to stress the *making* of socio-natures whose intricate geographies form

tangled webs of different length, density, and duration." Braun focuses on the diverse, uneven, and heterogeneous objects and practices that make up these socio-natural networks. Assemblages are "imbroglios that mix together politics, machines, organisms, law, standards and grades, tastes and aesthetics, even in the production of sovereign territory and the politics of scale" (2006: 647). Robbins and Marks (2009) offer four typologies of assemblage thought, inspired by Latour, Haraway, Marx, and Mitchell, that all share normative goals for advancing a "more-than-human" social geography, without having to bestow a pre-packaged "agency" on the nonhuman that it often does not have, or a constructed naturalness to social practice that does not necessarily exist.

DeLanda (2006: 5), furthermore, places relations at the core of an assemblage, "whose properties emerge from the interactions between parts." Like, a site ontology, DeLanda's assemblage is neither the sum of individual actions, nor the product of purely structural forces. Instead, DeLanda (building on Deleuze) offers an ontology that examines assemblages in their situated wholeness, through relations of objects that tend toward relative degrees of stability or instability. People, lobsters, elm trees, microscopes, rivers, etc. are always involved in active relation with one another – though some at more transformative moments than others. There is no independent arena of "social" and "natural" things, only relational moments between objects and people, humans and nonhumans. Others have shown that socio-natural assemblages can be understood as the product of a type of "dwelling" – or the "the intimate ongoing togetherness of beings and things which make up landscapes and places, and which bind together nature and culture over time" (Cloke and Jones 2001: 651). By emphasizing the relational moments of human and non-human intimacy, dwelling works to illuminate the role of nature in producing and transforming social landscapes and the agency of the non-human as an integral part of the spatial and temporal complexities of place-making.

In that case of Columbus, for example, conflicts between environmental groups and resource management agencies, rushing river flows and gushing rainstorms, and broken pipes and aging infrastructure produce a distinctly "urban" nature in the heart of the city. In Tijuana, a mix of socioeconomic disparity, state neglect, and hydrologic dynamics produces surface runoff, even as slum citizens organize and harvest renegade flows for material resource gains (e.g., water for backyard irrigation) and rights (e.g., access to infrastructure). In both cases, an explanation of that which is "natural" and that which is "social" in negotiations of sewage and development becomes increasingly difficult to discern. Instead, these proliferations of social natures require that we examine the intimate places within people and objects produce and reproduce social order: on the banks of the river bed, in the dirt trenches of informal settlements, in the conference rooms of state agencies, and in community centers where fed-up residents demand new services and clean water sources.

This approach, while fresh and exciting, also poses big dilemmas for research: if we consider social natures as sites or assemblages, how do we engage, study, and make sense of them? Where are site ontologies in everyday life? How do we analyze an assemblage? What methods are needed to investigate a site ontology of social natures?

The wrong tools for the right questions?

Research on social natures will inevitably run the gamut of methodological tools and techniques, but we offer three observations. First, rather than focus exclusively on *social* relations – as the productionists and constructionists inevitably do – a site ontology places *situated encounters* at the heart of analysis. Here, the primary methodological goal is to gain understanding of how "objects and practices ... continuously draw each other into relation" as they constitute the emergent and site specific configurations of social life (Marston et al. 2005: 425). For social natures, this involves a sustained look at how human-nonhuman relations, objects, and everyday practices converge in a site to produce necessary political outcomes. This approach does not contend that processes and practices are unique in every site, but instead demands attention to the *connectedness* and *extensiveness* of people and objects, both within *and* between multiple sites. Researchers should be less concerned with the repeatability or validity of their data (as it "accurately" reflects the world), and more interested in understanding the events and relations that make up socio-natural assemblages.

Second, alongside Braun (2006), we cite the need for nuanced, reflexive, ethnographic methods to examine sites and assemblages. Rather than abstracting social and political forces, an ethnography of situated encounters requires active observation of and participation in the production and connection of social sites. So too, it involves observation of the relationship between formal systems of knowledge and the everyday realities though which they play out (DeVault 2009). Braun (2006: 652) suggests that to achieve this, "ethnography itself has to be reworked, no longer focused on discrete places, but instead moving 'back and forth' between ... communities and landscapes and all the places 'implicated in the chains.'"

Third and finally, methodological creativity in social natures is often wanting. Given that heterogeneous social natures "are simultaneously salient and incapable of being squashed into isomorphic slots" (Haraway 1988: 586), it seems reasonable to engage multiple methods of understanding to engage situated encounters: quantitative and qualitative, social and physical. Many researchers in cultural and political ecology have pioneered the use of mixed methods to unpack environmental narratives (e.g., Bassett and Bi Zuéli 2000; Zimmerer 2000). A quick glance at social geography journals also reveals a bevy of innovative, cutting-edge ways to think and do social assemblages. Is it possible, we wonder, to import this imagination into methodological frameworks for understanding social natures?

In the case of wastewater, a more creative approach might use the "wrong" tools to speak to the "right" questions: such as employing focus group interviews with Columbus fishermen to chart the temporal fluxes of sewer overflows and river management, hydrologic modeling of Tijuana runoff to understand the spatiality of urban development in the global south, or discourse analysis of engineering texts to locate the power relations embedded in the hydrologic cycle. While utilizing biophysical methods are probably unfeasible for most social geographers, crosscutting collaborations with physical geographers and ecological scientists – a trend endorsed by many funding institutions (e.g., the NSF's Dynamics of Coupled Natural and Human Systems) – might provide richer ontological and methodological accounts of the production of social natures.

Material Worlds, Emerging Politics, and Concluding Thoughts

As environmental activists work to bring Columbus in compliance with the Clean Water Act, leaky pipes and sewer overflows evoke visceral responses from the community. State workers are set into motion as they assess infrastructure capacity and monitor water quality, producing and reproducing formal systems of expertise about ecological worlds in everyday state practice. In Tijuana, stormwater runoff from poor settlements flows downstream into protected areas and San Diego, transforming local estuarine ecologies and triggering questions of "what kind" of development, and for whom. While our stories of sewage offer striking similarities and differences, their complexity accentuates the analytical challenges facing researchers in making sense of entangled socio-natural relations. In both cases, nature and society never truly exist in isolation from one another, nor is their construction and final punchline known in advance. Furthermore, we animate the complexity of these socio-natural assemblages, while spotlighting on their deeply political, insistent and sometimes surprising materialities.

Drawing on the insights provided by scholars through decades of critical theory on the relations between nature and society, we provide an examination of social nature that recognizes the intimacy of human and non-human worlds, while rejecting the dichotomy of nature and culture that often underlies explanation in human–environment research. The site ontology approach discussed here, as a way of thinking though social nature, moves us away from questions about the ways that humans produce the physical environment, or they ways in which nature affects social and cultural practice, to questions about how social and physical landscapes are always in dynamic *co-constitution*. Also, it demands attention to the actual spaces where social natures are created and contested. From crumbling slum settlements to urban watersheds, none of these socioecological conundrums can be understood outside of context-specific situations.

Furthermore, while recognizing that similar forces and relations play out in multiple places, the site ontology remains continually focused on how these processes are negotiated in the *micro* spaces of social life, which serve as the potential avenue for new socio-ecological formations (Woodward, Jones & Marston forthcoming). In Columbus, for example, exposure to contaminants is minimized due to numerous treatment plants, but Tijuana has extended the sewerage grid to only 70% of its citizens. Environmental groups in Columbus have successfully challenged the city using scientifically based monitoring techniques and legal action under the US Clean Water Act (Rice 2007), while in Tijuana, neighborhood residents tackle stormwater with a cement mixer, building pervious pavers to absorb runoff on-site and provide concrete streets (Meehan 2008). In each of these two cities, the uniqueness of the social sites provides distinct and varying possibilities for alternative political practice. Sustained attention to both the similarities *and* differences of social natures within and between social sites, we argue, provides the best avenue for thinking and researching social natures in geographic research.

Geographers and social scientists now grapple with some of the most pressing issues the world has ever faced: human-induced climatic changes, serious shortages of food and freshwater, and newly mutating and fast spreading diseases that travel between human and animal hosts. Why, in the context of all this complexity, is it so important to rethink social natures? The proof is in the sewage. The world's social and environmental injustices clearly point to a tangled socio-natural web of life and the profoundly important and urgent political implications of understanding social natures. To remedy some of the stubborn pitfalls and stalemates of nature–society theory, we provide a framework, anchored by use of a site ontology, that attempts to uncover the socio-spatial assemblages of humans and nonhumans that connect to other emergent sites and make and remake social life. So too, we recognize the need for rich ethnographic accounts and diverse methodological techniques. This approach, we hope, may offer a deeper porthole into the complex, contradictory, and hybrid world around us.

Note

1 "Environmentality," first used by Timothy Luke (1999), is based on Foucault's notion of "governmentality," which was developed in a series of lectures about the nature of state power in France and the United States (see also Whitehead 2007).

References

Agrawal, A. (2005) *Environmentality: Technologies of Government and the Making of Subjects.* Durham, NC: Duke University Press.

Bakker, K. (2004) *An Uncooperative Commodity: Privatizing Water in England and Wales.* Oxford: Oxford University Press.

Bakker, K. and Bridge, G. (2006) Material worlds? Resource geographies and the "matter of Nature." *Progress in Human Geography* 30 (1): 5–27.

Bassett, T. J. and Zuéli, K.B. (2000) Environmental discourses and the Ivorian Savanna. *Annals of the Association of American Geographers* 90 (1): 67–95.

Benton, E. (ed.) (1996). *The Greening of Marxism.* New York: Guilford Press.

Braun, B. (2002). *The Intemperate Rainforest: Nature, Culture, and Power on Canada's West Coast.* Minneapolis: University of Minnesota Press.

Braun, B. (2006). Environmental issues: global natures in the spaces of assemblage. *Progress in Human Geography* 30 (5): 644–54.

Braun, B. and Castree, N. (2001) *Social Nature: Theory, Practice and Politics.* London: Wiley-Blackwell.

Castree, N. (2002) False antitheses? Marxism, nature, and actor-networks. *Antipode* 34: 111–46.

Cloke, P. and Jones, O. (2001) Dwelling, place and landscape: an orchard in Somerset. *Environment and Planning* A33 (4): 649–66.

Cloke, P. and Perkins, H.C. (2005) Cetacean performance and tourism in Kaikoura, New Zealand. *Environment and Planning D: Society and Space* 23 (6): 903–24.

Delanda, M. (2006) *A New Philosophy of Society: Assemblage Theory and Social Complexity.* New York: Continuum.

Deleuze, G. and Guattari, F. (1987) *A Thousand Plateaus: Capitalism and Schizophrenia.* Minneapolis: University of Minnesota Press.

Demeritt, D. (1998) Science, social constructivism, and nature. In B. Braun and N. Castree (eds.), *Remaking Reality: Nature at the Millennium.* New York: Routledge, pp. 173–93.

DeVault, M. (2009) Institutional ethnography: information about IE, http://faculty.maxwell. syr.edu/mdevault/Information_about_IEhtm. Accessed July 13.

Engler, M. (2004) *Designing America's Waste Landscapes*. Baltimore: Johns Hopkins University Press.

Foster, J.B. (2000). *Marx's Ecology: Materialism and Nature*. New York: Monthly Review Press.

Frow, J. (2003) Down the drain: shit and the politics of disturbance. In G. Hawkins (ed.), *Culture and Waste: The Creation and Destruction of Value*, Hawkins. Lanham: Rowman & Littlefield, pp. 39–52.

Gandy, M. (1999) The Paris sewers and the rationalization of urban space. *Transactions of the Institute of British Geographers* 24: 23–44.

Gibson-Graham, J.K. (1996) *The End of Capitalism (As We Knew It): A Feminist Critique of Political Economy*. Malden: Blackwell Publishers.

Gidwani, V. (2008) *Capital, Interrupted: Agrarian Development and the Politics of Work in India*. Minneapolis: University of Minnesota Press.

Hacking, I. (1999) *The Social Construction of What?* Cambridge, MA: Harvard University Press.

Haraway, D. (1988) Situated knowledges: the science question in feminisms and the privilege of partial perspective. *Feminist Studies* 14: 575–99.

Haraway, D. (2007) *When Species Meet*. Minneapolis: University of Minnesota Press.

Harvey, D. (1996) *Justice, Nature and the Geography of Difference*. Oxford: Blackwell.

Laporte, D. (2000) [1978] *History of Shit*. Cambridge, MA: MIT Press.

Latour, B. (2005) *Reassembling the Social: An Introduction to Actor-Network Theory*. Oxford: Oxford University Press.

Lorimer, J. (2007) Nonhuman charisma. *Environment and Planning D: Society and Space* 25 (5): 911–32.

Lorimer, J. (2008) Counting corncrakes: the affective science of UK corncrake census. *Social Studies of Science* 38 (3): 377–405.

Luke, T.W. (1999) Eco-managerialism: environmental studies as knowledge/power knowledge formation. In F. Fischer and M. Hajer (eds.), *Living with Nature*. Oxford: Oxford University Press, pp. 103–20.

Marston, S.A., Jones, III, J. P., and Woodward, K. (2005) Human geography without scale. *Transactions of the Institute of British Geographers* 30 (4): 416–32.

Marston, S.A., Woodward, K., and Jones, III, J.P. (2007) Flattening ontologies of globalization: the Nollywood case. *Globalizations* 4: 1–19.

Meehan, K. (2007) Wastewater. In P. Robbins (ed.), *Encyclopedia of Environment and Society*. Thousand Oaks: Sage.

Meehan, K. (2008) Power tools for justice. *Inside Mexico* May: pp. 31.

Melosi, M.V. (2008) *The Sanitary City: Environmental Services in Urban American from Colonial Times to the Present*. Pittsburgh: University of Pittsburgh Press.

Mitchell, T. (1999) State, economy, and the state effect. In G. Steinmetz (ed.), *State/Culture: State Formation after the Cultural Turn*. Ithaca and London: Cornell University Press, pp. 76–97.

Mitchell, T. (2002) *Rule of Experts: Egypt, Techno-Politics, Modernity*. Berkeley: University of California Press.

Moore, S.A. (2008) The politics of garbage in Oaxaca, Mexico. *Society & Natural Resources* 21 (7): 597–610.

Nightingale, A. (2006) *Caring for Nature: Subjectivity, Boundaries, and Environment*. Institute of Geography Online Paper Series: GEO-021. School of Geosciences, University of Edinburgh.

O'Connor, J. (1997) *Natural Causes: Essays in Ecological Marxism*. New York: Guilford Press.

Rice, J.L. (2005) Contested natures: sewer overflows, environmental activism, and ecomana-gerialism in Columbus, Ohio. Master's thesis, Ohio State University.

Rice, J.L. (2007) Ecomanagerialism. In P. Robbins (ed.), *Encyclopedia of Environment and Society*. Thousand Oaks: Sage P.

Robbins, P. (2004) *Political Ecology: A Critical Introduction*. Malden: Blackwell.

Robbins, P. (2007) *Lawn People: How Grasses, Weeds, and Chemicals Make Us Who We Are*. Philadelphia: Temple University Press.

Robbins, P. and Marks, B. (2009) Assemblage geographies. In S. Smith, R. Pain, S.A. Marston, and J.P. Jones, III (eds.), *The Sage Handbook of Social Geography*. London: Sage, pp. 176–94.

Schatzki, T.R. (2002) *The Site of the Social: A Philosophical Account of the Constitution of Social Life and Change*. University Park: Penn State Press.

Schatzki, T.R. (2003) A new societist social ontology. *Philosophy of the Social Sciences* 33 (2): 174–202.

Scott, J. (1998). *Seeing Like a State: How Certain Schemes to Improve the Human Condition Have Failed*. New Haven: Yale University Press.

Shildrick, M. (1997) *Leaky Bodies and Boundaries. Feminism, Postmodernism and (Bio) ethics*. London: Routledge.

Smith, N. (1984) *Uneven Development: Nature, Capital, and the Production of Space*. Oxford: Blackwell.

Smith, N. (1996) The production of nature. In. G Robertson, M. Mash, L. Tickner, J. Bird, B. Curtis, and T. Putnam (eds.), *Future Natural: Nature/Science/Culture*. New York: Routledge, pp. 35–54.

Stripple, J. (2008) Governing the climate, (b)ordering the world. In L. J. Lundqvist and A. Biel, (eds.), *From Kyoto to the Town Hall: Making International and National Climate Policy Work at the Local Level*. London: Earthscan, pp. 137–54.

Swyngedouw, E. (1999) Modernity and hybridity: nature, regeneracionismo, and the produc-tion of the Spanish waterscape, 1890–1930. *Annals of the Association of American Geographers* 89 (3): 443–65.

US Environmental Protection Agency (2009) Combined sewer overflow demographics, http://cfpub.epa.gov/npdes/cso/demo.cfm?program_id=5. Accessed June 29.

Wainwright, J. (2008) *Decolonizing Development: Colonial Power and the Maya*. New York: Wiley-Blackwell.

Walker, P.A. (2005) Political ecology: where is the ecology? *Progress in Human Geography* 29 (1): 73–82.

Whatmore, S. (2002) *Hybrid Geographies*. London: Sage.

Whitehead, M. (2007) Environmentality. In P. Robbins (ed.), *Encyclopedia of Environment and Society*. Thousand Oaks: Sage.

Whitehead, M., Jones, R., and Jones, M. (2007) *The Nature of the State: Excavating the Political Ecologies of the Modern State*. Oxford: Oxford University Press.

Woodward, K., Jones, III, J.P., and Marston, S. (2010) Of eagles and flies: orientation toward the micro. *Area* 42 (3): 271–80.

Zimmerer, K.S. (2000) The reworking of conservation geographies: nonequilibrium land-scapes and nature–society hybrids. *Annals of the Association of American Geographers* 90: 356–69.

Zimmerer, K.S. and Bassett, T.J. (eds.) (2003) *Political Ecology: An Integrative Approach to Geography and Environment-Development Studies*. New York: Guilford Press.

Chapter 5

Economie$

Geoff Mann

Introduction

There is currently a common sense circulating among many human geographers that the idea of "the economy" as an object of analysis and policy only arose after World War I in Europe and North America (Mitchell 1998). There are, I think, reasons to doubt the romantic notion that modern sensibility can be so cleanly periodized (Poovey 1998: 341 n. 54; Hart 2004: 94–5; Gidwani 2008: 280 n. 17), but the problem of "the economy" – let's set aside for a moment what exactly that means – is nevertheless a principal obsession of the twentieth century, and its discursive centrality is certainly historically unprecedented. No modern social scientist can safely bracket the "economic" (again, setting aside what that means for now).

This chapter considers the ubiquitous questions of "economies" and the economic from the perspective of social geography. I work from the broad question of what social geographies of economies might be, to more specific problems of scale and the meaning and function of money, to get at some of the ways broadly "economic" theoretical and categorical concerns matter in the work social geographers do. I am particularly interested in a crucial challenge for social geographers confronting the dynamics that constitute those related processes usually labeled globalization, neoliberalism, and financialization, problems posed by the "macro" quality of these processes. Below, I argue that they operate at a scale, and with distinct social geographies, irreducible to the aggregate of their component parts: neither the macroeconomy nor the social geography of the macroeconomy is merely an aggregate bunch of local economies all operating at once. If so, the macro has a first-order ontological status to which contemporary social geography, with its "particularist" emphasis on the space/place specificity of practice, is not always readily amenable.

A Companion to Social Geography, First Edition. Edited by
Vincent J. Del Casino Jr., Mary E. Thomas, Paul Cloke, and Ruth Panelli.
© 2011 Blackwell Publishing Ltd. Published 2011 by Blackwell Publishing Ltd.

That particularist orientation is certainly unsurprising, since showing that the "macro" is in fact the product of complex relations between historically and geographically specific places, spaces, and people is often the whole point of doing social geography. Most social geographers greet macro-dynamic claims with skepticism – when someone says "globalization," our instinct is to ask what geographies it papers over. Still, important, powerful and fine-grained studies often do not wrestle with macro-dynamics not usefully reduced to their historically-geographically particular parts. One could perhaps justify this by arguing that all macro-dynamics are in fact no more than the set of lived practices and relations that sustain them "on the ground": this is partly how we get to the valuable insights that neoliberalism is really neoliberal*izations* (Larner 2003; Prudham 2004; Roff 2007; Smith and Rochovská 2007), capitalism is capitalism*s* (Watts 2002), and the commodity is a bundle of dynamic and contingent relations (Cook and Crang 1996; Whatmore 2002). Yet it is also the case – or at least I argue below – that there are relations like money, and processes like financialization that, while having constituent particulars, always also have macro-scalar relevance that is more than the sum of their parts – i.e. they have a relative scalar autonomy – that often disappears or is obscured by a tightly focused lens.

What follows is divided into four sections. Section 1 considers the matter of economies in the context of contemporary geographic research, and argues that the expanding reach and blurring boundaries of social geography (and other subfields) with regard to economies should be celebrated. Section 2 suggests that these various researches are usefully differentiated by their approach, implicit or explicit, to the category of the "economic" itself. It identifies two broad categorical approaches to the economic in human geography, decentering and recentering. To show why the categorical question matters so much, and the different ways in which it can matter, section 3 discusses how differing conceptions of the economic as category shape the understanding of the "macro" qua scale of economic relations. The discussion contains a critique of the emergent "flat ontologies" literature in social geography in light of the problem of money. The conclusion returns to the issue of categories to suggest some work social geographies of "economies" might do in the critical production of different economies. It argues that thinking hard about the category of the economic demands we also think hard about the social, and that recentering the economic in social geography can open its horizons in meaningful ways.

The "Social Geography" of "Economies"?

Today, as geography's established economic, social, political, and cultural subfields stretch themselves to account for an increasing range of human relations, it is often impossible to tell them apart. This is a welcome development; I sincerely hope that identifying to which subfield a research question or program belongs will one day be impossible. I hasten to add this is not a comment on method, a wish that all economic geography, for example, will abjure quantitative analysis and "go cultural." The point, rather, is that there is no important geographic question I can think of whose consideration would be in any way improved by endorsing a disciplinary specialization in political, economic, cultural, or social relations. Indeed, I would argue that it is human geographers' willingness to reject these divisions, far more than the concepts of space or place, that defines the discipline's contemporary contribution. In combination with theoretical and methodological catholicism, the increasing meaninglessness of subdisciplinary borders is evidence that geography, as my friend Joel Wainwright says, is more properly an "anti-discipline." The subject of this chapter, the "social geography of economies," is a case in point. What human geographic question is not always "social"? What social question is not also "economic"? If social geography is, as the *Dictionary of Human Geography* puts it, "the study of social relations and the spatial structures that underpin those relations," what lies beyond its purview? The answer – that basically nothing is off-limits – is something we should celebrate, evidence of an enviable, if sometimes anxious, analytical potential.

Nevertheless, the continuing necessity of studies of the "social geography of economies" – not to mention how awkward the phrase sounds to our disciplined ear – is evidence that much remains to be done in the anti-disciplinary effort, and I would suggest that of all the fronts in that struggle, economies are among the most important. Economies – note the politics implicit in the plural – are a central question for geography, at least partly because modern economic analysis, and the hegemonic "market" category through which economies are presently understood, render them asocial and aspatial. It is no exaggeration to say that many find in economics' analytical shortcomings economic geography's ongoing reason for being, and any social geography of economies is virtually indistinguishable from much economic geography after the "cultural turn" of the 1990s. More recently, it is true, the so-called new economic geography tried to rescue "the economy" from aspatiality, but its achievements are by no means universally celebrated: its logical "space" is unrecognizable to most human geographers, and it leaves the economy's assumed asociality unchallenged.

This is an old problem. Long before Polanyi (1944, 1957), Habermas (1989), or Bourdieu (1990) demonstrated the social embeddedness of economic relations, Marx and others challenged the separation of the economic from the social at the very moment that separation was gaining acceptance. But a century and a half later, with "the economy" and "globalization" fundamental to popular notions of what the world is all about, the task of illuminating their social and spatial fabric remains a pressing critical task (Lee 2006).

Our premises – that everything has a social geography, and that economies are an essential subject of social geographical research – force the matter: what is the social geography of an economy? If you set out to examine it, what are you looking for? What should you pay attention to, or is there anything you can afford not to pay attention to? I doubt any human geographer would answer these questions definitively, saying instead something like "That depends" – which, true as it is, does not help that much. As it stands, a perusal of existing literature indicates that social geographies of economies take two main forms: (1) an examination of all the "social" spaces that "underpin" economic qua market phenomena like production, distribution, exchange, or consumption; and (2) an examination of the "economic" spatiality of nominally "social" phenomena like household reproduction or kinship networks.

Yet however compelling the work on these problems is, most of it takes only an implicit stance on a fundamental question, perhaps *the* fundamental question for the social geographic study of economies: in what qualities does the "economic" consist? What is it that makes something count as (part of) an "economy"?

The Economic as Category

In the words of Moishe Postone (1993), this is a question of category, essential to critique insofar as it is precisely through categories that discursive power is realized in intellectual practice. For Postone, for example, the meaning that "traditional Marxism" ascribes to the categories of value and labor misapprehends their historical particularity. Value, says Postone, is not a transhistorical category, but the specific form of social wealth in capitalism. The object of Marx's famous "labor theory of value" was not a fair distribution of value to the workers who produce it, but a world in which labor was not condemned to produce wealth as value. Postone's categorical analysis constitutes one of the most important contributions to critical theory in decades. I raise it here because similar attention to the category of the economic can help social geography extend and deepen our understanding of economies.

For the question of what the economic denotes is a priority in any social geographic study of economies. The common-sense answer is merely a conflation of "the economy" with "the market" (a concept with its own imprecisions). More precisely, the vaguely Weberian category "the economic" contains those phenomena which determine or are determined by the interaction of supply and demand in the market (Weber 1946: 181); for example, prices count, national borders do not. The problems with this definition, which reproduces the untenable distinction between exogenous and endogenous central to the discipline of economics, are legion, and well rehearsed (Barnes 1996; Mitchell 1998; Thrift 2000; Cameron; Gibson-Graham 2003; Lee 2002). Few geographers would endorse it; virtually all demand a more "embedded" concept.

Unfortunately, however, social geographers rarely take the time to explain what the economic means in their research. Much of the time it would seem to describe those activities that enable production and reproduction of livelihoods. To the extent that these are mediated by markets, formal or informal, there is a close but not exact correspondence to the modern colloquial meaning of economic. For example, in two compelling recent articles in *Social and Cultural Geography*, the economic is implicitly figured as either a more socially complicated, but fundamentally similar version of the markets familiar to contemporary economics (Round et al. 2008), or as regionally specific sets of productive and exchange relations basically commensurable with the "economic" in "economic development" (Lawson et al. 2008). I cite neither paper to identify some flaw, only as evidence that for the most part social geographers make common unexamined assumptions about the realm of "the economic" and the scales at which it can be examined. The question of why some things count as economic, while others do not, is rarely on the table.

There are helpful exceptions however. Here I discuss two ways social geographers, and those in related fields like economic sociology, have developed a critique of the tacit "the economy = the (capitalist) market" equation, so as to consider what the economic as category means. Though not entirely incompatible, a rough distinction between these approaches is possible. The first, which takes several forms, tries to decenter or even dissolve the category of the economic. This is the often implicit approach behind those studies described above as focusing on the unappreciated "social" spaces that "underpin" nominally economic phenomena. The second categorical approach to the economic goes the other way, recentering the category via a critical re-incorporation of the vast economies conventionally considered non-economic. This view motivates research that highlights the "economic" quality of supposedly "social" phenomena.

Decentering the economic

The decentering approach is perhaps most closely associated with Polanyi, long influential in economic sociology (Block 1990) and anthropology (Godelier 1974), but increasingly important to geographers (Hart 2004; Prudham 2005). For Polanyi, the social trumps the economic, or at least overdetermines its form. He insists on the "embeddedness" of the economic in historically particular social fabrics from which, even in advanced capitalism, it can never be entirely divorced. In a paper most economic anthropologists cite from memory – because it defines the terms of the field's most acrimonious debate – Polanyi argues that the "term economic is a compound of two meanings that have independent roots," the "substantive" and the "formal":

> The substantive meaning of economic derives from man's dependence for his living upon nature and his fellows. It refers to the interchange with his natural and social environment, in so far as this results in supplying him with the means of material want satisfaction.
>
> The formal meaning of economic derives from the logical character of the means-ends relationship, as apparent in such words as "economical" or "economizing." It refers to a definite situation of choice, namely, that between different uses of means induced by an insufficiency of those means. (Polanyi 1957: 243)

According to Polanyi, the two meanings "have nothing in common. The latter derives from logic, the former from fact ... The cogency that is in play in the one case and in the other differs as the power of syllogism differs from the force of gravitation" (Polanyi 1957: 243–4). (You can see how, when anthropology was going through a cold war positivist "revolution" akin to geography's, though less successful, remarks like this might start a debate.) Polanyi's point is that the relation of the hegemonic concept of the economic to the "real world" is tenuous as best: "Outside a system of price-making markets economic analysis loses most of its relevance as a method of inquiry into the working of the economy ... The human economy, then, is embedded and enmeshed in institutions, economic and noneconomic" (Polanyi 1957: 247, 250).

This framing is central to much of contemporary anthropology and sociology. Especially influential is Polanyi's (1957: 250) typology of "forms of integration," through which "empirical economies" acquire "unity and stability": reciprocity, redistribution, and exchange. Any echo of a modernist developmental hierarchy – reciprocity-tribal, redistributive–sovereign, exchange–market – is unintended; Polanyi explicitly notes the forms often coexist (1957: 256). Instead, he argues that economies' "appropriative movements" follow three general patterns, each of which is organized toward the same end: the supply of "the means of material want satisfaction." Insofar as this is what remains of the economic, it is either dissolved or decentered, its categorical independence destroyed, or its explanatory power greatly reduced, since Polanyi rejects any functionalist theory of the social.

Although different in many respects, a few more recent and perhaps equally influential thinkers have also taken a decentering approach to the category of the economic. For example, in his theory of practice and the concept of habitus, Bourdieu shares with Polanyi an effort to deconstruct the explanatory primacy and independence many otherwise quite incompatible theoretical approaches accord the economic, from some versions of Marxism to liberal rationalism to structural functionalism. A decentering approach also resonates with the work of Laclau and Mouffe (1985), who, perhaps unintentionally following Weber, situate the economic as always subordinate to the broad field of the political. They suggest that the economic (qua production) may once have constituted the central antagonism in capitalism, but that class, the positionality constituted by that antagonism, has been superseded by other modes of collective identification. The economic as field of human relations as such enjoys neither autonomy nor strategic priority.

Recentering the economic

The second approach to the economic as category has important affinities with the first, but derives from Marx, for whom the economic is comprised of the moments of circulation: production, distribution, exchange, and consumption. The relations that constitute this circulatory system, which includes dynamics like family structure (Marx and Engels 1967: 100, 1998: 52), the articulation of non-capitalist and capitalist modes of production (Marx 1977: 414), and modes of domination like race and imperialism (Marx 1973: 97–8, 326), are the fundamental bases of social life. His few remarks on the relation between the economic basis and social superstructure sometimes encourage an accusation of "economism," but if Marx is economistic, it is only on the terms of this porous definition of the economic. The caricature cannot stand a reading of Marx's work, and unwittingly confuses his concept of economies with the modern idea of "the" economy qua "the" market.

A broadly Marxian attempt to recenter the economic by respecifying its scope motivates Gibson-Graham's influential work on the variety and dynamism of non-capitalist economies, founded in feminist political economy and post-structural challenges to totalization and fixity (see also Thrift and Olds 1996). Like Polanyi, Gibson-Graham (Gibson and Graham) tackle the problem of the economy head-on, since, like him, they see its determination as of utmost political and analytical importance. But, despite the fact that they generally endorse categorical "destabilization," Gibson-Graham's project is definitively not to decenter the economic, but rather to challenge the discursive-ideological construction of the economic as overdetermined by, or coextensive with, capitalist exchange relations. They affirm the centrality of the economic, but insist that our conventional understanding of the category – the realm of capitalist exchange and associated imperatives – is impoverished. They offer a typology of economic relations which *may* be organized by the logic of capital, but are often or usually organized in other ways, even in "capitalist" societies. They suggest that the economic consists of three sets of practices:

- different kinds of *transaction* and ways of negotiating (in)commensurability;
- different types of *labor* and ways of compensating it; and
- different forms of economic *enterprise* and ways of producing, appropriating, and distributing surplus. (Gibson-Graham 2006: xii; emphasis in original)

Elsewhere (2005: 97), they offer an inspiring, if less categorically precise assessment of the content of the "economies" social geographers might consider:

> All these innovations are attempts to expand the boundaries of "economy" to include that which has been prohibited – the household, voluntary and community sectors, non-capitalist enterprises, and ethical judgments related to the future, the environment, and social justice. All are attempts to wrest "economy" back from the reductionism of the market and perhaps assert that an economy is, after all, what we make it.

These two approaches to the economic overlap in important ways. Indeed, both provide a sharp rebuke to orthodox liberal economics and to so-called orthodox Marxism. As Gibson-Graham note, "an economy is, after all, what we make it," and on this Marx (if not all Marxists) and Polanyi agree. Both would suggest the obscurantism that enables the contemporary economy = market equation is the product of the particular form of the economic we live and reproduce, capitalism. The point of both critiques is to demonstrate that the workings of capitalism neither exhaust nor fix the category of the economic. It is as historically and geographically particular as the relations to which it refers.

Also, both approaches complement geographers' emphasis on the importance of place and space in social life. Both Marx's and Polanyi's concepts of the economic have been refined so as to illuminate the geographical specificity of economic relations. David Harvey most famously "spatializes" the explanatory centrality of the economic, emphasizing the essentially geographical being of accumulation and crisis, evident in phenomena like transformations in urban form, governance, and residential patterns wrought by economic change in capitalism. A host of less self-consciously "political economic" Marxian work also uses interesting connections with the Marxian category of the economic to generate social geographic insight. Research associated with Gibson-Graham's community economies project comes immediately to mind (Gibson-Graham 2008; www.communityeconomies. org), as does brilliant recent work by Wainwright (2008) on the problem of development in Belize, and Ruth Wilson Gilmore (2007) on California's prison-industrial complex.

Polanyian embeddedness is also a theme is much recent economic and social geography (e.g. Yeung 1998; Taylor 2000; Hsu 2004), and more sophisticated engagements with his work are put to fascinating purpose in Scott Prudham's (2005) account of the Oregon forest industry. Prudham shows how the "disembedding" conceits of free-market capitalism are forever frustrated by the materiality of nature. Less explicitly, Vinay Gidwani's (2008: 100–1) discussion of the ways "development" intensifies the discursive power of the formal meaning of the economic, i.e. "market calculativeness," effects a sort of postcolonial Polanyian critique, while simultaneously relying on a postcolonial Marxism not entirely incommensurable with Gibson-Graham.

The decentering work of both Bourdieu and Laclau and Mouffe has also been influential in geography, but mostly in ways that are roughly compatible with the basic Polanyian critique of the economic (e.g. Bourdieu: Painter (1997) and Bauder (2004); Laclau and Mouffe: Massey (1995) and Rasmussen and Brown (2002)). It is important not to overstate the links, however. An essential difference between Polanyi and either Laclau and Mouffe or Bourdieu is Polanyi's avowed "institution-alism," and his "agnosticism" concerning the social and political forces – i.e. power – that shape the struggle over the relation of the economic to the social (Hart 2001: 650). The last thing one could say about Bourdieu or Laclau and Mouffe is that they are agnostic about power. Nevertheless, while the concept of "embeddedness" has been subjected to incisive criticism recently (Krippner 2001; Peck 2005; A. Jones 2008), the basic decentering move common to these accounts is what matters here, and it has undoubtedly had an important impact on social geographic work.

Part of the reason each of these categorical approaches is productive for many human geographers is that despite their differences, both are compatible with an emphasis on practice. For many contemporary social geographers, the whole point of studying economies is to understand the ways the economic is (re)produced, destabilized, and performed in everyday life. In both a Polanyian and a Gibson-Graham line, the economy's fundamental sociality consists in the fact that it is lived, in ordinary and extraordinary practice, by people doing and thinking their way through their days. A categorical approach, i.e. one that refuses to take the meaning of the economic for granted, but rather sees the continued production of the cate-gory itself as a social process, is thus well-suited to the analytical and political goals of many present-day social geographers.

Consider, for example, how this might help us think about something like the "pink pound," i.e. the gay community's consumer spending in the United Kingdom. Since the 1990s, when UK corporations, especially in entertainment and "luxury" goods sectors like electronics, finally recognized its purchasing power, the pink pound has presented an interesting set of problems for any attempt to understand the extent of the sphere of the economic (Johnson 2006). On the one hand, the recognition of, and commitment to, a specifically gay consumer on the part of mainstream corporations is arguably to be applauded. Indeed, for what they are worth, surveys indicate that gay consumers are very likely to patronize businesses that are explicitly gay-positive (BBC July 31, 1998). On the other hand, it is clear that the "packaging" of a relatively homogenous "gay consumer" in the interest of profit is problematic, and might potentially constrain more transformative political efforts on the part of the gay community to move beyond an "identity politics" that merely confirms the right to be exposed to appropriately sensitive advertising (Bell 1995; Lewis 1997; Reading 1999).[1] Either way, the category of the economic, and its interrelation with dimensions of social life, like sexuality, conventionally consid-ered non-economic is a crucial question posed by the dynamics of the pink pound. How are we to understand a dollar spent simultaneously as a statement regarding the politics of sexuality and as the more mundane reproduction of capitalist rela-tions of production and exchange? Among other questions, it is the content of the category of the economic at issue here.

Yet despite their similarities and points of agreement, there is one crucial point on which the decentering and recentering approaches differ (although that difference is given more or less emphasis by different accounts): one attempts to "disempower" the economic, even to dissolve it in the socio-cultural and/or political; the other refuses to let go of the centrality of the economic, even if it rejects an assumption that it is somehow the sole locus of causation in social life. I believe the latter is the more promising direction for social geography, for two reasons. First, because it is more readily amenable to the idea that the economic is a performative site, a set of relations that constitutes, both symbolically and materially, a "real" aspect of human communities. Decentering the economy is an important political move, but it also makes possible an elision of powerful forces at work in the economic per- formance. Laclau and Mouffe's silence on matters "economic" is exemplary: it is possible to read their account of socialist strategy and forget the power of capital and the relations, like finance, through which it functions. The point of the recenter- ing accounts is made clear by something like the question of the pink pound: the economic is real, if not properly bounded, and it matters enormously; the struggle is to perform it differently: "an economy is, after all, what we make it." Take, for example, the powerful work of McDowell (1997, 2003, 2009), which brings "culture and economy into a productive conversation in an attempt to understand the changing relationships between gender, employment and identity" in Britain (McDowell 2003: 19); for her, the "economy" is a crucial site of the production of gender. The second reason the recentering critique is more powerful, as I will try to show in the next section, is it is better equipped to deal with the question of money, a social force which, while not the be-all-and-end-all of the economic, must surely be accounted for in any approach to the economic and economies.

Money and the Challenge of the "Macro"

Another advantage of a recentering approach is its power to think the macro in the realm of the economic, arguably an especially challenging task for contemporary social geographers. Taking up that task does not mean we must abjure a commit- ment to local lived places and practices; again, with fellow-traveling anthropologists and sociologists, the elaboration of the material texture of economic life is one of human geographers' signal contributions to the study of human relations. Yet I believe social geography is unnecessarily constrained if it focuses mainly on its subjects' everyday social practices. That may seem wacky, given that it is *social* geography we are talking about, but my concern arises from the fact that I think there is something more than lived space, place and instance that matters here. Insofar as we write social geographies of economies, we are confronted with the fact that economic relations in their "macro" incarnation have both discursively and materially slipped their "lived" bonds. Of course they remain "lived," but they are also something else, something irreducible to the sum of their practices (Mitchell 1997: 107–10). This is one dimension of the problems of scale and the global-local relation hotly debated in recent years (Callon and Law 2004; Law 2004; Lee et al. 2004; Thrift 2004).

It is certainly not the case that social geography has neglected concepts that fall under the macro rubric; globalization and neoliberalism, for instance, have attracted much attention, and it is hard to imagine phenomena more macro. An excellent example is the work of Leyshon and Thrift (1997), who have helped destroy any notion that modern finance capital, via its technical wizardry, functions in some ethereal globe-space where agency is subordinated to the macro operations of "neo-liberalism." Their studies of financial firms show how "neoliberalism" is in fact at least partly constituted by the very contingent and embedded relations and spaces of, say, a London trader at Lloyd's Bank named Stanley. It turns out that where Stanley works, the history of his firm, and who Stanley knows and does not know have important implications for finance capital, and presumably there are innumerable Stanleys that help make up the supposed monolith of "global capital" (MacKenzie 2004). Yet a paradoxical outcome of social geographers' powerful demonstrations of the ways particular spaces, places, and social lives always matter, no matter how dematerialized or abstract some economic operations might appear, is an increasingly explicit assumption that macro-dynamics or processes have no independent ontological or epistemological status, no specific qualitative effects, but are instead mere sums of their constituent parts (see also Callon and Latour 1981).

The theoretical bases for this assumption, motivated partly by actor network theorists' insights into scalar "noncoherence" (Callon and Latour 1981; Law 2004), are still under construction, but are best developed in a growing literature – much of it penned by social geographers – advocating "flat ontologies." This reorientation would allow us, it is argued, to move beyond the vertical ontology inherent in the concept of "scale," which is framed as irremediably hierarchical (Marston et al. 2005). Flatness – not "horizontality," which implies as structural an imaginary as "verticality" – characterizes a world of assemblages of "self-organizing systems, or onto-genesis, where dynamic properties of matter produce a multiplicity of complex relations and singularities that sometimes lead to the creation of new, unique events and entities, but more often to relatively redundant orders and practices" (Marston et al. 2005: 422). In other words, there is no utility, and potentially only obscurantism, in the idea that there exist a series of increasingly comprehensive scales as we move "up" and "out" from the body: home, local, urban, regional, national, etc. This is a radical reorientation, one which exposes our tendency to hierarchy, and undermines the status of geography's holy trinity, space-place-scale. The alternative spatial frame proposed is the "site," "something that is materially emergent within its unfolding event relations. ... [A] social site ... inhabits a 'neighborhood' of practices, events and orders that are folded variously into other unfolding sites" (Marston et al. 2005: 426).

This is not the place to give a full account of the "flat ontology" position, but the goal of these efforts is partly to escape the ordering binaries with which most social science works, and which tend to minimize the meaning and political significance of the "lesser" element in the pair. Of particular concern is the micro–macro distinction, the "Trojan horse" which endows the idea of the "global economy" with its ontological basis and causal priority, "thereby eviscerating agency at one end of the hierarchy in favor or terms like 'global capitalism'" (Marston et al. 2005: 421). Although it does not exhaust its potentials, it is fair to say the point is to remember the "everydayness of even the most privileged social actors who, though favorably anointed by class, race and gender, and while typically more efficacious in spatial *reach*, are no less situated than the workers they seek to command" (Marston et al. 2005: 421) – think of Stanley at Lloyd's.

There are several ways to engage and critique the suspicion or hatred of scale (see, for example, the responses to Marston et al.: Collinge 2006; Hoefle 2006; Jonas 2006; Escobar 2007; Leitner and Miller 2007). Still, although many such critiques react to what they see as a rather radical break, there are some links here with earlier but undeniably scalar ideas. For example, the concept of "glocalization," or the now oft-repeated exhortation to recognize the apparent monolith of neoliberalism as in reality a set of historically and geographically specific "neoliberalisms," suggest important reasons to look past claims regarding "universal" or "global" wholes or modes to their particular, local, and embodied parts. There is a long tradition of thought justifying such a theoretical orientation and its methodological implications. As Marx argued regarding the analysis of production and its social relations, production is always "definite"; "there is no general production," only particular productions, or the totality of all productions (which is very different than general production). To imagine that each historical case is a variation on a universal macro-theme like "neoliberalism" is to smuggle in "inviolable natural laws," which is definitively non-Marxian (despite common mischaracterizations to the contrary). In economic geography, this commitment finds its best known formulation in Harvey's demand for "militant particularism."

Still, Harvey's, Smith's, or Swyngedouw's emphasis on scale in economic relations is definitively nested, i.e. hierarchical: "glocalization," for example, highlights the interpenetration of global and local scales, but is nonetheless based upon a scalar structure whose ultimate form is the global (Swyngedouw 1997, 2004). In the less structural frame of contemporary social geography, however, scale itself – and especially the idea of a "macro" scale – is the object of skepticism. "Particularism" here, even if it is not underwritten by an explicit anti-scale ontology or epistemology, more often operates as a sensitivity to the contingencies and constraints of practice, discourse, and performance that combine to construct, reproduce, and frequently destabilize the relations, institutions, and spaces that are collectively called "the economy" (e.g. Gibson-Graham 2003). As such, whether or not it denies that scale exists, much of social geography participates in what we might call "provincializing the macro." In this sense, the break proposed by flat ontology's advocates is a logical extension of the path many contemporary social geographers are already on: on the question of the macro, it is merely a step from skepticism to atheism.

That path, before and after flat ontologies, is traced by a successful effort to differentiate, specify, and de-universalize economic relations. Social geography beginning from this decentering approach to the economic focuses mostly on the historical-geographic production and reproduction of relations at the local or more immediate "scale" (psyche, body, household, firm, shop-floor, etc.). We learn, for instance, of the layers of human-landscape attachments enmeshed in deeply "placed" production systems and labor (e.g. O. Jones 2008), of the psychoanalytics of racial slavery in the United States (Nast 2000), or of the embodied time-geographies of parents' childcare responsibilities (Schwanen 2007).

This struggle to de-totalize, which began at least as far back as the advent of feminist and poststructural knowledge production in the 1960s, has shaped a social geographic practice that rightly questions the hegemony of the economic and economic reason. The facts that "universal" human propensities don't turn out to be universal, and that "objective" production functions and their constituent networks are not all that objective, has changed the way we think about economies. But the principal problem with this orientation away from the macro – which reaches its purest form in the flat ontology literature's denial of its existence – is that it limits social geographic study of actually existing economies, ruling out the examination of macrodynamics that are not readily amenable to this more "sited" approach, however theoretically sophisticated it is (and contemporary social geography is quite sophisticated). Consequently, crucial and indisputably (but not only) "economic" relations having an enormous impact on questions of interest to all human geographers, from liberals to Lacanians, are either ignored or taken at face value, spared the penetrating, denaturalizing critique of "economies" social geography has made its bread and butter.

Of course it is possible to say, as Marston et al. do (2005: 426), that social geography should not focus on macrodynamics because they do not exist, and their scalar foundations are indefensible: "we suggest an approach that begins with the recognition that scale and its derivatives like globalization are axiomatics: less than the sum of their parts, epistemological *trompes l'oeil* devoid of explanatory power." The logical corollary of this position, of course, is that the immediate, lived, or "local" (a term Marston et al. cannot avoid using; 2005: 425) is all there is, and anything we once called "macro" is merely an intricate and contingent "assemblage" of it. Such assemblages definitively do not attain independent ontological status or power: as Massey (2004: 8), the most eloquent advocate for "relational space," puts it, "[i]f space is to be thought relationally ... then 'global space' is no more than the sum of relations, connections, embodiments and practices."

Few have contributed more to social and human geography as we know it than Marston and Massey, but in this case I believe they are mistaken, unnecessarily constraining social geography politically and analytically. For in assuming away the independent force of the macro, they effectively affirm a tendency to tacit empiricism – since we do not live the macro or the global, they are "no more than the sum of relations," "epistemological *trompes l'oeil* devoid of explanatory power" – while putting some very real problems out of reach (Collinge 2006: 250). If relations are always the sum of their parts, then aspects of phenomena absolutely essential to many contemporary economies are off limits to social geography. They must, and often are, basically accepted as given – or, if not as given, then as beyond explanation, at least of the social-geographic kind.

The key to my argument is money. One can perhaps imagine a social geography of money that dismisses scale. Such an analysis might examine the socio-spatiality of money in financial centers like London, uncovering the various practices and discourses that organize finance capital today. This might be augmented by a study of central banks' monetary policies and the unquestionably geographically specific institutional knowledges and relations mobilized. Alternatively, one might undertake a study of the monetary practices of a specific community, like the relatively affluent gay community behind the pink pound, examining the meaning of money and its operation in exchange and other relations. This is of course impossible in any comprehensive manner, even for only one money or monetary practice, but comprehensiveness is not the issue. Little of the excellent recent social geographic work on derivatives markets (Pryke and Allen 2000; Tickell 2000), or newer work on local exchange trading systems (Maurer 2003; Lee et al. 2004; Collom 2005;

Evans 2009) for example, aims to be comprehensive; indeed, the fact that there is significant differentiation, making comprehensive claims regarding "global" social practices next to impossible, is part of the point.

Rather than pointing out some shortage in the case study inventory, my argument is that such a study, if it can be accomplished, prioritizes "unveiling" how money "really" works in everyday life's sites, by illuminating its unfolding constitutive practices and discourses. It would not, because it has decided it cannot, ask these same critical questions of those features of money that are, for much of the planet, for all intents and purposes imposed from outside, somewhere else. Indeed, unless one imagines that the best place to study the assemblage that constitutes money is where it is supposedly "produced," like the Federal Reserve offices in Washington or the Royal Canadian Mint, or "used," like among gay urban consumers in the UK, then the problem arises of exactly where and how to study it, since for any site much of what matters is always imposed from outside. Money operates from outside everywhere, the extra-local as such; or, alternatively, it operates everywhere it matters at the same time. The point is that in either case – as the product of relations particular to nowhere, or relations particular to everywhere – there is an essentially "macro" character to money, a crucial relation in the lives of many, though not all, contemporary economies. The only reason dynamics like those described by the pink pound are even imaginable is the fact that money has these characteristics. The work it does, the power it wields, the independent being money has consists in large part in its simultaneous nowhere-ness and everywhere-ness: money's very being "presupposes general recognition" (Marx 1973: 143). This, I believe, should be the definition of "macro," and it is, it seems to me, qualitatively different from and absolutely irreducible to its parts – a "relative autonomy," if you will. Nor is it a Trojan horse: to claim that money is a macro relation is not to say it is not also a lived or local relation, nor is it to grant it causal "priority," as if it were a universal phenomenon merely "manifested" in various places, overdetermining local relations. It is, rather, to recognize that there are some relations which go on basically everywhere, but their everywhereness is not constitutive of their being – love, for example, is still love if only one person feels it – and there are things, like money, that go on basically everywhere whose very being is constituted in the everywhereness that is the macro scale.

Consequently, insofar as we approach the economic in an attempt to recenter the category, and insofar as the economic, while not exhausted by the monetary, is incomplete without it, the macro is inseparable from the fact of modern money. Indeed, it may be that the macro found scalar life only with the advent of modern fiat money, or perhaps national money in general. That is a case to be made elsewhere, if it can be made at all. But money is the way in which the macro in "macroeconomic" lives, it is the absolutely essential player in the performance of the macroeconomic, a role it cannot shirk. In fact, I would go so far as to suggest not only that there *are* phenomena that are irreducible to the less-than-macro scalar, spatial and social practices of the individuals and institutions that produce them, but that money is the principal link between macro dynamics and lived practice.

I would also argue that any social geographic understanding of other crucial "macroeconomic" relations that work through money, like inflation, distribution, and the interest rate, is impossible unless the macro qua scale of the economic exists and enjoys a relative autonomy from its constituent parts. Even if one could tell a compelling story of their "production" – imagine, in the case of the interest rate, a "sited" ethnographic account of the US Federal Reserve's Open Market Committee – an emphasis solely on institutional practice and the everyday, if coupled with a general suspicion concerning the relative autonomy of the "macro," leads to either ignoring these macro dynamics despite their critical weight in many economic relations, or, more often, to unwittingly accepting them as "given," matters of economic "fact," without an equally complex, if differently scaled, social geography of their own. Neither is adequate to the problem of contemporary economies, for either way crucial questions remain unaskable.

Conclusion: From Economies to the Economic

What we need, I believe, is not to stop asking questions about embodied, emplaced, and historically particular human lives, but to ask them while *also* theorizing the macro relations they help produce and with which they interact. The problem, of course, is how to do that. If forced to provide a quick solution, I would probably say "Read Melissa Wright and Gillian Hart," who pull this off as well or better than any other geographers of whom I am aware (e.g. Hart 2002; Wright 2006).

If given a little more time, however, I would suggest we return to the challenges and questions with which we began: If you set out to examine the social geography of an economy, what are you looking for? What should you pay attention to? Or, perhaps, is there anything you can afford not to pay attention to? Like most others, I would say "That depends," but now I can be more specific about the nature of that dependence. For the task of any social geography of economies today, it seems to me, is by definition to recenter the economy in the fashion outlined in section 2. The task is not to dissolve or decenter the economic, but to put it squarely in the center of our analyses, to undertake its respecification in the spirit of Gibson-Graham. This is not exactly a call for a "new economism," in which the economy is discursively and materially remade as something more than the set of capitalist market-exchange relations with which it is currently conflated, although it is not far from that. Instead, social geography of economies must recenter the economic not to displace or determine other "realms" of social life – the common condemnation of "economism" – but to put the economic as category squarely at the center of the analysis. Social geographies of economies should invoke the economic, and seek it out, wherever it operates and wherever it matters, which will of course be historically and geographically specific (Gibson-Graham 2008).

The key to this, and the lesson an emphasis on category teaches us, is to make social geographies of economies the study of the myriad, sometimes contradictory, forces involved in the performance and production of the economic as category, including, among other things, the discipline of economics itself (Callon 1998; MacKenzie 2008; Mitchell 2008). This need not entail an attempt to fix the boundaries of the economic; on the contrary, it involves the constant tracing and retracing of the category throughout the social world, the incessant search to find how and why the social operates so as to produce the economic. Surely the emergence and persistence of the category in modern thought, improperly constrained by a market-oriented common sense or not, should not be accepted as a natural outcome, but as the result of social dynamics that are themselves illuminated by the power of the economic. In other words, the problem is not to show that the economic is reducible to the social, but to understand the form and content of the economic as a real realm of social experience, rendered "distinct" from the social according to the social's own rules.

This is another reason to take the macro seriously. For not only is it important because its social (re)construction demonstrates people "believe" it is important, but also because the ways in which the economic as category comes into being are always "real" processes that produce a "real," if abstracted, world. Indeed, "real abstraction" – the actual collective human performances through which categories come to matter – may be the defining condition of possibility for the market-capitalist elements of "economy" as it currently operates (Colletti 1975; Toscano 2008; Mann 2009). If so, this only reinforces the fact that the supreme real abstraction – money qua value – is both an essential piece of the puzzle in any study of economies (even in its absence, which must always be confronted), and irreducibly tied to the existence of the macro scale of the economic category.

Wright (2006), for example, illustrates the power of this approach to the social geography of the economic. She is interested in the ways in which the economic is discursively and materially consolidated by capital as the dis-interested sphere of value, and the vast array of power relations – founded in the power of capital or not – narrated, performed and sometimes violently imposed in its name. In a study of factory production in Mexico and China, she shows how the "myth" of female workers' disposability "actually creates the material embodiment of the disposable third world woman that houses this labor" (2006: 12). On the terms I have laid out here, the point of her work is an understanding of how the contemporary constitution of the social helps consolidate the economic as category through which women's bodies, even their very deaths, are spoken and performed. Moreover, Wright poses the fine-grained, geographically specific question of factory workers' lives in confrontation with an uneven, discursively variegated, but nonetheless macro "global capitalism" whose origin is part of the problem with which she wrestles. Although more convinced of capital's hegemony than Gibson-Graham, and less sanguine than their research on community economies, the message resonates: "an economy is, after all, what we [and they] make it."

Wright's research is enormously compelling, and puts us, I think, where we want to be. It demonstrates how, by elaborating a social geography that recenters the economy, we force ourselves to rethink not only the economic, but also the social (Lee et al. 2008). It also demonstrates that understanding the relative autonomy of the macro is essential to an analysis of capitalism – surely a prerequisite for any reasonable claim to an understanding of modern economies – but it never diminishes the import of the lifeworld of everyday practice. In contrast to the empiricism lurking behind "flatness," it shows the constitution of the economic not as an additive process, but one that unfolds on many scales at once. This is a social geography of the economic that fetishizes neither "local" presence nor "global" metaphysics.

Note

1   For an example of this "packaging," see Clear Channel's (2009) description of the uses to which its billboards can be put to attract the gay consumer.

References

Barnes, T.J. (1996) *Logics of Dislocation: Models, Metaphors, and Meanings of Economic Space*. New York: Guilford.

Bauder, H. (2004) Habitus, rules of the labour market and employment strategies of immigrants in Vancouver, Canada. *Social and Cultural Geography* 6 (1): 81–97.

BBC News (July 31, 1998) The pink pound, http://news.bbc.co.uk/2/hi/business/142998.stm.

Bell, D. (1995) Pleasure and danger: the paradoxical spaces of sexual citizenship. *Political Geography* 14 (2): 139–53.

Block, F. (1990) *Postindustrial Possibilities: A Critique of Economic Discourse*. Berkeley: University of California Press.

Bourdieu, P. (1990) *The Logic of Practice*. Tr. Richard Nice. Stanford: Stanford University Press.

Callon, M. (1998) Introduction: the embeddedness of economic markets in economics. In M. Callon (ed.), *The Laws of Markets*. Oxford: Blackwell, pp. 1–57.

Callon, M. and Latour, B. (1981) Unscrewing the big Leviathan: how actors macro-structure reality and how sociologists help them do so. In K. Knorr-Cetina and A. Cicourel (eds.), *Advances in Social Theory and Methodology*. London: Routledge and Kegan Paul, pp. 277–303.

Callon, M. and Law, J. (2004) Guest editorial. *Environment and Planning D: Society and Space* 22 (1): 3–11.

Cameron, J. and Gibson-Graham, J.K. (2003) Feminising the economy: metaphors, strategies, politics. *Gender, Place and Culture* 10 (2): 145–57.

Clear Channel (2009) Harnessing the pink pound, www.clearchannel.co.uk/content.aspx?ID=276&ParentID=93&MicrositeID=0&Page=1. Accessed October 2.

Colletti, L. (1975) Introduction. In K. Marx, *Early Writings*. London: Penguin, pp. 7–56.

Collinge, C. (2006) Flat ontology and the deconstruction of scale: a response to Marston, Jones and Woodward. *Transactions of the Institute of British Geographers* 31: 244–51.

Collom, E. (2005) Community currency in the United States: the social environments in which it emerges and survives. *Environment and Planning A* 37 (9): 1565–87.

Cook, I. and Crang, P. (1996) The world on a place: culinary culture, displacement and geographical knowledges. *Journal of Material Culture* 1 (2): 131–53.

Escobar, A. (2007) The "ontological turn" in social theory: a commentary on "Human geography without scale" by Sallie Marston, John Paul Jones and Keith Woodward. *Transactions of the Institute of British Geographers* 32: 106–11.

Evans, M. (2009) Zelizer's theory of money and the case of local currencies. *Environment and Planning A* 41 (5): 1026–41.

Gibson-Graham, J.K. (2003) An ethics of the local. *Rethinking Marxism* 15 (1): 49–74.

Gibson-Graham, J.K. (2005) Economy. In A. Bennett, L. Grossberg, and M. Morris (eds.), *New Keywords: A Revised Vocabulary of Culture and Society*. Oxford: Blackwell, pp. 94–7.

Gibson-Graham, J.K. (2006) *The End of Capitalism (As We Knew It)*. Minneapolis: University of Minnesota Press.

Gibson-Graham, J.K. (2008) Diverse economies: performative practices for other worlds. *Progress in Human Geography* 32 (5): 613–32.

Gidwani, V. (2008) *Capital, Interrupted: Agrarian Development and the Politics of Work in India*. Minneapolis: University of Minnesota Press.

Gilmore, R.W. (2007) *Golden Gulag: Prisons, Surplus, Crisis, and Opposition in Globalizing California*. Berkeley: University of California Press.

Godelier, M. (1974) Une anthropologie économique est-elle possible? In M. Godelier (ed.), *Un Domain Contesté: L'Anthropologie Économique*. Paris: Mouton Éditeur, pp. 285–345.

Habermas, J. (1989) *The Structural Transformation of the Public Sphere: An Inquiry into a Category of Bourgeois Society*. Tr. Thomas Berger. Cambridge, MA: MIT Press.

Hart, G. (2001) Development critiques in the 1990s: *culs-de-sac* and promising paths. *Progress in Human Geography* 25 (4): 649–58.

Hart, G. (2002) *Disabling Globalization: Places of Power in Post-Apartheid South Africa*. Berkeley: University of California Press.

Hart, G. (2004) Geography and development: critical ethnographies. *Progress in Human Geography* 28 (1): 91–100.

Hoefle, Scott W. (2006) Eliminating scale and killing the good that laid the golden egg? *Transactions of the Institute of British Geographers* 31: 238–43.

Hsu, J. (2004) The evolving institutional embeddedness of a late industrial district in Taiwan. *Tijdschrift voor Economische en Social Geografie* 95: 218–32.

Johnson, B. (2006) Pink pound flexes technological muscle. *Guardian* February 7, www.guardian.co.uk/technology/2006/feb/07/news.retail.

Jonas, A. (2006) Pro scale: further reflections on the "scale debate" in human geography. *Transactions of the Institute of British Geographers* 31: 399–406.

Jones, A. (2008) Beyond embeddedness: economic practices and the invisible dimensions of transnational business activity. *Progress in Human Geography* 32 (1): 71–88.

Jones, O. (2008) Of trees and trails: place in a globalised world. In N. Clark, D. Massey, and P. Sarre (eds.), *Material Geographies: A World in the Making*. London: Sage, pp. 213–56.

Krippner, G. (2001) The elusive market: embeddedness and the paradigm of economic sociology. *Theory and Society* 30: 775–810.

Laclau, E. and Mouffe, C. (1985) *Hegemony and Socialist Strategy: Towards a Radical Democratic Politics*. New York: Verso.

Larner, W. (2003) Neoliberalism? *Environment and Planning D: Society and Space* 21: 509–12.

Law, J. (2004) And if the global were small and noncoherent? Method, complexity, and the baroque. *Environment and Planning D: Society and Space* 22 (1): 13–26.

Lawson, V., Jarosz, L., and Bonds, A. (2008) Building economies from the bottom up: (mis) representations of poverty in the rural American northwest. *Social and Cultural Geography* 9 (7): 737–53.

Lee, R. (2002) Nice maps, shame about the theory'? Thinking geographically about the economic. *Progress in Human Geography* 26 (3): 333–55.

Lee, R. (2006) The ordinary economy: tangled up in values and geography. *Transactions of the Institute of British Geographers* 31 (4): 413–32.

Lee, R., Leyshon, A., and Smith, A. (2008) Rethinking economies/economic geographies. *Geoforum* 39 (3): 1111–15.

Lee, R., Leyshon, A., Aldridge, T., Tooke, J., Williams, C., and Thrift, N. (2004) Making geographies and histories? Constructing local circuits of value. *Environment and Planning D: Society and Space* 22 (4): 595–617.

Leitner, H. and Miller, B. (2007) Scale and the limitations of ontological debate: a commentary on Marston, Jones Woodward. *Transactions of the Institute of British Geographers* 32: 116–25.

Lewis, R. (1997) Looking good: the lesbian gaze and fashion imagery. *Feminist Review* 55: 92–109.

Leyshon, A. and Thrift, N. (1997) *Money/Space: Geographies of Monetary Transformation*. London: Routledge.

MacKenzie, D. (2004) Social connectivities in global financial markets. *Environment and Planning D: Society and Space* 22 (1): 83–101.

MacKenzie, D. (2008) *An Engine, Not a Camera: How Financial Models Shape Markets*. Cambridge, MA: MIT Press.

Mann, G. (2009) Colletti on the credit crunch: a response to Robin Blackburn. *New Left Review* II/56: 119–27.

Marston, S., Jones, J.P., and Woodward, K. (2005) Human geography without scale. *Transactions of the Institute of British Geographers* 30: 416–32.

Marx, K. (1973) *Grundrisse*. New York: Vintage.

Marx, K. (1977) *Capital*, vol. I. New York: Vintage.

Marx, K. and Engels, F. (1967) *The Communist Manifesto*. London: Penguin.

Marx, K. and Engels, F. (1998) *The German Ideology*. Amherst: Prometheus.

Massey, D. (1995) Thinking radical democracy spatially. *Environment and Planning D: Society and Space* 13 (3): 283–8.

Massey, D. (2004) Geographies of responsibility. *Geografiska Annaler* 86 B (1): 5–18.

Maurer, W. (2003) Uncanny exchanges: the possibilities and failures of "making change" with alternative monetary forms. *Environment and Planning D: Society and Space* 21: 317–40.

McDowell, L. (1997) *Capital Culture: Gender at Work in the City*. Oxford: Blackwell.

McDowell, L. (2003) *Redundant Masculinities: Employment Change and White Working Class Youth*. Oxford: Blackwell.

McDowell, L. (2009) *Working Bodies: Interactive Service Employment and Workplace Identities*. Oxford: Wiley-Blackwell.

Mitchell, K. (1997) Transnational discourse: bringing geography back in. *Antipode* 29 (2): 101–14.

Mitchell, T. (1998) Fixing the economy. *Cultural Studies* 12 (1): 82–101.

Mitchell, T. (2008) Rethinking economy. *Geoforum* 39 (3): 1116–21.

Nast, H. (2000) Mapping the "unconscious": racism and the oedipal family. *Annals of the Association of American Geographers* 90 (2): 215–55.

Painter, J. (1997) Regulation, regime and practice in urban politics. In M. Lauria (ed.), *Reconstructing Regime Theory: Regulating Urban Politics in a Global Economy*. London: Sage, pp. 122–43.

Peck, J. (2005) Economic sociologies in space. *Economic Geography* 81: 129–76.

Polanyi, K. (1944) *The Great Transformation: The Political and Economic Origins of Our Times*. Boston: Beacon Press.

Polanyi, K. (1957) The economy as instituted process. In K. Polanyi, C. Arenberg, and H. Pearson (eds.), *Trade and Market in the Early Empires: Economies in History and Theory*. Glencoe, IL: Free Press, pp. 243–70.

Poovey, M. (1998) *A History of the Modern Fact: Problems of Knowledge in the Sciences of Wealth and Society*. Chicago: University of Chicago Press.

Postone, M. (1993) *Time, Labor, and Social Domination: A Reinterpretation of Marx's Critical Theory*. Cambridge: Cambridge University Press.

Prudham, W.S. (2004) Poisoning the well: neoliberalism and the contamination of municipal water in Walkerton, Ontario. *Geoforum* 35 (3): 343–59.

Prudham, W.S. (2005) *Knock on Wood: Nature as Commodity in Douglas-Fir Country*. London: Routledge.

Pryke, M. and Allen, J. (2000) Monetized time–space? Derivatives: money's "new imaginary"? *Economy and Society* 29 (2): 264–84.

Rasmussen, C. and Brown, M. (2002) Radical democratic citizenship: amidst political theory and geography. In E. Isin and B.S. Turner (eds.), *Handbook of Citizenship Studies*. London: Sage, pp. 175–90.

Reading, A. (1999) Selling sexuality in the lesbian and gay press. In J. Stokes and A. Reading (eds.), *The Media in Britain: Current Debates and Developments*. London: Palgrave Macmillan, pp. 265–72.

Roff, R.J. (2007) Shopping for change? Neoliberalizing activism and the limits to eating non-GMO. *Agriculture and Human Values* 24 (4): 511–22.

Round, J., Williams, C.C., and Rodgers, P. (2008) Everyday tactics and spaces of power: the role of informal economies in post-Soviet Ukraine. *Social and Cultural Geography* 9 (2): 171–85.

Schwanen, T. (2007) Matter(s) of interest: artefacts, spacing and timing. *Geografiska Annaler: Series B, Human Geography* 89 (1): 9–22.

Smith, A. and Rochovská, A. (2007) Domesticating neoliberalism: everyday lives and the geographies of post-socialist transformations. *Geoforum* 38 (6): 1163–78.

Swyngedouw, E. (1997) Neither global nor local: "glocalization" and the politics of scale. In K. Cox (ed.), *Spaces of Globalization: Reasserting the Power the Local*. New York: Guilford, pp. 115–36.

Swyngedouw, E. (2004) Scaled geographies: nature, place, and the politics of scale. In E. Sheppard and R. McMaster (eds.), *Scale and Geographic Inquiry: Nature, Society and Method*. London: Wiley-Blackwell, pp. 129–53.

Taylor, M. (2000) Enterprise, power and embeddedness: an empirical investigation. In E. Vatned and M. Taylor (eds.), *The Networked Firm in a Global World*. Aldershot: Ashgate, pp. 199–233.

Thrift, N. (2000) Pandora's box? Cultural geographies of economies. In G. Clark, M. Feldmann, and M. Gertler (eds.), *The Oxford Handbook of Social Geography*. Oxford: Oxford University Press, pp. 689–702.

Thrift, N. (2004) Remembering the technological unconscious by foregrounding knowledges of position. *Environment and Planning D: Society and Space* 22 (1): 175–90.

Thrift, N. and Olds, K. (1996) Refiguring the economic in economic geography. *Progress in Human Geography* 20 (3): 311–37.

Tickell, A. (2000) Dangerous derivatives: controlling and creating risks in international money. *Geoforum* 31 (1): 87–99.

Toscano, A. (2008) The open secret of real abstraction. *Rethinking Marxism* 29 (2): 273–87.

Wainwright, J. (2008) *Decolonizing Development: Colonial Power and the Maya*. Oxford: Blackwell.

Watts, M. (2002) Capitalisms, crises, and cultures I: notes toward a totality of fragments. In A. Pred and M. Watts (eds.), *Reworking Modernity: Capitalisms and Symbolic Discontent*. New Brunswick: Rutgers University Press, pp. 1–19.

Weber, M. (1946) *From Max Weber: Essays in Sociology*. In H.H. Gerth and C. Wright Mills (eds.), New York: Oxford University Press.

Whatmore, S. (2002) *Hybrid Geographies: Natures, Cultures, Spaces*. London: Sage.

Wright, M. (2006) *Disposable Women and Other Myths of Global Capitalism*. London: Routledge.

Yeung, H. (1998) The socio-spatial constitution of business organization: a geographical perspective. *Organisation* 5: 101–28.

Chapter 6

Community

Marcia England

At this historical moment we are awash in communities. (Watts 2004: 197)

Introduction

Social theorists have been examining "community" and "communities" for over a century. The term is espoused often in colloquial speech, but just what it actually means is difficult to pin down. This chapter outlines the lineage of the concept of community as geographers and other prominent social theorists have used it since the late nineteenth century. Community intrigues scholars as it is an often-contested term. Some seek to find meaning in the term, others want to examine the consequences of the concept and others dismiss it as meaningless. Within this chapter, I do not attempt to solve the problem of what community is/means, but instead document the work of those who have explored the term and show the utility/futility in its employment. The term is convoluted and nebulous,[1] but holds meaning for a number of people in multiple ways. This chapter addresses the notion of community with regard to collective action and collective identity and synthesizes the literatures surrounding community. Drawing on works from sociology, philosophy, and geography, the following examines how community affects the organization of social life and the spaces that emerge from those social interactions.[2]

The spaces of community influence social relations and vice versa. Community affects not only social geographies, but also political, economic, historical, and cultural geographies. These geographies that surround community can be both "real" and "imagined." For example, the nation is a community space that is imagined, but feels very real to many. Another example is the neighborhood, often touted as the "space of a community," which may have boundaries grounded in reality, but

A Companion to Social Geography, First Edition. Edited by
Vincent J. Del Casino Jr., Mary E. Thomas, Paul Cloke, and Ruth Panelli.
© 2011 Blackwell Publishing Ltd. Published 2011 by Blackwell Publishing Ltd.

can involve imagined sentiments of neighborliness. Within these real and imagined communities, people have multiple identities as a result of belonging to, and identifying with, multiple communities.

Community is further clouded when looking at larger social issues that surround it. The concept of community involves issues of inclusion and exclusion, of unity and division, and of heterogeneity and homogeneity. Many of the debates connected with the term are rooted in these binaries.[3] In order to discuss community, one must address the dualisms and debates that surround the notion. Many questions surround the notion, and characteristics of community are often discussed within these conversations: Are communities inclusive? Exclusive? A source of division? Or one of unity? Are they heterogeneous or homogeneous? Even, do these questions matter? This is not to say that communities are one or the other or that binaries hold true when theorizing and discussing community, but that there are sides of the debates on community that see the term as possibly conflict-ridden.

Scales of Community

Community is a concept that resonates on multiple scales from the global to the local. Globally, one might imagine a community based on fair trade or the enactment of community based on charity or aid, e.g. the Indian Ocean tsunami of December 2004 or the Haitian earthquake of 2010. At the local level, one could consider the production of rural communities constructed as bucolic ideal. Because there are so many scales involved in shaping communities (as people can belong to multiple communities at the same time), there are many ways to frame community and communities. Scholars in the social sciences and humanities have been studying the different forms communities take. Below are a few examples that show the variation in community scales.

Global communities can be formed in numerous ways. The cosmopolitan community is one that, while imagined and socially constructed, can have very real effects. Think of times when there have been global movements – boycotts, protests, and economic aid are just a few examples. Chilean grapes boycotts, dolphin-safe tuna endorsement and the South African boycotts and embargoes (due to the South African apartheid regime) are famous global movements that have spurred people across the world to mobilize. Other famous examples are advertising (Coca-Cola's campaign of "I'd like to buy the world a Coke" or the Visa's Olympics' campaign: "Go, world"). The processes of globalization have had two primary effects on community. One is that it has formed a global community; the other is a desire to cling to the local and a rise in sentiment regarding the local community.

A meso-level of imagined community can be found in the state. Community at the state level is often complemented by a sense of nationalism. The nation ideal is touted in times of state crisis. Take for example the case study of September 11, 2001. Inhabitants of the United States rallied around a cry of "United We Stand" and formed a community not only with their neighbors and co-workers, but with strangers across their cities, states and the U.S. as a whole. People proudly wore their patriotic iconography in order to symbolize their allegiance to the state and to its people. As Billig (1995: 74) argues, "Nationhood ... involves a distinctive imagining of a particular sort of community rooted in a particular sort of place" (quoted in Dittmer 2005: 633). This is not to say that all felt the welcoming spirit of the national community. Many minorities (religious or otherwise) felt alienated in this time of bonding. They formed their own communities of exile in some cases. The emergence of the nation as a community also does not discount local diversity.

Local communities, depending on their locale, are often seen as sites of homogeneity. For instance, rural communities are regularly touted as the ideal of the concept of community and, as such, have been the source of social science research for decades. If one pictures the countryside, idyllic images of rural communities are often conjured. A place where everyone knows everyone else, people say "hello" to everyone and there is a smile on every face. Of course, this is not the reality of many rural communities, but many people romanticize the rural as space of common values and morals. Anthropologists, sociologists, and geographers have proven that differences exist in rural communities, especially along gender lines (Little and Leyshon 2003; Little and Panelli 2003). Drawing on the work or Dempsey (1990, 1992), Little and Panelli (2003: 282) note research has argued that "men experienced substantial authority and control in communities, while women were expected to engage in activities and behaviours that would nurture, service and maintain traditional values, practices and relations within the community." Campbell and Phillips (1995) demonstrate that key communal spaces and practices propagate unequal gender relations. The rural community has been impugned by some due to the social constraints put on its inhabitants. For example, Hillier et al. (1999: 93) found that "rural young women lack a sense of embodied sexuality" and that "they are constrained by their own ideas of their sexuality and the restrictions placed upon them by their communities" (Little and Leyshon 2003: 266).

Rural places often tout the notion of community as an ideal. Valentine (2001: 255) points out that rural space "has been understood to have a particular form of rural society: community." But rural ideas of community are not monolithic, but instead are rooted in one's communal identity. For example, one's status or time spent in the community can affect one's sense and definition of belonging. Gender, lifestyle choices and class can all be markers of difference in a rural community, as they are with urban communities.

Community as Relationship

Ferdinand Tonnies (2001 [1887]) is often cited for his work on distinguishing "community" from "society." For Tonnies, *gemeinshaft* (community) was epitomized by close and personal ties, while *gesselschaft* (society) was centered more on anonymous and impersonal relationships. For example, Tonnies (2001: 18) explains: "We go out into *Gesellschaft* as if into a foreign land. A young man is warned about mixing with bad society: but "bad community" makes no sense in our language." Tonnies furthermore spatializes his argument by stating that country people know little of *gesellschaft*, as he argues that society is an urban phenomenon.

One of the most influential schools of thought on community came from the Chicago School of Human Ecology (hereafter referred to as simply the Chicago School), which developed a theory of "natural communities" in the first decades of the twentieth century. Headed by Robert Park, Ernest Burgess, and Roderick

McKenzie, the Chicago School examined Chicago neighborhoods using plant and nature analogies. In Park's analysis of urban communities, he compared human social relationships with plants through three key concepts: competition, ecological dominance and invasion and succession. Park's theories were advanced by Burgess and McKenzie, who continued to use the framework of nature to describe urban human activity. These theories are still widely circulated in introductory urban geography texts.

Critiques of the Chicago School emerged shortly after as other urban scholars found the theory did not match up well to research in other landscapes. The Chicago School has also been criticized theoretically, most notably by scholars who did not believe the ecological analogy compared significantly with urban human social relations (see a discussion of Firey 1945, 1947; Sjoberg 1960 and Harvey 1973 in Valentine 2001). Although widely critiqued, the Chicago School opened up discussions and theorizations of "community" and promoted research into neighborhoods for subsequent geographers and other urban scholars. The Chicago School has decidedly marked history and created a body of work upon which urban scholars build.

The Chicago School focused on the scale of the neighborhood and left a legacy of examining community at that level. As such, neighborhoods are often envisioned as the spaces of community. Yet neighborhoods are specifically contingent on location (Martin 2002) and communities are not necessarily (see Anderson 1991). Neighborhoods change and people and businesses move in and out over time. But there is a constant of location. When the idea of the neighborhood is mentioned, a sense of nostalgia is often evoked. There is almost a utopian vision of the area that is "the neighborhood." When the characters on Sesame Street ™ sing of the "people in your neighborhood," they sing of business owners and police, school children and postal workers, not marginalized members of society like the homeless. This idealizes the community in such a way that can be polarizing and exclusive.

Cox (2002: 148–9) argues that there is a creation of a moral hierarchy of "good" and "bad" neighborhoods, arguing that "[w]ithin this moral socio-spatial hierarchy residents jostle further to redefine their spaces, their neighborhoods, in some way which will further enhance their sense of social worth." Within these neighborhoods comes the possibility for the formation of a community, but that is not necessarily so. Many neighborhoods are not communities. Drawing on the Cater and Jones (1989: 169) definition of a neighborhood community as "a socially interactive space inhabited by a close-knit network of households, most of whom are known to one another and who, to a high degree, participate in common social activities, exchange information, engage in mutual aid and support and are conscious of a common identity, a belonging together" (quoted in Valentine 2001: 112), one can see that this does not hold true in many neighborhoods. Anonymity and socio-spatial isolation can be found in many neighborhoods, leading some to question whether or not community is on the decline.

Cater and Jones's definition of community may seem contradictory to definitions posed by other scholars on the subject. There has been a recent move away from prior approaches to community, which treated it as a homogenous entity (Panelli and Welch 2005, see EPA 1999). Multiple identities interact and intersect within communities, and scholars from various disciplines often debate the notion of the homogeneity of communities. Homogeneity may exist within a community, but it is not necessarily a precursor.

Defining Community

How to define community has been a source of struggle for many social theorists, yet many definitions abound. For some, the definition of community may be as simple as Johnston's (1994: 80), which defines community as "a social network of interacting individuals, usually concentrated in a defined area." Furthermore, Silk (1999: 6) argues that community "suggests any or all of the following: common needs and goals, a sense of the common good, shared lives, culture and views of the world, and collective action" (quoted in Panelli and Welch 2005: 1590). Watts (2004: 196) sees community as "constitutive of modern politics, a keyword whose meaning turns on questions of membership, shared meanings, identity and imagination."

For others, community is a more complicated concept with various definitions and perceived notions. As an example, Selznick's (1992: 361–4) "fully realized community" shares a mixture of the following key values:

> *historicity* (interpersonal bonds fashioned in a shared history and culture), *identity* (a sense of community manifest in loyalty, piety, and a distinctive identity), *mutuality* (relations of interdependence and reciprocity), *plurality* (persons engaging in intermediate associations or group attachments), *autonomy* (the flourishing of unique and responsible persons), *participation* (in different roles and aspects of society), and *integration* (via political, legal, and cultural institutions). (quoted in Smith 1999: 21)

The combination of characteristics can be a useful tool in defining communities, but community is more than the sum of its parts. Community has certain connotations that can frame it theoretically in a number of ways.

Community is often defined positively. Schofield (2002: 663) writes, "[t]o think community is to enter a world without enemies." Etzioni (1995: 31) explains that "[w]hen the term community is used, the first notion that typically comes to mind is a place in which people know and care for one another" (quoted in Smith 1999: 23). Williams (1973: 76) comments on the affect of community stating that it,

> can be the warmly persuasive word to describe an existing set of relationships or the warmly persuasive word to describe an alternative set of relationships. What is important is that unlike all other terms of social organization (state, society, etc.) it never seems to be used unfavourably and never to be given any positive opposing or distinguishing term. (quoted in Schofield 2002: 664)

Although community has often been conceptualized at times as a "warm," "welcoming," and "positive" idea (Putnam 2000; Valentine 2001; Williams 1973), it can have an exclusionary aspect as well.

Another definition of community is that of a politically/economically/culturally motivated group that excludes those that do not have the same intended outcomes or goals. This process of exclusion can help to form the community in a way. For instance, I have argued elsewhere (England 2006), as have other scholars, that communities can have common goals. One definition of community holds that a common objective is necessary to communal identity. The community as such is motivated politically, socially, economically, culturally to achieve some purpose. This purpose can be of an inclusive or exclusive nature, can be polarizing or unifying.

For Selznick (1992: xi), "a proper understanding of community ... presumes diversity and pluralism as well as social integration" (quoted in Smith 1999: 23). This conceptualization of community accommodates both the inclusionary and exclusionary, the cohesive and devisive forces present in communities and collective identities. The exclusion of others from the community provides a uniting moment for the collective group, it provides a sense of identity in that it forms a dialectic moment of identification by naming that which it is not.

Communal Identity

Identity is formed by difference, which can be theorized through alterity. One asserts an identity through the process of negation of that which one is not (e.g., A, not A). This process denies an essential identity, since identification takes place only through a relation to another that is different. The outside of the category is therefore already embedded within the category. Boundaries between categories are not stable and need each side to exist. This relationship marks a "trace" of the other onto the self, onto one's identity for the self cannot be formed without the other (Derrida 1991a, 1991b, 1991c; Dwyer and Jones 2000; Isin 2000; Laclau and Mouffe 1985; Natter and Jones 1997).

One's communal identity can gloss over differences between members of a community and exacerbates difference between the community and those who are not accepted as part of the fold. Young has written about the "ideal of community" and argues that the notion of community implies intolerance of difference (Young 1990a, 1990b). Valentine (2001: 106) states that for Young, community "privileges unites over difference, that it generates social exclusions and that it is an unrealistic vision." The ideal of community "expresses a desire for the fusion of subjects with one another which in practice operates to exclude those with whom the group does not identify" (Young 1990a: 227). For Young (1990a: 303), "The ideal of community ... denies the difference between subjects. The desire for community relies on the same desire for social wholeness and identification that underlies racism and ethnic chauvinism on the one hand and political sectarianism on the other."

Being part of a community can produce a feeling of belonging and acceptance. Being left out of a community can cause one to feel ostracized and alone. Both the inclusion and exclusion of people(s) from a community provide moments of definition for the "community." When one is set apart from a community, whether it be internal or external definition that causes that differentiation, a myriad of sentiments can result ranging from pride to resentment. The marking of an individual or set of individuals as those that do not belong produces distinction between the insiders (often theorized as the "community") and the outsider.

One of the key debates of scholars who examine communities surrounds the degree of homogeneity in a community as mentioned earlier. For most common purposes, a community is defined as a group that shares a set of common interests.

In this understanding, communities involve interacting individuals that form a group with some common characteristic or goal. But just how heterogeneous that group is depends on the community.

Some argue that in order for community to exist, microdivisions are ignored and commonality is promoted (Fischer and Poland 1998; Schofield 2002; Sennett 1970). Communities can produce a feeling of cohesiveness. For Sennett (1970: 31): "[t]he bond of community is one of sensing common identity, a pleasure in recognizing 'us' and 'who we are.'" Sennett distinguishes between a social group and a community: "a community is a particular kind of social group in which men believe they *share* something together" (1970: 31, original emphasis). Those who are seen as being outsiders do not belong to the community, they do not "share something together." This has spatial outcomes.

For some scholars, in order to recognize a community, an "us," there must be a "them," an other. Discourses on community inform conceptions of communal/public space and who has access to that space. Fear of the social other leads to exclusion from public space of those who are seen as threatening. Fischer and Poland (1998: 193) argue that community "has become a critical resource as well as a product of effective self-selection of norms, stakeholders and resources" which often results in the "exclusion of the ones disrupting the order striven for."

Those constructed as disorderly are policed (whether by the community itself or by the state) and surveilled to uphold local government, police, and community definitions of public space and of who belongs in public space and to maintain public order and public health (Chauncey 1996; Daly 1998; Fyfe and Bannister 1998; Herbert 1996a, 1996b; Hunt 2002; Lupton 1999; Lyon 1994; Ogborn 1993). Those who are considered disorderly are often considered unworthy of being in public space because they evoke feelings of unease or even dis-ease.[4]

Purified Communities

The mobilization of social and spatial separation can be viewed as a purification attempt. A seminal work on urban community is Sennett's *The Uses of Disorder* (1970), in which he argues that there is a sense of community in suburban areas (though the flight to the suburbs is motivated by a desire of isolation), but that it is more difficult to create social cohesion in an urban environment. Sennett's notion of a "purified identity" comes into play when theorizing "community." Arguing that most people live in an adolescent stage of maturity when it comes to social relations, Sennett discusses the myth of a purified community. Within this purified community, there is a push to exclude that seen as different based on "adolescent" fear.

The process of separation serves two functions: one, it maintains an idea of social purity and two, it compartmentalizes society into categories of pure and defiled (Sibley 1995). Hubbard (2004) states that abjection marks the boundary between pure and polluted and drawing on Sibley (2001) suggests that desires to prevent boundary violation thrive on stereotypical images of repulsion which become mapped onto particular social groups. Moral panics are reflective of fears about belonging and not belonging, about the purity of territory and the fear of transgression. Sibley (1995: 69) states,

Feelings of insecurity about territory, status and power where material rewards are unevenly distributed and continually shifting over space encourage boundary erection and the rejection of threatening difference.

Deviance is often reinforced geographically though the spatialization of social boundaries. Spatial distance facilitates social distance. When proximity occurs, the categorical boundaries are challenged and can lead to unease. According to Hubbard (2002: 371), the "potential for abjection is thus present when spatial orders are called into question, blurring the distinction of pure and polluted." In the formation of communities, order is established to provide a clearer distinction between the pure and the polluted, the marked and unmarked, the community and its other.

Territoriality

Territoriality can be a spatial manifestation of communities. Gottman (1973: x) stated that examinations of territories show the "internal" relationships between communities and space, and the "external" relationships between communities and their neighbors. He argued that,

> the significance of territory, at least in "western" history, has not been simply in the routine of political processes but also as a `psychosomatic device ... [whose] evolution ... [is] closely related to the human striving for security, opportunity, and happiness.

Johnston (2001) argues that territoriality can be useful in studies of group and individual behavior at multiple scales, although it is commonly used to describe state power, leading to criticism. For example, Agnew (1994) and others believe that political geography suffers from a "territoriality fetish" (see Johnston 2001) and need to focus on territoriality on smaller scales due to a fragmentation of the state. Pacione (1983) in an examination of definitions of space and territory as they relate to community cohesion outlines six factors that create a sense of territoriality: "personal attachment to the neighborhood, friendships, participation in neighborhood organizations, residential commitment, use of neighborhood facilities, and resident satisfaction" (quoted in Valentine 2001: 112–13).

Yet Valentine (2001) argues that territoriality is not evoked in many cases unless the territory is under threat, which can result in NIMBYism. For example, "community watch" groups are predicated upon the notion of exclusion. They monitor those who are outsiders or strangers. Police often encourage citizens of "communities" to use strategies of territoriality to mark their spaces and to discourage those who are not citizens of the community, especially those seen as deviants such as prostitutes, drug users, vagrants, etc., to move on or to avoid the community (and the space) all together. Spaces, including public spaces, are controlled and patrolled in order to purify those spaces and reduce risk of "contagion" of the community (socially and physically) in the name of safety and health. States Fischer and Poland (1998: 191), " 'community policing' has come to entail governance of local space by targeting 'problem' hosts or carriers of 'disorder'." The boundaries and areas that are created through territoriality mark areas of community and belonging.

Here, it is important to highlight that the formation of communities is a social construction. The construction of communities, though, takes many forms.

Imagined Communities

For Anderson (1991), communities are constructs. Using the nation as example, he discusses how members of a nation often feel bonded to one another even though the totality of the membership will never be known and inequality may exist between members of that nation. Anderson posits the framework of "imagined communities" in which the community may be at a larger scale than the local or may be even simply a concept, but still has very real, material effects (Valentine 2001). Imagined communities take place at multiple scales from the local to the global. For example, various diasporas have created very close relations between those have migrated (whether forced or not) and the region from which they came. Diasporic groups can form global imagined communities. Although they are distanced from their national territory, diasporic migrants feel very strong connections to that place. They form a community with other migrants as well as those that may remain in the home(land) region.

For Rose (1990), communities may be constructed concepts, but are grounded in the material as well. Rose (1990: 426) defines imagined communities as "a group of people bound together by some kind of belief stemming form particular historical and geographical circumstances in their own solidarity" (quoted in Valentine 2001: 124). Valentine (2001) demonstrates using Dwyer (1999) and Rose (1990) how communities are constructed and imagined through discourse and debate. Valentine states that "[a]lthough these structures of meaning are fluid and contested, they are important to their "members" and have wider political meaning" (2001: 127).

Other aspects of community involve socio-political movements and imagined communities that span scales from the local to the global. Community organizations[5] can change power relations within a geographic area, hence changing social relations since communities are built through a social interaction between invested individuals (Martin 2002). With this interaction comes a power, participating in a community can also mean deciding who gets to participate. Sadd and Grinc (1994) argue that those who greater social capital participate to a greater extent than those with fewer social resources. They argues that sometimes this leads policies that "target of members of the community who do not [or even cannot] participate" (quoted in Bass 2000: 151).

Communitarianism

For certain scholars who examine the concept of community, a sense of morality is tied to the term. For example, Smith (1999) questions the moral dimensions of community and for Etzioni[6] (1995: ix), "communities are social webs of people who know one another as persons and have a moral voice." At this point, it is of use to highlight the role of communitarianism in discussions of community. Smith (1999: 21) explains that communitarianism is "expressed as a focus on 'we' rather than 'I.'" For Watts (2004: 196), communitarianism is the "meaning and possibility of justice or democracy turns, in this view, on the identification of individuals with their community and its values."

Etzioni (1995: ix) discusses the role of morality in regards to his conceptualization of communitarianism: "Communitarians call to restore civic virtues, for people to live up to their responsibilities and not merely focus on their entitlements." Furthermore, Etzioni (ibid.) states that communities "draw on interpersonal bonds to encourage members to abide by shared values ... Communities gently chastise those who violate shared moral norms and express approbation for those who abide them. They turn to the state (courts, police) only when all else fails" (quoted in Smith 1999: 22).

Of course, there are multiple and different types of communitarianism. Bader (1995) outlines the tension between conservative/protective and liberal-democratic in Smith (1999), stating that the conservative version, which is the often the subject of myth, is when:

> internal homogeneity of communities is postulated and cross-cutting communal allegiances and collective identities are forgotten; in a kind of retrospective nostalgia, communities are thought to be harmonious (traditional) *Gemeinschaften* and confronted with conflict-ridden (modern) strategic *Gesellschaften*; cultural communities are constructed without any analysis of structural antagonism and conflict, particularly class antagonism and conflict; the idea of shared meaning, of shared cognitive and normative frames and interpretations is very much overstressed. (Bader 1995: 217 in Smith 1999: 25)[7]

In the examination of the collective actions and identity associated with community, scholars also looked at the ethical conflicts that can arise in the formation of community–self-named community organizations or organizations that represent the community as a whole, ignore and sometimes harm members of their own communities, often without a second thought to the consequences of their actions. This can create a displacement of those that are deemed as different, as outside of the community, which can have deleterious social consequences.

Some argue that in order to address this social effect, "communities" should operate from a place of empathy and ethical concern. As the neoliberal state cuts social services more and more, the community becomes the site of care in many cities and states. Communal goals should not be the "displacement" of those who are in need of community or community services, but to help those who are *in* the community, but not necessarily seen as such. Sypnowich (1993: 106–7) argues that,

> Resolving the tension between difference and sameness involves understanding that the rationale of the politics of difference is for those "others" to become part of a "we" which is a source of social unity, as a community or nation but potentially including all of humankind. (quoted in Smith 2000: 1151)

Decline of the Community

Some scholars have stated that community as a way of life is in a state of decline. Some challenge the very existence of "community" at all. These theorizations of community only add to the debates that have been discussed for generations. Valentine

(2001) classified the debates into three categories: community lost, community saved and community liberated (for a thorough discussion, see pp. 115–16). As a example of community lost, Tonnies (2001 [1887]) argued that the shift from community (*gemeinschaft*) to mass society (*gesellschaft*) was a result of industrialization, urbanization and mass communication. For Tonnies, there has been a loss of emotional social ties that bring people together into close contact. The loss of these ties, for Tonnies and others, signals the onset of social alienation.

Putnam's (2000) *Bowling Alone* is an influential work that is often recognized as the story behind the decline of community in the modern era. Discussing the disintegration of social structures like church, Putnam argues that social ties have devolved. In his discussion of social capital, Putnam examines civic and social life in American communities. Discussing another community controversy, Putnam quotes Wellman (1988):

> It is likely that pundits have worried about the impact of social change on communities since human beings ventured beyond their caves ... In the [past] two centuries many leading social commentators have been gainfully employed suggesting various ways in which large-scale social changes associated with the Industrial Revolution may have affected the structure and operation of communities ... This ambivalence about the consequences of large-scale changed continued well into the twentieth century. Analysts have kept asking if things, have, in fact, fallen apart. (quoted in Putnam 2000: 25)

This debate as to whether or not communities are in a state of decline has played out in scholarship for decades. In *Bowling Alone*, Putnam quotes figures that state how Americans lament the loss of community and feel that its loss is a signal of a breakdown in American society. He shows how many feel that more emphasis should be put on the community, even if it means further demands on the individual.

Yet there are counterarguments that conflict with the notion that community is in a state of decline. Scholars like Gans (1962) have posited that strong networks exist in the urban environment. These networks may be along socio-economic lines (race, class, ethnicity) and show that there is still a sense of belonging or at least a desire to belong to a larger social entity. Following that, urban dwellers live and operate in multiple networks that are diverse and wide-ranging in scale (Wellman 1979; Valentine 2001). Multiple networks can sometimes cause confusion in one's identity as allegiances can be split. Wellman (1979) goes on to argue that the ambiguity people feel as they engage multiple networks can feel as if one has "lost" community (Valentine 2001).

A resurgence of community perhaps can be found in the concept of neotribalism. Using the terms *tribus* or neo-tribes (there are debates as to which is the better term – see Bennett 1999). For Maffesoli (1996: 98) the tribe is "without the rigidity of the forms of organization with which we are familiar, it refers more to a certain ambience, a state of mind, and is preferably to be expressed through lifestyles that favour appearance and form." Maffesoli's tribes are heterogeneous remainders of a mass consumption society and they are formed through, and distinguished by the choices, both lifestyle and consumptive, of its members. The key to neo-tribes is lifestyle choice and consumptive practices, where a lifestyle is " 'a freely chosen game' and should not be confused with a 'way of life,' which is 'typically associated with a more-or-less stable community'" (Kellner 1992: 158; Chaney 1994: 92, quoted in Bennett 1999: 608). Hetherington (1992: 93) points out that Maffesoli's tribalization involves "the deregulation through modernization and individualization of the modern forms of solidarity and identity based on class occupation, locality and gender … and the recomposition into 'tribal' identities and forms of sociation" (quoted in Bennett 1999: 606).

Tribal identities are more ephemeral as tribalists constantly reconstruct themselves. They are seen as more transitory than prior generations. These are not long-lasting bonds necessarily, but instead, tribalists move from group to group. Shields (1992b: 108) argues that: "Personas are "unfurled" and mutually adjusted. The performative orientation toward the other in these sites of social centrality and sociality draws people together one by one. Tribe-like but temporary groups and circles condense out of the homogeneity of the mass" (quoted in Bennett 1999: 606).

Modern networks have replaced the traditional forms of kin and neighbors with friends and co-workers (Valentine 2001). In very recent times, the rise of social networking sites like Facebook and MySpace have reinforced and reestablished contacts between people who live both in proximity and at a distance to each other. These websites, along with other websites that seek to bring people/ideas together, can be see as the basis for virtual communities (refer back to Anderson's imagined communities).[8]

Conclusion

In Valentine's (2001: 105) discussion of community, she states that since the 1980s "the notion of community has been retheorized as a structure of meaning or imagining. This marks a major shift in understandings of community." For Day (2006: 1), the "idea of community continues to grip people's imaginations, and even grow in significance as it takes on new applications." But measuring and locating "community" is a difficult process. Perhaps community as a concept, as a way of life, means nothing. Yet many would argue to the contrary. Community, the feeling that one engages with society meaningfully, is something that people often seek out. Whether it be through neighborhood associations, the PTA or the internet, there is often a longing or yearning to be part of something larger than oneself. Over and over again, a lament for the loss of community can be heard throughout modern Western society due to many feeling alienated and disconnected from others on a number of scales.

Scholars debate how communities accommodate difference in the attempt to form collective identity and this conversation has been in effect for many decades (Sennett 1970; Tonnies 2001 [1887]; Young 1990a, 1990b). Does community allow for difference within or does it ignore difference for the greater whole? The same debate resounds for collective identities. For Niethammer (2003), the "construction of collective identities arises out of broader practices of defining and delimiting communities. As a rule … these dissolve internal differentiations within any given collectivity in favor of a common external demarcation" (quoted in Watts 2004: 198).

It is, of course, possible to have fidelity to more than one community (Watts 2004). These sometimes intersecting, sometimes conflicting, communities can play off one another to form one's identity. Often based on social markers and/or political goals, one can belong to a number of communities at the same time as community is not necessarily tied to territory (Anderson 1991; Knox 1995; Davies and Herbert 1993).

Many scholars have used the neighborhood or even the scale of the local as the spaces of community, but there is also a proliferation of work that shows that community on a much larger scale, including the national and global scales since communities are both place-based and imagined. Communities bond through a sense of shared identity or goals or actions regardless of where its members actually are.[9]

What community means is subjective. As this chapter has discussed, community is a concept, process, and effect. It can lead to both social harmony and social discord. It is difficult to define, to see, and to study. It can be both exclusionary and inclusionary at the same time. It has been debated for generations and communities and their members will continue to be a source of interest for scholars. There is not a solution to the "problem" of community, but it is a subject worthy of examination and debate in the future.

Panelli and Welch (2005) ask: "is there some condition that drives human engagement with community?" and furthermore question: "why community?" These questions are difficult to both pose and answer. They raise the idea that there may be an innate need for social cohesion by humans. Yet, community is a social construct and much effort has been put forth in its deconstruction. The breakdown of its construction has been a source of much geographic work, yet some question whether there is an inherent need for community. Whether the question of an innate sense of community may be controversial, it is a question that is both interesting to community and social scholars and one that calls for further exploration of why people form communities and seek to join them.

Notes

1   The term is so debated with geography that a colleague of mine once said that he would not touch the term "with a ten-foot theoretical pole."

2   The basis of this chapter comes from a comprehensive literature review of scholarly work on community and collective action and from fieldwork conducted that looked into community policing and its affects on communities, both those that were included as part of the "community" and those that were seen as "outsiders."

3   It is hard to discuss community and all that it encompasses without evoking binaries on some level. Within this chapter, I use binaries as a way of simplifying debates on the term. This is not to sidestep the complexities of community, but instead to try to create a framework from which to understand the term better. Binaries as themselves can be problematic for a myriad of reasons, but their utility in this case outweighs the inherent problems found in employing dualisms.

4   Abjection, for some scholars, is a concept by which exclusion is justified. Those who are constructed as abject (such as the homeless) are not seen as having the same "moral" fortitude as those who are not seen as engaged in some form of illicit behavior. When action is perceived to be disorderly, both the activity and the person performing the deed are seen as abject. It is their transgression of morality and lack of orderly actions in space that instigates their production as the social other of the individual and of certain

communities. Mobilizations of abjection involve the discourses, effects and problems associated with this casting out and stigmatizing of social groups as others. This social othering can be a key component in the formation of community. For those labeled as "abject" use of communal space is circumscribed. Those who are seen as transgressive and who contradict notions of order are not constructed as part of the "community."

5 Within understandings of community, the community organization must be defined as well. Elwood (2006) in her examination of community organizations and urban spatial politics defines a community organization as "nongovernmental agencies whose activities are geographically specific within a locality – directed at fostering change within a defined area in a city – with no implication that the social community represented is singular, unchanging, or uncontested" (p. 338, n. 1). This definition is important to note as community organizations often effect power on multiple geographical scales.

6 Etzioni is a key figure in theorizations of communitarianism. The political implications of his work have been of interest to geographers (Smith 1999) as they show the inherent power relations involved in formulations of community.

7 Social theorists such as Emile Durkheim, Max Weber, and Karl Marx have taken on the concept, but one of the most referenced is Ferdinand Tonnies due to his distinction between society (*gesellschaft)* and community (*gemeinschaft).* It is necessary, following Tonnies ([1887] 1998), to differentiate between a community and a group.

8 Another example, YouTube, posts videos that go "viral" and global within days, constructing a global audience.

9 Virtual communities have formed through various forms of media. For example, online communities have existed via the internet since the 1980s, television has united people since the mid-twentieth century, and newspapers have created collective identities since the nineteenth century.

References

Agnew, J. (1994) The territorial trap: the geographical assumptions of international relations theory. *Review of International Political Economy* 1: 53–80.

Anderson, B. (1991) *Imagined communities: Reflections on the Origin and Spread of Nationalism.* New York: Verso.

Bass, S. (2000) Negotiating change: community organizations and the politics of policing. *Urban Affairs Review* 36 (2): 148–17.

Bennett, A. (1999) Subcultures or neo-tribes? Rethinking the relationship between youth, style and musical taste. *Sociology* 33 (3): 599–617.

Billig, M. (1995) *Banal Nationalism.* Thousand Oaks: Sage.

Campbell, H. and Phillips, E. (1995) Masculine hegemony in rural leisure sites in Australia and New Zealand. In P. Share (ed.), *Communication and Culture in Rural Areas.* Wagga Wagga: Centre for Rural Social Research.

Cater, J. and Jones, T. (1989) *Social Geography: An Introduction to Contemporary Issues.* London: Edward Arnold.

Chauncey, G. (1996) Privacy could only be had in public: gay uses of the streets. In J. Sanders (ed.), *Stud: Architectures of Masculinity.* New York: Princeton Architectural Press, pp. 244–67.

Cox, K. (2002) *Political Geography: Territory, State and Society.* Oxford: Blackwell.

Daly, G. (1998). Homeless and the street: observations from Britain, Canada and the United States. In N.R. Fyffe (ed.), *Images of the Street.* London: Routledge, pp. 111–28.

Davies, W.K.D. and Herbert, D. (1993) *Communities within Cities: An Urban Social Geography.* London: Belhaven Press.

Day, G. (2006) *Community and Everyday Life*. London and New York: Routledge.

Dempsey, K. (1990) *Smalltown: A Study of Social Inequality, Cohesion and Belonging*. Melbourne: Oxford University Press.

Dempsey, K. (1992) *A Man's Town: Inquality between Men and Women in Rural Australia*. Melbourne: Oxford University Press.

Derrida, J. (1991a) Speech and phenomena. In P. Kamuf (ed.), *A Derrida Reader: Between the Blinds*. New York: Columbia University Press, pp. 6–30.

Derrida, J. (1991b) Of grammatology. In P. Kamuf (ed.), *A Derrida Reader: Between the Blinds*. New York: Columbia University Press, pp. 31–58.

Derrida, J. (1991c) Difference. In P. Kamuf (ed.), *A Derrida Reader: Between the Blinds*. New York: Columbia University Press, pp. 59–79 .

Dittmer, J. (2005) Captain America's empire: reflections on identity, popular culture and geopolitics. *Annals of the Association of American Geographers* 95 (3): 626–43.

Dwyer, C. (1999) Contradictions of community: questions of identity for young British Muslim women. *Environment and Planning A* 31: 53–68.

Dwyer, O. and Jones, III, J.P. (2000) White socio-spatial epistemology. *Social and Cultural Geography* 1 (2): 209–22.

Elwood, S. (2006) Beyond cooptation or resistance: urban spatial politics, community organizations, and GIS-based spatial narratives. *Annals of the Association of American Geographers* 96 (2): 323–41.

England, M. (2006) Citizens on patrol: community policing and the territorialization of public space in Seattle. PhD dissertation. Washington: University of Kentucky.

Environment and Planning A (EPA) (1999) The dynamics of community, place and identity. *Theme issue* 31 (1).

Etzioni, A. (1995) *The spirit of community: rights, responsibilities and the communitarian agenda*. London: Fontana.

Firey, W. (1945) Sentiment and symbolism as ecological variables. *American Sociological Review* 10: 140–8.

Firey, W. (1947) *Land Use in Central Boston*. Cambridge, MA: MIT Press.

Fischer, B. and Poland, B. (1998) Exclusion, "risk," and social control: reflections on community policing and public health. *Geoforum* 29 (2): 187–97.

Fyfe, N.R. and Bannister, J. (1998) The eyes upon the street: closed circuit television surveillance and the city. In N.R. Fyffe (ed.), *Images of the Street*. London: Routledge, pp. 254–67.

Gans, H. (1962) *The Urban Villagers*. New York: Free Press.

Gottman, J. (1973) *The Significance of Territory*. Charlottesville: University of Virginia Press.

Harvey, D. (1973) *Social Justice and the City*. London: Edward Arnold.

Herbert, S. (1996a) *Policing Space: Territoriality and the Los Angeles Police Department*. Minneapolis: University of Minnesota Press.

Herbert, S. (1996b) The normative ordering of police territoriality: making and marking space with the Los Angeles Police Department. *Annals of the Association of American Geographers* 86 (3): 567–82.

Hillier, L., Harrison, L., and Bowditch, K. (1999) Neverending love and blowing your load: the meanings of sex to rural youth. *Sexualities* 2: 69–88.

Hubbard, P.J. (2001) Sex zones: intimacy, citizenship and public space. *Sexualities* 4 (1): 51–71.

Hubbard, P.J. (2002) Sexing the self: geographies of engagement and encounter. *Social and Cultural Geography* 3 (4): 356–81.

Hubbard, P.J. (2004) Cleansing the metropolis: hiding vice in the contemporary city. *Urban Studies* 41 (9): 1687–1702.

Hunt, A. (2002) Regulating heterosexual space: sexual politics in the early twentieth century. *Journal of Historical Sociology* 15 (1): 1–34.

Isin, E. (2000) Introduction: democracy, citizenship and the city. In E. Isin (ed.), *Democracy, Citizenship and the Global City*. London: Routledge, pp. 1–21.

Johnston, R.J. (2000) Community. *The Dictionary of Human Geography*. Oxford: Blackwell, pp. 101–2.

Johnston, R.J. (2001) "Out of the 'moribund backwater'": Territory and territoriality in political geography. *Political Geography* 20 (6): 677–793.

Kellner, D. (1992) Popular culture and the construction of postmodern identities. In S. Lash and J. Friedman (eds.), *Modernity and Identity*. Oxford: Blackwell, pp. 141–77.

Knox, P. (1995) *Urban Social Geography: An Introduction*, 3rd edn. Harlow: Longman.

Laclau, E. and Mouffe, C. (1985) *Hegemony and Socialist Strategy*. London: Verso.

Little, J. and Leyshon, M. (2003) Embodied ruralities: developing research agendas. *Progress in Human Geography* 27 (3): 257–72.

Little, J. and Panelli, R. (2003) Gender research in rural geography. *Gender, Place & Culture* 10: 281–9.

Lupton, D. (1999) Crime control, citizenship and the state: lay understandings of crime, its causes and solutions. *Journal of Sociology* 35 (3): 297–311.

Lyon, D. (1994) *The Electronic Eye: The Rise of Surveillance Society*. Minneapolis: University of Minnesota Press.

Martin, D. (2002) Constructing the "neighborhood sphere": gender and community organizing. *Gender, Place and Culture* 9 (4): 333–50.

Maffesoli, M. (1996). *The Time of the Tribes: The Decline of Individualism in Mass Society*. London: Sage.

Natter, W. and Jones III, J.P. (1997) Identity, space and other uncertainties. In G. Benko and U. Strohmayer (eds.), *Space and Social Theory*. London: Blackwell, pp. 141–61.

Ogborn, M. (1993) Ordering the city: surveillance, public space and the reform of urban policy in England 1835–56. *Political Geography* 12: 505–21.

Pacioni, M. (1983) The temporal stability of perceived neighbourhood areas in Glasgow. *Professional Geographer* 35 (1): 66–73.

Panelli, R. and Welch, R. (2005) Why community? Reading and difference and singularity with community. *Environment and Planning A* 37: 1589–1611.

Putman, R. (2000) *Bowling Alone: The Collapse and Revival of American Community*. New York: Simon and Schuster.

Rose, G. (1990) Imagining Poplar in the 1920s: contested concepts of community. *Journal of Historical Geography* 16: 425–37.

Schofield, B. (2002) Partners in power: governing the self-sustaining community. *Sociology* 36 (3): 663–83.

Selznick, P. (1992) *The Moral Commonwealth: Social Theory and the Promise of Community*. Los Angeles: University of California Press.

Sennett, R. (1970) *The Uses of Disorder: Personal Identity and City Life*. New York: W.W. Norton.

Sibley, D. (1995) *Geographies of Exclusion*. London: Routledge.

Sibley, D. (2001) The binary city. *Urban Studies* 38 (2): 239–50.

Silk, J. (1999) The dynamics of community, place and identity. *Environment and Planning A* 31: 515.

Sjoberg, G. (1960) *The Pre-industrial City, Past and Present*. New York: Free Press.

Smith, D.M. (1999) Geography, community, and morality. *Environment and Planning A* 31: 19–35.

Smith, D.M. (2000) Social justice revisted. *Environment and Planning A* 32: 1149–62.

Tonnies, F. (2001) [1887] *Community and Civil Society*, ed. J. Harris. Cambridge: Cambridge University Press.

Valentine, G. (2001) *Social Geographies*. Harlow: Prentice Hall.

Watts, M.J. (2004) Antimonies of community: some thoughts on geography, resources and empire. *Transactions of the Institute of British Geographers* 29: 196–216.

Wellman, B. (1979) The community question: the intimate networks of East Yorkers. *American Journal of Sociology* 84: 1201–31.

Wellman, B. (1988) The community question re-evaluated. In M.P. Smith (ed.), *Power, Community and the City*. New Brunswick: Transaction.

Williams, R. (1973) *The Country and the City*. London: Chatto and Windus.

Young, I.M. (1990a) The ideal of community and the politics of difference. In L.J. Nicholson (ed.), *Feminism/Post-modernism*. London: Routledge, pp. 300–23.

Young, I.M. (1990b) *Justice and the Politics of Difference*. Princeton: Princeton University Press.

Chapter 7

Belonging

Caroline Nagel

Introduction

The concept of belonging has become increasingly salient in the lexicon of social geographers and is often seen in conjunction with concepts relating to boundaries, borders, "sense of place," and identity. The term conjures up a variety of meanings – some quite positive, reflecting feelings of warmth, security, and being at home; some perhaps more ambiguous, hinting at exclusion, conformity, and struggle. Dictionary definitions of the term belonging capture some of this ambivalence. Belonging, for instance, can mean "to fit in" and to be a "member of a group," but also to be "proper," "appropriate," and "suitable," as well as "to have in one's possession." The term thus conveys a paradoxical notion that to be a "natural" part of something involves following a particular set of rules and norms that are set by actors more powerful than oneself. In this chapter, I wish to examine the ambivalences surrounding belonging, and how these are structured and experienced spatially.

Most of the discussion that I present in this chapter revolves around migration and migrants. Belonging certainly is relevant beyond the issue of migration, and geographers have addressed belonging in relation to young people (e.g., Vanderbeck et al. 2004), nationalist movements (e.g., Hakli 2001), urban planning (e.g., Fenster 2004), and historical memory (e.g., Mills 2006). While I draw on insights from all of these literatures, I give special attention to migrants because questions of belonging seem to be particularly salient in migrant-receiving contexts, especially in a modern nation-state system that presupposes cultural homogeneity within fixed territorial boundaries. Migrants by their very nature transgress boundaries and force a rethinking of who belongs and on what terms. The rethinking and reconfiguration of belonging that accompanies migration takes place at multiple levels and geographical scales, from the nation-state itself to the everyday spaces of cities and neighborhoods where migrants and non-migrant encounter, confront, and interact with one another.

A Companion to Social Geography, First Edition. Edited by
Vincent J. Del Casino Jr., Mary E. Thomas, Paul Cloke, and Ruth Panelli.
© 2011 Blackwell Publishing Ltd. Published 2011 by Blackwell Publishing Ltd.

This chapter will draw out four interrelated dimensions of belonging that reflect the complex meanings of this concept and the different spatialities and spatial practices associated with belonging. In doing so, this chapter also engages with some of the ontological tensions and epistemological contestations present in geographical approaches to belonging. The first dimension I will consider is the *emotional dimension*, which refers to people's attachments to places and the ways they construct a sense of belonging in, and in relation to, particular places. Second, I will examine the *formal dimension* of belonging, and specifically how citizenship, immigration, and border control policies create formal parameters of membership in nationally defined communities. Third, I will explore the *normative dimension* of belonging that often accompanies legal frameworks of belonging and that encompasses the everyday, "informal" practices and narratives through which social actors construct particular places as belonging (or not belonging) to certain groups. Finally, I will examine the *negotiated dimension* of belonging, by which I mean the ways that marginalized groups actively contest their exclusion from places and engage with dominant groups to widen the boundaries of belonging.

This exploration of four dimensions of belonging, I hope, will provide a multi-layered understanding of belonging, and one that brings to light the myriad ways that dominant and subordinate groups spatialize belonging. While multilayered, this account is in no way complete, and I emphasize that by focusing on examples from the field of migration studies, I am touching only superficially on many axes of belonging, including gender and sexuality. This chapter should therefore be viewed as an entry point into a critical engagement with the concept of belonging for geographers working in a variety of subject areas.

Dimensions of Belonging

I want to begin my overview of the four dimensions of belonging (as I have conceived them) by looking at a case that speaks to the multiple layers of belonging that people experience in their everyday lives. The case is that of an Arab-American community activist in Dearborn, Michigan, that my colleague, Lynn Staeheli, interviewed in 2005 for a study we conducted on immigration and citizenship. This activist, Muna, is a member of a Palestinian family who sought refuge in Jordan after Israel declared its independence in 1948. In the 1970s, Muna immigrated to the United States and settled in the Detroit-Dearborn area in Michigan, which has one of the largest and most concentrated populations of Arab-origin people in the United States. Muna had worked for many years in a large social service organization in Dearborn that serves mainly Arabic-speaking immigrants, and in this capacity, led the effort to establish a purpose-built Arab-American museum. The museum – one of the few "ethnic" museums in the country to have partnership status with the much-revered Smithsonian Institution – opened in 2005 across the street from Dearborn's city hall, thus situating the Arab-American community physically and symbolically at the center of the city.

The museum was built, in part, to bring pride to Arab Americans and to teach younger generations about their heritage, and its collection and displays firmly place Arab Americans in a multicultural, immigrant-centered narrative of American nationhood. The museum, in this regard, is also intended to serve as a space in which non-Arabs can learn about Arab culture and Arab Americans. As Muna states, "[We] believe that art is a good bridge between people, between ethnic groups. It's a place where people feel not threatened, to learn more about each other."

Neither the presence of the museum, nor its validation by "mainstream" institutions and political actors, however, has eliminated the tensions experienced by Arab Americans as a result of the 9/11 terrorist attacks and the US government's response to them, which included the mass deportation in 2001–2 of "Middle Eastern" men whose immigrant paperwork was not in order. For many Arab Americans, the politics of inclusion in American society have played out in ways that are decidedly less celebratory than the creation of the Arab-American museum – as seen, for instance, in the perceived need among many Arab Americans to display American flags in their businesses and homes, and in the tendency for many Christians of Arab origin to dissociate themselves from the "Arab American community" (see Howell and Shyrock 2003). Muna expresses a deep sense of being alien in America when she states,

> We are always constantly under attack for one reason or another … There's always some Arab country that is being targeted … And that sense of being constantly attacked, I really don't think it has been communicated … If *you* criticize America, if you criticize Bush, it's a sign of how great our democracy is working. If *I* criticize Bush, I'm un-American, I'm not loyal. We get this all the time.

The issue of belonging is therefore very complicated for Muna and others activists, reflective as it is of multiple experiences of acceptance, rejection, and exile both in the United States and in places of origin. Thus, when asked where she feels her home is, Muna responds, "No home." And when asked if this makes her uncomfortable, she replies,

> As I get older it becomes more uncomfortable. You start thinking about retiring, where you will be in your older age. You need some kind of support and you don't find it. It's becoming with time more uncomfortable because I'm getting there [*laughs*] … Deep in my heart, I feel like somebody had done lots of injustice to my family and nobody understands … Every day I feel like, How does the world think of Israel as democracy and Palestinians as terrorists? How? Most of the time I don't even get in discussion with people, but that is the overpowering thing in my life.

This vignette illustrates different layers of what it means to belong to a place. Experiences of belonging (or not belonging), as Muna's story indicates, are shaped at different scales (e.g., locally, nationally, and transnationally) and they are enacted in and in relation to multiple spaces (e.g., Dearborn, the Arab-American Museum, the United States, Palestine/Israel). Belonging is partly a matter of the heart – it is a feeling of being part of something and some place, and its absence is felt very keenly. It is structured through laws, policies, and norms, and it involves negotiations between dominant and subordinate groups that can lead to a reformulation of the terms of membership. Belonging, it seems, is often an incomplete process – individuals can belong in some ways but not others; and it can be tenuous and conditional, as many Arab Americans have learned. In the following sections, I wish to draw out different dimensions of belonging and to highlight the ambiguities that surround it. I also wish to draw out the different kinds of spatialities associated with each dimension of belonging, including the spatialities of nation-states, public spaces, cities, and "home."

The Emotional Dimension of Belonging

Geographers have long made belonging central to their analyses of space and place. In the 1970s, geographers began to advocate the discipline's engagement with humanistic philosophies, largely as a response to the sterile "spatial science" approaches that dominated the discipline in the post-war period. Yi-Fu Tuan (1976), arguing that "scientific approaches to the study of man [sic] tend to minimize the role of human awareness and knowledge," proposed a mode of geographic enquiry that would appeal "to such distinctively humanistic interests as the nature of experience, the quality of the emotional bond to physical objects, and the role of concepts and symbols in the creation of place identity" (1976: 269). Similarly, Anne Buttimer (1976) suggested the aim of geography is to uncover how and why people endow certain places with special significance, and she proposed a phenomenological approach that would bring into focus people's perceptions and awareness of place, their everyday experiences of place, and the meanings and identities they give to the places they inhabit – in short, their "sense of place."

The humanistic perspective informed numerous geographical studies in the 1970s and 1980s that explored the ways in which places evoke deep feelings of belonging among their inhabitants. One example is Ben Marsh's (1987) study of declining coal-mining towns in eastern Pennsylvania. This region, as Marsh described it, "by conventional economic or demographic measures, and by the normal standards of landscape esthetics [sic], is the least attractive part of Pennsylvania" (1987: 337). Yet, "the people remaining in these towns … have a powerful sense of belonging just where they are, with such ties to these tired old places that they are reluctant to move under any circumstances" (1987). In exploring this deep attachment to place, geographers began to engage with notion of historical memory, as well. Tuan (1976) remarked that humanistic geography required not just a historical understanding of place, but an awareness of the role that history and group memory play in creating territorial claims. Similarly, Knight (1982: 514), chastising political geographers for ignoring the "emotional bonds of groups to political-territorial identities," urged that greater attention be paid to the historical myths that underpin collective consciousness and loyalty to place.

Humanistic geography's concerns with belonging, place-based meanings, memories, and emotional attachments, in some ways can seem naive to contemporary readers. While some, like Tuan (1976) and Knight (1982), recognized the fabricated nature of place-based identities (especially those espoused by nationalists), humanism had relatively little to say about the role of gender, race, and class relations, much less capitalism, in structuring people's relationships and identifications with places. Some descriptive accounts in the humanistic vein, moreover, border on the patronizing – Marsh's account of Pennsylvania's coal mining towns, for instance, focuses more on the foibles of local residents, including their "charmingly" misguided efforts to attract tourists, than it does on the complex political-economic processes that rendered this a "broken landscape." Still, humanistic concerns with meaning, memories, and emotional attachment remain important impulses in social and cultural geography and, indeed, have gained new resonance as geographers have fixed their attention on identity and difference, psychogeographies (e.g., Sibley 2003), and "affect" (e.g., Thrift 2004). These contemporary approaches vary in content and in the particular vocabularies they use to describe geographical phenomena, but they tend overall to situate attachments to and identifications with place within broader discussions of cultural domination and subordination, inclusion and exclusion, and the politics (and politicization) of place. Place-based meaning and "sense of place," from this vantage point, are viewed through the lens of power relationships, performativity, discourse, and contestation.

This re-engagement with notions of belonging has come at a time when the intensification of globalization processes, marked by rapid technological changes, high levels of human mobility, and global flows of images and commodities, has raised important questions about what it means to belong to place in a world characterized, it seems, by placelessness and deterritorialization (Appadurai 1996; Massey 1993). Such questions are especially (though not solely) relevant to the tens of millions of migrants in the world today – guestworkers, refugees, exiles, and transnational elites – who, for varying reasons and in various ways, have been rendered displaced or placeless. How do such groups construct places of belonging? How do places come to have meaning for groups who are defined by mobility and whose lives are lived, in many cases, across borders and in multiple spaces?

One particularly rich vein of scholarship has used the term *diasporic* to describe contemporary migrants' experiences and spaces of belonging. The term diasporic suggests that migrants' point of reference is "elsewhere," and that they are, in some way, using place to reproduce a collective identity based on homeland. Diasporic practices, in this sense, involve the production and reproduction of collective memories of places and the inscription of these memories into new places. Alison Blunt (2003) has described the process of spatializing belonging among displaced or placeless people in terms of "productive nostalgia." Productive nostalgia, she elaborates, involves often gendered (re)enactments of home through social practices, ritual, and the actual work of constructing domestic and community spaces. Mavroudi (2008) describes such a process in her account of Palestinian exiles in Athens who use the space of the community house, or parikia, to socialize with one another and to convey their political and emotional commitment to Palestine. Another example is Fortier's study of Italian émigrés in Britain and their various performances (including outdoor processions and distinctive religious liturgies) that serve to reiterate the Italian identity of their former London neighborhood (Fortier 1999: 50).

A key point made in these and other accounts is that such performances and enactments of belonging are not simply carried over and reproduced intact from the place of origin. Even if not outright inventions, all such enactments of communal belonging – whether in relation to diasporic groups or to nationalist ideologies – involve the selection of particular memories to be commemorated and the manufacturing of particular histories to be honored. They can be seen, moreover, as "regulatory practices that produce social categories and the norms of membership within them" (Fortier 1999: 43). It is through rituals and cultural practices, in other words, that community leaders set the terms of belonging in the group. In these ways, performances of belonging must be understood in terms of the politics of establishing authenticity – politics that are experienced and practiced very differently by the old and young, men and women, elites and non-elites within any given "community" (see Houston and Wright 2003).

At the same time, the politics of belonging within migrant and minority groups become enmeshed in wider political struggles vis-à-vis dominant, "host society" groups. Dominant groups interpret minority cultural traditions and performances in relation to their own narratives of identity and their own norms. For instance, the annual procession of British Italians described by Fortier (1999), which winds its way through an old Italian neighborhood, is significant not only for what it means to the British Italian community, but also for what it means to a British mainstream that wavers between the "celebration of multiculturalism" and the demonization of allegedly unassimilable minorities. The ambivalence that surrounds visible displays of cultural difference – i.e. Are particular practices quaint remnants of exotic cultures, or are they dangerous forms of deviance that threaten the integrity of the nation? – continuously informs mainstream responses to migrants and minorities and the disciplinary actions imposed upon them, a point to which we will return shortly when we address normative dimensions of belonging. Before this, though, I want to change tack and look at belonging not in terms of emotional investments in place-creation, but, rather, in terms of the power of states to formally structure social membership.

The Formal Dimension of Belonging

In revealing the practices and performances through which migrant groups produce and reproduce their identities in places of settlement, and the tensions and ambivalences that surround these practices and performances, scholars of diaspora take a decidedly *political* view of belonging – one that regards place-based meanings as both products and reflections of unequal and contested power relationships. But there are other equally important angles from which we can approach the politics of belonging, and I want to explore here the territorialization of belonging in the nation-state. Political geography as a sub-discipline has, until recently, tended to view the nation-state as the main locus of power in the world and, therefore, as the main unit of analysis. This perspective has been rightfully criticized for reifying state power – that is, for treating the nation-state as a monolithic entity that possesses agency and "acts" of its own accord – as well as for sidelining the multiple relationships and spatialities of power that people, and especially marginalized people, experience in their everyday lives (see Staeheli and Kofman 2004). Yet it would be a mistake to ignore the formal, territorial dimensions of belonging and the ways in which powerful actors at the nation-state level implement and enforce the boundaries of social membership.

The defining feature of modernity, Peter Taylor (1994) has argued, has been the congruence (or assumed congruence) of political membership, cultural belonging, and territory within a system of sovereign nation-states. Such congruence, in turn, has required the assertion of control over human mobility (Sassen 1999). In the nineteenth and early twnetieth centuries, states exerted ever more stringent control over cross-border movements of people through passports and visas, border check-points and patrols, and through immigration laws restricting the entry of certain unwanted groups, such as the Chinese Exclusion Act of 1882 in the United States and the Aliens Act of 1905 in Britain (Rystad 1992). Likewise, states devised legal codes that set the terms of settlement for "aliens" and that spelled out the proce-dures for attaining full membership, most of which reflected gendered and racialized suppositions about who could "assimilate." Citizenship in the nation-state, as Jacobson (2002: 167) notes, signified more than simply a legal status; it marked at a very fundamental level "belonging-in-space," answering the essential questions of "Who am I?" and "To whom do I belong?" (also Goodwin-White 1998).

Almost as soon as it had imprinted itself on the globe, however, the nation-state system was being compromised by rapid and intense globalization. While very few scholars claim that the nation-state has become irrelevant or obsolete, many have demonstrated that the nation-state no longer serves as an exclusive container of power, and that social, political, and economic processes are increasingly organized at different scales (Taylor 2000; Purcell 2003; Brenner 2004). Political membership and belonging, as well, seem to be "deterritorialized" vis-à-vis the nation-state, and reterritorialized in new political spaces that are simultaneously local and global. Scholars have adopted a new vocabulary to indicate the shifting nature of political membership and the proliferation of modes of political belonging in a "post-Westphalian" world – a vocabulary that includes terms like post-national citizen-ship, flexible citizenship, global civil society, transnationalism, the global public sphere, and the diasporic public sphere.

Migrants have experienced this unhinging of territory, citizenship, and polity more directly than other groups, just as they did the consolidation of territory, citi-zenship, and polity in the nineteenth and twentieth centuries. A great deal of litera-ture has explored, on the one hand, how shifting political structures within and beyond nation-states enable or constrain citizenship practices among migrants, and how, on the other hand, migrants enact new, more complex modes of political belonging by situating themselves both "here" and "there." One prominent theme has been the growing salience of international human rights discourses and norms and the incorporation of these norms into national-level policies (and, concomi-tantly, a decoupling of formal citizenship and ethnicity – see Joppke 1999). Soysal (1994) has argued that proliferation of human rights norms has given rise to post-national modes of citizenship, whereby those lacking formal citizenship in the countries where they reside have many if not most of the rights of full citizens. These "denizens" are able to press their political claims – many of which revolve around cultural rights – within new political-institutional parameters, including municipali-ties (see Kofman 1995) and supranational entities like the European Union (see Soysal 1997).

Intersecting with conceptions of post-national citizenship (as well as with ideas of diaspora described earlier) are discussions of migrant transnationalism, which speak to migrants' simultaneous participation in the social, political, and economic life of two or more nation-states. The phenomenon of transnationalism covers an array of social patterns, behaviors, and forms; what interests us here is the idea that migrants' political activity across national borders signals the multiplication of political memberships (Baubock 2003). This process has been fed by the growing tolerance among host and sending societies of dual citizenship. In a number of examples, "tolerance" of dual citizenship has led to the virtual incorporation of émigré communities into the political life of the sending society (Itzigsohn 2000). Migrants' actual ability to remain engaged across long distances is typically attributed to rapid advances in communication.

Overall, these literatures suggest that belonging is no longer contained, bounded, or exclusive. Memberships appear to be proliferating, and people appear to be oriented toward multiple polities and political spaces. Normative readings of these trends vary widely in the literature. For some, they have led to new diasporic modes of belonging, in which deterritorialized nations come together within a virtual public sphere (Bernal 2006; Parham 2004), or to new forms of cosmopolitan belonging (Sassen 2000; also Clifford 1994; Gilroy 1994). Others offer a less sanguine analysis, pointing to a "thinning out" and devaluation of citizenship, the fragmentation of polities, and an inability to create the cohesion necessary for a functioning public sphere (for elements of these discussions, see Renshon 2001; Soysal 1997).

But if the formal structures of belonging clearly are undergoing significant changes, it is important to recognize that state-based citizenship has not been entirely emptied of its content. The nation-state, first of all, has not relinquished its capacity to structure belonging among its inhabitants. If anything, states have an unprecedented capacity to regulate borders and to place under surveillance those inside and outside of its borders, and they are acting upon this capacity with gusto. State bureaucracies increasingly control and manage who can enter and under what terms they are able to do so. The proliferation of complex point systems, visa requirements, and migrant categories in countries of the global north, combined with the prevalence of biometric methods of tracking foreigners within borders, point to the power of state actors to shape mobility according to their own wishes and needs. As Sparke (2005) notes, state power is brought down more heavily on some groups than others: while transnational business elites are often granted quick and easy access to state territory, those deemed undesirable, unclean, or unassimilable face dangerous border crossings despite, in many cases, clear demands for their labor.

Increasingly for "unwanted" migrants, these border crossings do not end at the border itself. Rather, migrants face a multiplication of borders within host societies. In the United States for instance, federal immigration enforcement officers have deputized local law enforcement officers to carry out the work of immigration control, while local jurisdictions have passed an array of punitive legislation targeting Latinos and undocumented workers (Coleman 2007; Winders 2007). Laws and ordinances designed to enforce English language usage, to deny immigrants access to public services, to regulate migrant workers' presence in public spaces, and to hinder their access to affordable housing, reflect and reproduce clear distinctions between who belongs and who does not belong. Such phenomena are not unique to the United States, of course. The ban on the headscarf in France, the emergence of the "social cohesion" agenda in Britain, and a new policy in the Netherlands that requires foreign-born brides and grooms of Dutch residents to pass an assimilation test before receiving a visa, all point to the power of the state to control "foreigners" (see Joppke 2007). Discussions of multiple memberships, post-national rights, transnational citizenship, and diasporic public spheres, therefore, must be tempered by a recognition of states' ongoing prerogative to set the parameters of membership in the polity.

The Normative Dimension of Belonging

Having just provided an account of belonging that focuses on the regulation of territorial boundaries by powerful and pervasive state entities, I want to shift attention once again to the articulation and contestation of belonging in the more intimate spaces of everyday experience and interaction – though discussions of the "everyday" and of the state are by no means unrelated. Earlier in this chapter, when describing emotional dimensions of belonging, I spoke of the tensions that can emerge between diasporas and dominant host society groups, who may look at diasporic practices as quaint and colorful or as subversive and deviant. My discussion here expands on this observation and considers how dominant groups enforce the boundaries of membership through the production and reproduction of social norms.

Norms operate within and through a multitude of spatial and discursive practices and are internalized in such a way that they are seldom recognized or acknowledged by those who abide by them. Indeed, norms often are not apparent to group members until they are challenged by the visible and felt presence of others, such as immigrants, minorities, and other marginalized or "deviant" groups. These others bring norms into sharp relief and give rise to efforts by dominant groups to defend norms and to define more clearly what it means to belong. Such efforts often involve setting rules of appropriate behavior in certain spaces.

As suggested earlier, humanistic geography brought attention to the emotional attachments that people have to places, but they gave less attention to the ways in which place-based attachments create feelings of being "out of place" for those who are deemed not to belong. The feeling of being out of place – of not belonging in or to a place – reflects the fact that places are inscribed not simply with meanings, but with *dominant* or hegemonic meanings that include some while simultaneously excluding others (see Kinsman 1995; Jenkins 2003). Those who are "out of place" – i.e. who transgress space through their inappropriate behaviors, attitudes, or appearance – are subject to sanction, discipline, and expulsion (Cresswell 1996). I am mainly interested here in the informal mechanisms of control and the enforcement of norms through everyday social practices. But I also want to demonstrate that such norms are often expressed and enforced through state action and authority, especially at the neighborhood or municipal level.

Katharyne Mitchell's account of the debates surrounding the settlement of wealthy Chinese immigrants in Shaughnessy, an affluent Vancouver suburb, in the 1980s illustrates well the deep entrenchment of norms and the discourses through which different groups express – and contest – these norms. Shaughnessy's exclusivity, Mitchell notes, had long been maintained through a variety of informal procedures, from the cultivation of a particular image of the area by developers and real estate agents to the use of restrictive covenants and exclusionary municipal zoning codes.

Well into the 1970s, while other parts of Vancouver were becoming increasingly ethnically diverse, this neighborhood remained implicitly coded as "white." During the 1980s, in response to the demolition of several old properties and the construction of new "monster houses" by Chinese newcomers, homeowners' associations mobilized and worked closely with city planning agencies to establish strict guidelines – many of them written in minute detail – to preserve the area's architectural character. As Mitchell (1997: 168) describes it, homeowner mobilization reflected not only anxiety about the construction of aesthetically displeasing homes, but a "general fear concerning the possible diminution, deprivation and dispossession of [the] way of life" embodied by Shaughnessy – a way of life consciously modeled on a pre-war British idyll – due to the arrival of moneyed Chinese immigrants. In the ensuing controversy about house demolition, which was aired in the media and in public hearings about proposal zoning restrictions, a strong discourse about appropriateness emerged, in which established white residents posited the Chinese and their homes as fundamentally out-of-place, as quite simply not belonging to the neighborhood, or indeed, to Canada.

Another illustration of the normative and discursive dimension of belonging can be seen in Daniel Trudeau's (2006) account of a controversy in Hugo, Minnesota – a small town with an almost entirely white population on the suburban fringes of Minneapolis – that erupted over a slaughterhouse operated by a Hmong immigrant named Lee. Starting in the late 1990s, Lee's slaughterhouse began to draw growing numbers of Hmong immigrants who used the facility for ritual animal sacrifice. Lee's operation also attracted Muslim customers – most of them from Somalia and Eritrea – in need of fresh, halal meat. The slaughterhouse, and the traffic congestion associated with it, provoked complaints by Lee's neighbors, who asked for the town council to intervene. As in the case of the large houses built by Chinese immigrants in Vancouver, the controversy surrounding the slaughterhouse was posed as a struggle between the right of an established community to maintain the essential character of place and the right of newcomers to practice their culture freely. For opponents of Lee's venture, the slaughterhouse was an affront to Hugo's idyllic, rural, and agricultural landscape (somewhat paradoxically, given that one might expect to find slaughtering activities taking place on farms and in grazing areas); for Lee's supporters, the slaughterhouse was cast as a central element of the Hmong community's spiritual life. Despite the swirl of emotions, however, much of the discussion at the level of municipal government (as was the case of Shaughnessy) was framed in putatively "neutral" terms – that is, as a matter of complying with existing codes and uses (see also Nelson's (2008) account of conflict in a small Oregon town over the construction of farmworker housing, and Naylor and Ryan's (2002) account of local opposition to the expansion of a mosque in a London suburb).

These cases illustrate that "transgressions are moments in which landscapes are (re) constructed in order to fix a particular meaning of place" (Trudeau 2006: 434) – in other words, transgressions become opportunities for dominant groups to assert and to enforce a particular conception of place and what and who belongs within it. Two more specific points emerge from these cases. The first point relates to the intersection between norms and formal structures or legal systems. Norms circulate and are conveyed through discourse, but they also underlie and are given a degree of fixity through legal systems. In these cases, particular ideas of appropriateness have been established and reinforced through legal means, and especially through municipal planning authorities who set zoning regulations and who determine whether certain uses are "in keeping" with the local environment. It is crucial, therefore, to understand how informal, discursive practices of inclusion and exclusion, belonging and rejection, intersect with state or quasi-state structures. The second point is that those seen to be in violation of norms can actively challenge these norms. In the case of Vancouver's Shaughnessy neighborhood, for instance, Mitchell emphasizes that Chinese residents were not passive through the debate about Chinese houses. On the contrary, they responded to white angst quite vigorously by invoking notions of rights, liberties, and well as the superiority of Chinese values. In doing so, they challenged the terms of belonging articulated by white residents and attempted to re-work notions of the "public" and the common good. We will delve more deeply into such negotiations in the following section.

The Negotiated Dimension of Belonging

We have seen in the previous three sections that belonging is necessarily relational: it involves the construction of boundaries that distinguish between "us" and "them." Marking our place in the world means the creation and enforcement of multiple borders and barriers, some of them informal (as with norms of conduct) and others of them formally encoded in legal systems and even enforced through military means. But while these boundaries often achieve a degree of fixity, especially insofar as they become legally encoded, they can also be altered, sometimes dramatically so. The politics of belonging, in this regard, involve negotiations between dominant and subordinate groups through which the latter attempt to shift the terms of belonging as defined by dominant groups.

These negotiations are central to interactions between immigrant and host society groups and between ethnic minorities and majorities. In some cases, these negotiations shake societies to their core and signal a major shift in the ways societies imagine themselves and structure relationships between different groups. Such was the case with the Civil Rights movement in the United States, in which black activists and their white allies mobilized against the elaborate quasi-legal system of racial segregation that had been in place in the South since the 1870s. The Civil Rights movement resulted in the creation of an entirely new set of laws and policies centered on non-discrimination and affirmative action, and in a new set of discourses and images that placed African Americans more visibly in American national narratives (for a fuller discussion, see Dwyer 2000). This is not to say that inequalities have disappeared, that social and spatial segregation have ceased to exist, or that "race" no longer matters in American society. But it is to say that African American political activism challenged a deeply entrenched system of exclusion and altered, if only partially, the terms of belonging in the United States.

The Civil Rights movement – or, at least, how it has been remembered – played out on a national stage through a dramatic series of events. But negotiations of belonging also take place in the mundane interactions between dominant and subordinate groups on an everyday basis and in everyday spaces. For those in subordinate groups, negotiations may involve subtle changes in voice, comportment, or appearance either to assert sameness or to mark difference as a challenge to existing norms. Such subtle actions may not be noticed or remarked upon by onlookers, but they infuse common interactions with meaning and significance. In my own research on British Arab communities in London, for instance, I recall discussions with my Muslim interviewees about the politics of drinking beer at the pub. For some, abstaining from alcohol was a crucial part of being Muslim, and they found themselves avoiding pubs – a quintessential British space – though they also rejected the notion that their abstention made them somehow less "British." Others, however, insisted on the need to consume alcohol and to go to pubs precisely to fit in with what they saw to be a core cultural practice of British society. The point is that such mundane, everyday actions (at least in the British context) of drinking beer at a pub are loaded with meaning and become part of British Arabs' efforts both to "be British" and to alter what it means to be British. Some scholars speak of these negotiations in terms of "micropolitics" – acts of compliance, accommodation, or defiance in the spaces of daily interaction through which power relationships between groups are produced and reproduced (cf. Amin 2002). While indeed small, they are part of much wider sets of politics through which subordinate groups attempt to re-work existing norms and discourses and generate new, counter-narratives about societal membership.

As indicated in the example of British Arabs and beer drinking, negotiations of belonging are often fraught with contradiction and disagreement. There is rarely total consensus within a group about how they should engage with dominant groups, present themselves in the public sphere, or incorporate themselves into mainstream spaces and narratives (see, for instance, Veronis 2007). Group members might attempt, for instance, to "preserve" cultural differences, to make them visible, and even to bring wider societal recognition and validation to distinctive cultural practices, such as language, dress, or religious rituals (as seen in the case of the Italian processions in London described by Fortier 1999). In contemporary societies, these sorts of claims are often folded into discourses of multiculturalism and cultural diversity, and at times are looked at favorably by elements of dominant societies who see cultural differences and exotic spaces as marketable commodities (Goonewardena and Kipfer 2005). Others, in contrast, might attempt to reduce the sense of foreignness that surrounds them by eliminating visible differences or by sequestering these differences in "private" cultural spaces. Such is the case with many of the British Arabs I have interviewed, who have tended to view visible differences, especially those linked to Islamic practice, as stigmatizing Arabs and perpetuating the association between Arabs and religious extremism (Nagel and Staeheli 2008). Still others may embrace markers of difference but attempt to re-work the meanings and discourses surrounding these differences and to re-construe them in terms of mainstream norms. This can be seen with some young Muslim women in France who have justified the wearing of headscarves in terms of liberal values of liberty and equality, thus setting up a debate over the meaning of these terms, and, indeed, of "liberalism" itself. Different views, strategies, and actions can reflect the particular experiences and outlooks of generational, class, or gender groups, and can lead to politics that are highly fragmented (see Secor 2003).

For contemporary migrant groups, negotiations of belonging are situated in multiple spaces and are shaped by political processes and narratives operating within and beyond nation-state borders. Patricia Ehrkamp (2005), in her work on Turkish immigrants in Germany, sees neighborhood spaces as part of a wider "negotiated reality" that is simultaneously local, national, and transnational. In Marxloh, a working-class neighborhood in the industrial city of Duisburg, neighborhood spaces are constituted through residents' ongoing connections with Turkey evident in the plethora of local shops selling Turkish goods, in the ubiquity of satellite dishes, and the use of Turkish language by residents. The creation of a "Turkish" landscape in Marxloh, in turn, becomes central to the struggles of Turkish immigrants to achieve full membership in German society. In one instance, the request by local mosque to amplify calls to prayer sparked widespread opposition among German residents of Marxloh, who argued that this would be both bothersome and out-of-place. The issue, however, went far beyond Marxloh, and it quickly was taken up by national media as an example of the supposed Islamicization of Germany. This episode and other like it have heightened Turkish residents' sense that Germans are unwilling to accept cultural differences (Ehrkamp 2006). While some long for Turkey and see their time in Marxloh as temporary, others are keen to use local spaces, including various municipal forums, to voice their views and to question homogenous ethnocultural understandings of nationhood and citizenship (ibid.).

In focusing on the negotiated dimension of belonging, I have reiterated the idea that belonging is, above all, a political process through which different groups continuously produce and reproduce the boundaries of membership. The dynamics of inclusion and exclusion, and the struggles to alter these dynamics, are experienced very differently within and across groups, reflecting diverse positionalities of gender, class, and generation (among others). Negotiations of belonging are often highly localized and focus on the use of, and/or access to, spaces by particular groups. But while localized, these negotiations are often connected to wider sets of political concerns, reflecting the fact that "localities" themselves are situated in multiple, overlapping spaces of interaction, identity, and power relations.

Conclusion

Belonging as a state of being is fraught with ambivalence. Surely, we all want to feel that we belong. Belonging imparts a tremendous sense of security and comfort and a sensation that many associate with being home or, at least, feeling "at home." It is no wonder that migrant groups attempt to re-create home in the places where they have settled and to reproduce the cultural practices associated with their places of origin. Yet practices of belonging, as much as they create solidarity and emotional well-being, also involve following rules, enforcing norms, and policing the boundaries between "us" and "them." Belonging excludes as much as it includes, and it disciplines as much as it sustains and nurtures.

The ambivalence at the heart of belonging prompts us to think about the different dimensions through which people experience belonging (or not belonging) in their everyday lives. I have identified four dimensions – emotional, formal, normative, and negotiated – but I do not wish to suggest that this is an exhaustive list. The main point I have tried to convey in exploring these particular dimensions is that belonging is necessarily *political*. In common parlance, the term political is typically associated with the workings of states and governments – the realm of legislation and public policy. This understanding of political is certainly relevant to understanding belonging, as I discussed in the section on the formal dimension of belonging. The political, though, also has a broader meaning that refers to the exercise of power in interpersonal and societal relationships, whether in the "private" realm of family and "community" or the "public" realm of work, citizenship, and the state (realms which are not readily distinguishable in "real life"). Any geographically oriented discussion of belonging – of emotional attachments to place and a sense of ownership vis-à-vis place – inevitably leads us to an investigation of power: power to claim and to control space, to inscribe space with particular meanings, and to regulate who and what can be fully part of any given place.

Locating the dynamics of belonging requires that we consider the role of states and governing elites, as well as those whose authority is far more circumscribed and spatially limited. Belonging is structured through nation-states, but it also takes shape in the multitude of interpersonal encounters that one finds in schools, workplaces, neighborhoods, places of worship, cities, and so on. Such encounters reflect, inform, and reinforce societal understandings of who belongs and where they belong. They provide opportunities, as well, for challenging the exclusions inherent in belonging.

I think here of an Arab-American activist named Hayder that Lynn Staeheli and I interviewed in 2004, who, in responding to a question about citizenship, spoke of an incident at his son's school in Anaheim, California. After an event in the school gym, a teacher came up to Hayder's family and repeatedly asked his American-born son where he was from.

> [My son] said "I'm from Anaheim." She said "No, no, originally, originally, like, where were you from?" He said, "Oh yeah, I was born in La Habra," because that's where the hospital was [*laughs*]. So I guess the teacher was kind of frustrated; he said, "No, I notice your mom was wearing Islamic dress. Where's she from?" He said "She's from Texas." She got really frustrated and said "Okay, what about your dad?" He said "Oh, my dad's from Lebanon"; So she said "You're from Lebanon." He said "No I'm not from Lebanon; *He's* from Lebanon." Not that he's ashamed of it, but it's a sense of belonging.

For Hayder, this incident – which undoubtedly went unnoticed by most of the people in the gym that day – was indicative of the struggle for belonging that many Muslims and Arab Americans experience every day. Hayder has embraced this struggle, committing himself to an active role in an Islamic organization where he spearheads public relations efforts. His aim, as he describes it, is to enhance Muslims' sense of belonging in the United States by making Muslims more visible and more active in community life and in their cities. "I don't want anybody to look at my son or my daughter and say 'Your name is Omar, you must not be an American'," he told us, "And I know it's not going to happen because we're going to be able to make people accept Arabs and Muslims the way they finally accepted the Irish and Italians." Hayder's remarks make plain the politics of belonging. For those excluded from the boundaries of social membership, belonging is something to be achieved and something to fight for; its presence, as well as its absence, makes every place – even a school gym in Anaheim, California – meaningful and consequential.

Note

1   This study, entitled "Community, Immigration, and the Construction of Citizenship," involved interviews with Arab-origin activists in 4 US cities (Washington, DC, Los Angeles, San Francisco, and Detroit/Dearborn) and 4 British cities (London, Sheffield, Liverpool, and Birmingham). The research was funded jointly by the Economic and Social Research Council (UK) and the National Science Foundation (US). See www. arab-communities.org for details.

References

Amin, A. (2002) Ethnicity and the multicultural city: living with diversity. *Environment and Planning A* 34: 959–80.

Appadurai, A. (1996) *Modernity at Large: Cultural Dimensions of Globalization*, Minneapolis: University of Minnesota Press.

Baubock, R. (2003) Towards a political theory of migrant transnationalism, *International Migration Review* 37 (3): 700–23.

Bernal, V. (2006) Diaspora, cyberspace, and political imagination: the Eritrean diaspora online. *Global Networks* 6 (2): 161–79.

Blunt, A. (2003) Collective memory and productive nostalgia: Anglo-Indian homemaking at McCluskieganj. *Environment and Planning D: Society and Space* 21: 717–38.

Brenner, N (2004) *New State Spaces: Urban Governance and the Rescaling of Statehood.* Oxford: Oxford University Press.

Buttimer, A. (1976) Grasping the dynamism of the lifeworld. *Annals of the Association of American Geographers* 66 (2): 277–92.

Clifford, J. (1994) Diasporas. *Cultural Anthropology* 9 (3): 302–38.

Coleman, M. (2007) Immigration geopolitics beyond the Mexico–US border. *Antipode* 39 (1): 54–76.

Cresswell, T. (1996) *In Place/Out of Place*, Minneapolis: University of Minnesota Press.

Dwyer, O. (2000) Interpreting the Civil Rights movement: place, memory, and conflict. *Professional Geographer* 52 (4): 660–71.

Ehrkamp, P. (2005) Placing identities: transnational practices and local attachments of Turkish immigrants in Germany. *Journal of Ethnic and Migration Studies* 31 (2): 345–64.

Ehrkamp, P. (2006) "We Turks are no Germans": assimilation discourses and the dialectical construction of identities in Germany. *Environment and Planning A* 38: 1673–92.

Fenster, T. (2004) Belonging, memory and the politics of planning in Israel. *Social and Cultural Geography* 5 (3): 403–17.

Fortier, A.-M. (1999) Re-membering places and the performance of belonging(s). *Theory, Culture and Society* 16 (2): 41–64.

Gilroy, P. (1994) Diaspora. *Paragraph* 17 (1): 207–12.

Goodwin-White, J. (1998) Where the maps are not finished: a continuing American journey. In D. Jacobson (ed,), *The Immigration Reader: America in a Multidisciplinary Perspective.* Oxford: Blackwell, pp. 415–29.

Goonewardena, K. and Kipfer, S. (2005) Spaces of difference: reflections from Toronto on multiculturalism, bourgeois urbanism, and the possibility of radical urban politics. *International Journal of Urban and Regional Research* 29 (3): 670–8.

Hakli, J (2001) The politics of belonging: complexities of identity in the Catalan borderlands. *Geografiska Annaler* 83 (3): 111–19.

Houston, S. and Wright, R. (2003) Making and remaking Tibetan diasporic identities. *Social and Cultural Geography* 4 (2): 217–32.

Howell, S. and Shyrock, A. (2003) Cracking down on diaspora: Arab Detroit and America's "War on Terror." *Anthropological Quarterly* 76 (3): 443–62.

Itzigsohn, S. (2000) Immigration and the boundaries of citizenship: the institutions of immigrants' political transnationalism. *International Migration Review* 34 (4): 1126–54.

Jacobson, D (2002) *Place and Belonging in America*. Baltimore: Johns Hopkins University Press.

Jenkins, W. (2003) Between the lodge and the meeting house: mapping Irish Protestant identities and social worlds in late Victorian Toronto. *Social and Cultural Geography* 4 (1): 75–98.

Joppke, C. (1999) How immigration is changing citizenship: a comparative view. *Ethnic and Racial Studies* 22 (4): 629–52.

Joppke C. (2007) The transformation of immigration integration in western Europe: civic integration and antidiscrimination policies in The Netherlands, France, and Germany. *World Politics* 59 (2): 243–73.

Kinsman, P. (1995) Landscape, race, and national identity: the photography of Ingrid Pollard. *Area* 27 (4): 300–10.

Knight, D.B. (1982) Identity and territory: geographical perspectives on nationalism and regionalism. *Annals of the Association of American Geographers* 72 (4): 514–31.

Kofman, E. (1995) Citizenship for some, but not for others: spaces of citizenship in cotemporary Europe. *Political Geography* 14 (2): 121–37.

Marsh, B. (1987) Continuity and decline in the anthracite towns of Pennsylvania. *Annals of the Association of American Geographers* 77 (3): 337–52.

Massey, D (1993). Power geometry and a progressive sense of place. In J Bird, B Curtis, T.P., Robertson, G., and Tickner, L. (eds.), *Mapping the Futures: Local Cultures, Global Change*. London and New York: Routledge, pp. 60–70.

Mavroudi, E. (2008) Palestinians in diaspora, empowerment and informal political space. *Political Geography* 27 (1): 57–73.

Mills, A. (2006) Boundaries of the nation in the space of the urban: landscape and social memory in Istanbul. *Cultural Geographies* 13 (3): 367–94.

Mitchell, K. (1997) Conflicting geographies of democracy and the public sphere in Vancouver BC. *Transactions of the Institute of British Geographers* 22 (2): 162–79.

Nagel, C. and Staeheli, L. (2008) Being visible and invisible: integration from the perspective of British Arab activists. In C. Dwyer and C. Bressey (eds.), *New Geographies of Race and Racism*. Aldershot: Ashgate, pp. 83–94.

Naylor, S. and Ryan, J.R. (2002) The mosque in the suburbs: negotiating religion and ethnicity in South London. *Social and Cultural Geography* 3 (1): 39–60.

Nelson, L. (2008) Racialized landscapes: whiteness and the struggle over farmworker housing in Woodburn, Oregon. *Cultural Geographies* 15 (5): 41–62.

Parham, A.A. (2004) Diaspora, community, and communication: internet use in transnational Haiti. *Global Networks* 4 (2): 199–217.

Purcell, M. (2003) Citizenship and the right to the global city: reimagining the capitalist global order. *International Journal of Urban and Regional Research* 27 (3): 564–90.

Renshon, S. (2001) *Dual Citizenship and American National Identity*. Washington: Center for Immigration Studies.

Rystad, G. (1992) Immigration history and the future of international migration. *International Migration Review* 26 (4): 1168–99.

Sassen, S. (1999) *Guests and Aliens*. New York: New Press.

Sassen, S. (2000) The global city: strategic site/new frontier. In E Isin (ed.), *Democracy, Citizenship, and the Global City*. London and New York: Routledge, pp. 48–61.

Secor, A. (2003) Citizenship in the city: identity, community, and rights among women migrants to Istanbul. *Urban Geography* 24 (2): 147–68.

Sibley, D. (2003) Geography and psychoanalysis: tensions and possibilities. *Social and Cultural Geography* 4 (3): 391–9.

Soysal, Y. (1994) *Limits of Citizenship: Migrants and Postnational Membership in Europe*. Chicago: University of Chicago Press.

Soysal, Y. (1997) Changing parameters of citizenship and claims-making: organized Islam in European public spheres. *Theory and Society* 26 (4): 509–27.

Sparke, M.B. (2005) A neoliberal nexus: economy, security and the biopolitics of citizenship on the border. *Political Geography* 25: 151–80.

Staeheli, L. and Kofman, E. (2004) Mapping gender, making politics: toward feminist political geographies. In L. Staeheli, E. Kofman, and L. Peake (eds.), *Mapping Women, Making Politics: Feminist Perspectives on Political Geography*. London and New York: Routledge, pp. 1–13.

Taylor, P (1994) The state as container: territoriality in the modern world system. *Progress in Human Geography* 18 (2): 151–62.

Taylor, P. (2000) World cities and territorial states under conditions of contemporary globalization II: Looking forward, looking ahead. *GeoJournal* 52: 157–62.

Thrift, N. (2004) Intensities of feeling: towards a spatial politics of affect. *Geografiska Annaler* 86B (1): 57–78.

Trudeau, D. (2006) Politics of belonging in the construction of landscapes: place-making, boundary-drawing and exclusion. *Cultural Geographies* 13: 421–43.

Tuan, Y.-F. (1976) Humanistic geography. *Annals of the Association of American Geographers* 66 (2): 266–76.

Vanderbeck, R.M., Dunkley, C.M., and Morse, C. (2004) Introduction: geographies of exclusion, inclusion, and belonging in young lives. *Children's Geographies* 2 (2): 117–83.

Veronis, L. (2007) Strategic spatial essentialism: Latin Americans' real and imagined geographies of belonging in Toronto. *Social and Cultural Geography* 8 (3): 455–73.

Winders, J. (2007) Bringing back the (b)order: post-9/11 politics of immigration, borders and belonging in the contemporary US South. *Antipode* 39 (5): 920–42.

Part II   Thinking and Doing Social Geographies

Introduction

Vincent J. Del Casino Jr.

There is no doubt that the reinvigorated interest in social geography as a subdiscipline has been tied to a more general concern with how geographers go about "thinking" and "doing" social geography. Indeed, methodological conversations, which engage with the substance of how social geographers frame questions, think about data, or engage research participants, are commonly found in geography

A Companion to Social Geography, First Edition. Edited by
Vincent J. Del Casino Jr., Mary E. Thomas, Paul Cloke, and Ruth Panelli.
© 2011 Blackwell Publishing Ltd. Published 2011 by Blackwell Publishing Ltd.

journals, research manuscripts, and textbooks. This excitement for all that is "meth-odological" is prompted by the concerns that have been raised over the last three to four decades related to the politics and ethics of research, activist research tied to social justice movements, and feminist concerns related to reflexivity, subjectivity, and identity in the research process. These concern are not simply about the nuts-and-bolts of on-the-ground research or method, but are tied to the theory of method – how geographers conceptualize the various stages of the methodological process, from how to think about the "social fields" of research to how to analyze, write, and use research findings.

In conceiving this part, we thought it best to offer comment on the meta-processes of methodology, avoiding chapters dedicated solely to particular data collection approaches – structured interviews verses life history interviews verses participant observation and ethnography. It is not that different ways of approaching data col-lection are uninteresting. Rather, what we hope to suggest with this part is that before one even thinks about what data are it is important to understand how and why social geographers ask the questions they do. This demands a conversation about the "phases" of the research process – from thinking about how to frame social geography to how to conceive of one's place in field (both the field of research and the field of geography) to how to analyze and write about one's findings once you have them to how to use one's findings to do work that is engaged with one's subjects and with one's politics. While one certainly need not read these chapters in order, by doing so it is possible to see how research "happens" within the differ-ent spaces of the research process – office, home, field, or even the coffee house. By the end of this part, it is clear to see that social geographic research does not take place in a "black box" or the "ivory tower," it is an engaged set of practices operat-ing across a myriad number of spaces.

It is appropriate in the context of how this part was conceptualized, that it begins with Richard Howitt's chapter (Chapter 8) titled, "Knowing/Doing." In this pro-vocative chapter, Howitt suggests that it is naïve to do social geography research as if it is located "somewhere else," a place outside of ourselves as both researchers and political subjects. Instead, he argues that social geography is always about the "experiences and practices of coexistence – the being-together-in-place" (p. 132). Social geographers must begin by engaging with the context of how they know what they know and why they do what they do. In constructing a relational approach to knowing and doing social geography, Howitt thus offers a radical agenda that appreciates the constitutive nature of social geographic research, using his own career and research experiences to suggest that his knowledge of his research and his research process is contingent not only on how the research was done and with whom but by when he is reflecting on that work – immediately following his field-work or twenty years later. In sum, he argues, that "expertise...is experiential as well as theoretical" (p. 142). As he traces these issues, he illustrates that the bounda-ries are quite blurred between knowledge (theory) and practice (application), a theme that resonates throughout this part.

Chapter 9, titled "Framing the Field" by Joanne Sharp and Lorraine Dowler, next engages one of the core "objects" of social geographic research, the field. Arguing that the field is a historically contested site of knowledge production, Sharp and Dowler tease out the various meanings that geographers have brought to the field and their fieldwork over time. The chapter rightly begins with a critical look at how fieldwork has historically been situated within a "view from above," a vision informed by a masculinist tradition through which the practices in the field are captured and objectively reproduced in written form. This distanciated perspective has been scrutinized by scholars, including feminist and postcolonial thinkers, who have challenged the notion that researchers and research subjects are somehow separated by a barrier of objectivity. Instead, as Sharp and Dowler argue, the field is better conceived of as a messy and complex set of socio-spatial relations constituted by a myriad number of subjects, including researchers. Put more directly, the field is not an easily defined and clearly articulated space; it is a site through which the research process is embodied through "connections as well as differences" (p. 156). Social geographers must thus consider how the field site and fieldwork is defined not simply by what is there "in the field," but by what is produced by their own "field practices" (p. 156).

Following on this theme of the embodiment of knowledge and knowledge production in social geographic research, van Hoven and Meijering (Chapter 10) further interrogate the important questions of positionality, subjectivity, and institutional context in their chapter titled "On the Ground." Situating their own work within the context of Dutch geography, van Hoven and Meijering argue that research interactions are structured by processes that extend beyond the spoken or visual exchanges that take place between researcher and respondent (in an interview setting). These interactions are also structured by the institutional contexts in which that knowledge is generated as well as through the processes of linguistic and emotional translation. It is logical, then, that van Hoven and Meijering engage with theories of the body and body politics in their discussion of the grounded experiences of social geographic research. They suggest that new methodological concerns around body politics, reflexivity, and positionality have pushed for new methods and approaches to data collection and analysis. What their chapters suggests, therefore, is that a methodology that conceptualizes data as something more than the spoken word of an interviewee demands new approaches that take into consideration the wide array of actors involved in any research process – from the emotions of both researcher and respondent to the non-human subjects involved in research such as the "environment."

If the challenges of marking out how social geographers define knowledge, the field, or data are not provocative enough, social geographers also have to consider what to do upon "Leaving the Field" (Chapter 11). As Carolyn Gallaher rightly avers, while "field research is often a lonely venture … [l]eaving the field can also be frustrating" (p. 181). This is particularly true for those new to fieldwork and data analysis. What does one do with all those data? How does one prepare and then use the data in her analysis? In addressing these questions, Gallaher offers an important set of guiding practices, asking social geographers to avoid being "tempted to make conclusions that go beyond what their data will allow" (p. 194). This may sound simple but it is one of the greatest methodological challenges for many researchers – what can I or can I not say with the data I have. Drawing, then, from her own research, Gallaher offers a set of insightful and practical questions (and answers) to what happens once you have your data. "In academic work," Gallaher argues, "the translation of expertise into written work is accomplished when the researcher cum writer is faithful to her methodology" (p. 195). And, as this chapter illustrates that means engaging the intimate relationship between social theoretical framework and methodological approach in and through the entire process from defining the data, to collecting the data, to analyzing the data.

In Chapter 12, Lieba Faier further expands on the question of what happens with all those data in her chapter titled, "The Worldly Work of Writing Social Geography." Drawing on her own research experiences, Faier examines "ethnography as a means for writing social geography, and, specifically, social geographies of enounter" (p. 198). In particular, she draws our attention to the messiness of everyday encounter and the value that ethnography as both a research and a writing practice can have when exploring the contingencies of engagement. Her own work on Filipina women in Japan highlights what she sees as the theoretical and methodological value of writing about/through encounters. As she argues, "ethnographic writing shows us how sociospatial processes play out on-the-ground, illuminating the nuances and contradictions that studies of large-scale political economic forces often miss" (p. 209). It is these nuances that many social geographers have now turned their attention, asking not only how the sociospatial organization of the economy looks from a distance but how the economy operates in the daily lives of individuals who encounter each other within different frames of reference and authority. Writing about those encounters, then, offers a wider lesson for social geographic research in general – asking us to consider how through writing we can "narratively" remap places as heterogenous sites of social engagement.

The remaining two chapters by mrs c kinpaisby-hill (Chapter 13) and Sarah Johnsen (Chapter 14) extend the conversation about what social geographers do in and beyond the field. And, both ask how social geographers engage the world in and through their research. Methodologically, these two chapters further collapse the boundaries between researcher and researched, academic and activist. In writing about *participatory praxis and social justice*, mrs c kinpaisby-hill provide an important methodological intervention in/for social geography by suggesting that participatory research approaches "not only explain socio-spatial injustices" but provide research that can actually help bring about change. This methodological approach fits well into the already politicized landscape of social geography, which has long been a subfield dedicated to engagement with issues related to social (in)justice. In moving toward what she argues are "more fully social geographies," mrs c kinpaisby-hill thus force social geographers to reconsider their own elite place as those who "produce theory." Their approach to participatory research "views theorizing as ontologically distinctly social: it is a citational process, a relational bricolage of one's own and other's ideas, and it is an embedded and constantly ongoing praxis of coming to knowing through iterative cycles" (p. 223). It is, as they argue, an ongoing iterative process of knowing and becoming, one in which the boundaries between theory and practice are not only blurred but also directly contested.

The arguments of mrs c kinpaisby-hill segue nicely with the final chapter in this part by Sarah Johnsen (Chapter 14), who engages the methodological concern of social relevance in social geographic research. In her chapter, "Using Social Geography," she challenges readers to consider how social geographers might maintain an "ethical and political commitment to people" (p. 236) with whom they work while also rethinking relevance as more than the "application" of geographic tools to everyday problems. In short, relevance is reflected in the realm of both everyday social geographic practice and action as well as in the various policy debates that structure the lives of individual people. Working through this concern about relevance in social geography through her discussion of homeless geographies, Johnsen powerfully argues that "projects designed to evaluate the outcomes of specific policies (i.e., apparently 'shallow' analysis) need not preclude researchers from reflecting on broader parameters, presumptions and premises of such policies (i.e., conducting 'deep' analysis) (p. 240). Indeed, as many other geographers have argued about the epistemological distinctions that often structure research – hard and soft science, objectivity and subjectivity, society and space – Johnsen argues that so-called policy research need not be cleaved apart from so-called theoretical work. They are, actually, constitutive. That said, theorizing social geographic research as constitutive is one thing, doing the on-the-ground social work to engage policy-makers is another. To that end, Johnsen remains committed to pushing for a social geographic research agenda that is engaged and conscious of its political consequences in and beyond our own academic lives.

Overall, then, this part illustrates that social geographers cannot simply remove themselves from the research process. And, it also shows that social geographers must appreciate that methodological work is theoretical work – it demands understanding that the research process is always already informed by the epistemological and ontological assumptions that geographers bring to the field. Embracing the notion that methodology matters encourages social geographers to imagine how and what they do and why they do it. This is, in the end, a good thing.

Chapter 8

Knowing/Doing

Richard Howitt

A Manifesto for "Radical Contextualist" Approaches to Social Geography

Social geography is neither a unified sub-discipline nor a singular field of practice. It is characterized more by diversity, pluralism, and multiplicity than coherent agreement on methods for doing social geography, conceptual frameworks for understanding social geographies or politics of engagement in response to these challenges. It has long been thus.[1] While the various cultural, economic, environmental, and various other "turns" in geography have threatened to marginalize the "social," the value of socially informed readings of geographies and geographically informed readings of social relations suggest that questions of social connection and its implications should continue to underpin how geographers deal with the world.

There are significant challenges in thinking about social geographies and doing social geography. Lowenthal (1961) noted the epistemological foundations of such pluralism in his account of the experiential basis of geographical thought. His influential paper considered the interplay of personal geographies and consensual discourses about the world by geographers and others. In considering the spatiality of inequality, the uneven geographies of access to resources and opportunities, and the importance of connection (to both society and place) in understanding social geographies, the sub-discipline has a strong thread of concern with justice, relevance, participatory methods, and ethical responsibility in social geography. This chapter is situated firmly within those traditions.

A Companion to Social Geography, First Edition. Edited by
Vincent J. Del Casino Jr., Mary E. Thomas, Paul Cloke, and Ruth Panelli.
© 2011 Blackwell Publishing Ltd. Published 2011 by Blackwell Publishing Ltd.

Awkward Sticky Messes in Social Geography

The changing influence of different philosophical movements and academic fashions in social geography has been well reviewed by others (e.g., Jackson and Smith 1984; Kitchin 2007; Del Casino 2009; Smith et al. 2010) and need not be further rehearsed here. Rather this chapter invites readers to consider how relationships between knowing and doing geography might be illuminated by the light generated around the margins of the field.

There is much of interest and value to social geographers in the margins of discourses; the edges that overlap and blur the apparent certainties of particular theoretical positions; the awkward, even uncomfortable juxtapositions that occur across the frontiers, borders, edges, and boundaries of places, peoples, and ideas (Howitt 2001a). Such margins often challenge claims to certainty, privilege, and superiority. Australasian geographers, for example, have come to understand the huge significance of Indigenous geographies only relatively recently (e.g., Howitt and Jackson 1998; Louis 2007; Panelli 2008). The ways disciplines engage with these edges, how one's work is situated in these often hotly contested and awkward geographies, actually offers a fine place to think, to come to know, to be challenged, and to act. Indeed, my own experience has been that working within and across such margins has been an appropriate place to both engage with the mechanisms of marginalization and exclusion intellectually and practically and to understand and challenge the economic, political, and cultural dynamics of social change at various places and across various scales.[2]

These messy complexities shape interactions within and between peoples and places. They create a stickiness that adheres to the relationships that geographies create and the connections between people and places. Tsing (2005: 10) suggests that this stickiness and the friction it affords simultaneously gives purchase to the universal appeal of some ideas (consider for example transformative ideas such as nationalism, environmentalism, and human rights[3]), while also ensuring that "engaged universals are never fully successful in being everywhere the same" (Tsing 2005: 10).

This chapter, then, considers how social geographers might think about the awkward sticky messes that characterize the experiences and practices of coexistence – of being-together-in-place. It discusses how we might deal with the ways in which specificities of place and culture adhere to the relationships and processes of social interaction and change to create the mosaics of coexistence, betweeness, and possibility that mark the best of contemporary social geographical writing. In the same sense that it is commonly argued that "geography matters" (e.g., Massey and Allen 1984), this chapter argues that context matters! It advocates the importance of context in knowing, doing, and responding to social geographies against approaches based on universalist theory and structure.

The chapter is subtitled as a "radical contextualist" manifesto. It advocates an approach to geographical knowledge that is responsive to and aware of the context(s) in which knowledge is formed, debated, and applied. It also advocates recognition of multiple contexts influencing the social geographies in which our knowledge is constructed, tested, and applied. This probably requires some explanation as everyone runs off to Google "radical contextualism"! Like many terms that we wish were our original contributions, radical contextualism already has a meaning. Interestingly, the term – like much of the language of structuralism – comes from linguistics and the philosophy of language more than social science. In those discourses, radical contextualism is often opposed to "moderate contextualism." In drawing the term into social science, however, it is better seen as opposed to the sort of reductionist claims of universalizing self-important and self-referential theories that so often render social, cultural, and environmental context marginal in discussions of social relations, social policy, and social justice in favor of some universal claim to truth that is independent of the context in which it delivered.[4]

In my own lexicon, the idea of radical contextualism points to an epistemological, political, philosophical, and aesthetic orientation to the importance of the material, transactional, and relational connections of history, geography, and society (of time, place, and social process) as influential on how things unfold, and how we come to understand and respond to the events, places and people around us – the sticky materialism of experience and being-together-in-place. It points also to the priority of ethical connection as a basis for understanding sociality and responding to the world of social relationships, social process, and of being-together-in-place (see also Levinas 1998, 1999; Eskin 1999; Visker 2003). Radical contextualism is in tension not only with the grand theoretical claims of structuralist thinking but also the superficiality of postmodernist collage and the naïve specificity of parochial and exclusive localisms.

Radical contextualism is advocated here as a foundation for thinking that is manifested in work that values the "complexities of agency" (Laurie 2005), the power of "weak theory" (Stewart 2008), and the significance of "humble theory" (Noyes 2008). In this sense, radical contextualism parallels Geertz' methodological advocacy of "thick description" (1973) and Hall's emphasis of the importance of contextual knowledge in intercultural domains (1977). It points to the value of field-based research in which observation, experience, and engagement with the processes of everyday life and the need make sense of gesture, symbol, and signal on the basis of the context in which they are delivered and received as the basis for knowing and doing social geographies. Massey (1993) championed of Geertz' thick description as fundamental to the theory and practice of geography. This chapter champions radical contextualism as a fundamental to how we do, know, and respond to social geographies.

The conceptually and empirically rich narratives that tell stories of belonging, alienation, loss, movement, and the experience of change narrate both material and imagined geographies. They are central to the concerns of contemporary social geography (Gibson-Graham 2005). Dealt with sensitively, these narratives can nurture social theory that is situated, engaged, and based on the relationships and processes that occur in the lived experience of places (at multiple scales from the interpersonal to the cosmological (Howitt 2002). They provide building blocks (or perhaps the conceptual and narrative material) for "hopeful geographies" (Lawson 2005) engaged by and responsive to "ontological pluralism" (Howitt and Suchet-Pearson 2003, 2006) and the "social embeddness of action" (Curry 2005: 130).

So, this chapter explores and advocates the fundamental significance of context in social geographical research (and social actions in response to social geographical research) as fundamental to and constitutive of how we come to know and do social geographies.

Weaving Social Geographies from Personal Narrative, Social History, Cultural Biography

This is not to abandon the power of big stories. Indeed, I want to use stories to shape my small commentary here, starting with Hugh Brody's quite brilliant reflection (2000) on his career as a cultural anthropologist. Brody considers where his love of wild landscapes and places arose. At one level, Brody offers a compelling spatial story, carrying his readers from Canada's High Arctic to the Pennine Hills around Sheffield in the post-World War II era of his childhood. His autobiographical sketch identifies tramping in Sheffield's rather-less-than-wild Pennines as a motivation of his lifelong engagement with a much wider and wilder world.

Brody's narrative, however, is not limited to a personal spatial chronology. It also weaves the inevitably smallish stories of personal biography into a wider and altogether more startling storytelling. His narrative fabric uses not only personal biography, but also larger scale elements of social history and cultural narrative. His text beautifully encompasses not only the intimate family stories of belonging, escape, rejection, difference, loss, and refuge that link his mother's and father's stories to the cataclysmic violence of the twentieth century's devastating European wars, but also the wider, larger, longer story of cultural identities and human history.

Brody offers his readers a social geography that links disparate worlds. In one sense, it is perhaps an archetypal postmodern social geography, juxtaposing urban and remote places and different times through the accidents of connection that arise from personal experience. There is, however, more at work here than a simplistic postmodern pastiche. Brody's work beckons us towards an approach to social geography that would weave together different spatial and temporal scales, not simply by allowing us to map spatial stories and connections, but also by encouraging us to begin building a more nuanced understanding of our places-in-the-world, and a capacity to do social geography in new, ethically engaged ways; to be more responsive to the building blocks of our understanding – people, place, and environment.

I follow Brody's lead and discuss how contemporary work within the broad communities of practice that constitute social geography might engage with ideas of building knowledge about the social geographies we create and inhabit – and how to act responsibly in the light of the knowledge we construct. Let me first take three narrative glimpses into the development of my own knowing and doing.

Sticky Moments: Narrating Geographies of Coexistence

1 Collett's Crossing, New South Wales, 1976

As an undergraduate, I juggled summer holidays between poorly paid laboring jobs, life on a surfboard, and time spent helping a beloved cousin on his farm on the north coast of New South Wales. One afternoon, after a long morning spent fencing in the heat, my cousin took me to a shallow river crossing named after my great grandfather. My maternal ancestors had been early settlers in the region and were deeply implicated in denuding the landscape and displacing and dispossessing the country's Yeagl and Gumbangirr traditional owners. My cousin drove to the river and arranged for me to meet him further downstream after spending a little time looking at the old farmstead area. I watched him drive away into the forest and immersed myself in the solitude of this place that connected me to both family and social history.

After an hour or so of contemplation and exploration, I stepped into the river to begin walking to our rendezvous. Within a few steps, however, found myself unable to move. My legs were paralyzed in the warm, shallow water that was tugging around them as the tide moved seawards. The land asked who I was. The place had recognized me – although I thought myself a stranger here. My great grandparents' stories of possession (and dispossession) of this place had passed to me in childhood stories. At university I was already studying geography and seeking to respond to the dominant politics of racism, injustice, and dispossession that litter Australian landscapes and which were already too familiar from my own experience. I had begun to hear other versions of similar stories to those of my own childhood[5] – but not of this place. In accepting the generous mentoring of Joyce Clague, a Bundjulung elder from Maclean, a little further north, I had begun to hear a different set of stories to those of my grandmother's childhood. I was told of the blood that had stained the lands and waters around me. But this place, too, had stories I had to hear.

I stood knee-deep in the river, compelled to bear witness to this place. I was compelled to speak – out loud – to the place that held me. And in the sunlight and warm water, my audience was so much more than place. This place was alive with presence and power; here was a place at the "edge of the sacred" (Langton 2002). I spoke to the particular everywhere and everywhen, the here-and-now *and* the "before" and "to be," of that place[6] and acknowledged my family and the complex stories of belonging, possession, and violence that were then beginning and have since continued to motivate my work. I also acknowledged the stories of a local farm, the frontier violence it embodied, and the later stories of abandonment, change, and loss. I spoke of the things I was doing – in my studies, in my political activity, in my heart – to move my own story beyond the history of racist violence that I was beginning to understand framed my family history in this place. And I acknowledged that if this was not enough, I would accept the judgment of this place and whatever that might imply.

And the place listened.

Cicadas and birds chorused around me. A slight breeze moved the branches of the casuarina trees along the banks. The outgoing tide pulled at my legs – and I could not move.

But slowly, the place allowed me to move. I stepped forward, having been heard and both released and held by that place – and have endeavored to do so ever since.

This is not a story I've told often. More than thirty years after the experience, however, it remains a raw and compelling moment for me. It defies rational explanation – but it was real. My Aboriginal colleagues with whom I've occasionally discussed it, nod in a simple affirmation that a truth has been spoken; and that I am perhaps a little more comprehensible to them than before. Whatever happened at that place, my own place in the world, my own sense of place and connections to "country" were changed as a consequence. In this moment, I was opened to a landscape of being, belonging, and responsibility in a way that, like Brody's re-reading of Genesis, revealed my family stories as partial and self-defensively partisan. In that long and continuing fragment of time and place, I began my preparation to accept and work towards understanding Levinas' ideas of ethical availability and the priority of and one's responsibility to the "other" (e.g., Levinas 1998, 1999).

In more recent times, through my work as a professional geographer, I have understood what it means to be accountable to the Dreaming – the environmental, cultural, and cosmological relationships that many Aboriginal people understand to be deeply embedded in places, to be true to oneself, to take responsibility for what one does. And I've come to understand what might be characterized as the "sentient landscape" (Rose 1996) that "spoke" and "listened" to me on that day. But in this chapter, I recount this moment as a window onto the ideas of belonging, knowing, and doing that are the focus of the chapter – a window on what it is to act in response to knowing.

2 Argyle exploration area, East Kimberley, Western Australia, 1980

During my PhD fieldwork, I worked closely with Aboriginal people in the Kimberley region of Western Australia. My experience included singing for Yungngora traditional owners at Noonkanbah Station, where the state government had escorted a drilling rig to desecrate a scared site with a paramilitary convoy that had met with Aboriginal and trade union protests as it traveled north the Noonkanbah. It also included a visit to inspect damage to registered sacred sites at the exploration site that has since become the Argyle Diamond Mine (see Howitt 1989, 2001b: ch. 8).[7] In many ways the experience opened my eyes, my mind, and my heart to the simultaneous realities that constitute coexistence and pluralism in social and cultural landscapes. It was a transformative experience that I recorded in my fieldnotes as follows:

> Like a fish gasping for water, the hill lays upon the land. It lies there in front of me, an exhausted fish which has struggled against enormous forces to retain its freedom. The mountain range from which I view this spectacular site lies broken by the struggle – an unsuccessful trap set by wily hunters to catch the fish.
>
> Sitting in the back of the four wheel drive utility with a small group of old men, I hang on to avoid falling out as we negotiate the half-made track up the hillside. I strain to hear the words of the old men whose country this is. I am a stranger here – the only person in the truck whose first language is English. All the other men speak one, or two, or more languages whose sounds have been heard in this country for centuries. As they speak, they struggle to open my eyes to another way of seeing. Company geologists have been here, looking for diamonds. Their explanation of the landscape involves volcanoes and tectonic forces. These old men understand that. But for them the mountain was both mountain and fish. It is not that the mountain is a geological symbol of a spiritual reality. Rather it is both. For them, the two realities comfortably occupy the same space. Neither is more or less real. Both are real. The barramundi struggled here more than two-hundred kilometres from the coast, chased all the way by birds which threw up traps of spinifex grass to catch it. Finally, at Barramundi Gap,

Site K1098 on the geologists' and anthropologists' maps, the fish escaped, leaving a gap in the range which ensured the future of all barramundi, and the people who relied on them; and leaving a mountain whose belly is full of precious stones.

For me, this visit to the Argyle Diamond site in 1980 was an "ah-ha" time. After working with Aboriginal people for several years, I finally felt I had reached an understanding of why the Land Rights movement was simultaneously political and spiritual. The land itself was simultaneously a political and spiritual entity; simultaneously a reflection of geology and cosmology. As a stranger to this country of new dreamings, created by fish I had never seen, and birds I had never sat and watched, I could not reach a genuine understanding of the complexities and subtleties of the stories and the realities which lay within them, but in this moment I felt I had glimpsed something profound and important. It was on that hillside on that day that I began to understand the geographies of coexistence; the social geographies of contemporary Australia. I began to understand that seeing the landscape is always mediated by culture, power, values, and contested and evolving meanings. Country and human relationships with it are always becoming. There could be no clearer demonstration than this of the integral connections between the ecological, the economic, and the social dimensions of life.

3 Native title negotiations meeting, Hahndorf, South Australia, 2000

In 1999, I was invited to join a small team at the Aboriginal Legal Rights Movement in South Australia that was exploring possibilities for a negotiated settlement of Native title claims in that state. We were involved in discussions with the government of South Australia, Native title claimants from across most of the state,[8] and a range of industry groups. This work has been widely reported (e.g., Agius et al. 2002, 2004, 2007), but I want to reflect briefly on incidents at a meeting of about 200 Native title claimants at Hahndorf in 2000, which had a profound effect on our processes – and on my understanding of what it means to "do" social geography.

Negotiating about complex political issues is difficult in any circumstances. But in South Australia in 2000 many issues were beginning to boil over. We were dealing with a great diversity of cultural and historical experience in the Native title claimant groups. While the ALRM team was ably led by Parry Agius and had considerable expertise at its disposal, we were not always able to generate trust and understanding in our critical audience. We had managed to keep lawyers away from most of the process – indeed, we had managed to avoid the government's proposal that we get the lawyers to define key terms such as Native title, coexistence, and extinguishment before the negotiations began. But there were deep divisions and tensions between and even within the Native title claimant groups, and increasing frustration at the complexity and slowness of the process.

At the beginning of the meeting, Parry was asked (again) if we had secured payment for the claimants' time in the meetings. This was a serious issue for many of the participants as the meetings were three-day meetings, and many people had more than a day's traveling to and from each meeting. We had secured travel costs, accommodation, and food – but we had not secured any personal allowance for people's time. This heightened frustration and anger within the group, who felt we were not hearing their demands. We were equally frustrated that we were already using up the funds government had made available for the process and could not see how we could meet this demand. In frustration, Parry explained again what he had done to try to secure the funds, but offered to call the government's lead negotiator and invite him to talk directly to the claimants without our involvement.

The government negotiator arrived, and we left to wait outside. Two hours later, a government vehicle was sent from Adelaide with a strongbox full of cash to make payments for the meeting, and an agreement that claimants would be paid for their attendance at future meetings. We had learnt an important lesson – and the claimants realized that we were serious about recognizing them as the principal negotiators rather than taking that role for ourselves.

Later in the meeting, however, things again began to unravel as tensions exploded when a woman whose family had been removed from the areas under their claim several generations ago referred to young initiated men from the desert areas as "boys" – a profound insult in customary legal terms. Anger boiled over and serious implications were beginning to develop. By early evening, death threats had been issued, and prospects for a "statewide Native title agreement" seemed to be fading by the minute.

Parry and I had long discussed the need for our process to be held accountable to The Dreaming. We felt the negotiations must be accountable in customary legal process rather than authorized by the government against whom claims were being mounted. The government was still opposing the claims in the court proceedings and allowing their power to be the source of negotiating authority would have been unacceptable. Late into the night, there were diplomatic discussions between representatives of the two groups, meditated by senior men and women from each group and from across the whole claimant community.

In the morning, the air was electric with anticipation. The woman who had spoken the insult unintentionally rose to her feet and apologized. It was a powerful speech in which she referred to her own frustration at her loss of access to language, to cultural knowledge and to customary legal knowledge that was a product of her family's displacement and the possession of their country by others. She spoke with great humility and a powerful dignity that reminded everybody in the room that history rode with us in every step, every word, and every moment. She had narrated a spatial story that was simultaneously deeply personal, profoundly social and powerfully cultural in its scope. She sat, and slowly one of the older men, a respected elder from the insulted group rose to his feet. There were some grumbles from the young men who felt defamed, but their respect for the new speaker was apparent and they settled as he spoke slowly and with great authority. He accepted the apology and offered his own sorrow to the other group for their loss. He offered support and friendship, and said that this was the end to the dispute. If any of the young men still had issues – they had issues with him and would need to speak to him, to deal with him.

If the air was electric at the opening, at this point it was alive. The older people – the ancestors whose law, culture, and being were the foundation of the claims under discussion in the negotiations – were a palpable presence. These were people from the hardest of times, from times when survival depended on knowledge, accountability, and toughness. These ancestors were beyond the edge of death (Langton 2002), but had set a tough hurdle to test the resilience and reliability of the leaders and followers in this process. And against the odds, this incident built a level of trust and mutual recognition, an acceptance of coexistence, of being-together-in-country amongst the claimants. It also pushed the ALRM team further into a relationship of trust with the claimants. It was not simply our willingness to recognize the claimants and their own governance processes as the key drivers of the negotiations. It was also that we realized that we trusted them. We saw clearly that we were becoming an accountable element of a much bigger set of relationships and processes than simply transforming the state of South Australia. We were producing a new form of political accountability to customary law – and that had been tested and accepted in this moment.

Radical Contextualist Orientations to Engagement with Place, Scale, and Process

These personal accounts of knowing and doing social geographies narrate moments of being and becoming that have profoundly shaped my understanding of the awkward sticky messes of human–human, human–nature, and human–cosmos relationships that engender contemporary social geographies. Each suggests a window on wider methodological, epistemological, and conceptual issues. Each points to an openness, a plural and sticky simultaneity of connections within, between and across spaces, places, times, and scales that challenge geographers trying to know, write, and do social geographies.

In writing the first of my personal narratives, I was struck by the sentence "I stepped forward ... and have endeavored to do so ever since." Where had that come from? This sentence connected a teenage moment, this treasured private moment, to a lifetime of aspiration and motivation. It pointed to a link between knowing and responsibility; between thought and action; between the multiple narratives of family and place; belonging and possession; history and geography; culture and identity. That sentence situates my engagement with questions of social and environmental justice and theoretical engagement with notions of scale, human rights, and coexistence within both a personal narrative, a social history and a cultural biography of alienation and belonging in Australia. And it possibly connects the discussion of knowing and doing social geographies to the question of responsibility, of ethical availability to others.

Others have commented on the significance of the radical and relevance turns in social geography (e.g., Blunt and Wills 2000; Swyngedouw 2000), but I note the particular significance of the link to struggles for justice in various forms as particularly important in the development of thinking about social geographies. Geographical knowledge is seen as a purposeful, valuable element in actions not just to describe or explain the world, but also to change it. While that might echo Marx's theses on Feuerbach, it also echoes my conclusion to my PhD that suggested that the structural analysis provided of transnational corporation in North Australia that it provided had been concerned with "the perspectives of local interests and counter strategies at that scale (which) meant the thesis (had) been more influenced by Bhuddist values ... and anarchist hopes ... than Marxist theory" (1986: 509).

In the case of the Argyle Diamond Mine, I have revisited the learning and knowing achieved there many times over my professional life in geography. The juxtaposition of the sacred site and the diamond resource at Argyle seemed such a powerful metaphor of the relationship between Aborigines and miners that was not just a focus of my work, but which also cast such clear light onto so many aspects of contemporary Australian society. In the mid-1990s, along with a group of my Aboriginal colleagues, I was invited to contribute to high-level intercultural training for Rio Tinto plc, the owner of the Argyle mine. In retelling the story, I invited the miners themselves to revisit the exploration phase of the project, and begin rethinking what relationships had been constructed there; what assumptions had been made; what had been learned and what had been forgotten. Ten years later, I had the rare privilege of supervising Kim Doohan's extraordinary re-investigation of the social geographies of the Argyle mine (Doohan 2006), in which she drew on more than twenty years' ethnographic engagement with traditional owners and their communities and their tireless efforts to renegotiate the social geography of the mine site. In their words, they needed to "make things come good"; to re-order the way the miners related to the sacred geography that had created the landscape that was so important to both Aboriginal people and diamond miners.

In Doohan's re-narration of the Argyle story (2006, 2008), it is not a simple binary between sacred and economic landscapes. In her narration, the cultural values of place that adhere to the dreamings that motivate the landscape in Aboriginal traditions begin to adhere to mining company practices. Cross-cultural training programs have traditional owners welcoming new mining company staff to the mine site as part of induction. Traditional ceremonies are offered and accepted as part of safety training. In customary law, the traditional owners accept responsibility for introducing the miners to the ancient fish whose scales are the prized diamonds. After an industrial accident in which a young man died, many staff requested traditional owners perform a cleansing and healing ceremony that reconnects them to the place and each other. Despite the very real senses in which local Aboriginal people had been displaced from their country by the mining operation, their persistent presence in the landscape, their profound commitment to "making things come good" – to setting relationships between people and place in order – had created an extraordinary social geography in which customary ceremony, cultural landscape, global commodity trade and property rights had become intertwined in new and profoundly powerful ways.

Here was no simple hybrid landscape, but a new narrative of belonging, being together and place. This is not a narrative that is easily accessible to the naive observer. It is not a geography to be found by research that brings together clever quantitative and qualitative techniques to reveal some pre-existing state of things. Rather it needs to seen as a narrative of place that is built and understood by the processes of connection. The narration makes it accessible to a wider audience; but is not able to be produced as if the analysis were reducible to a set of dry facts. In the interplay between my own early experience and Doohan's prolonged engagement, our conversations in supervision sessions nurtured a very different understanding of the social geographies developing at (and around) Argyle – the place, the company and the idea.[9] Clearly, the context that matters is similarly irreducible to merely the material phenomena. In constructing the awkward, sticky juxtapositions of being-together-in-time-and-place at Argyle, circumstances exposed both traditional landowners and corporate operators to each other. In each taking the "other" seriously, in engaging with rather than reducing the other to a political issue or externality of some kind, the evolving relationships created new social geographies which simultaneously reflect, challenge, and disrupt the dominant narratives of possession, dispossession, and belonging in North Australia. There is no singular and neat methodological solution available to allow research to reveal these emergent geographies. The narrative reinforces the initial proposition in this chapter that there is no singular field of practice available to reveal a singular, representative truth. Doing social geography as an academic endeavor inescapably confronts plurality, difference, otherness, and questions of ethical engagement, partial and partisan knowledge, and simultaneous but incompatible truths.

The third narrative of applied geographical research and advocacy in the South Australian Native title processes raises the stakes in terms of academic accountability – the consequential implications of "doing" social geography in such a complex and sticky world (also see Johnsen, Chapter 14, this volume). My role as a researcher was under constant scrutiny. Was the knowledge I contributed relevant? Useful? Powerful? Would it stand up in legal and political process? Did it make sense to those whose rights were at stake in the negotiations and would they be able to mobilize it in negotiations? How might such research-based knowledge be drawn into a relationship of accountability to the customary legal imperatives of the Dreaming? These are questions that have echoed in much of my work in various times and places across Australia. The activity I was involved in was mandated to change the world – not simply to describe or explain it. In other activities, such as applied research in social impact studies, similar imperatives have governed my work and that of my students. While the university sector and our disciplinary colleagues have elevated peer reviewed publications as the pinnacle of achievement in our profession as geographers, in the places created by our human experience of being-together-in-place, it is the value of our knowledge in facilitating engagement; its robustness when challenged across ontological difference and its facility in fostering new ways of doing the material and social geographies at various scales from the inter-personal to the intercultural, from the local to the global that seems to matter most to most people.

This does, I think, present some significant challenges in "doing" social geography. It is certainly not a basis for rejecting any particular set of methodological or conceptual tools that we might use to examine and interpret social relationships and processes in places. But it does challenge the authority of expert knowledge as if expertise can be constructed without social and ethical accountability. I am not suggesting that some new form of political correctness should be adopted to insist on accountability to local values – indeed, I have already argued that the radical contextualism advocated here is inescapably conscious of and opposed to the prejudices and privileges of parochial localism as surely as it is opposed to the reductionist generalizations of grand narrative and the specious fragmentation of postmodernist collage. It is not my accountability to and engagement with Indigenous groups that is the point. I could construct another set of narratives about similar engagements with state agencies and transnational resource companies – and others could present narratives of engagement with many other sorts of places, institutions, and values.

The point is that the social geographies that engage us as geographers are precisely that – they are social. We overlook that methodologically and conceptually at our peril – and at great cost to the sorts of contributions that geographers might make to describing, understanding, and changing the circumstances that we find ourselves in.

Conclusion

Context matters – the historical, geographical, social, and cultural context in which social geographers undertake research fundamentally shapes what we come to know and how we come to represent it to our various audiences. The relationship between "doing" social geography and "knowing" social geographies is never a simple matter if identifying a research site and a set of methods for collecting and analyzing data. Our choices on such matters not only reflect various contingencies of circumstance, but also have implications for what becomes knowable, and what is erased or marginalized in the geographies and representations that we are involved with.

Radical recognition of the significance of context pushes social science towards "weak" theory and ethics of engagement and availability. It emphasizes what Flyvbjerg (2001) refers to as "phronesis" – expertise that is experiential as well a theoretical. It pushes us towards accepting that the conventional academic boundaries between "basic" and "applied" research, between "research" and "teaching" and between "academy" and "society" are much more blurred, messy, and awkward than university administrators and research funding agencies acknowledge. It diffuses the source of authority away from academic privilege and quite literally replaces it into a process of socially constructed discourse and argument. It renders the influential presentation to a lay audience just as significant as the acceptance of a peer publication. And it allows the co-construction of knowledge through research, action, and education to become an important part of the being-together-in-place that represents the foundations for social geographies.

Notes

1   In the early 1980s, Jackson and Smith recognized that social geographers were "being urged to consider a wide range of alternatives to the positivistic assumptions which formerly guided the great majority of geographical research (1984: 1). By the early 1990s Gregson felt that social geography was a fragmenting field "seen as including anything and everything beyond 'the economic'" (Gregson 1992: 387). Since then, a lot of the conventionally "economic" has also moved within the scope of social geography as separate conceptualizations of social and economic theory have been challenged by more integrative approaches. Most recently this volume, Del Casino (2009) and Smith et al. (2010) have remapped the scope and consequence of both social geographies and social geographical thought.

2   Howitt (1993) draws a parallel with Eric Wolf's well-known discussion of the "people without history" (Wolf 1982) and refers to Indigenous peoples as "people without geography": "Dispossessed of their traditional country, and marginalized from the economic, political and social mainstreams of colonial and neo-colonial life, Indigenous Peoples have, in the eyes of the discipline it seems, become "people without geography" – mar-

ginalized and excluded from the production of geographic knowledge, and simultaneously marginalized and excluded from their traditional country, and the social fabric woven from it" (Howitt 1993: 1).

3 Exhaustive consideration of these ideas is beyond the scope of this chapter, but see, for example, Anderson (1983) on nationalism, Tully (1995) and Ignatieff (2000) on rights and O'Riordan (2005) on environmentalism and sustainability.

4 Radical contextualism in linguistics is contrasted with moderate contextualism as well as structuralism. In philosophy more generally, it has been associated with debates about the extent to which the meaning of an utterance is dependent upon its context (Norman 1999; Brendel and Jäger 2004).

5 As an undergraduate I was profoundly affected by the spatiality of Henry Reynolds' account of Indigenous Australians' experience on what he referred to as "The Other Side of the Frontier" (Reynolds 1976: 1981).

6 The image of the everywhen is drawn from Stanner's (1969) description of Aboriginal cosmology that is all-too often glossed as "The Dreaming." The cosmological order that is conceptualized as "The Dreaming" in Aboriginal Australian ontologies is profoundly implicated in the relationships of connection between human and non-human entities in the world of lived experience. A brief overview that draws on several powerful Aboriginal accounts is offered by Rose (1996: particularly 28–33).

7 For a powerful recent reinterpretation of Aboriginal experience at Argyle, see Doohan 2008.

8 Pitjantjatjara and Maralinga-Tjaratja lands in the western part of the state were already held as Aboriginal land and had not been claimed under the Commonwealth's *Native Title Act 1993*.

9 See Macquarie Human Geography Group (2001) for discussion of the idea of graduate supervision as a "nourishing conversation."

References

Agius, P., Davies, J., Howitt, R., and Johns, L. (2002) Negotiating comprehensive settlement of Native title issues: building a new scale of justice in South Australia. *Land, Rights, Laws: Issues of Native Title* 2 (20): 1–12.

Agius, P., Davies, J., Howitt, R., Jarvis, S., and Williams, R. (2004) Comprehensive Native title negotiations in South Australia. In M. Langton, M. Teehan, L. Palmer, and K. Shain (eds.), *Honour Among Nations? Treaties and Agreements with Indigenous People.* Melbourne: Melbourne University Press, pp. 203–19.

Agius, P., Jenkin, T., Jarvis, S., Howitt, R., and Williams, R. (2007) (Re)asserting Indigenous rights and jurisdictions within a politics of place: transformative nature of Native title negotiations in South Australia. *Geographical Research* 45 (2): 194–202.

Anderson, B. (1983) *Imagined Communities: Reflections on the Origin and Spread of Nationalism.* London and New York: Verso.

Blunt, A. and Wills, J. (2000) *Dissident Geographies: An Introduction to Radical Ideas and Practice.* Harlow: Prentice Hall.

Brendel, E. and Jäger, C. (2004) Contextualist approaches to epistemology: problems and prospects. *Erkenntnis* 61 (2–3): 143–72.

Brody, H. (2000) *The Other Side of Eden: Hunters, Farmers and the Shaping of the World.* Vancouver: Douglas and McIntyre.

Curry, G. (2005) Reluctant subjects or passive resistance? *Singapore Journal of Tropical Geography* 26 (2): 127–31.

Del Casino Jr., V.J. (2009) *Social Geography: A Critical Introduction*. Oxford: Wiley-Blackwell.

Doohan, K. (2006) "Making things come good": Aborigines and miners at Argyle. PhD thesis. Macquarie University: Department of Human Geography.

Doohan, K. (2008) *Making Things Come Good: Relations between Aborigines and Miners at Argyle*. Broome, WA: Backroom Books.

Eskin, M. (1999) A survivor's ethics: Levinas's challenge to philosophy. *Dialectical Anthropology* 24 (3–4): 407–50.

Flyvbjerg, B. (2001) *Making Social Science Matter: Why Social Inquiry Fails and How It Can Succeed Again*. Cambridge: Cambridge University Press.

Geertz, C. (1973) *The Interpretation of Cultures*. New York: Basic Books.

Gibson-Graham, J.K. (2005) Surplus possibilities: postdevelopment and community economies. *Singapore Journal of Tropical Geography* 26 (1): 4–26.

Gregson, N. (1992) Beyond boundaries: the shifting sands of social geography. *Progress in Human Geography* 16 (3): 387–92.

Hall, E.T. (1977) *Beyond Culture*. Garden City: Anchor Press.

Howitt, R. (1986) Transnational corporations in North Australia: strategy and counter strategy. PhD dissertation. Sydney: School of Geography, University of NSW.

Howitt, R. (1989) A different Kimberley: Aboriginal marginalisation and the Argyle Diamond Mine. *Geography* 74 (3): 232–38.

Howitt, R. (1993) People without geography? Marginalisation and Indigenous peoples. In R. Howitt (ed.), *Marginalisation in Theory and Practice*. Sydney: Economic and Regional Restructuring Research Unit, Departments of Economics and Geography, University of Sydney, pp. 37–52.

Howitt, R. (2001a) Frontiers, borders, edges: liminal challenges to the hegemony of exclusion. *Australian Geographical Studies* 39 (2): 233–45.

Howitt, R. (2001b) *Rethinking Resource Management: Justice, Sustainability and Indigenous Peoples*. London: Routledge.

Howitt, R. (2002) Scale and the other: Levinas and geography. *Geoforum* 33: 299–313.

Howitt, R. and Jackson, S. (1998) Some things do change: Indigenous rights, geographers and geography in Australia. *Australian Geographer* 29 (2): 155–73.

Howitt, R. and Suchet-Pearson, S. (2003) Ontological pluralism in contested cultural landscapes. In K. Anderson, M. Domosh, S. Pile, and N. Thrift. *Handbook of Cultural Geography*. London: Sage, pp. 557–69.

Howitt, R. and Suchet-Pearson, S. (2006) Rethinking the building blocks: ontological pluralism and the idea of "management." *Geografiska Annaler: Ser B, Human Geography* 88 (3): 323–35.

Ignatieff, M. (2000) *The Rights Revolution*. Toronto: Anansi.

Jackson, P. and Smith, S.J. (1984) *Exploring Social Geography*. London: George Allen & Unwin.

Kitchin, R. (2007). *Mapping Worlds: International Perspectives on Social and Cultural Geographies*. Abingdon: Routledge.

Langton, M. (2002) The edge of the sacred, the edge of death: sensual inscriptions. In B. David and M. Wilson (eds.), *Inscribed Landscapes: Marking and Making Place*. Honolulu: University of Hawaii Press, pp. 253–69.

Laurie, N.D. (2005) Putting the messiness back in towards a geography of development as creativity: a commentary on J.K. Gibson-Graham's "Surplus possibilities: postdevelopment and community economies." *Singapore Journal of Tropical Geography* 26 (1): 32–5.

Lawson, V. (2005) Hopeful geographies: imaging ethical alternatives – commentary on J.K. Gibson-Graham's "Surplus possibilities: postdevelopment and community economies." *Singapore Journal of Tropical Geography* 26 (1): 36–8.

Levinas, E. (1998) *Entre Nous: Thinking-of-the-other*. New York: Columbia University Press.

Levinas, E. (1999) *Alterity and Transcendence*. New York: Columbia University Press.

Louis, R.P. (2007) Can you hear us now? Voices from the margin: using Indigenous methodologies in geographic research. *Geographical Research* 45 (2): 130–9.

Lowenthal, D. (1961) Geography, experience, and imagination: towards a geographical epistemology. *Annals of the Association of American Geographers* 51 (3): 241–60.

Macquarie Human Geography Group (2001) Nourishing conversations in the co-construction of knowledge. In A. Bartlett and G. Mercer (eds.), *Postgraduate Research Supervision: Transforming (R)Elations*. New York, Peter Lang, pp. 261–75.

Massey, D. (1993) Questions of locality. *Geography* 78 (2): 142–19.

Massey, D. and Allen, J. (eds.) (1984) *Geography Matters! A Reader*. Cambridge: Cambridge University Press.

Norman, A.P. (1999) Epistemological contextualism: its past, present, and prospects. *Philosophia* 27 (3–4): 383–418.

Noyes, D. (2008) Humble theory. *Journal of Folklore Research* 45 (1): 37–44.

O'Riordan, T. (2005) Beyond environmentalism: towards sustainability. In J.A. Matthews and D.T. Herbert (eds.), *Unifying Geography: Common Heritage, Shared Future*. London and New York: Routledge, pp. 117–43.

Panelli, R. (2008) Social geographies: encounters with Indigenous and more-than-white/Anglo geographies. *Progress in Human Geography* 32 (6): 801–11.

Reynolds, H. (1976) The other side of the frontier: early Aboriginal reactions to pastoral settlement in Queensland and Northern New South Wales. *Historical Studies* 17 (66): 50–63.

Reynolds, H. (1981) *The Other Side of the Frontier: An Interpretation of the Aboriginal Response to the Invasion and Settlement of Australia*. Townsville: James Cook University of North Queensland.

Rose, D.B. (1996) *Nourishing Terrains: Australian Aboriginal Views of Landscape and Wilderness*. Canberra: Australian Heritage Commission.

Smith, S.J., Pain, R., and Jones III, J.P. (eds.) (2010) *The Sage Handbook of Social Geographies*. London: Sage.

Stanner, W.E.H. (1969) *After the Dreaming: Black and White Australians – an Anthropologist's View*. Sydney: Australian Broadcasting Commission.

Stewart, K. (2008) Weak theory in an unfinished world. *Journal of Folklore Research* 45 (1): 71–83.

Swyngedouw, E. (2000) The Marxian alternative: historical-geographical materialism and the political economy of capitalism. In E. Sheppard and T.J. Barnes (eds.), *A Companion to Economic Geography*. Oxford, Blackwell, pp. 41–59.

Tsing, A.L. (2005) *Friction: An Ethnography and Global Connection*. Princeton and Oxford: Princeton University Press.

Tully, J. (1995). *Strange Multiplicity: Constitutionalism in an Age of Diversity*. Cambridge: Cambridge University Press.

Visker, R. (2003) Is ethics fundamental? Questioning Levinas on irresponsibility. *Continental Philosophy Review* 36: 263–302.

Wolf, E. (1982) *Europe and the People without History*. Berkeley: University of California Press.

Chapter 9

Framing the Field

Joanne Sharp and Lorraine Dowler

It is still the case, we would argue, that field research involves an encounter that confronts, engulfs and even overwhelms us, and that this is true whether we are working on a glacier, through a book, in a library or a city-street ... But it seems to us curious at best that so much of what informs our work can be so succinctly written out of the account, or, even where the performative is acknowledged, turned into a sort of doing as self-control. (Dewsbury and Naylor 2002: 246)

Introduction

Despite – or, perhaps, because of – the discipline's tradition of field expeditions and exploration, there has been surprisingly little discussion in geography about the actual practices of fieldwork. Since the 1990s there has been a significant amount of literature devoted to critical approaches to qualitative methodologies (Rose 2001; Limb and Dwyer 2001; Shurmer-Smith 2002; Hay 2005; Delyser et al. 2009) but there is much less attention paid to where we enact our methods, "the field," and the myriad experiences, emotions, and practices which make up "fieldwork" (but see *Professional Geographer* 1994; *Geographical Review* 2001). Conventionally, the field has been understood to be something away from the researcher, somewhere that s/he has to go to. It is imagined as a place separate from the home and the university. However, as the opening quote suggests, there have been critiques of this from radical perspectives, feminists, and postcolonial scholars in particular, who have emphasized connection to the field and analyzed the politics and ethics of fieldwork, while still others have presented the idea that the field should be understood as constituted through practice rather than being a pre-existent and stable place awaiting discovery by the field researcher. In this chapter, then, we trace meanings of "the field" from the period of the discipline's establishment through to more

A Companion to Social Geography, First Edition. Edited by
Vincent J. Del Casino Jr., Mary E. Thomas, Paul Cloke, and Ruth Panelli.
© 2011 Blackwell Publishing Ltd. Published 2011 by Blackwell Publishing Ltd.

recent feminist and postcolonial critiques of this fieldwork tradition. Examining such debates highlights issues of power and representation, in addition to challenging notions of where the field is, what comprises it, and how we as researchers engage with it.

Stout Boots and Hairy Chests: Fieldwork Traditions

Geography as a discipline has its roots in the field more than in the library. The discipline was established as an "aid to statecraft" (Mackinder 1904) to support European expansionary and colonial ambitions and so the origins of geography are based in colonialist exploits of discovery and exploration. Thus, there has been a particular hue to the celebration of geography's founding fathers, as heroic explorers struggling against the elements and natives, "solid hunks of British manhood" as Stoddart (1986: 143) put it, quoting Freeman approvingly. The Royal Geographical Society (established in 1830) sponsored many of the great colonial explorers such as Burton, Speke, and Stanley in their attempt to cross Africa and map the sources of its great rivers.[1] Put rather bluntly, the role of colonial geographical fieldwork was to discover new lands, to mark out and catalogue resources, and to map the territory to support the effective and efficient capture and control of new territories (see Godlewska 1994).

Philips (1997) and others have argued that exploration was marked by the convergence of nationalism, masculinity, and romance of far off places. It was a kind of muscular, individualistic independent masculinity, not a bookish intellectual version. The narrative excludes women, unless they represent ties to home (wives and mothers who urge them not to go, who try to domesticate them, loved ones, and safety longed for when far away) or as prizes or things to protect (Sharp 2008). The first-person accounts articulated through the explorer's journal ensured that he was at the centre of the story, creating him as the hero of the story:

> The mythology of exploration insists that the explorer be pitted against the vicissitudes of nature, hounded by inconsiderate indigenes and worn out by hunger in the service of his country. Above all, the explorer is an heroic *individual*: in exploration hagiographies there is rarely mention of the other members of the party. (Ryan 1996: 21)

The imperial project clearly had the field marked as "over there," the blank spaces on the map to be filled with information useful to colonial governance and trade. While any fieldwork was regarded as valuable, influential US geographer Carl Sauer and his students tended to work in societies distant from their own. As Stoddart (1986: 146) put it, "whatever its educational advantages, "fieldwork" at home was but a pale substitute for active exploration overseas."

Nevertheless, geography's obsession with the field also included more local fieldwork. Not for geographers was the dry and lifeless library (Stoddart 1986), but for "the geographer the ground is the primary document" (Wooldridge in Stoddart 1986: 144). The object of geographical education, Wooldridge (1955) argued, "is to develop 'an eye for country'" (in Stoddart 1986: 144). Similarly, Sauer insisted on the pedagogical value of learning by doing fieldwork, a rite of passage with those who had already mastered the techniques of seeing what eluded the non-trained eye.

> Being afoot, sleeping out, sitting about camp in the evening, seeing the land in all its seasons are proper ways to identify the experience, of developing impression into larger appreciation and judgment. I know no prescription of method; avoid whatever increases routine and fatigue and decreases alertness. (Sauer 1956: 296)

The trained geographer, Stoddart (1986: 56) argues, has an eye "to discern what others cannot." The "eye" was, for many geographers, the main tool for fieldwork. This naturalistic approach to fieldwork suggested that there were no specific methodologies to be taught in the classroom; rather, the fieldwork skill was imparted through an apprenticeship, a rite of passage with a great master who could guide his [sic] students. Both Sauer and Stoddart suggested that this was a skill that good geographers were born with though, rather than achieving through education:

> One of the rewards of being in the field with students is in discovering those who are quick and sharp at seeing. And then there are those who never see anything until it is pointed out to them. (1956: 290)

Sauer insisted on going to promontory to get the best view over a new landscape. This highlights one of the effects of the privileging of vision, which is that it distances the viewer from what is being viewed and, as a result, objectifies it. While researchers such as Sauer were *in* the field, they were not *of* it. In other hands this language became almost God-like:

> From my loft eyrie I can see herds upon herds of cattle, and many minute specks, white and black, which can be nothing but flocks of sheep and goats. I can also see pale blue columns of ascending smoke from the fires, and upright thin figures moving about. Secure on my lofty throne, I can view their movements, and laugh at the ferocity of the savage hearts which beat in those thin dark figures … As little do they know that human eyes survey their forms from the summit of this lake-girt isle as that the eyes of the Supreme in heaven are upon them. (Stanley 1878)

Feminist and Postcolonial Challenges

The fieldwork tradition in geography has conventionally worked to produce a sense of the field as something fixed and bounded and a separate space from that of the researcher. The field was a place the researcher left home to go to. However, various critical engagements with fieldwork by feminist, postcolonial and other scholars has challenged this notion of the field, highlighting the complex and entangled nature of the different spaces of research. Driver has argued that the field is "a region undergoing continual processes of construction by the fieldworker, inhabitants of the field, and those elsewhere" (Driver 2001 as cited in Powell 2002: 264). From this perspective, and contrary to the likes of Sauer and Stoddart, the field is not just "there" awaiting the arrival of the researcher to be known, "it is produced and reproduced through both physical movement across a landscape and other sorts of cultural work in a variety of sites" (Driver 2000: 267). Thus, "the field" is cross cut by relationships of power/knowledge, representation, and practice.

One of the first challengers to the traditional approach to the field was in William Bunge's founding of the Society for Human Exploration in 1968. Bunge called for "rediscovering geographical skills of exploration and using them for new purposes" (Peet 1998: 73). His Detroit Geographical Expedition was set up with this radical approach to fieldwork:

> For Bunge (1969: 3), "the tyranny of fact compels that geographers go into a state of rationally controlled frenzy about the exploration of the human condition" by forming "expeditions" to the poorest areas, contributing rather than taking, planning with (rather than for) people. Geographers, he said, should become people of the regions they explore, should discover the research people need doing, and address themselves energetically to these problems. Local people should become trained in geographic skills so they could become part of the solution rather than being objects of study. (Peet 1998: 73).

Bunge's expeditions lasted for only a few years in the early 1970s, however, their spirit of inclusion and cooperation, challenging rather than acting on behalf of dominant powers, continues through the fieldwork politics and ethics of feminist geography.

As already noted, the heroic traditional figure of the field researcher was gendered so as to exclude women. This was the case figuratively but also materially, as institutions such as the Royal Geographical Society barred women as members until the twentieth century (1913 in the case of the RGS). The field was seen as no place for a woman. Feminist geographers have suggested that it is this fieldwork tradition that has helped to marginalize women in the field (see Rose 1993). Domosh (1991) took Stoddart (1986) to task for ignoring women in his account of the history of geography – particularly in his narration of the place of the field in the establishment of the discipline – and thus reproducing a heroic, masculinist account of the discipline emerging from the deeds of early European explorers. Stoddart (1991) replied that his account included only men because women had not been part of scientific field expeditions due to their exclusion from professional societies, and had thus not contributed to the scientific development of the discipline. His vision of the role of fieldwork in the establishment of this discipline was one run through with science – not any travel could be considered good geography: for this accolade, some form of scientific study must also be included. This meant that many male travelers were also excluded from Stoddart's account of the discipline. However, Domosh countered this by insisting that his account perpetuated a partial, masculinist account of the disciple by virtue of his privileging of the scientific gaze. As a result of their exclusion from formal education and institutions of higher learning, women could not travel as part of scientific expeditions, and nor had they generally had extensive education in science. The knowledge that they would collect from their journeys would inevitably be different and Domosh and others argued it would necessarily be more partial, qualitative, interpretive and based around discussion with marginalized figures (such as Mary Kingsley's reliance on the knowledge of old women in her *Travels in West Africa* (see Blunt 1994)) rather than chiefs and community representatives who would meet official explorers. While in the past, this more subjective and partial knowledge had been devalued, Domosh (1991) argued that with the qualitative turn in geography, the field knowledge of women travelers was as valuable – if not more so – than that collected through formal scientific explorations.

The question of how we define the field had developed from the wider concerns of feminists from both within and outside of the discipline, which has universally impacted the research process. As Ekinsmyth (2002) argues, the feminist critique of positivism and the notion of the impartial detached researcher have changed the way we think about the practice of research. She outlines what she refers to as the building blocks of a feminist methodology as: "acknowledgement of the partiality of knowledge, a sensitivity to power relations, faith in 'everyday knowledges', openness to diversity of approaches and emancipatory goals for research outcomes" (Ekinsmyth 2002: 178). Feminist researchers are committed to destabilizing power relationships in the research process, and nowhere would this be more important than during the process of fieldwork. Critical to framing a feminist field are questions of subjectivity and reflexivity. Although rather commonsensical ideas in terms of understanding one's position within the field there has been a reduction of these concepts, which unfortunately has at times led to some egotistical diatribes or naval gazing by scholars that tend to focus more on the experiences of the researcher rather than understanding the complex power dynamics that can occur in the field. For example, many feminists work on issues of violence against women, and although working with women who have been the victims of violent attacks will undoubtedly affect us emotionally, the bigger question about "writing violence" is the establishment of power relationships when researchers try to "speak" for those who have experienced the violence of war, domestic violence, or violent crimes. Kobayashi (1994) thoughtfully engages with issues of subjectivity when she questions the dilemma of being both a field researcher and activist, working with communities marginalized by racism and sexism, when she herself had also been marginalized by racism and sexism. England (1994) and Gibert (1994) also point to the paradox of being feminist activists who want to destablize power relationships while recognizing that historic techniques of field research were laden with imbalanced power relationships. As Nast (1994) argues, the simple act of choosing a particular field is a political act. Robben and Nordstrom (1995), taking their cue from Spivak (1994), argue that fieldwork will always be tied to questions of power and politics. To this end, they contend that western researchers must continually engage in the process of auto-critique of privilege. This critique would also include feminist research. As Staeheli and Lawson (1994) warn us, in a rush to critique masculine domination of the discipline, feminists need to be vigilant and not quickly define fieldworks as qualitative in nature. At the same time, they remind us that feminist scholars who work with large data sets to uncover the marginalization of women at larger scales are also practicing a form of feminist fieldwork (Staeheli and Lawson 1994).

Wolf (1996) expands on the politics of fieldwork and contends that feminist methodologies will always be evolving but she also posits that all researchers need to be aware of the power dynamic of fieldwork in three important ways. First, and perhaps most obviously, power differences are always existent as they relate to identities of the researcher and the researched, such social categories as race, class, gender, age, urban versus rural, and so on. Second, is the need to examine the power exerted through the research process, in the form of defining the research relationship, unequal exchange of information, and issues of exploitation. Third, power is often exerted in the "post-site," after the period of fieldwork has concluded, in terms of writing and representation. Wolf (1996: 3) warns us that even though feminists tend to be at the forefront of "experimenting with strategies of co-authoring, poly-vocality, and representation" as a way of challenging power dynamics, academic feminists still maintain control over research projects and knowledge creation. Wolf argues that there is only so much feminists can do in terms of changing the power dynamic inherent in fieldwork when the field is part of the larger institutional structures of academia, which determines what intellectual projects are acceptable and should be rewarded by way of publications, tenure, and promotions. Therefore, it is critical to feminist research not to simply think about the power relationship between the research and the researched but to understand the field as part of a much larger institution in which there will be on-going power struggles.

Similarly, the feminist field is broader than a single site and researcher. As Hesse-Biber (2006) argues, feminist researchers may share some common goals, such as social justice and gender equality; however, not all feminists share the same talents and perspectives. In a similar way, third world feminists have highlighted differences between women (Mohanty 1997) and opened up space for postcolonial critiques of field work. Discomfort at the colonial history of white, often male, western research-ers studying black, non-western, feminized others, has led some to question the right of researchers to do work in societies outside of their own. Responding to Gayatri Spivak's (1994) claims that "the subaltern" cannot speak, some feel that it is an appropriation of voice for outsiders to represent communities. This fear of imposing inappropriate judgements on different voices has led some to suggest relativism, and it seems that the fear of appropriating the voice of others has led some researchers to question their abilities to say anything about communities of which they are not a member (see, for example, England 1994), creating a general anxiety around questions of representation.

However, Radcliffe (1994) argues that this is not a solution to problems of power and (mis)representation. She argues that "disclaiming the right to speak about/with Third World women acts ... to justify an abdication of responsibility with regard to global relations of privilege and authority which are granted, whether we like it or not, to First World women (and men)" (Radcliffe 1994: 28). Spivak similarly regards a withdrawal of western academics from such discussions as too easy an option, and instead insists on a tracing of histories and geographies of connection between researcher and research participants. In New Zealand and the United States, some Indigenous communities have taken the initiative over this issue, requir-ing that potential researchers are questioned about their work and must gain the community's permission before the research can take place.

Bennett argues that the theoretical "hunches" one brings with them to the place where one works "shapes the field" along with one's respondents. Calling on Katz (1994) and Crossley (1996), she argues that the field is constructed from spaces of "betweenness" or "interworlds" and is not simply the fixed space of the other (Bennett, 2002: 141). In a world made up of complex interrelationships and depend-encies, to talk of coherent communities, within which some are members (and therefore somehow able to represent their community) and others are outsiders (and therefore cannot), is simplistic and misleading (see Jones 2000). Moreover, this view of discrete communities is not one that most postcolonial theorists would be willing to adopt when analyzing the identities of groups other than academics and their research participants (Briggs and Sharp 2004). The whole nature of "the field" as a coherent place to be examined and captured in print is challenged by the entangled nature of our globalized world. The traditional anthropological idea of spending extended periods in one place is rendered problematic when we acknowledge that the boundaries of place are porous or are extroverted (Massey 1991, 2009). This has led feminist geographers and other to examine the necessity for multiple sites for fieldwork to emphasize connection through global process:

> By displacing the field and addressing the issue in rural Sudan and East Harlem, New York – settings that on the surface appear to have little in common – I am able to tell a story not of marginalization alone where "those poor people" might be the key narrative theme, but of the systemic predations of global economic restructuring. (Katz 1994: 68)

Katz's work goes beyond notions of multi-site ethnographies and reflects a shift whereby ethnographic sites are viewed as fluid, crossing scales from the local to the global, thereby contesting the notion of a bounded study site. Ethnographers, such as Katz, Wright, and Pratt to name a few are redefining place as it relates to broader definitions of community. Consequently, some ethnographers are examining the notion of "living the global," whereby "ethnographers can reveal how global processes are collectively and politically constructed, demonstrating the variety of ways in which globalization is grounded in the local" (Gille and Ó Riain 2002: 271). The notion of "living the global" is important in understanding how the exploitation of women, such as factory workers in Juarez, Mexico (see Wright 2006) and Filipina domestic servants in Vancouver, Canada (see Pratt 2004) are part of a transnational phenomenon which focuses on the exploitation of women on a world scale in the context of global capitalism.

This emphasis on connection rather than separation also allows feminist researchers to address postcolonial challenges to the problem of representation. In addition, many geographers, including feminists, have sought to challenge the notion of distance between researcher and researched through collaborative research. By recognizing the knowledges held by community members about their own situation, and incorporating this into the research design, methodology, and outputs of research, research participants are involved at every stage rather than being othered by the process.

Moreover, while the field presents the potential for misrepresentation and the inappropriate performance of colonizing power relations, it also presents at the same time the possibility of meeting and opening a true dialogue with those so often marginalized and silenced by dominant discourses and representations (Dowler and Sharp 2001; Hyndman 2001; Sundberg 2003). Despite the feminist critique of fieldwork for its masculinist tradition, the field is an ambivalent site for feminist work within which relations of collaboration and the dependency of the researcher on her/ his research participants does not allow for any simplistic exercise of power. Vincent Del Casino (2001) thus situates the fluidity of relationships in the field when he describes how relationships with respondents are temporal and how, at times, a relationship can be central to the project and at other times, distant. In this way, relationships change as the project matures. Clearly, power relations are inescapable but there are times when this operates beyond the control of the researcher – it is not always the researcher who is in a powerful position, driving forward the development of the research. One can feel entirely powerless and dependent upon others and their knowledge of how to challenge local power relations as well as the very different environments, languages, and customs of a particular place. Both authors of this chapter have been in field situations where they have felt very vulnerable and entirely dependent upon the local knowledge of the research participants, whether when negotiating the desert (see Sharp et al. 2003) or a war zone (see Dowler 2001a, 2001b). Chacko (2004) recognizes that despite her connections with the women she worked with in fieldwork in India, in some ways she was "the other" to her respondents, and it was she who was out of place and in a relatively powerless position, particularly so when she was invited into the homes of her respondents. The relationship – and the lines of power and dependency – changes over the course of any field experience, as researchers become familiar with their field sites and their research participants become familiar with them. Friendships are sometimes formed. Hapke and Ayyankeril (2001) discuss the multiple negotiations they had to make to address their respondents' expectations of a white female researcher working alongside a male Indian research assistant, and how these negotiations changed once they married.

Performing the Field

> It is also important to note that the field sites themselves shape our research outcomes as much as our own research *graphs* them. Materialities come into play in the field that proffer their own delimiting agency upon us. Here we are thinking not only about the situated and embodied nature of fieldwork … but also of the ways in which the materiality of the field site itself affects our actions. (Dewsbury and Naylor 2002: 256)

Feminist geographers have long challenged the masculinism of the field. But understanding the *representations* of "the field" is not enough; there is also the need to see the field as a set of *practices*. As Driver (2000: 267) explains, "Where field-work has been the subject of debate amongst historians of geography, it has too often been considered as an unproblematic expression of ruling ideologies or institutional projects." Instead, he continues, we need to consider the much more complex sets of experiences, practices, and emotions that make up what "the field" is to geographers. Following this focus on the creative effects of practice, some have considered the ways in which field practices have constituted the discipline and disciplinary identity in different places (e.g., Powell's (2008) work on Arctic fieldwork and Lorimer's (2003) consideration of the "small stories" of geography made in student fieldwork and other "marginal" field sites), and how serendipities of fieldwork location can lead to changed ways of thinking (e.g., Briggs and Sharp 2009 also see Chapter 8, this volume).

It is the network of bodies and things, actors and actants, human and more-than-human, "who come into contact with researchers in the field" (Dewsbury and Naylor 2002: 257) and make knowledge. And, yet, these knowledges are often written out of the final product of the fieldwork. The complex, emotional, and ambiguous experience of the field is fixed into a singular text or report. Most clearly we can see how the great industry of exploration in geography's early days was hidden in narratives of individual heroes. However, even now, the apparatus and networks of entry and support that facilitate fieldwork, and the acts, emotions, and practices that go into the doing of it, are still generally silenced in our accounts of research that are circulated through academic publication. Recent feminist work, which privileges the body and the geographies of emotions (see Bondi et al. 2002), help to bring to the fore a whole array of experiences and emotions, traditionally silenced in our field accounts, which has encouraged critical reflection over the process of research methodologies. The significance of the embodied challenges of the field (often physically or emotionally overwhelming at the time) is often silenced in the written and presented accounts of methodology, which tend to focus on the formal aspects of project design and implementation, and yet everyone who has spent time in the field has felt emotions linked to such things as displacement and disorientation from being away from home, frustrations at the failure of meticu-lously planned research methodologies to work out in practice, loneliness and joy. Mountz et al. (2003) highlight the fact that whereas there is much guidance on the *technicalities* of fieldwork and research methodologies, researchers have much less help on the everyday performance of fieldwork – the embodied realities of *being* in the field and *doing* work there – and call for greater support for fieldworkers often isolated in the field (Monk et al. 2003; but see Scheyvens and Storey 2003).

Recognizing the fact that, for example, the "field and the fieldworker are co-constructions, and the knowledge produced between them reflects the materiality and mutability of this relationship" (Katz 2009: 252), highlights the inescapable embodiment of the field researcher. It is impossible to ignore the embodied experi-ence of the field, especially where the spaces of fieldwork are distanced and different from the spaces of home and everyday experience. Sharp (2005: 3) noted that anyone "who has spent time in the field will have stories of the challenges of day-to-day work, whether the difficulties of accessing respondents or simply boredom at the new routine." Mandel (2003) reflects that the field can be a disorientating and exhausting place. She illustrates how fatigue from a long period in the field led her to react badly to a challenging situation causing her to have a "lost day" of work. "Often," she says (Mandel 2003: 12), "I think, when we are most tired we rely on what is most comfortable to us, making it difficult to accommodate others' views on how things should be done." She emphasizes the need for greater attention to the personal costs of fieldwork and the need to ensure there is space for the "care of the self" in addition to paying close attention to the ways in which research subjects are treated.

Valentine (2002: 124) suggests that one aspect missing from this openness to the research process, however, is recognition of desire and lust in the field. Perhaps because of the colonial associations of desire and the actualities of sexual relations between male travelers and female natives played out over again in the archives of fieldwork (and now repeated in tourism), feminists have been unwilling to consider this emotional and embodied element of the field experience. One exception is Cupples' (2002) remarkably frank discussion of her sexual engagement with and in the field. She argues that the failure to acknowledge this process is demonstration of "the ongoing power of the myth of the researcher as detached and objective" (Cupples 2002: 382).

While the majority of recent work has also focused upon the importance of truthful, open encounters between fieldworker and others in the field, some have reflected upon the necessity for dishonesty in order to get at real processes. Some are keen not to offend hosts with different values and so, for example, female researchers don appropriate dress for research in Islamic societies and hide views which they know will cause offence to the populations being researched. Parr (2001) highlights the embodied effects of fieldwork on herself as she adjusted her behavior, presentation of self, and even levels of intoxication, in order to fit in and put her respondents at ease.

Bennett (2002) contends that the reason there is nothing natural about the field is that there is nothing natural about our behaviors when we enter the field and just as geographers are attuned to the multiplicity of place, so should we also be aware of the complexities of our own characters. She contends that even when conducting participant observation, "people behave differently in different settings according to where they are, who they are with and their agenda" (Bennett 2002: 143). She likens researchers to actors who are capable of a number of different performances depending on with whom they are interacting, and points to how fieldwork can be physically challenging and researchers often call upon their bodies in unfamiliar ways when encountering new experiences. She refers to other research- ers embodied field experiences such as Parr adopting a more authoritarian cadence in certain situations and Coffey's appearing more self-assured by adopting business attire when interviewing professionals (Bennett 2002). Herbert (2001: 309) assures us that although informants might act in unnatural ways owing to the observer's presence, the researcher will also fall sway to the behavioral and cognitive patterns of the field. However, Dowler (2001b: 161) has questioned whether this is fair or appropriate:

> While in West Belfast, I represented myself in a way that would minimize my status as an outsider. For example, even though I am a staunch advocate of a woman's privilege to choose, I signed anti-abortion petitions. When I was asked how I felt about the issue of divorce, I was evasive. I promoted my Roman Catholic upbringings, even though I do not consider myself as a practising Catholic. I attended weekly services at the local Catholic church and made sure I was seen at them. In hindsight, these actions did make my respondents feel more comfortable with me. However, I think it is important to ask ourselves whether, if we adopt participant observation in order to promote an open relationship with our respondents, the respondents should not have the same benefit of a candid relationship with the researcher?

For others, the issue of presentation of self is entirely – and consciously – a political one based around an ethical evaluation of power relations. While being "himself" with villagers in Goa, Routledge (2002) adopted the persona "Walter Kurtz" – a smartly dressed, UK-based tour agent – when speaking with hotel owners so as to get a more honest assessment of the environmental impact of tourism in the area than he would have got as himself. Routledge's decision to reveal his true identity, or to project an image more acceptable to his interviewees, was a political one linked to his belief in the value of the outcomes of his research.

Familiar Fields

We cannot assume, however, that being in a familiar location or situation will necessarily remove the complications outlined above. Browne (2003) has highlighted the multiple power relations running through research with friends. Here, rather than having to deal with constructions of otherness, Browne has had to negotiate apparent relationships of sameness; instead of having to establish relations of trust she had to negotiate the challenge of attempting to carve out a position within already-existing trust from which to enact her research critically (see also Avis 2002; Valentine 2002). Illustrative of this type of re-positioning is Karen Till's (2001) discussion of the field as a number of research territories that cross scale, which leads to an identity formation whereby the researcher cannot separate themselves into their research and personal selves. In this way Till (2001) contends that, in this respect, when someone enters the field they are going home. Stan Stevens (2001) points to the difficulties of designating a space as the field when as in his case he has spent a quarter of his life in the Nepal Himalayas. For Stevens (2001: 66) the field is "a way of life." In a similar way, Dydia Delyser (2001) questions the boundaries of the field when studying as an insider in a community. The insider/outsider status is a slippery slope and although there has been a plethora of writing on recognizing one's status as an outsider there has been relatively little written on the researcher as insider. As Delyser argues some researchers start out as insiders and conduct research "close to home, or close to our hearts – topics so compelling we can't leave them alone – and we try to find way to use our 'insider' status to help, not hinder, insights" (Delyser 2001: 441). Delyser, a long time staff member of a Northern California ghost town, was attracted to how both the visitors and park staff drew meaning from this historic landscape. As a result of her insider status she had to develop ways of inhabiting many communities simultaneously which was far more complex than traditional views of the insider/ outsider binary assumed as part of the process of participant observation. Most importantly, the questions these scholars raise, poses a challenge to existing notions of detachment from the particular communities and places they live and work.

Conclusions

The trajectory of geography's relationship with the field could be characterized as one of increasing complexity. From the discipline's starting point where scholars produced detailed tomes on their chosen place of study, we have moved to a position today where the field is regarded as multiple and porous, we can regard places as unique but only through unique sets of connections rather than because of immanent characterization. For some, the field is the product of field practices rather than being a pre-existing place.

Our methodologies for fieldwork have reflected this, taking in interpretative and collaborative methods, tracing connections as well as differences. New sites for fieldwork are emerging, most notably the virtual spaces of online communities, from imagined landscapes of fantasy massive multiplayer online games to the safe spaces of internet chatrooms for those othered and excluded by dominant society. At the same time, conventional fieldwork is facing challenges from environmental concerns, raising questions about the importance of long haul flights in the context of global warming. Calls for carbon neutral fieldwork will require new imaginative and collaborative methodologies. Thus, in conclusion, the field site is considered more than a fixed place and is the constituted by global economies, new technologies and broader ecologies. When considering the field site as multidimensional, inclusive of many different experiences, it is important to consider the field researcher also transcending fixed notions of identity embodying multiple standings within the field.

Note

1   And, at the time of writing (summer 2009) the role of the RGS in large-scale expeditions of discovery (rather than the support of a wider range of geographical scholarship and activity) was once again subject of heated debate at the society.

References

Avis, H. (2002) Whose voice is that? Making space for subjectivities in interviews. In Bondi et al. (eds.), *Subjectivities, Knowledges, and Feminist Geographies: The Subjects and Ethics of Social Research*. Lanham: Rowman and Littlefield, pp. 191–207.

Bennett, K. (2002) Participant observation. In P. Shurmer-Smith (ed.), *Doing Cultural Geography*. Thousand Oaks, London, and New Delhi: Sage, pp. 139–44.

Blunt, A. (1994) *Travel, Gender and Imperialism: Mary Kingsley and West Africa*. New York: Guilford.

Bondi, L. (2003) Empathy and identification: conceptual resources for feminist fieldwork. *ACME: An International E-journal for Critical Geographies* 2 (1): 64–76.

Bondi, L., Avis, H., Bankey, R., Bingley, A., Davidson, J., Duffy, R., Einagel, V.I., Green, A.-M., Johnston, L., Lilley, S., Listerborn, C., Marshy, M., McEwan, S., O'Connor, N., Rose, G., Vivat, B., and Wood, N. (2002) *Subjectivities, Knowledges, and Feminist Geographies: The Subjects and Ethics of Social Research*. Lanham: Rowman and Littlefield.

Briggs J. and Sharp J. (2004) Indigenous knowledges and development: a postcolonial caution. *Third World Quarterly* 25 (4): 661–76.

Briggs, J. and Sharp, J. (2009) One hundred years of researching Egypt: from rule of experts to Bedouin voices? *Scottish Geographical Journal* 3–4: 256–72.

Browne, K. (2003) Negotiations and fieldworkings: friendship and feminist research. *ACME: An International E-journal for Critical Geographies* 2 (1): 132–46.

Bung, W. (1969) The first years of the Detroit Geographical Expedition: a personal report published by Detroit, Society for Human Exploration. LCCN: 72180053. Dewey:910/7/11 LC:G74.

Chacko, E. (2004) Positionality and praxis: fieldwork experiences in rural India. *Singapore Journal of Tropical Geography* 25 (1): 51–63.

Crossley, N. (1996) *Intersubjectivity: The Fabric of Social Becoming*. Thousand Oaks, London, and New Delhi: Sage.

Cupples, J. (2002) The field as a landscape of desire: sex and sexuality in geographical fieldwork. *Area* 34: 382–90.

Del Casino, V. (2001). Decision-making in an ethnographic context. *Geographical Review special issue on Doing Fieldwork* 91 (1–2): 454–62.

Delyser, D. (2001) "Do you really like it here?" Thoughts on insider research. *Geographical Review special issue on Doing Fieldwork* 91 (1–2): 158–67.

Delyser, D., Herbert, S., Aitken, S., and Crang, M. (2009) *Sage Handbook of Qualitative Geography*. Thousand Oaks. London, New Delhi: Sage.

Dewsbury, J. and Naylor, S. (2002) Practising geographical knowledge: fields, bodies and dissemination. *Area* 34 (3): 253–60.

Domosh, M. (1991) Towards a feminist historiography of geography. *Transactions, Institute of British Geographers* 16: 95–104.

Dowler, L. (2001a) The four square laundry: participant observation in a war zone. *Geographical Review* special issue on *Doing Fieldwork* 91 (1–2): 414–22.

Dowler, L. (2001b) Fieldwork in the trenches: participant observation in a conflict area. In C. Dwyer and M. Limb (eds.), *Qualitative Methods for Geographers*. London: Arnold, pp. 153–65.

Dowler, L. and Sharp, J. (2001) A feminist geopolitics? *Space and Polity* 5 (3): 165–76.

Driver, F. (2000) Editorial: field-work in geography. *Transactions, Institute of British Geographers* 25: 267–8.

Driver, F. (2001). *Geography Militant: Cultures of Exploration and Empire*. Oxford: Blackwell.

Ekinsmyth, C. (2002) Feminist cultural geography. In P. Shurmer-Smith (ed.), *Doing Cultural Geography*. Thousand Oaks, London, New Delhi: Sage pp. 53–66.

England, K. (1994) Getting personal: reflexivity, positionality, and feminist research. *Professional Geographer* 46 (1): 80–9.

Gibert, M. (1994) The politics of location: doing feminist research at "home." *Professional Geographer* 46 (1): 90–6.

Gille, Z. and O'Rian, S. (2002) Global ethnography. *Annual Review of Sociology* 28: 271–95.

Godlewska, A. (1994) Napoleon's geographers (1797–1815): imperialists and soldiers of modernity. In A. Godlewska and N. Smith (ed.), *Geography and Empire*. Oxford: Blackwell.

Hapke, H. and Ayyankeril, D. (2001) Of "loose" women and "guides," or, relationships in the field. *Geographical Review* special issue on *Doing Fieldwork* 91 (1–2): 342–52.

Hay, I. (2005) *Qualitative Methods in Human Geography*. Oxford and New York: Oxford University Press.

Herbert, S. (2001) From spy to ok guy: trust and validity in fieldwork with police. *Geographical Review* special issue on *Doing Fieldwork* 91 (1–2): 304–10.

Hesse-Biber, S. (2006) Feminist research, exploring the interconnections of epistemology, methodology and method. In Hesse-Biber *Handbook of Feminist Research: Theory and Praxis*. Thousand Oaks, London, New Delhi: Sage, pp 1–28.

Hyndman, J. (2001) The field as here and now, not there and then. *Geographical Review* special issue on *Doing Fieldwork* 91 (1–2): 262–72.

Jones, P. (2000) Why is it alright to do development "over there" but not "here"? Changing vocabularies and common strategies of inclusion across the "first" and "third" worlds. *Area* 32 (2): 237–41.

Katz, C. (1994) Playing the field: questions of fieldwork in geography. *Professional Geographer* 46 (1): 67–72.

Katz, C. (2009) Fieldwork. In D. Gregory, R. Johnston, G. Pratt, M. Watts, and S. Whatmore (eds.), *The Dictionary of Human Geography*, 5th edn. Oxford: Wiley-Blackwell.

Kobayashi, A. (1994) Coloring the field: gender, "race," and the politics of fieldwork. *Professional Geographer* 46 (1): 73–80

Limb, M. and Dwyer, C. (2001) *Qualitative Research Methods for Geographers*. London: Arnold.

Lorimer, H. (2003) Telling small stories: spaces of knowledge and the practice of geography. *Transactions, Institute of British Geographers* 28 (2): 197–217.

Mackinder, H. (1904) The geographical pivot of history. *Geographical Journal* 23: 421–37.

Mandel, J. (2003) Negotiating expectations in the field: gatekeepers, research fatigue and cultural biases. *Singapore Journal of Tropical Geography* 24 (2): 198–210.

Massey, D. (1991) A global sense of place. *Marxism Today* June: 24–9.

Massey, D. (2007) *World City*. Cambridge: Polity.

Mohanty, C.T. (1997) Feminist encounters: locating the politics of experience. In L. McDowell and J. Sharp (eds.), *Space, Gender, Knowledge: Feminist Readings*. London: Arnold.

Monk, J., Manning, P., and Denman, C. (2003) Working together: feminist perspectives on collaborative research and action. *ACME: An International E-journal for Critical Geographies* 2 (1): 91–106.

Mountz, A., Miyares, I., Wright, R., and Bailey, A. (2003) Methodologically becoming power, knowledge and team research. *Gender, Place and Culture* 10 (1): 29–46.

Nast, H. (1994) Women in the field: critical feminist methodologies and theoretical perspectives: opening remarks on "women in the field." *Professional Geographer* 46 (1): 54–66.

Parr, H. (2001) Feeling, reading, and making bodies in space. *Geographical Review special issue on Doing Fieldwork* 91 (1–2): 158–67.

Peet, R. 1998. *Modern Geographical Thought*. Oxford: Blackwell.

Philips, R. (1997) *Mapping Men and Empire*. London: Routledge.

Powell, R. (2002) The siren's voices? Field practices and dialogue in geography. *Area* 34 (3): 261–72.

Powell, R. (2008) Becoming a geographical scientist: oral histories of Arctic fieldwork. *Transactions, Institute of British Geographers* 33 (4): 548–65.

Pratt, G. (2004) *Working Feminism*. Philadelphia: Temple University Press.

Professional Geographer (1994) 46 (1): 80–9.

Radcliffe, S. (1994) (Representing) colonial women: authority, difference and feminism. *Area* 26 (1): 25–32.

Robben, A. and Nordstrom, C. (1995) The anthropology and ethnography of violence and sociopolitical conflict. In A. Robben and C. Nordstrom (eds.), *Fieldwork Under Fire, Contemporary Studies of Violence and Survival*. Berkeley and Los Angeles: California University Press, pp. 1–23

Rose, G. (1993) *Feminism and Geography*. Thousand Oaks, London, and New Delhi: Sage.

Rose, G. (2001) *Visual Methodologies*. Thousand Oaks, London, and New Delhi: Sage.

Routledge, P. (2002) Travelling east as Walter Kurtz: identity, performance and collaboration in Goa, India. *Environment and Planning D: Society and Space* 20: 477–98.

Ryan, S. (1996) *The Cartographic Eye: How Explorers Saw Australia*. Cambridge: Cambridge University Press.

Sauer, C. (1956) The education of a geographer. *Annals of the Association of American Geographers* 46 (3): 287–99.

Scheyvens, R. and Storey, D. (2003) *Development Fieldwork: A Practical Guide*. Thousand Oaks, London, and New Delhi: Sage.

Sharp J. (2005) Geography and gender progress report 1: feminist methodologies in collaboration and in the field. *Progress in Human Geography* 29 (3): 304–9.

Sharp, J. (2008) *Geographies of Postcolonialism: Spaces of Power and Representation*. Thousand Oaks, London, and New Delhi: Sage.

Sharp, J., Briggs, J., Hamed, N., and Yacoub, H. (2003) Doing gender and development: understanding empowerment and local gender relations. *Transaction, Institute of British Geographers* 28 (3): 281–95.

Shurmer-Smith, P. (2002) *Doing Cultural Geography*. Thousand Oaks, London, and New Delhi: Sage.

Spivak, G.C. (1994) Can the subaltern speak? In P. Williams and I. Chrisman (eds.), *Colonial Discourse and Post-colonial Theory*. New York: Columbia, pp. 66–111.

Staeheli, L. and Lawson, V. A. (1994) Discussion of "women in the field": the politics of feminist fieldwork. *Professional Geographer* 46 (1): 96–102.

Stanley, H.M. (1878) *Through the Dark Continent*. London.

Stevens, S. (2001) Fieldwork as commitment. *Geographical Review* special issue on *Doing Fieldwork* 91 (1–2): 66–73.

Stoddart, D. (1986) *On Geography*. Oxford: Blackwell.

Stoddart, D. (1991). Do we need a feminist historiography of geography: and if we do, what should it be like? *Transactions, Institute of British Geographers* 16: 484–7.

Sundberg, J. (2003) Masculinist epistemologies and the politics of fieldwork in Latin Americanist geography. *Professional Geographer* 55 (3): 181–91.

Till, K. (2001) Returning home and to the field. *Geographical Review* special issue on *Doing Fieldwork* 91 (1–2): 158–67.

Valentine, G. (2002) People like us: negotiating sameness and difference in the research process. In P. Moss (ed.), *Feminist Geography in Practice*. Oxford: Blackwell, pp. 116–26.

Wolf, D. (1996) Feminist dilemmas in fieldwork. In D. Wolf, *Situating Feminist Dilemmas in Fieldwork*. Jackson: Westview, pp. 1–55.

Wright, M. (2006) *Disposable Women and Other Myths of Global Capitalism*. New York and London: Routledge.

Chapter 10

On the Ground

Bettina van Hoven and Louise Meijering

Thinking Through the Production of Knowledge

Whilst we do not wish to discredit ways of conducting research in the past, there is no denying that thinking about the research process in social geography has changed … a lot! In this chapter we aim to explore some of the developments and issues related to doing research by focusing particularly on "the ground," i.e. the interaction with people, institutional contexts and elements in and through various spatial practices/processes. To do this, the chapter begins with a sketch of our own institutional contexts, exploring how knowledge production in The Netherlands has been perceived and valued over time and illustrating the impact of these roots on current research (and teaching). In so doing, we also clarify our own positionality, at least to some degree, in approaching the themes throughout this chapter. In the second section, titled "situating the politics of knowledge construction," we consider the role of the body, social contexts (e.g., food and language), emotion and performance in the interaction between researcher and respondent. We draw attention to the fact that interviewing is, by no means, simply "extracting information" from another person but that knowledge generated in this context relies on more than questions and answers. In that context, we briefly address the ethics of disseminating knowledge. Whilst much of the social research literature has drawn on interviews as main method of data collection, we also examine the impact of "newer" methods of data generation, such as walks and video, to draw attention to the different types of knowledge that different methods may generate. We illustrate that such performance-based methods produce different power relations, especially where respondents generate data on their own terms. The final section of the chapter focuses more on other "elements" that impact the production of knowledge,

A Companion to Social Geography, First Edition. Edited by
Vincent J. Del Casino Jr., Mary E. Thomas, Paul Cloke, and Ruth Panelli.
© 2011 Blackwell Publishing Ltd. Published 2011 by Blackwell Publishing Ltd.

in particular those in our natural environment. We revisit a physical geographical study on ecoforestry on Vancouver Island to discuss the possible roles of non-human elements that were previously ignored. In so doing, we draw on actor-network and to a lesser extent on non-representational theory to illustrate, for example, how agency can be viewed differently thus assigning more active roles not only to humans but also to non-humans, i.e. trees in our example. The discussion encourages readers to think in new ways about the roles of different kinds of participants in doing geography.

Institutional Contexts of Knowledge Production

Researchers cannot be seen as detached from the institutional and national contexts in which they have been educated and are employed. In this section we turn to our own context to provide a background to what facilitates, restricts, and motivates our research practice and politics. Firstly, we are writing this chapter as we are embedded, at least in part, in the Dutch Geography tradition.[1] Secondly, we have worked in Geography and have grown to "become geographers" in the Department of Cultural Geography at the University of Groningen (Netherlands). Thirdly, we have our own personal backgrounds (and funding sources) that have helped shape the kinds of geographies we do and geographers we are. All of these issues have been noted elsewhere, too, as influencing geographical projects (see, for example, Cloke et al. 2004) and they are worth noting in the context of a discussion on people, elements, and relations involved in the production of knowledge.

Since we are most likely to address an audience that "grew up" in an Anglo-American context, a few remarks on the way in which geography has been practiced elsewhere, and in this case the Netherlands, are worth making. Louise (MSc social/regional geography, PhD cultural geography) received her entire training in the Netherlands, Bettina diverted through Germany (to study biology) and England (to obtain a BSc in physical geography and a PhD in social-feminist geography) before coming to a Dutch Geography Department as a member of staff. Both of our research has been affected by how geography and the production of geographic knowledge has been valued and perceived in the Netherlands.

Like geography's history in the United Kingdom (see e.g., Cloke et al. 2004), the advent of Dutch geography is related to Dutch colonialism (until the independence of the Dutch Indies in 1949), with the aims to explore, map, and categorize unknown territories in order to exploit or claim its human and non-human resources. Geography was more akin to the natural than the social sciences, it utilized the same approaches to ordering the world and was judged by the same parameters. Since the 1930s, the role of geography and geographers remained prominent in the Netherlands as they played an important part in large-scale planning projects in the struggle against water and development of new land (van der Vaart et al. 2004). When creating new land, such as the *Noordoostpolder*, planners made a point of including people that they believed were representative of Dutch society, incorporating culture as a variable in planning. This manifested, for example, in establishing three churches and three schools in the new villages to account for the different, dominant religions in the Netherlands.

Musterd and de Pater (2003) similarly emphasized that geography has long been an applied and practical science, strongly imprinted by spatial planning and regional-economic policy, while largely undervaluing the significance of social and cultural issues. With the notable exception of gender, Dutch human geography has begun to include issues of power and identity only relatively recently.[2] Gender was a topic in the early 1980s when courses in "women studies" and later a network for gender studies were set up based in Amsterdam. But in spite of opportunities to make a profound contribution to feminist geography (see also Peake 1989), these were marginalized due to institutional constraints. Both initiatives ceased to exist as staff involved was often on part-time and/or temporary contracts and financial support became problematic (see van Hoven et al. 2010). The most promising avenue to recognizing the significance of personal experience and difference to the advancement of geographic theories (see also McDowell 1992) was threatened to become a cul-de-sac. In spite of "a recognition of alternative knowledges," overall in geography, "there is still a tendency towards what could be seen as 'masculinist' knowledges/ways of knowing" (Widdowfield 2000: 199) in Dutch geography.

Looking at our institutional context more locally, the Department of Cultural Geography in Groningen has been interested, predominantly, in "Making Places," which implies an interest in relations between people and places and the role of difference in establishing such relations. Although several different "entry points" are used, i.e. quantitative and qualitative, most members of the Department became geographers in a more positivist milieu, they came from a more quantitative background and, as Musterd and de Pater (2003: 555) noted for Dutch geography in general, "refus[e] to get carried away with new [UK] trends." This leads to continuous discussions about what and how geography should be done, and when this geography is valid (or even if it is actually geography). For instance, in our faculty (comprised of cultural and economic geography, planning and population studies) many Bachelor and Master thesis supervisors instruct students to perceive of their role as researcher as being an "objective instrument." This translates into resistance (by supervisors) to the use of the first person in reporting, as "it is not relevant who the researcher is."[3] The implication is that the institution remains somewhat ignorant of the impact of interaction between the researcher and researched as embodied persons, as well as the recognition that *unique contexts* are created in which equally *unique knowledge* is constructed.[4]

As Droogleever-Fortuijn (2004) further avers, the lack of diversity (and experimentation in thought and practice) in Dutch geography is tied to the broader national context, which includes that fact that: (1) many Dutch geography graduates find employment in the civil service or as policy consultant; (2) geographers are often involved directly in policy making through their role as advisory board members and media experts; and (3) much research is *government funded*. Working in a Geography Department in the Netherlands, this means that the choice of topics reflects these constraints. Faculty draw on their applied expertise and students often want to be equipped for the aforementioned labor market.

The implications of the above reflections for writing this chapter are that we are more experienced with some ways of thinking about and collecting data than others and perhaps more positive about certain approaches compared with others. This is not to say that we feel our work falls into the "traditional" Dutch way of doing geography but rather that we have some reservations toward some of the literature emerging from UK geography. For example, although we find current theoretical innovations in social geography intriguing as it "positively challenged us to recast aspects of our own research" (Laurier and Philo 2006: 354), we do wonder whether it may be "more tightly bound to becoming philosophy?" (p. 359). In addition, we concur with Staddon (2009: 163) who notes that "after all, the point of critical analysis must be ... not merely to understand the world but to change it."

Situating the Politics of Knowledge Construction

When considering knowledge as situated, the construction of knowledge is always "political." Which knowledge is constructed is a process of negotiation between researcher, researched, and institutional and societal contexts, such as universities (Purcell 2007), funding agencies (Cloke et al. 2004), journal editors (Kitchin and Fuller 2003), policy makers (Lairumbi et al. 2008), and the general public (Garvin 2001). In the context of this chapter, we focus on the researcher-respondent relation, as that provides the most tangible illustration of how knowledge is produced on the ground. We discuss the (overlapping) identities of the researcher: the impact of his/ her body, social identities, and non-visual and unspoken elements.

Bodies play an important role in generating knowledge. In an interview situation, the ways in which bodies are read (both by researcher and respondent) can impact rapport as well as (the nature and extent of) topics discussed. Such bodily markers, e.g., gender, race, age, affiliation with cultural subgroups (through clothing, tattoos, piercings etc), or economic status, can work to the benefit of an interview situation or not (see also DeVerteuil 2004). For example, in a study on the experiences of (male) Indian information technology (IT) professionals in Germany (Meijering and van Hoven 2003), Louise found that her respondents had preconceived notions of her bodily markers based on her ascribed status as academic/researcher. One of the respondents noted he had expected to be interviewed by a middle-aged, male, gray-haired professor instead of by a woman his own age (early twenties). Finding more commonalities than expected turned out to be helpful in establishing rapport.[5]

As an embodied and cultural experience, the preparation of food can also play a role in the production of knowledge. The (male) Indian IT professionals in Louise's study enjoyed to cook and eat Indian dishes together, even though preparing food is traditionally a female task. The activities that took place around the food gave everyone involved a sense of being part of India, and allowed them to share a sense of home (see also Law 2001). Some respondents let Louise also experience this "sense of India" through letting her try Indian food and drink, such as *samosas* and *lhassi*. Such invitations to experience the Indian culture of the respondents occurred after an interview had been conducted. This experience evolved from the rapport established between Louise and the IT professionals in her study. Taking part in these social events thus helped Louise to make more sense of the data generated through the previous interviews.[6]

Social identities, too, can impact the researcher–respondent relation; they can serve as boundaries or bridges. Since these identities are not always visible and people subscribe to many different positions (which can cause a researcher to be part insider and part outsider to the group s/he studies), boundaries or bridges between people are continuously renegotiated. For respondents, it can be important to explore the researcher's social identities before agreeing to participate in a study; for many respondents finding some kind of common ground leads to more trust. In a study by Louise on intentional communities in Western Europe, such as a Hare Krishna group, a hippy commune, a vegetarian collective, and a cohousing group (Meijering, 2006), many respondents liked to establish the extent to which Louise identified with their communal ideals before engaging in an interview. They asked, for instance, whether she lived in a community herself, whether she was a vegetarian, and what religious beliefs she had. It seemed that respondents felt less inhibited to talk about their personal experiences and ideas when these coincided with the researcher's. At the same time, they tended to explain everyday life in more detail when Louise's experiences did seem more distant from their own community life.[7]

In interactions such as the above, the researcher may start to feel a sense of "responsibility" for the group her respondents represent, evoking ethical questions with regards to disseminating knowledge.[8] In a personal conversation with Louise and Bettina, David Sibley commented on his ethical doubts about the power of academic knowledge, which had developed through his work with gypsies from the early 1970s onwards in a paper called "Problematizing exclusion" (1998). In the interview, he discussed how his ideas of the purpose of knowledge production changed:

> [The gypsies] were kind of ... on the borderline in terms of their relationship with the law ... I knew what was going on, right? But I thought, well, OK, these people have invited me in. There's a trust between us. So the worry that made me stop writing about gypsies ... was that I was becoming increasingly uneasy about the possible uses that would be made of things I wrote, to work against their interests ... And I have to say that initially, I wasn't really thinking about this too much, I felt that OK it must be good to write, because you're providing knowledge, which is good, because there was a lot of myth and misunderstanding ... But then I began to think on reflection that knowledge is not always good. Not because it's untrue necessarily, but because of the ways in which it might be applied against the interests of the group that you're supposed to be committed to. (Sibley 2009)

As a result of his reflections, David stopped writing about gypsies: when telling the "real story" the potential harm for the gypsies as a group outweighed the advantages of uplifting some of the prevailing prejudices.

Returning to levels of difference and similarities, language is another relevant practice. In Louise's intentional communities study, interviews were held in Dutch, English, and German. In some cases, either Louise or the respondent could not express him/herself in his/her native tongue. This affected the interaction during the interview since questions needed to be rephrased and, more importantly, responses were simplified, losing some of their depth and richness. In addition, when disseminating the results, respondents' views had to be translated into English. The translation process itself is highly political; interpretations may be changed, lost, or misunderstood when translating interviews from one language into the other (Müller 2007). Not only in translating, but also in recording, transcribing and quoting, the politics of doing interviews becomes apparent. Through analyzing the non-verbal in (telephone) interviews, Cook (2009: 178) analyzed "whether and how the textual orders mobilized in the achievements of qualitative interviews attune to non-verbal realism." She concluded that interviews produce not only texts, but also conversationalists, and that thus power imbalances between researcher and respondent may be overcome.

Other invisible and unspoken elements that play a role in knowledge construction are emotions. Considering the role of emotions in academic research has become part of a reflexive approach in which the subjective nature of doing research is recognized (see Widdowfield 2000; Bennett 2004, 2009. But see also Gaskell 2008 for experiences in an unsupportive academic environment).[9] In particular in Bettina's prison research, emotions played a role throughout the research process, i.e. from generating ideas about conducting a study, through data collection and data analysis/ presentation (i.e. at international conferences). From 2002 to 2004, she kept a research diary, which was published in her faculty's student magazine *Girugten*, and emotions were often expressed in this "informal" setting.[10] In an entry on October 31, 2002, for example she recalls a letter by prisoner Richard (on Death Row) who describes how he tries to feel something. He had begun to withhold part of his lunch tray in order to make prison authorities send in a riot team who would, upon further refusal to comply, physically harm him. Richard wrote that this experience made him aware that he *can* still feel, that he is still alive: "I thrived on this which is scary. I loved the stress relief, the actual physical contact with another human being, negative is as good as positive at this point." In her notes, Bettina wonders how she should feel about this. She speculates whether all of this was true, whether Richard was describing his emotion or tried to portray himself as tough and manly, whether she should feel sorry for him or whether, as a convicted criminal, he was "deserving"[11] (van Hoven 2003a). In the Dutch case study, too, there are several accounts of the role of emotion in the interaction with the correctional officers, the prisoners and the materiality of the prison. The most memorable event she describes is: "When I arrived at D-block today, two correctional officers met me, took me to the team room and closed the door. "Did you interview S. ?," they ask. "He's wasted." I don't understand, wasted- he's depressed or sick? No, they mean he's dead, he was shot shortly after his release, liquidated. As they leave, the officers say "Well, it's his own fault." I am shocked. S. was no innocent, he had grown up in a criminal environment. But he had two daughters and had been leading a double life. My heart went out to his girls" (van Hoven 2003b).

In the above examples, the ways in which the body and social identities impacted on data generation were not premeditated. There was no particular effort on the side of the researcher to appear one way or the other in order to influence an interview. However, researchers and respondents can (and sometimes have to) make strategic use of their identities to effect the production of academic knowledge (see Crang 2005), for example through deciding which parts of their identities they expose or thinking about how to modify their bodies to adapt to the research situation. For example, in her study on prison spaces in the Netherlands and the United States (see e.g., van Hoven and Sibley 2008; Sibley and van Hoven 2009) Bettina considered her appearance more carefully. In the Netherlands, she was advised not to wear "revealing clothes" whilst in the United States, visitor regulations state that visitors need to dress modestly (e.g., no short skirts). She found that, in the Dutch case, prisoners were very attentive to her clothing, commenting on details like the shoes she was wearing.

She also "dressed up" as a correctional officer in a Dutch case in order to do participant observation prior to the interview phase, to get a "feel" for the prison environment and culture. Here, too, she noted that some of her research participants responded to her "costume" more quickly than to "herself." During her week as "officer," she conducted many of the tasks assigned to this group: unlocking and locking cells, accompanying prisoners to sports or other activities, monitoring prisoners during their yard hour, during visitation and on D-block itself. She returned after the summer break as "researcher" which confused some of the prisoners who (eventually) recognized her. She now mingled freely with inmates and some approached her and requested an explanation for her role change. One prisoner later recalled he had been somewhat suspicious of Bettina after this but was swayed to agree to an interview after positive accounts by fellow-prisoners who seemed to enjoy being interviewed and listened to (see also Gaskell (2008) on the role and importance of "listening").

The Effect of Methods on Knowledge Production

So far, we have focused on knowledge construction largely as it occurs in the context of an interview situation. However, recent developments in methodology show that a range of methods have been adopted in geographic research. In this section, we explore how the use of different methods affects the knowledge that is produced. We take the example of a study on youth and belonging on Vancouver Island (Trell et al. 2009; Trell and van Hoven forthcoming), and show how interviews, mental mapping, walks, and video revealed different facets of place experiences of school.

Interviews as a method of data collection take place as in interaction between the interviewer and respondent, but the interviewer often determines the questions and directs the discussion. Thus, the power relations between interviewer and respondent are seldom in balance and this impacts the information given by the respondent. As the "traditional" interview is an "indoors-method" (but see Anderson 2004; Hitchings and Jones 2004, for exceptions) there is no direct interaction between the respondent and the place/event/object s/he is talking about. The information revealed is based only on one's mental image of the place, or one's memories. Cele (2006) argues that in interviews, respondents reveal the use and experience of place but that details about what the place looks (feels, smells, sounds) like can often be forgotten or diminished. Instead, discussions often focus on he social and cultural dimensions of place. The following quote from Trell et al. (2009: 11) illustrates the above:

> There's not really much to say about the environment [at Cedar school] because it's just a normal, relaxed, easy-going environment ... I like pretty much everything about this school ... It's a really nice comfortable place. (Kevin, male, 17 years, peer-led interview on belonging and school)

In comparison, information about (the meaning of) places in mental maps is based on respondents' view of relative importance of places in their daily lives (see also

Matthews 1984a, 1984b; Young and Barrett 2001). Although there is no direct interaction between the objects, places, events, and the respondent, (undirected) mental mapping allows for more creativity and freedom to express oneself with less influence from the researcher. The respondent chooses which elements to include and exclude from the map which means that the places a researcher chooses to focus on in an interview may be absent altogether. If done in a group context, the group can influence the places respondents add to their mental maps and the ways in which they talk about these places. This process can also trigger spontaneous discussion about daily places, activities, and people with whom the respondents spend time. The discussions based on such maps are comparable to unstructured interviews, as the researcher asks questions about elements on the maps that are unclear. In Trell et al.'s study (2009), mental maps of respondents' everyday places were drawn by each of the four research participants in a group context. The following quote by Kevin shows that he did include school as important place in his map (see Figure 10.1) but did not elaborate on its meaning:

> Anyway, I got the school right here because I like coming to school, although I don't really do much but it's social, climate. (Kevin, male, 17 years old, discussion about mental maps).

Walks (or "go-along interviews," see Carpiano 2009 for discussion) can also reveal information about seemingly ordinary aspects of place, on details, context and sensual aspects, that are nevertheless important in everyday interaction. Walks

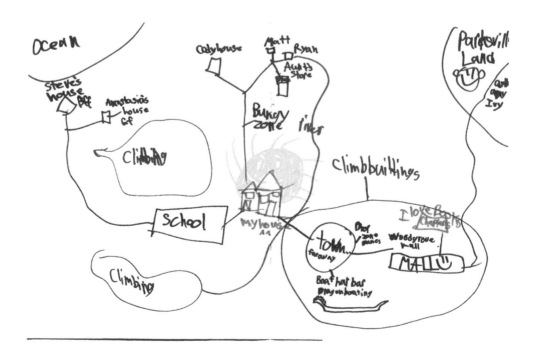

Figure 10.1   Mental map by Kevin. Source: Trell et al. 2009.

allow the respondent to be "in charge," as the researcher can be bracketed outside the data generation process or observe it. Put simply, observation by the researcher during a walk or by means of video, captures an event *as it happens*. Hall et al. (2006: 3) commented:

> [W]e have felt [the] walks to be three-way-conversations, with interviewee, interviewer and locality engaged in an exchange of ideas; place has been under discussion but, more than this, and crucially, under foot and all around, and as such much more of an active, present participant in the conversation, able to prompt and interject.

The knowledge produced in this context is tied to the physical experiences of place. Cele (2006) calls walking (with children) a performance-based research method where being active and on the move, while constantly seeing, hearing, feeling, and smelling the place, triggers conversations and reflections that probably would not otherwise occur (also see Katz 2004). In Trell et al. (2009) Ryan, whilst filming the outside of their classrooms, focuses on birds as he happens to see them:

> A bird just came out of the school! Oh yeah, we should get the birds! We have birds in our rafter. I can hear them, but I can't see them right now, they're up in there. There's a bunch of ... yeah, the dung [*laughing*]. (Ryan, male, 17 years, video-recorded walk)

The possibility of being outdoors and hearing the birds in school rafter had a direct influence on the way Ryan represented his school; he was able to reveal a new dimension of his school experiences that was not talked about in the interview. In fact, it is quite possible that Ryan would hear the birds during class and that they affect his in-class experience (e.g., provide a welcome distraction during boring moments or prove to be annoying during exams) but this is likely the kind of information forgotten during an interview.

Videos made by respondents can also provide information similar to walks, as the respondents are in charge, and the videos are a product of the multi-sensual encounter with human and non-human participants and the material environment. In addition, video may be empowering as respondents get to represent things they choose, when and from whichever angle they want (see also Gomez 2003; van der Sloot 2005). In Trell et al. (2009), the respondents were given a camera to film their places (and homes) in their own time, i.e. unrestricted by schedules imposed by the researcher. In his video, Kevin, revealed a special place that he would not normally have taken other people.

> And just like that I'm at the school. In order to get to my favorite place, up there, I have to GET up there [*filming the school roof*] ... And just like that I'm up ... And this, this is my favorite spot to be [the school roof]. In fact I even come up and read right there, right in the shade. I just love this place. (Kevin, male, 17 years, unguided video)

The school rules do not allow students to be on the roof thus Kevin did not reveal it as his special place in the in-class interview, nor did he mention it during the walk with the rest of the group. Only when he was alone did he reveal the importance and his use of this place. In Kevin's case, the video method proved particularly useful since he found "sitting still" and focusing in an interview situation difficult. In a situation and at a time of his own choosing, however, he revealed important experiences about "school" that would have remained hidden if the research had been confined to using interviewing only. In addition, due to the relative freedom the method provides, as well as the absence of the researcher, the power imbalance is less pertinent to results obtained.

The above discussion has illustrated how different methods generate different types of knowledge, amongst others through the fact that the interaction between researcher and respondent is different for each method.

Actor-networks, the Non-representational, and Changing Relations in Doing Geography

In the previous paragraphs, our attention has focused mainly on people and relations involved in doing geography whilst largely ignoring "the elements" – those nonhuman participants that are often critical to personal experience. We have also emphasized important issues that affect the production of knowledge, such as subjectivity, positionality, and various politics and methods. In so doing, we have gone some way, perhaps, in breaking down dualisms and looking at research from a relational perspective.[12] However, and importantly, we have remained "stuck" to an anthropocentric way of doing research, thus excluding a number of elements that are vital to, even though underexplored in, the construction of knowledge. For example, we noted that walking can be regarded as a performative, "embodied" method which actively engages the elements of the world around – buildings, trees, rooftops, etc. But these elements have remained static and silent, "dead" even, in our discussion. Drawing on the conceptual language of actor-network theory (ANT) and to a lesser extent non-representational theory (NRT), in the remainder of this chapter we want to attempt to "breathe life" into these elements – thinking about them as more than passive recipients of human agency. Since neither of us work within the framework of ANT or NRT, and we have had to "undo" some of our "knowledge" (e.g., about what agency is, or intentionality) this constitutes somewhat of a thought experiment.

One important limitation of this chapter has been the way in which we have implicitly treated agency as "belonging" to human actors. Using ANT,[13] we can view agency differently, namely as decentred and an outcome of networks and interactions. In so doing, we include more explicitly the role of non-humans and can begin to theorize nature and society from a relational (but less anthropocentric) perspective. In order to think through some contributions from ANT, we take as a point of departure an earlier research project in the field of physical geography (conducted by Bettina). We first describe, briefly, the original study and use this to contrast the way in which nature was objectified with "newer" conceptions of nature.

In 1996, Bettina undertook a study aimed at assessing the impact of management on old growth forest ecosystems on Southeast Vancouver Island (Canada). The study consisted of a comparison of (visible, above-ground) plant life as well as soil chemistry and microbiology in selected quadrants in a forest area where ecoforestry was practiced and an area that had been clearcut about ten years prior to the study. In order to collect data, a selection of "suitable" locations to set up quadrants was made in which plants were to be identified, counted, and compared and in which soil samples were analyzed. In a nutshell, the knowledge generated consisted of an archive of plant species and abundance, bacteria and fungi present, and various soil data. Bettina had "take[n] the world [study site] apart conceptually to find out what it [was] made of (via taxonomy) and then put it back together to explore interconnections among the parts (causal relations)" (Rhoads 1999: 763). In this case, the key causal relation to be confirmed was between ecoforestry and the "quality" of the ecosystem as determined by ecology/conservation standards. As a physical geographer, Bettina had "subscribe[d] to the general metaphysical sentiment that 'there is one world susceptible of scientific investigation, one reality amenable to scientific descriptions'" (Hacking (1996: 44) as cited in Rhoads (1999: 765)). Hence, in this study, the natural elements were treated as autonomous objects and their function was fit into classifications and models in a scientific network of biodiversity, and a policy, management, or forest economy. However, the relative role of a plant species in each of these networks differs,[14] their value becomes magnified or diminished, their existence essential or negligible. A plant species' "being" relies on its networks and the actors, events and processes within (see also Whatmore 2002).

We now want to look at the forest in the above study in a more symmetrical way,[15] thus focusing more on the natural elements as "*subjects* in a much more fully dynamic and fascinating set of relationships" (Staddon 2009: 163, original emphasis). Considering more explicitly the "things" Bettina studied as *subjects*, opens up opportunities for looking more in depth at the ways in which humans and non-humans relate to each other. In so doing, we need to consider non-human actors (even if they are not "conscious" (Bear and Eden 2008)) as possessing the ability to exercise agency. This implies that "the arrows of causality and intentionality [in networks] can run in *all* directions" (Staddon 2009: 165), i.e. both humans and non-humans are capable of affecting something, of changing something.

In order to think through the multi-directionality of these arrows of causality and intentionality in the case of the forest, we draw on Thrift (2005). Thrift (2005: 464) argues that "organisms extend beyond the obvious integuments of their "internal physiology" in persistent and systematic ways and adaptively modify their environment. Environments, in turn, can be thought of as a myriad of "external physiologies" that have been adapted to act in roles substitute or accessory organs, means of communication, or even microclimates." In the case of the forest study, examples are forms of collaboration that enhance the vitality and longevity of trees such as those with lichens or mycorrhizal fungi. These organisms both function as accessory organs and modify their environment, ultimately encouraging the cohabitation of some plant and animal species and denying access of others. In this case one could, perhaps, speak of trees exercising agency, possessing intentionality, or following Cloke and Jones (2003), purposive agency. With this in mind, we can look with different eyes at ecoforestry (than was done in the above example). For example, purposive agency can become important in defining the relationship between the (trees in the) forest and its "manager." We use here the example of Merve Wilkinson, the ecoforester/manager of "Wildwood"[16] (one of Bettina's study locations in British Columbia) to illustrate how we imagine this. The decision for

Merve to harvest a tree takes into consideration this tree's coming into "being." When harvesting a tree, he (implicitly) takes into account a tree's purposive agency, its external physiology and its linkages with various networks, since cutting down one tree may cause a "ripple effect" affecting other networks of which this tree is a part (see also Stalder 1997). Merve considers the benefits of the tree to soil fertility, the penetration of light to the forest floor as a result of the density of the tree's foliage, both of which result in the future success or failure of saplings. He considers the tree's contribution to a microclimate that may benefit insect life, which in turn promotes birdlife, etc. In addition, he is concerned with the place of this tree in an economic network, which renders it valuable or not. Last but not least, the place of the tree in a network that includes the emotion and identity of Merve himself also plays a role in how the tree acts upon his person. This may include visual, audible or olfactory aspects of the presence/absence of the tree, in particular since the forest is also a part of the foresters' home, or where the preservation of such a forest may rely, at least in part, on its pleasing nature to its visitors, who are willing to spend money and time to preserve it. In other words, Merve "attempt[s] to enlist the interest or action of another 'so that [his] own desired performance can take place'" (Power 2005: 42). The trees, in turn "recommend themselves ... through a variety of characteristics, [drawing] the person down into their world and make for an understanding of their concerns and a commitment to their care" (Power 2005: 48).[17] In Bettina's study, none of these relations and considerations played a role as they were not classifiable as a visible and measurable outcome of a particular way of forest management.

In terms of purposive agency in the context of this ecosystem, however, it is not so difficult to see how one act may benefit another and result in a better life' for a tree. The question arises perhaps to what extent we may assume that purposive agency may also be targeted at or include humans? One reason for denying this possibility can be found in progressive thought, which has turned the relationship between humans and non-humans into a defensive one with strict boundaries (see also Davies and Dwyer 2007). Clarke (2002) notes (drawing on Horkheimer and Adorno) that "civilization, the modern world, has slowly and methodically prohibited instinctual behaviour" and "touching, feeling, smelling [is] something unhomely, uncanny." As a result, "the [human] body ... as a vessel of consciousness, [has become] ontologically sealed off from the world it is conscious of" (Carolan 2009: 1). This would make it difficult for us to employ our senses to interconnect with the different worlds which animals (and plants) inhabit, as do animals (see Thrift, 2005).[18] Thrift (2005) mentions, for example, the spider who creates a web that is attuned to the visual (in)capacity of the fly.[19] If all these systems exist and make perfect sense, why then should we assume there is no such purposive link between the natural worlds and humans?[20] Whilst we would not go as far as stating that the forest deliberately creates a comfortable, pleasant or economically feasible environment for humans, we would subscribe to Gibson's view that "the biophysical ecology of space does make possible an array of doings, doings that in turn "tune" bodies for certain understandings of the natural world" (Gibson 1986 as cited in Carolan 2009: 4), which may trigger certain responses.

Perhaps Jennifer Lea's (2008: 95) article on retreats, or "healing places" illustrates this, too, as she describes the way in which retreat participants "could sense their (specifically interconnected) place in the world" and through their connectivity with nature were helped in going "deeper" in a therapeutic experience. Lea's is an example of "new subjectivities" as well as "spaces [that] emerge through the enactment of practices that explicitly attempt to facilitate a kind of transformation in awareness, thinking, feeling and relating" (McCormack in Rycroft 2007: 616). It is here that we can begin to see that geographers are addressing the kind of thinking that "lies in the body, understood not as a fixed residence for "mind" but as a dynamic trajectory by which we learn to register and become sensitive to what the world is made of" (Carolan 2009: 3). Carolan (2009) pointed out that this thinking through the body, or "unconscious thought makes up approximately 95% of all cognition"[21] (p. 3). Some recent geographic work, then, tries to capture this richness that has been missed in current social science practice by turning to "creativity, specificity, openness, fluidity, risk, uncertainty, and pluralistic views of knowledge as practice in/of/for the world" (Jones 2008: 1603). This way of addressing the world has begun to demand different ways of carrying out research, for example through dance (McCormack 2003) or walking in nature (Wylie 2005). At the same time, as Davies and Dwyer (2007: 261) point out, the challenge "to report back from such embodied experiences is never easy."

Whether these efforts will be labeled geography or philosophy eventually, as we have wondered in the above, may be left in the middle for now, as they "are an opportunity to re-enchant the present" (Fenton 2005, as in Davies and Dwyer 2007: 262). They offer new, creative and dynamic ways to explore life and living as it unfolds. They open up new methodologies and new ways of thinking about research, researchers and the subjects they study as well as ethics.

Conclusion

In this chapter, we explored some of the aspects of doing research and generating knowledge by looking specifically at the people, elements, and relations involved in the research process. We referred to the impact of our own Dutch backgrounds on the kind of knowledge that we produce and, in so doing, which perspectives we may exclude. Being embedded in Dutch geography, with its practical orientation and close relation to policy planning, implies that recognition of work of a more "abstract" nature as well as working with a more reflexive approach can be more of a challenge, even though it may be required in the international context (and when aspiring to publish in English peer-reviewed journals). We then explored the politics of knowledge construction in research based largely on interviewing and, in so doing, touched on the role of visible as well as unspoken and unseen bodily markers on research relationships. We argued that some of the "newer" (more participatory) methods have much to contribute in generating different aspects of knowledge that would remain hidden in interviewing. Furthermore, we thought through some of the implications of actor-network theory and, to a lesser degree, non-representational theory, on how we think about the connectivities among the people, elements, and relations involved in doing geography. We have seen that knowledge is partial, subjective and shaped in interaction between researcher, respondents and the rest of the world, rather than objective and universal. This implies a fundamental challenge to the epistemological basis of social geography historically.

The generation of knowledge encompasses more aspects, of course, many of which we have ignored in the context of this chapter. Importantly, we have neglected the distribution, valuing and use of knowledge generated. Researchers become entangled in highly complex and contradictory interests and agendas, such as those set by funding agencies, policy-makers, and research assessment exercises, as well as research participants and local populations. All actors involved impact, to various extents, the knowledge produced by academics, resulting in a new distribution of ownership of academic research, with a less prominent role for the researcher.[22]

Whilst we have seen that many actors are involved in the production of academic knowledge, we would argue that doing research is often a "narrow" exercise: when we go out to collect data we usually exclude a significant part of the networks in which (we and) our respondents are embedded in order to be able to focus on a research question and relate respondents' experiences to specific theories. Therefore, looking at research through an actor-network lens has been interesting in the context of this chapter. Paying attention to networks and how "things" become what they are puts the production of knowledge in perspective in that it highlights the fact that "the construction of knowledge [is] a product of *all kinds of things*" (Bosco 2006: 136, our emphasis), and extends beyond our usually narrow focus. Refocusing attention in this way also helps establish bridges when doing interdisciplinary research. Buller (2009: 6), for example, argues that when the boundaries of the individual disciplines have been established, "the 'objects' of our investigation begin to move around. Not only do we need to 'translate' them ourselves to endow them with communicable cross-disciplinary and cross-actor relevance, but the objects themselves act upon our disciplinary preconception."

In writing this chapter, we have enjoyed critically engaging with, and even dismissing, some of our own preconceptions, which has enabled us to think about knowledge production in ways that we had not done before. Having said this, we remain in doubt about the value or "usefulness" of some knowledge, for example in terms of helping change the world for the better. For example, we found Tolia-Kelly's (2006) critical look at "affect" useful for appreciating the "everyday value" of the concept rather than its philosophical. In our own, Dutch context we can convey her argument to students as it is applicable to their future work in governmental institutions, or teaching. However, Wylie's (2005) "single day walking" (to "pick on" just one text) would be less successful in achieving a positive appreciation of the concept "affect." In concluding, then, we would like to borrow from Massumi (in Jones 2008: 1603), who adequately represents our own sense of purpose of "new" forms of knowledge: "the question is not: is it true? But: does it work? What new thoughts does it make possible to think? What new emotions does it make possible to feel [...] and what new actions, politics and ethics can these new thoughts/feelings conjure?" We ourselves have already experienced this thinking of new thoughts in the ecoforestry example above and feel that our students may be equally inspired to think beyond their (institutional) horizon, thus working and reworking their identities as geographers.

Notes

1 The impact on national contexts on "doing geography" is illustrated in the series of Country Reports in Social & Cultural Geography, e.g., on the Netherlands (Musterd and de Pater 2003), South-East Asia (Bunnell et al. 2005), Greece (Vaiou 2005), and Norway (Berg 2007).

2 Rather than referring to individual authors, we refer to cumulative publications of the research groups here. See for Amsterdam: http://www.fmg.uva.nl/amidst/publications. cfm/A043C1D4–1A0A-4530–8141DCC008F468CF. See for Groningen: www.rug.nl/ursi/publications/publications2/index. See for Nijmegen: www.ru.nl/socialegeografie/onderzoek/publicaties/. See for Utrecht: http://researchgroup.geo.uu.nl/index.php?groupid=55.

3 Such an approach can be contrasted with some recent (English) handbooks, who teach the aspiring researcher that interviews can never be replicated or unaffected by the researcher's person and personality (see, for instance, Dunn 2005; Valentine, 2005a). Moreover, Valentine (2005a) draws on feminist research to illustrate that the researcher may not necessarily remain speechless: for example when respondents express racist, homophobic, or other offensive views, interviewers may find themselves in dialogue because silence can legitimize such views.

4 We need to emphasize that this is certainly not the case for all researchers in the department, some of whom have a more ethnographic or anthropological background.

5 Several other studies further suggest how the bodies of both researcher and researched can play a role in knowledge construction. For example, Bain and Nash (2006) described their experiences when "dressing up" for and observing a queer bathhouse event, while Worth (2008) shows how common experiences between researcher and respondent may contribute to establishing a relation of trust, for instance in disability geography. Hester Parr (1998) has also discussed the researcher's body when doing research in a mental health clinic, suggesting that the body can be used as a strategic research tool in interpersonal connections. Based on these studies, it can be suggested that "knowledge is articulated with embodied codes and memories that 'emerge as flashes' not following 'logical-rational knowledge'" (Ylönen (2003: 565) as cited in Crang 2005: 231).

6 Longhurst et al. (2008) similarly discuss the consumption of local foods in the context of researcher–respondent relations.

7 Respondents asking questions about the researcher's identity have been described by other authors, such as Hopkins (2009) in his study on Muslim youth. Both Hopkins and his respondents perceived him occupying a space of "betweenness": like his respondents, he was young and male, but he was not Muslim, "black," or "Asian." Perhaps in an attempt to establish more "sameness," his respondents were interested in where he was from, which schools he had attended, and which soccer team he did or did not support. Constructing levels of sameness and difference helped them to make sense of each other.

8 In the context of the moral turn in geography, ethical considerations have become a part of the discipline, especially since the late 1990s. See, for example, Cloke (2002), Proctor (1998), Smith (2001), and Valentine (2005b).

9 Widdowfield (2000), for instance, described her own emotional response to doing research on lone parents in Newcastle. Bennett (2004) focused on the emotions of the researched, instead of the researcher, and discussed how emotional responses provided more depth in the context of her study on foot and mouth disease in England. In a subsequent paper, she included both respondent and researcher, through stressing the

relational nature of unspoken elements such as emotions: "my feelings cannot simply be explained through my self, but through my relationship with others and the context of our interactions and connections" (Bennett forthcoming: 2).

10 Bettina has not addressed emotions in doing the prison research explicitly in the form of an academic (peer-reviewed) publication for similar reasons addressed by Gaskell (2008), i.e. an academic setting unappreciative of an "emotional researcher." In 2006, she presented some of the more "emotional" research at the Annual Meeting of American Geographers in Chicago ("Abandon hope, all ye who enter here": Discussing everyday experiences of H/hope in prison) but the paper never made the special issue which focused more on the non-representational aspects of hope.

11 Further exploration, i.e. the accounts of other prisoners of the riot incident, confirmed Richard's account. In addition, Richard was one of the prisoner's who claimed innocence (see also: http://en.wikipedia.org/wiki/Richard_Cartwright_%28murderer%29).

12 What we mean here is that identities are fluid and dynamic: they are affected by the respective contexts in which they are expressed and interwoven with humans, non-humans, and materialities of their everyday lives.

13 In our argument, we have drawn on Fernando Bosco's excellent discussion of networks. He notes that "ANT is about uncovering and tracing the many connections and relations among a variety of actors (human, non-human, material, discursive) that allow particular actors, events and processes to become what they are" (2006: 136). He exemplifies this using his own identity as a geographer: "From the perspective of ANT, I would no longer be a geographer with the ability to write papers and produce knowledge if my computer, my colleagues, my books, my job, my professional network, and everything else in my life that allows me to act as what I am were taken away from me ... if that were to happen, I would become something different" (2006: 137).

14 Even where the same representation is used. A Western Red Cedar may be perceived as "good" in ecological, cultural, and economic terms. Yet, the consequences of such labeling differ significantly resulting in preservation, moderate use, or harvest.

15 Michael Shanks (2007) provides an interesting account of thinking symmetrically in archaeology. He maintains: "Symmetry draws attention to mutual arrangement and relationship. Symmetry, in this mutuality, implies an attitude, that we should apply the same measures and values to ourselves as to what we are interested in ... A consonance of past and present, individual and structure, person and artifact, biological form and cultural value, symmetry is about relationships ... Symmetry here also holds that we are not essentially different from those ... we study. We are all bound up in different kinds of relationship with the materiality of the world" (591).

16 See also: www.conservancy.bc.ca/content.asp?sectionack=wildwood.

17 Lorimer (2005: 85) further notes that "these entangled relationships, that are found to incorporate love, care, need and (commercial) demand, are also a means to consider place-making agencies and therapeutic feelings of dwelling." Elsewhere this has been discussed in the context of ethics, e.g., that "cultivating affective attachment to the world is valuable for ethical life" (Bennett (2001) as cited in Davies and Dwyer 2007: 260; but see also Cloke and Jones 2003; Popke 2009).

18 Thrift (2005: 465) argued that "animals are bound up with different and diverse spaces ... they also live in very different times, in terms of metabolic rates, reaction times and forms of foresight, lifespans and memories ... animals exist in spaces and times which mean that the relation that they have to things in an environment may be radically different from ours and each others."

19 Other instances of an attunement between animals and their environment or food sources are ultraviolet light seen by birds, infrared light seen by insects, or the sensitivity of some fish to electric and magnetic fields (Thrift 2005).

20 Sarah Whatmore (2006: 603) argues that recent work has begun to address this, in a way, by "focusing attention on bodily involvements in the world in which landscapes are co-fabricated between more-than-human bodies and a lively earth ... interrogat[ing] 'the human' as no less a subject of ongoing co-fabrication than any other socio-material assemblage."

21 "While '95 percent of academic thought [tends to be] concentrated on the cognitive dimension of the conscious I'" (Thrift (2000: 36) as cited in Carolan 2009: 3).

22 For instance, Underhill-Sem and Lewis (2008), who worked with indigenous people, argued that within the complex whole of politics, research *can* help to abandon inequalities and mechanisms exploiting historically oppressed groups.

References

Anderson, J. (2004) Talking while walking: a geographical archaeology of knowledge. *Area* 36: 254–62.

Bain, A.L. and Nash, C.J. (2006) Undressing the researcher: feminism, embodiment and sexuality at a queer bathhouse event. *Area* 38 (1): 99–106.

Bear, C. and Eden, S. (2008) Making space for fish: the regional, network and fluid spaces of fisheries certification. *Social & Cultural Geography* 9 (5): 487–505.

Bennett, K. (2004) Emotionally intelligent research. *Area* 36 (4): 414–22.

Bennett, K. (2009) Challenging emotions. *Area* 41 (3): 244–51.

Berg, N.G. (2007) Country report. Social and cultural geography in Norway: from welfare to difference, identity and power. *Social & Cultural Geography* 8 (2): 303–30.

Bosco, F.J. (2006) Actor-network theory, networks and relational approaches in human geography. In: S. Aitken and G. Valentine (eds) *Approaches to Human Geography*. London: Sage, pp. 136–46.

Buller, H. (2009) The lively process of interdisciplinarity. *Area* 41 (4): 395–403.

Bunnell, T., Kong, L., and Law, L. (2005) Country reports: social and cultural geographies of South-East Asia. *Social & Cultural Geography* 6 (1): 135–49.

Carolan, M.S. (2009) "I do therefore there is": enlivening socio-environmental theory. *Environmental Politics* 18 (1): 1–17.

Carpiano, R.M. (2009) Come take a walk with me: the "go-along" interview as a novel method for studying the implications of place for health and wellbeing. *Health & Place* 15: 263–72.

Cele, S. (2006) *Communicating Place, Methods for Understanding Children's Experiences of Place*. Stockholm: Almqvist & Wiksell International.

Clarke, S. (2002) From aesthetics to object relations: situating Klein in the Freudian "uncanny," www.btinternet.com/~psycho_social/Vol1/JPSS1-SHC1.htm. Accessed July 15, 2009.

Cloke, P. (2002) Deliver us from evil? Prospects for living ethically and acting politically in human geography. *Progress in Human Geography* 25 (5): 587–604.

Cloke, P. and Jones, O. (2003) Grounding ethical mindfulness for/ in nature: trees in their places. *Ethics, Place and Environment* 6 (3): 195–214.

Cloke, P., Cook, I., Crang, P., Goodwin, M., Painter, J., and Philo, C. (2004) *Practising Human Geography*. London: Sage.

Cook, N. (2009) It's good to talk: performing and recording the telephone interview. *Area* 41 (2): 176–85.

Crang, M. (2005) Qualitative methods: there is nothing outside the text? *Progress in Human Geography* 29 (2): 225–33.

Davies, G. and Dwyer, C. (2007) Qualitative methods: are you enchanted or are you alienated? *Progress in Human Geography* 31 (2): 257–66.

DeVerteuil, G. (2004) Systematic inquiry into barriers to researcher access: evidence from a homeless shelter. *Professional Geographer* 56: 371–80.

Droogleever Fortuijn, J. (2004) Gender representation and participation in Dutch human geography departments. *Journal of Geography in Higher Education* 28 (1): 133–41.

Dunn, K. (2005) Interviewing. In I. Hay (ed.), *Qualitative Research Methods in Human Geography*, 2nd edn. Oxford: Oxford University Press, pp. 79–105.

Garvin, T. (2001) Analytical paradigms: the epistemological distances between scientists, policy makers, and the public. *Risk Analysis* 21 (3): 443–55.

Gaskell, C. (2008) "Isolation and distress"? (Re)thinking the place of emotions in youth research. *Children's Geographies* 6 (2): 169–81.

Gomez, R. (2003) Magic roots: children explore participatory video. In S. White (ed.), *Participatory Video: Images That Transform and Empower*. London: Sage, pp. 215–34.

Hall, T., Lahua, B., and Coffey, A. (2006) Stories as sorties. *Qualitative Researcher* 3: 2–4.

Hitchings, R. and Jones, V. (2004) Living with plants and the exploration of botanical encounter within human geographic research practice. *Ethics, Place and Environment* 7: 3–19.

Hopkins, P.E. (2009) Women, men, positionalities and emotion: doing feminist geographies of religion. *ACME* 8(1): 1–17.

Hoven, B. van (2003a) "Imprisoned geographies" (research diary). *Girugten* 35 (1): 34–5.

Hoven, B. van (2003b) "Imprisoned geographies" (research diary). *Girugten* 35 (3): 34–5.

Hoven, B. van and Sibley, D. (2008) "Just duck": the role of vision in the production of prison spaces. *Environment and Planning D: Society and Space* 26 (6) 1001–17.

Hoven, B. van, Been, W., Droogleever Fortuijn, J., and Mamadouh, V. (2010) Teaching feminist geographies in the Netherlands: learning from student-led fieldtrips. *Documents d'Anàlisi Geogràfica* 56 (2): 305–21.

Jones, O. (2008) Stepping from the wreckage: geography, pragmatism and anti-representational theory. *Geoforum* 39 (4): 1600–12.

Katz, C. (2004) *Growing Up Global: Economic Restructuring and Children's Everyday Lives*. Minneapolis: University of Minnesota Press.

Kitchin, R. and Fuller, D. (2003) Observation: making the "black box" transparent: publishing and presenting geographic knowledge. *Area* 35 (3): 313–15.

Lairumbi, G.M., Molyneux, S., Snow, R.W., Marsh, K., Peshu, N., and English, M. (2008) Promoting the social value of research in Kenya: examining the practical aspects of collaborative partnerships using an ethical framework. *Social Science & Medicine* 67: 734–47.

Laurier, E. and Philo, C. (2006) Possible geographies: a passing encounter in a café. *Area* 38 (4): 353–63.

Law, L. (2001) Home cooking: Filipino women and geographies of the senses in Hong Kong. *Ecumene* 8 (3): 264–83.

Lea, J. (2008) Retreating to nature: rethinking "therapeutic landscapes." *Area* 40 (1): 90–8.

Longhurst, R., Ho, E., and Johnston, L. (2008) Using "the body" as an "instrument of research": kimch'i and pavlova. *Area* 40 (2): 208–17.

Matthews, M.H. (1984a) Cognitive mapping abilities of young boys and girls. *Geography: Journal of the Geographical Association* 69 (4): 327–36.

Matthews, M.H. (1984b) Cognitive maps: a comparison of graphic and iconic techniques. *Area* 16 (1): 33–40.

McCormack, D.P. (2003) An event of geographical ethics in spaces of affect. *Transactions of the Institute of British Geographers* 28 (4): 488–507.

McDowell, L. (1992) Multiple voices: speaking from inside and outside "the project." *Antipode* 24 (1): 56–72.

Meijering, L. (2006) *Making a Place of Their Own: Rural Intentional Communities in Northwest Europe.* Utrecht/Groningen, Netherlands: Royal Dutch Geographical Society/ Faculty of Spatial Sciences, Geographical Studies (NGS), 349.

Meijering, L. and B. van Hoven (2003) Imagining difference. The experiences of "transient" Indian IT-professionals in Germany. *Area* 35 (2): 174–82.

Müller, M. (2007) What's in a word? Problematizing translation between languages. *Area* 39 (2): 206–13.

Musterd, S. and de Pater, B. (2003) Country reports – eclectic and pragmatic: the colours of Dutch social and cultural geography. *Social & Cultural Geography* 4 (4): 549–63.

Parr, H. (1998) Mental health, ethnography and the body. *Area* 30 (1): 28–37.

Peake, L. (1989) The challenge of feminist geography, *Journal of Geography in Higher Education* 13: 85–121.

Popke, J. (2009) Geography and ethics: non-representational encounters, collective responsibility and economic difference. *Progress in Human Geography* 33 (1): 81–90.

Power, E.R. (2005) Human–nature relations in suburban gardens. *Australian Geographer* 36 (1): 39–53.

Proctor, J.D. (1998) Ethics in geography: giving moral form to the geographical information. *Area* 30 (1): 8–18.

Purcell, M. (2007) "Skilled, cheap and desperate": non-tenure-track faculty and the delusion of meritocracy. *Antipode* 39: 121–43.

Rhoads, B.L. (1999) Beyond pragmatism: the value of philosophical discourse for physical geography. *Annals of the Association of American Geographers* 89 (4): 760–71.

Rycroft, S. (2007) Towards an historical geography of nonrepresentation: making the countercultural subject in the 1960s. *Social & Cultural Geography* 8 (4): 615–33.

Shanks, M. (2007) Symmetrical archaeology. *World Archaeology* 39 (4): 589–96.

Sibley, D. (1998) Problematizing exclusion: reflections on space, difference and knowledge. *International Planning Studies* 3 (1): 93–100.

Sibley, D. (2009) Personal conversation with authors: interview held for use in this chapter on March 17.

Sibley, D. and van Hoven, B. (2009) The contamination of personal space–boundary construction in a prison environment. *Area* 41 (2): 198–206.

Sloot, M. van der (2005) I/eye Tibet: autovideographies depicting ethno-national identity in daily lives of young Tibetans in Dharamsala, India. Unpublished master's thesis. University of Amsterdam.

Smith, D.M. (2001) Geography and ethics: progress, or more of the same? *Progress in Human Geography* 25 (2): 261–8.

Staddon, C. (2009) Towards a critical political ecology of human-forest interactions: collecting herbs and mushrooms in a Bulgarian locality. *Transactions of the Institute of British Geographers* 34: 161–76.

Stalder, F. (1997) Actor-network-theory and communication networks: toward convergence, http://felix.openflows.com/html/Network_Theory.html. Accessed July 15, 2009.

Thrift, N. (2005) From born to made: technology, biology and space. *Transactions of the Institute of British Geographers* 30: 463–76.

Tolia-Kelly, D.P. (2006) Affect – an ethnocentric encounter? Exploring the "universalist" imperative of emotional/affectual geographies. *Area* 38 (2): 213–17.

Trell, E.M. and van Hoven, B. (forthcoming). Everyday geographies and place attachment: a participatory research project with youth in Cedar, British Columbia. *Proceedings. Re-exploring Canadian space. ACSN Canada Cahier.*

Trell, E.M., van Hoven, B., and Huigen, P.P.P. (2009) Everyday places and practices in photos, video, walks, mental maps and narratives of youth. Exploring the possibilities and limitations of different research techniques with youth in Canada. Paper presented at the 3rd Nordic Geographers Meeting. Turku, Finland, June 8–11.

Underhill-Sem, Y. and Lewis, N. (2008) Asset mapping and Whanau action research: "new" subjects negotiating the politics of knowledge production in Te Rarawa. *Asia Pacific Viewpoint* 49 (3): 305–17.

Vaart, R. van der, de Pater, B., and Oost, K. (2004) Geography in The Netherlands. *Belgeo* 1: 135–44.

Vaiou, D. (2005) Country report: the ambiguities of social and cultural geography in Greece. *Social & Cultural Geography* 6 (3): 455–69.

Valentine, G. (2005a) Tell me about … : using interviews as a research methodology. In R. Flowerdew and D. Martin (eds.), *Methods in Human Geography: A Guide for Students Doing a Research Project*, 2nd edn. Harlow: Pearson Education, pp. 110–27.

Valentine, G. (2005b) Geography and ethics: moral geographies? Ethical commitment in research and teaching. *Progress in Human Geography* 29 (4): 482–7.

Whatmore, S. (2002) *Hybrid Geographies: Natures Cultures Spaces*. London: Sage

Whatmore, S. (2006) Materialist returns: practicing cultural geography in and for a more-than-human worlds. *Cultural Geographies* 13: 600–9.

Widdowfield, R. (2000) The place of emotions in academic research. *Area* 32 (2): 199–208.

Worth, N. (2008) The significance of the personal within disability geography. *Area* 40 (3): 306–14.

Wylie, J. (2005) A single day's walking: narrating self and landscape on the South West Coast Path. *Transactions of the Institute of British Geographers* 30: 234–47.

Young, L. and Barrett, H. (2001) Adapting visual methods: action research with Kampala street children. *Area* 33 (2): 141–52.

Chapter 11

Leaving the Field

Carolyn Gallaher

"But they didn't do that, so I don't have data on it." (A frustrated author after returning from the field)
   "What they don't do is as important as what they do." (A patient professor, in response)

Field research is often a lonely venture, with days structured by work and little else. Leaving the field, then, can be exhilarating – a welcome return to friends and family. Leaving the field can also be frustrating. Having completed one difficult, time consuming task, an equally complex, protracted one awaits. What does one do with all the data collected in the field? Indeed, whether one is returning from a village in Oaxaca, an archive in London, a hospice center in Chiang Mai, or a refugee camp in Uganda, sorting through and making sense of one's data can be an intimidating affair. Harry Wolcott (1994) rightly observers "that the real mystique of qualitative inquiry lies in the processes of using data rather than in processes of *gathering* data" (p. 1, emphasis in original). The focus of this chapter is the transition period between data collection and analysis. In particular, it has two goals. The first is to help readers identify, arrange, and sort their data. As the quotes that start the chapter above indicate, sometimes a scholar is not sure what data she has. The second aim is to help readers determine how best to analyze data once it is organized. The chapter is organized in the following manner. The first section introduces the reader to the notion of methodology – a way of thinking about the organization of research that is greater than the methods employed to do it. The second section addresses issues related to identifying/preparing your data. The third section offers pointers for how to use data. Particular attention is paid to understanding the

A Companion to Social Geography, First Edition. Edited by
Vincent J. Del Casino Jr., Mary E. Thomas, Paul Cloke, and Ruth Panelli.
© 2011 Blackwell Publishing Ltd. Published 2011 by Blackwell Publishing Ltd.

differences between "description," "analysis," and "interpretation." Throughout, the author uses examples from her and others' research to illustrate important points.

Never Forget Why You Went in the First Place

One of the tried and true axioms of qualitative research is that what one proposes to do in the field rarely turns out to be what one does there. To be sure, few research projects morph into something entirely different, but little changes often accumulate, making the actual project somehow different from the proposed one. This happened in a research project I began in 2002 on Loyalist paramilitary activity in the wake of the 1998 Northern Ireland peace accord. My goal was to understand if the way Loyalists defined themselves vis-à-vis Republicans had changed since 1998.[1] When I started doing interviews, however, I discovered that Loyalists were increasingly defining themselves with reference to other Loyalists instead of Catholics. I frequently heard people say, for example, "We're the real Loyalists." What seemed to matter most to people who defined themselves as Loyalist was not opposition to a United Ireland (a given for most Loyalists) but one's affiliation with one of three paramilitary groups. In response, I changed my interview instrument to include questions about the importance of paramilitary affiliation to one's self-definition within the wider Loyalist fold. I kept earlier questions in my interview instrument, but I discovered that the most detailed responses were for my new questions. They reflected a reality of the field of which I was previously unaware.

Changes like the ones I encountered in the field can often create a sense of dissonance for a researcher when she returns home. In this period it is important to remember what one initially set out to do in the field. Indeed, though one's survey or interview instrument may have been modified, such changes normally reflect empirical realities to which a researcher must respond rather than a rethinking of a project's aims. In these instances, one's extant methodology can and should serve as a compass. And, although the terms method and methodology are frequently used interchangeably, they are not synonymous. Methods are tools one uses to collect data. A survey, an interview questionnaire, and participant observation are all data collection methods. How one uses these tools is methodology. Put another way, methodology is what a researcher thinks is the best way to obtain knowledge (Denzin and Lincoln 2005a). There are, of course, as many "best ways" to obtain knowledge as there are individuals wanting to get that knowledge. In academia methodological best practices are usually defined by one's theoretical approach, whether explicitly stated or not.

In general there are four broad theoretical frames that drive qualitative methodology in the social sciences: positivist, ethnographic, Marxist, and poststructuralist (adapted[2] from Denzin and Lincoln 2005b). These categories are broad and often include sub-theories and/or approaches. The poststructuralist frame, for example, is employed by scholars working within feminism, postcolonialism, and critical race theory. Each of these subfields has its own literature, debates, and empirical touchstones, but practitioners within all of them tend to agree on the "best practices" for obtaining knowledge (see Del Casino 2009).

A methodological set of "best practices" applies throughout the research process. It includes an approach to *posing questions* (e.g., does one use a hypothesis or an open ended question?) *collecting data* (e.g., does emotion count as data?), *evaluating data* (e.g., is external validity necessary?), and *presenting results* (e.g., can one make a truth claim about the topic studied?). The distinction between method and methodology means, for example, that two researchers studying the concentration of poverty, one working from a positivist theoretical perspective and the other from a poststructural one, could both use interviews, but would likely construct and administer interview instruments differently, and ultimately analyze results in distinct ways as well. The positivist likely would start with a hypothesis to explain the concentration of poverty. She would design an interview instrument to elicit factual information that could to help her decide whether to reject a null hypothesis. A poststructuralist, by contrast, would tend to ask exploratory questions. His interview instrument might include open-ended questions about how poverty is defined and understood by those so categorized.

Giving one's methodology a second look is a good starting point for anyone just returning from the field. Indeed, self-consciously returning to one's methodology before organizing and analyzing data will help provide continuity to a project. The next section discusses the first post-field step – identifying/preparing your data for analysis. Examples from actual studies are used throughout to illustrate how data is understood by the four main methodological traditions identified above.

Preparing Your Data

One of the hardest parts of getting data organized is determining what counts as data. There is little debate, for example, on whether an interview transcript, a completed survey, or a memo from an archive constitutes data. But, within these collection methods, wider questions loom. Does the emotion of the interview subject count as data – does it matter, for example, if one subject responds to a question in an angry fashion while another appears unemotional? Likewise, how does one count or categorize an inarticulate response to a question? And, what do we make of a controversial statement from an archival memo when the preceding clause is redacted? In short, organizing one's data presents conundrums the researcher must address before moving onto data analysis. How one addresses such conundrums depends in large part on one's methodology.

Positivist approaches

As a methodology positivism is usually based on hypothesis testing (for recent examples in geography, see Tierney and Kuby 2008 and Graves and Peterson 2008). Although it tends to be associated in geography (and other social science disciplines) with the quantitative revolution of the 1970s, positivist methodology can involve the use of quantitative and/or qualitative data. In geography, for example, scholars who do spatial analysis often use qualitative data to nuance statistical results. In a study about gender and child chauffeuring in two income households in Utrech, the Netherlands, for example, Schwanen (2007) supplements statistical data he collected from a sample of households with interview data. Schwanen asked parents to create narratives about the division of labor for child chauffeuring in their home and then spoke with parents about these narratives. Positivist studies can also use only qualitative data. In the discipline of international relations, for example, scholars working within realist traditions use policy positions, formal actions (e.g., sanctions), and international agreements (treaties) to test hypotheses about interstate relations. In general, positivist data that is qualitative is factual information that can be counted or categorized, and compared across cases.

The first step for scholars who have completed data collection using a positivist methodology is to develop a system for recording and arranging data. If surveys were used, for example, the researcher may create a spreadsheet that lists all responses by question. Doing so allows the researcher to more easily identify patterns. Scholars will also want to track the relationship between the main variables in the study hypothesis. If a scholar is looking at how climate change legislation affects state behavior, for example, then she will want to track state responses to the legislation.

Ethnographic approaches

Ethnographic methodology takes a different approach to data. Ethnographic methodology is designed to help researchers understand something that is either unexplained or poorly understood. Ethnographic methodology is often associated with participant observation.[3] A classic example of ethnographic methodology can be found in studies of "street corner society," a term coined in the 1960s to describe underemployed African-American men who spend their days hanging out and sometimes hustling on inner city street corners (Whyte 1955; Liebow 1967). At the time, street corner men were viewed as "losers" and their behavior as pathological. Elliott Liebow's *Tally's Corner* (1967) – a study of street corner life in Washington, DC, – was designed to move beyond the stereotypes of street corner men, to understand what their lives were really like. In an ethnographic study data is information that helps you understand the phenomenon you want to study. In this case, Liebow collected data on the types of activities street corner men engaged in on a day to day basis (playing craps, meetings with wives/girlfriends, etc.) as well as what these engagements meant for the structuring of their lives. In Liebow's study (as in other ethnographic studies) emotions and perspectives are important pieces of data. Liebow found that his informants often felt like failures, but they responded in different ways. Some reacted with anger, others with hopelessness. These emotions, placed in context, led Liebow to theorize that hanging out on the streetcorner provided structure and meaning to lives that had precious little of either. [4]

Such data is not, however, easily categorized or directly comparable across cases. Indeed, unlike positivist data, which is often readily organized, scholars using ethnographic data tend to have more difficulty organizing it. It is often a good idea to begin by organizing data in the way that it was collected. If a scholar organized her research notes by interview subject, for example, she might want to start by setting up a table with a column for each subject. She can then arrange emotional responses, anecdotes, or portions of interview transcripts into the appropriate column. Normally, however, the level at which ethnographic data is collected and recorded (at the individual level) is not the same level at which data is discussed (at the group level). As such, the researcher will need to do a second round of organization. If we use *Tally's Corner* as an example, a researcher might want to take data collected on individual subjects' emotional responses to relationships (i.e. with John, or Louise) and put them into bigger categories that cover all subjects, such as "relationships with friends" or "relationships with children."

Marxist approaches

Marxist methodology is designed to understand capitalism and the social relations that ensure it. The data that one could collect to understand capitalism (and its social relations) could certainly be quantitative. In geography and other social science disciplines, however, researchers guided by Marxist methodology have tended to forgo an econometric approach for a political economic one. Generally, a political economic approach entails looking at how political structures are organized to support capitalist accumulation and reinforce class structure (Peet 1985). Data collection in such a study would focus on political regulation with an eye to who benefits from it, and who does not.

Mohameden Ould Mey's analysis (1996) of the effects of International Monetary Fund (IMF) and World Bank structural adjustment programs (SAPs) in Mauritania provides a case in point. Ould Mey's data collection focused on international and national regulations. At the international level, he collected "documents and declarations coming out of the G-7 economic summits held between 1975 and 1995" (1996: 2). At the national level Ould Mey examined briefs, memos, and other government documents produced by the state of Mauritania as it negotiated with the IMF and the World Bank for loan money. These documents included factual information, such as specific policies, as well as perspectives, such as opinions by IMF officials on how an SAP should be structured.

Given that Marxist methodology is driven by concerns for emancipation, Marxist data collection can also include data on the effects of marcoeconomic policies on everyday people. In her study of SAPs in Jamaica, for example, Susan George (1990) collected data on the cost of basic staples like flour, cornmeal, and chicken before and after the country took out an IMF SAP. She also collected data on the calories supplied by standard units of the same staples in order to identify changes in caloric intake (assuming a fixed family budget) after an SAP was adopted. Indeed, a combination of macro and micro level data is a hallmark of Marxist methodology because it seeks to highlight the way macroeconomic policy can impinge on the lives of society's most vulnerable populations.

Poststructuralist approaches

As a methodology, poststructuralism is designed to understand how categories and the discourses that surround them come to be taken for granted, understood as "normal" or "commonsense." Poststructuralists believe common social categories, like "race," or economic categories, such as the "free market," are constructed rather than given. What "white" as a racial category meant in 1850, for example, is not what it means today. Nor does today's "free market" imply the exact same thing it did in the 1920s. Meanings are fixed (or become dominant) temporarily, but are always contingent and thus unstable. Poststructuralist methodology is designed to understand systems of meaning.

Most poststructuralist work in geography is based on the work of Ernesto LaClau and Chantal Mouffe. Their seminal work, *Hegemony and Socialist Strategy* (1985), lays out a number of methodological concepts to guide poststructuralist inquiry. The starting point for LaClau and Mouffe is *discourse*, which can broadly be understood as a system of meaning through which people, things, events, etc. are defined. An earthquake, for example, can be a "natural phenomenon" or "the wrath of God," depending on whether one understands the world through a scientific discourse or a religious one (LaClau and Mouffe 1985: 108). In the social sciences discourse is understood as political. It is built around *nodal points*, broad categories that "underpin and organize social orders" (Howarth 2000: 110). Within a societal level discourse, people occupy *subject positions* such as "black" and "working class." When people act out (or act out against) a particular subject position (and the wider discourse that frames them), they express *political subjectivity*. While many discourses are created and maintained by the powerful in a society, discourse can also be created by less powerful members of society. If a system of meanings seeps into popular culture, and becomes commonplace and recognizable, it is discourse no matter its source. LaClau and Mouffe also discuss antagonism – a situation where a social actor cannot fully occupy his subject position and constructs "others" for blame. Geographers have used LaClau and Mouffe to understand the identity politics of a variety of social movements. The author has used LaClau and Mouffe, for example to understand the identity politics of the far right in the United States (see next section for a detailed explanation). Such studies focus on very different groups, but they find common ground in tracking the emergence of political subjectivity, the creation of nodal points, and the presence (or lack) of antagonism in their respective groups.

The data that researchers working under a poststructuralist methodology collect, especially in early phases of research, is often "textual." Textual data is not, of course, confined to books. Rather, words (written or spoken) that name, identify, and place people/things/events within a discursive regime are considered data. Perspective and emotion are also considered data in poststructuralist methodology. Indeed, understanding how social actors define themselves and how they express or contradict the subject positions they are defined by are important for identifying political subjectivity and whether that political subjectivity is antagonistic.

How to Use Data

Every year in my qualitative methods class for PhD students I assign an "interview project." Students must select a professor, read a selection of his/her work, and then prepare an interview instrument with targeted questions. They must also arrange the interview and obtain permission to tape it. Once the interview is complete, students are asked to transcribe it and then write a 12-page paper about major themes in the professor's work. One of the most common questions I get about the assignment is a deceptively simple one – "what do I do with the interview material?" Indeed, students often treat the interview as a "side show," as if the "real" data comes from what the professor wrote, or what others have written about him. Once students realize that the interview and its transcript are also data, they turn to technical questions. They want to know if they should summarize what a professor says or quote her directly. If they choose to quote her, they wonder whether to use snippets or paragraph length quotes, and whether to include their questions in the text or leave them unstated. Students also want to know how to treat the information in an interview. Is interview information "factual" or is it opinion to be dissected?

There is, of course, no right answer to any of these questions. How one uses interview data (or other qualitative data) depends on the methodology, broadly, and more specifically what the data is supposed to help one do (i.e. what one was trying to find out in the first place). Wolcott (1994) identifies three general approaches to using data – description, analysis, and interpretation. For the most part, these three approaches apply to non-causal research designs. As such, these approaches would not apply to some positivist studies. Of the three, description is the most straight-forward. A descriptive study is designed to describe something unknown to the reader. An analysis uses description, but it also examines data systematically. An analysis focuses on how things/people in a system fit together or interact. An interpretive analysis addresses wider issues – what the working of this particular system says about society at large. These three approaches to using qualitative data are not rigidly distinct. As Wolcott (1994) notes, it is difficult to specifically identify where the borders between them lie. Likewise, these categories also contain elements of the others within them. A researcher who wants to use a descriptive approach to present interview data, for example, will still rely on wider references points to craft questions. Asking a gay activist, for example, if his activism is related to discrimination is based on an interpretive assumption that contemporary social movements are rooted in challenging social inequality. Finally, it is possible to use all three approaches. A researcher can describe a group's belief system while also analyzing how the various parts fit together and what all of it means in a wider context. In the remainder of this section I detail these three approaches drawing on examples from geography and other social science disciplines.

Descriptive approaches

Descriptive approaches can be used by any of the four methodologies detailed in the section above. The descriptive approach is designed "to stay close to the data as originally recorded" (Wolcott 1994: 10). Scholars using a descriptive approach tend to treat fieldnotes or interview transcripts as "facts." The goal is to describe something that is unknown or poorly known. Descriptive studies can flesh out what is known about a topic. Once a phenomenon is thoroughly described, other researchers can use that descriptive base to develop a hypothesis or test a theory.

Although description is rarely used by itself today, it was an important part of the early anthropological canon. In the first half of the twentieth century, for example, Anthropologists like Margaret Meade and Bronislaw Malinowski provided detailed accounts of societies unknown in and to western countries. They described the family structures, mating rituals, and belief systems of societies first encountered by westerners during the age of colonialism. The organization and structure of these societies was a complete mystery to western anthropologists and their job was to describe it, often in excruciating detail. In *The Argonauts of the Western Pacific*, for example, Malinowski (1922) described broad topics like the geographic bounds of the Kulu region of Papau New Guinea, social taboos, and the organization of ownership. He also described (in great detail) specific topics such as canoe making, the handling of yams, witchcraft, gardening, and attitudes towards wealth. In many respects early anthropology was as much about collecting data as understanding what it all meant.

While descriptive analysis (alone) began to fall out of favor within Anthropology and other social science disciplines by the 1950s, description remains an important element of qualitative research. Clifford Geertz, who did much to move anthropology away from its colonial roots, argues that "thick description" remains a valuable part of the anthropologist's tool box. Description also continues to be important for research focused on new phenomena. Loretta Lees' work (2003) on supergentrification in Brooklyn Heights, New York provides a case in point. Drawing on Marxist inspired theories of gentrification (Smith 2002; Hackwork and Smith 2001), Lees identifies a new type of gentrification, supergentrification, which entails "the transformation of already gentrified, prosperous and solidly upper-middle-class neighbourhoods into much more exclusive and expensive enclaves" (p. 2487). Lees documents supergentrification by using census and interview data. Census data between 1970 and 2000 indicate, for example, that while Brooklyn Height's racial composition remained relatively stable during the period, its economic character changed. Mean income, educational attainment, and levels of professional employment increased between 1970 and 2000. Lees also used interviews with extant residents in Brooklyn Heights to understand how they saw their new neighbors. She found that most extant residents viewed new neighbors with suspicion. They described their tastes as ostentatious and their efforts at remodeling rude and noisy. They also took issue with what they saw as the declining cohesiveness of their neighborhood; new residents, many of them Wall Street financiers, routinely failed to participate in community events. To demonstrate these tensions Lees uses lengthy quotes from her informants.

Lees's work provides a good example of how description can still be useful in contemporary social science. Her description (of a new phenomenon) provides a basis from which to test existing theories. Indeed, she concludes her article by noting that the supergentrification of already gentrified neighborhoods "presents something of a challenge" to traditional models of gentrification that presume "a final endpoint to the process" (p. 2506).

Although descriptive accounts (of communities, people, and/or places) can be fascinating, especially for outsiders, they have their limits. In particular, descriptive work tends to beg more questions than it answers. Lees' study, for example, poses a provocative conclusion – that classic gentrification models might be wrong – but it fails to definitively assert or prove it. Moreover, for practitioners and other specialists "on the ground" (e.g., a low income housing advocate in a gentrifying neighborhood), descriptive work can often appear redundant. Practitioners usually already know the contours of a problem; they tend to prefer analysis geared towards a solution. Descriptive work also can be criticized for being subjective. What details a researcher decides to include (or not include) in her work is not governed by a set standard. As Wolcott notes, "there is no consensus on what exactly constitutes the 'thick description' Clifford Geertz immortalized" (p. 14).

Analytic approaches

The analytic approach to research is designed to uncover the internal infrastructure of a phenomenon. In particular, a scholar using this approach asks questions about structure, patterns, and relationships. Unlike descriptive approaches, analytic ones aspire to be more "scientific." A researcher using an analytic approach wants outsiders to view his research as not only accurate in terms of the depiction presented, but also the mechanisms used to draw conclusions about it (Mitchell 1983 as cited in Wolcott 1994). Scholars who want to take an analytic approach must make this determination upfront because this sort of analysis is only possible if data is collected in a systematic way. Systematic data collection includes using random or systematic sampling (so you can use your sample data to make claims about a wider population), standardizing interview questions, and normalizing observation times (to ensure that observations don't vary across time and/or place). Kathy Blee's study on white supremacist women (2002) provides an excellent example of employing an analytic approach to data.

Blee's study is designed to understand how and why women in the United States join hate movements. Blee notes that developing a statistically random sample of women in the hate movement is impossible because "there is no comprehensive list of racist activists or even a reliable estimate of their numbers" (p. 198). Many groups are underground and those that are countable tend to inflate their numbers. However, Blee determined that it was still possible to develop a representative sample. She began by developing a comprehensive list of hate groups through personal contacts (based on prior research), secondary literature, lists maintained by watchdog groups, and archival collections at two universities. Blee collected and read/viewed/listened to anything produced by the groups (newspapers, videos, speeches, etc.) for one year. Doing this allowed Blee to create a subset of groups that had significant participation or leadership by women. From this list Blee selected thirty groups that represented the ideological, structural, and geographic diversity of the wider group. Blee then sought to interview at least one female leader or active participant in each group.

Blee's systematic approach to creating a sample of women to interview allowed her to make general statements about how and why women join white supremacist groups. She identified several characteristics most women in her sample had in common: they were educated, had good jobs, grew up in financially stable families, and did not suffer abuse.[5] Blee also identified the relationship between beliefs and movement participation. Contrary to stereotype, Blee discovered through her interviews that women did not join hate movements because they were looking for an outlet for well developed ideological beliefs. Rather, most developed or crystallized racist beliefs after joining a group. The recruitment process was more important than extant beliefs because it created a sense of community, identity, and belonging for Blee's interview subjects. Blee also discovered that affiliation with a racist group did not occur because of specific targeting by recruiters or a prior receptivity by the women in her sample. Rather, "simple happenstance is often an element of racist affiliation" – meeting someone who is a white supremacist and getting to know them over time.

Blee's work is a good example of an analytic approach to handling data. Indeed, she self-identifies her work as analytical. She notes, for example, that her study is "the only relatively systematic sample of racist group members in the contemporary United States" (p. 198). Analytic studies like Blee's study are useful for policy-makers and other practitioners who want to act on research, to make policies beyond the area or people studied. By collecting qualitative data in a systematic way, Blee's findings can be generalized to the wider population of females in the white supremacist movement. And, policy-makers and other practitioners can use this data to design prevention and intervention schemes that are most likely to work.

The analytic approach is not, however, without flaws. A primary problem with this approach is that it does not permit the researcher to address the wider meaning of the phenomena under study. By sticking so closely to the data, and the "rules" about what one can and cannot say about data, a messy world is often isolated, boiled down, and made unrecognizable. Blee's work, for example, provides readers with a plethora of detail on how women get into the hate movement, but it tells us very little about what female participation in the hate movement says about changing gender norms or the state of race relations in the United States. What does it mean, for example, that feminist ideals of independence and self-fulfillment can be achieved in a non emancipatory context – i.e. used to denigrate others? In short, an analytic approach allows us to understand how something works, but it does not allow us to put the phenomenon into relief, showing its place in a wider context.

Interpretive approaches

If description is about facts, and analysis about showing how they fit together, then interpretation is about saying what it all means (Wolcott 1994). Pinning down the meaning of something is often a subjective process. If one were to ask 50 people what they think a Picasso or Matisse painting means, for example, one would likely get 50 unique responses. Assessing meaning in a scholarly way, then, requires having a metric or reference point against which to make a statement about the meaning of something. Usually, in the social sciences, the reference point is theoretical. It can also be experiential – based on the life experiences of the subject being studied and in some cases those of the researcher. Interpretation is also about taking information that is particular (to a person, a group, a place) and using it to make a comment on something more general. Unlike the analytic approach, however, interpretation does not require that one develop a random or representative sample before making more general statements. Indeed, the goal of interpretation isn't to make declarative of definitive statements about a population based on a sample of it but rather to say what the condition of that population means in wider society.

The author's study of the militia movement in Kentucky provides a good example of an interpretive study (Gallaher 2003). The study was driven by a poststructural methodology and was designed to unpack the identity politics of the militia movement in Kentucky. While I wanted to say something about the particular militia I was studying, I also wanted to say something about what the resurgence of far right militarism in the United States meant for our political system. My project involved data collection from three sources. I began by collecting textual material produced by militia groups in Kentucky, including magazines, leaflets, and other pamphlets.

I also engaged in participant observation by attending meetings of a militia-affiliated group. At these meetings I identified key militia leaders and requested interviews with them. Like many interpretive studies, mine involved both description and some analysis. I described the emergence of the Kentucky militia movement, identified its key leaders, and discussed the key issues on which the movement had staked positions. I also took an analytic approach to some data. I compared descriptive data on the Kentucky movement (when the movement emerged, why/how it emerged, what its key issues were, etc.) to regional and national studies. I discovered that the movement I chose to study varied significantly from groups in the Great Lakes region and in the intermountain west. The Kentucky movement started, for example, after the 1995 bombing of the Murrah Federal building in Oklahoma City – a period in which other militias contracted in size or fell apart. In contrast to other groups, the Kentucky movement had also adopted an "above ground" policy, attempting to work with local and state representatives rather than plotting to overthrow them. The majority of my analysis, however, was interpretive. I wanted to understand what my informants' views on the world meant for wider issues of race and class.

To assess my informants' views for wider meanings I used my theoretical perspective as a measuring stick. My study was informed by poststructural understandings of social positionality. This framework views social positions such as race, gender, sexuality, etc. as socially constructed rather than given. In the contemporary United States, white, heterosexual, and male are hegemonic subject positions. Norms of behavior, social expectations, etc. are defined around what is normal for white, heterosexual males. People are "othered" if they differ from this tripartite norm in one or more ways. I used this theoretical insight to interpret why my informants – a group of white (presumably heterosexual) men – would feel the need to arm themselves for a showdown against a state in which they occupy hegemonic subject positions. I also wanted to know how the term "patriot" became a nodal point around which a variety of meanings, positions, views were collapsed.

To make my assessment, and bring the reader along, I used empirical chapters to first lay out my informants' views on a variety of topics, including the American political system, specific issues of concern in Kentucky, and the state of the local and national economies. I used long excerpts from my interviews and often included the back and forth of conversation between myself and an interview informant. I interpreted their comments vis-à-vis my theoretical framework. I argued that militia members stood on an identity fault line. As white men, they occupied a dominant place in society, yet as working class, they were "exploited (or exploitable) by virtue of their position within the global division of labor" (p. 19). The dual identity of militia men meant that they could be mobilized "through progressive outlets by focusing on their class position within a global capitalist system and their linkages with workers in other parts of the world, or…[through] regressive outlets, focusing on what they regard as assaults on their social dominance" (p. 20). I also argued that discourses of patriotism "deflect[ed] a sustained analysis of class-based concerns while buttressing notions of cultural and racial superiority through 'safe' nationalist coding."

The interpretive use of data allows scholars to offer insights that cannot be counted, measured, or otherwise verified. They allow one to make connections that a quasi-scientific[6] approach would never permit. This characteristic is both the strong and weak point of interpretation. On the positive side, my use of interpretation allowed me to make conclusions about the movement that neither a descriptive or analytic approach could allow. Creating a random sample of militia leaders in Kentucky (and a control sample of militia leaders across the country) would have been time and cost prohibitive. It would have also undermined the approach I wanted to take – one that allowed me to dig into detail and nuance because my study scope was small. Moreover, even if I had been able to do such a study, there is no scientific tool for measuring meaning. The kind of insights I was able to share were based on an experiential level – reflecting on my experiences studying the movement, those of my informants over time, and more broadly my roots growing up in a white, working-class family in the south. However, my study is limited in that it can be neither replicated nor verified. The lack of systemization in my data collection, for example, would make it impossible for someone to replicate my study in Pennsylvania or Ohio to test my findings. In an author-meets-critics review of my book on the movement in ACME, for example, one critic made just this point, noting that my study would be more useful in comparative context (i.e. designed to be replicated) because it could help explain why there are "very different [militia] politics despite similar circumstances" (Miller 2004: 6). Likewise, while some of my findings could be internally verified (by checking the public record for arrest warrants, reviewing my interview tapes, etc.), the wider interpretation cannot be tested vis-à-vis external reference points. Indeed, there is no comparative "test" of the meaning or significance of nodal points.

A combination of approaches

In many cases a researcher will choose to write up his data using a combination of approaches. A good example can be found in a now classic essay anthropologist Clifford Geertz (1973) wrote on Balinese cockfighting. The essay deftly combines all three approaches.

Geertz begins the piece descriptively, by recounting his and his wife's arrival in 1958 in a Balinese village they planned to study. They arrived, "malarial and diffident" (p. 412) and remained that way for more than a week. Ten days into their stay, and still feeling disoriented and unwelcome, Geertz and his wife decided to attend a cockfight being held in the village square. The location of the fight – in a public place – was not typical. As Geertz notes, cockfighting is illegal in Bali, so "fights are usually held in a secluded corner of a village in semisecrecy" (p. 414). This particular cockfight, however, was out in the open, in plain view. Geertz offered several speculations as to why – "perhaps because they were raising money for a school that the government was unable to give them, perhaps because raids had been few recently, perhaps, as I gathered from subsequent discussion, there was a notion that the necessary bribes had been paid, they thought they could take a chance on the central square and draw a larger and more enthusiastic crowd without attracting the attention of the law" (p. 414).

Despite the organizers' assumptions, the fight would attract the attention of the authorities. "Amid great screeching cries of 'Pulisi! Pulisi!' from the crowd, the policemen jumped out, and, springing into the center of the ring, began to swing their guns around like gangsters in a motion picture, though not going so far as actually to fire them" (p. 414). Geertz and his wife would find themselves fleeing uniformed officers with the rest of the attendees. They eventually ducked into the compound of a fellow fleeing man, whose wife quickly arranged "evidence" the three had been having tea and chatting all afternoon should a policeman knock on the door, as one shortly did.

The decision to run (like everyone else at the fight) would prove advantageous to Geertz and his wife. As Geertz recounted, "everyone was extremely pleased and even more surprised that we had not simply 'pulled out our papers' (they knew about those too) and asserted our Distinguished Visitor status, but had instead demonstrated our solidarity with what were now our covillagers. (What we had actually demonstrated was our cowardice, but there is fellowship in that too)" (p. 416).

Geertz supplements descriptive information about his acclimatization in the village with information pertaining to the rules of cockfighting. He pays particular attention to rules governing the placement of bets. He describes the two main types of bets, who can and cannot place them, how much a person may bet per type of bet, and the calling of odds. Geertz description of cockfighting rules was also guided by analytic principles. Indeed, he attended and collected data at 57 cockfights. And, he collected the data in a systematic way, allowing him to make generalization about all Balinese cockfights. As he explains, "of the fifty-seven matches for which I have exact and reliable data on the center bet, the range is from fifteen ringgits [Balinese currency] to five hundred, with a mean at eighty-five and with the distribution being rather noticeably trimodal" (p. 426). Indeed, one could imagine successfully placing a bet in a cockfight anywhere in Bali after reading Geertz' analysis of betting mechanics.

Geertz also wanted to understand what the cockfight meant in Balinese culture. He knew they were illegal, that they were bloody (chickens often fight to the death), and that people frequently lost large sums of money in bets placed at them. But, he did not understand what they meant to the Balinese that participated in them. Geertz uses Bentham's notion of deep play to ground his interpretation. Geertz describes Bentham's concept as a game "in which the stakes are so high that it is, from his utilitarian standpoint, irrational for men to engage in it at all" (p. 432). Geertz interpreted cockfights in Bali to be a symbolic act. Battles of social dominance (at the individual and kin group levels) within the village were given symbolic form in the cockfight. This interpretive measuring stick allowed Geertz to identify several unspoken rules or practices about cockfighting that seem counterintuitive from a financial perspective (i.e. for a bet placer that wants to win, not lose money). Geertz discovered, for example, that an individual "never bets against a cock owned by a member of his kingroup" (p. 437), even if the other group's cock is the better bird. Likewise, when there is no kin group bird to place a bet on, the bet placer will always choose the bird of an allied kin group over that of an unallied one, even if that means betting against the stronger bird. And, in a cockfight where a man could reasonably vote on either bird (because of competing or complex loyalties), "he tends to wander off for a cup of coffee or something to avoid having to bet" (p. 439). In short, cockfighting provides a safe outlet for the airing of "interpersonal and intergroup" rivalries (p. 440).

Dance with the One That Brung Ya: Being True to Your Methodology

Researchers are often tempted to make conclusions that go beyond what their data will allow. In many ways this is to be expected. After spending several months (or more) in the field, and several more sorting and organizing data, scholars often rightly feel they know a topic inside and out. One way to avoid this temptation is to remember what kinds of questions your theoretical perspective is designed to answer, and of importance here, what kinds of answers your methodology can provide.

One key mistake students and more accomplished scholars alike make is to assert causality when the methodology does not permit it. Postitivism is the only methodology discussed here that permits a scholar to make a statement of causality.[7] The most a scholar can assert using the other methodologies here is a link between "variables." For academic reviewers (whether they are members of a dissertation committee or a journal editorial board) asserting causality when your method does not allow it is often considered a fatal mistake. It can be enough to fail a dissertation or reject an article. For the young scholar eager to make a mark on his discipline, using less forceful language can feel wrong. After the great time investment research often takes, scholars rightly feel they want to say something definitive about the topic studied. However, following the rules of one's methodology will pay off in the long run. It is also worth noting that other methodologies do not view causality as uni-directional. Marxists, for example, view social change as dialectical, where causality is multi-directional and difficult to untangle in a standard regression model. Likewise, some methodologies, such as poststructuralism, tend to ignore questions of causality altogether. Understanding meaning is not about delineating what the effect of "y" is on "x," rather it is about understanding the complex, messy processes by which meaning is discursively cemented (if only temporarily) through iterative processes.

A second, common mistake is to make analytic statements from data that was not collected systematically. For people who do small "n"[8] studies (which are rarely random or representative), it can be tempting to make declarative statements about the larger group from which the "n" is derived (e.g., members of an ethnic group). However, if data is not collected systematically, the only statements that can be made definitively are those about informants part of the study "n." Scholars can, of course, indicate that they believe their finding may hold true for a wider group, but the language must be suggestive rather than definitive. As in the above mistake, it can sometimes be frustrating for a researcher who has spent so much time in the field to come back and use language that feels hedged or partial. In these circumstances it is best to remember that what small "n" or otherwise descriptive studies lack in scope, they make up for in detail. Indeed, *Tally's Corner* (which is discussed above) is a good example. Liebow never claimed his work was the definitive account of African-American streetcorners in the United States. It was a study of one street corner in one city. In that respect, it contained no scientific value. However, by providing detailed contextual information on just a few men, it highlighted the structural constraints that many other impoverished men faced at the time (Duneier 2007). Indeed, and perhaps ironically, Liebow's restraint in *Tally's Corner* allowed his work to spark a national conversation on race and poverty.

A final common problem can be found in interpretive approaches. Wolcott (1994) suggests that interpretation is the most difficult way to approach data. Indeed, he suggests that "novice researchers who feel uncertain about how far to go, how much interpretation to offer, should err on the side of too much description, too little interpretation" (p. 36). Skill at interpretation is something that comes with experience, and even then only for some. A key problem with interpretation is, as Wolcott suggests, that it is often "little more than a point of view in disguise" (p. 37). Of course, it is difficult to draw a fine line between opinion and interpretation precisely because the building blocks of interpretation are often experiential – whether the experiences are those of the informant or the researcher. Experience is often a crucial (though certainly not the only) element in the fashioning of opinions. It is probably best to recognize at the outset of any interpretive analysis that some will view interpretation as opinion no matter what one does. There are, however, some things an author can do to avoid this sort of criticism – a criticism, by the way, that can lead a reviewer to reject a work out of hand. Primary among them is keeping one's interpretive measuring stick close at hand, and making sure the reader knows it is there as well. Thus, for example, if one is a Marxist by political bent, but one's methodology is interpretive and one's study is about something not directly connected to class, one should resist the hesitation to bring it into the interpretation. If one feels compelled to do so, one should make sure the connection between class and one's theoretical measuring stick is clear and on point. Otherwise, one's analysis may appear as poorly costumed opinion, or worse as incoherent and tangential.

Conclusion

When a scholar leaves the field she can rightly be called an expert. Professors often tell nervous graduate students as much in the minutes before a thesis or dissertation defense. It is cliché but true advice. In most cases, a scholar will have spent more time in the field studying a given topic than anyone on his or her committee (it can be another story once you enter wider academe). Unfortunately, expertise is best used sparingly. In academic work, the translation of expertise into written work is accomplished when the researcher cum writer is faithful to her methodology. There are two ways to do this. The first is to develop a research design that is consistent with the theoretical position you are using to guide your work. The second is to make sure that the handling, organization, and ultimate analysis of one's data is consistent with one's methodology.

As noted earlier in the chapter, there are four broad theoretical approaches that guide qualitative methodology – positivist, interpretive, Marxist, and poststructuralist. Each of these theoretical frames takes a different view not only on what constitutes data, but how one should analyze that data. Keeping one's methodology in mind during time in the field, and once one gets home will provide consistency to a project. It also will help one avoid the common mistakes discussed in the section above.

Once field data is organized and ready for analysis, researchers have three ways to approach it – using description, analysis, or interpretation. No one approach is better than the others. Rather, it depends in large part on what type of data a researcher has and what he or she wants to do with it. A researcher who wants to "prove" something would be better off, for example, using an analytic approach while one who wants to spark a debate would be better placed using an interpretive approach. It is also useful to keep in mind that researchers can and do use all three approaches simultaenously, although doing so usually entails emphasizing one approach over the others.

In following these "rules," of course, it is also important to bear in mind an important fact about research. No matter what methodology one chooses, good research speaks to the reader. Researchers can speak to their readers with good writing, insight, and most of all humility.

Notes

1 The civil war (or "Troubles") in Northern Ireland was a conflict over the province's geopolitical status. Catholics wanted the province to become part of a united Ireland while Protestants wanted to remain part of Great Britain. During the war Catholics were "represented" by Republican paramilitaries such as the Irish Republican Army (IRA). Protestants were "represented" by the state and by Loyalilst paramilitaries, including the Ulster Volunteer Force (UVF) and the Ulster Defence Association (UDA).
2 Denzin and Lincoln (2005b) use the term interpretive rather than ethnographic. I use ethnographic here in order to avoid confusion with my discussion in the next section on the three ways to "use" data – description, analysis, and interpretation. Interpretive (or ethnographic) methodology is not the same thing as the interpretative use of data. While a person using an ethnographic methodology would likely decide to interpret (rather than describe or analyze) his data, he could also choose to analyze it. Likewise, a Marxist or a poststructuralist could interpret (rather than describe or analyze) data.
3 Participant observation is a method first employed by anthropologists. It has since been adopted and used by sociologists, geographers, and other social scientists.
4 In the foreword to *Tally's Corner* Hylan Lewis summarizes Liebow's findings, noting that "they use friendship and the 'up-tight' buddy system as resource and buffer and in their design to protect dignity and to rationalize and conceal failure if necessary" (p. x).
5 Although the financial and familial status of a given recruit is descriptive, when a sample is used, a researcher can identify a pattern of characteristics common in recruits as a whole (making it an analytical exercise).
6 I use the term quasi-scientific here to mean approaches that attempt to approximate the scientific method by developing and testing a hypothesis.
7 It is important to note that not all positivist studies are causal.
8 "n" is often used as a placeholder for sample size, i.e. a study can have an "n" of 20, 50, etc.

References

Blee, K. (2002) *Inside Organized Racism: Women in the Hate Movement*. Berkeley: University of California Press.
Del Casino, V.J., Jr. (2009) *Social Geography: A Critical Introduction*. Chichester: Wiley-Blackwell.
Denzin, N. and Lincoln, Y. (2005a) Paradigms and perspectives in contention. In D. Denzin and Y. Lincoln (eds.), *The Sage Handbook of Qualitative Research*, 3rd edn. London: Sage, pp. 183–90.

Denzin, N. and Lincoln, Y. (2005b) Introduction: the discipline and practice of qualitative research. In D. Denzin and Y. Lincoln (eds.), *The Sage Handbook of Qualitative Research*, 3rd edn. London: Sage, pp. 1–32.

Duneier, M. (2007) On the legacy of Elliot Liebow and Carol Stack: context-driven fieldwork and the need for continuous ethnography. *Focus* 25 (1): 33–8. Available online at University of Wisconsin–Madison Institute for Research on Poverty website, www.irp.wisc.edu/publications/focus/pdfs/foc251.pdf. Accessed July 13, 2009.

Gallaher, C. (2003) *On the Fault Line: Race, Class, and the American Patriot Movement*. Lanham: Rowman and Littlefield.

Geertz, C. (1973) *The Interpretation of Cultures*. New York: Basic Books.

George, S. (1990) *A Fate Worse Than Debt*. New York: Grove Weidenfeld.

Graves, S. and Peterson, C. (2008) Usury law and the Christian right: faith based political power and the geography of American payday loan regulation. *Catholic University Law Review* 57 (3): 637–700.

Hackworth, Jason, and Smith, N. (2001) The changing state of gentrification. *Tijdschrift voor Economische en Sociale Geografie* 92: 464–77.

Howarth, D. (2000) *Discourse*. Buckingham: Open University Press.

LaClau, E. and Mouffe, C. (1985). *Hegemony and Socialist Strategy*. London: Verso.

Lees, L. (2003). Super-gentrification: the case of Brooklyn Heights, New York City. *Urban Studies* 40 (12): 2487–509.

Liebow, E. (1967) *Tally's Corner: A Study of Negro Streetcorner Men*. Boston: Little, Brown, and Company.

Malinowski, B. (1984) [1922] *The Argonauts of the Western Pacific*. Prospect Heights: Waveland Press.

Miller, B. (2004) Thoughts on Carolyn Gallaher's *On the Fault Line*. ACME 3 (1): 4–6.

Ould-Mey, M. (1996) *Global Restructuring and Peripheral States: The Carrot and the Stick in Mauritania*. Lanham: Littlefield Adams Books.

Schwanen, T. (2007) Gender differences in chauffeuring children among dual-earner families. *Professional Geographer* 59 (4): 447–62.

Smith, N. (2002) New globalism, new urbanism: gentrification as global urban strategy. *Antipode* 34: 427–50.

Tierney, S. and Kuby, M. (2008) Airline and airport choice by passengers in multi-airport regions. *Professional Geographer* 60 (1): 15–32.

Whyte, W.F. (1955) *Street Corner Society: The Social Structure of an Italian Slum*. Chicago: University of Chicago Press.

Wolcott, H. (1994) *Transforming Qualitative Data*. Thousand Oaks: Sage.

Chapter 12

The Worldly Work of Writing Social Geography

Lieba Faier

In his now famous discussion of the poetics and politics of "writing culture," James Clifford tells us that "[t]he making of ethnography is artisanal, tied to the worldly work of writing" (1986: 6). In this chapter, I focus on ethnography as a means for writing social geography, and, specifically, social geographies of encounter. I consider how ethnography, not only as a research method but also as a narrative practice, enables us to engage in the worldly work of mapping the ways lives and perspectives unequally come together in the making of places and cultural meanings. My aim is to consider how we can use ethnography to represent those quotidian, unequal, and intimate processes of encounter through which places and cultures today emerge as, in Doreen Massey's words, "articulated moments in networks of social relations" (1993: 66).

Ethnography is useful for social geographers because it draws attention to the dynamics of everyday sociospatial processes. Methodologically, it is most frequently associated with participant observation, an approach that is usually traced back to Bronislaw Malinowski. Ethnographers engage in what Clifford Geertz has famously called "deep hanging out" (Geertz 2001: 107). They immerse themselves for extended periods of time in the day-to-day worlds of those whose lives they are studying, and they attempt to learn to see things through these people's experiences and perspectives. Ethnographers also conduct numerous formal and informal interviews in which they sometimes allow those with whom they are speaking to direct the conversation onto topics these people find most important. In analyzing their material, they make an effort to theorize social and cultural processes through the details of people's stories and everyday lives.

A Companion to Social Geography, First Edition. Edited by
Vincent J. Del Casino Jr., Mary E. Thomas, Paul Cloke, and Ruth Panelli.
© 2011 Blackwell Publishing Ltd. Published 2011 by Blackwell Publishing Ltd.

Long overlooked in geography, ethnography has of late been attracting attention in the discipline (e.g., DeLyser 1999; Herbert 2000; Katz 2004; McHugh 2000; Merrill 2006; Till 2005). In recent years, geographers have argued the utility of ethnographic methods for examining the ways that place and space are produced through ongoing social and political economic processes (e.g., Hart 2004; Herbert 2000; Katz 2004; Merrill 2006). They have focused on ethnography as an analytic tool for developing frameworks that can explain structures of social action (Herbert 1997). They have also suggested that ethnography can be helpful for understanding the articulations of different trajectories of sociospatial change in different parts of the world (Hart 2004).

However, most geographers focus on ethnography as a research and analytical method. Few have considered its significance as a *narrative genre*. Here I draw attention to ethnography as a writing practice – a means for writing social geographies, and, specifically, social geographies of encounter. I argue that because ethnography allows for the presentation of multiple voices and perspectives, it is particularly well suited for illustrating the intersecting social and political economic processes through which different social geographies emerge.

Since the mid-1980s, attention to ethnography as a narrative practice has gained ground in other disciplines, and particularly in anthropology, alongside the reflexive turn in studies of culture prompted by the work of James Clifford, George Marcus, and Michael Fischer, among others (see Clifford and Marcus (eds.) 1986; Marcus and Fischer 1986). These scholars stress the inseparability of the poetic and the political in ethnographic writing. They argue that relations of power lie at the center of such a writing practice. They tell us that how one writes can be as important as what one says.

These arguments have had two implications for ethnographers. First, they have pushed ethnographers to consider how their own positioning affects their relationships with the people they study and the data they gather (see Chapter 10, this volume). Feminist geographers, in particular, have productively embraced this point as they have explored how our research and writing practices are based in embodied, everyday, social, and political economic relations of power (e.g, Davis 1994; Gilmore 2007; Katz 1994; Mountz, Moore, and Brown 2008; Nagar 2006; Pratt 2004; Pulido 2006; Wright 2006). Second, critical attention to ethnography as a narrative practice has prompted ethnographers to develop new, experimental writing strategies that undercut claims of absolute knowledge and ethnographic authority, which have been identified with forms of domination (e.g., Abu-Lughod 1993; Kondo 1990; Pratt 2004; Stewart 1996; Steedly 1993; Tsing 1993; Visweswaran 1994). These writing strategies, which have included fragmenting narratives and blurring boundaries between fiction and ethnography, highlight the limitations of producing knowledge about others and stress the situated and partial nature of all theory. They also offer means for evoking some of the contingency, messiness, and surprises of everyday social life and the complex workings of power within it, which are often overlooked by (if not written out of) all-encompassing analytical frameworks.

In the remainder of this chapter, I consider one ethnographic narrative strategy: a focus on cultural encounters. We might think of an ethnography of encounters as a narrative map that charts how social and spatial formations take shape through the unequal convergences of people's paths. Ethnographies of encounters highlight different voices and perspectives as these unequally come together to remake social worlds (see Faier 2009). They consider how these convergences create new spatial and cultural forms not only through the articulation of discrepant stakes and desires, but also through their misalignments – the misunderstandings, tensions, and gaps that develop between different people's objectives and dreams as these are situated within relations of power.

In what follows, I first discuss some of my own research on Filipina-Japanese marriages in a mountainous region of rural Nagano that I call Central Kiso. I ethnographically focus on encounters among Filipina migrants and Japanese residents as a strategy for conveying some of the awkward and unequal ways that the alignments and misalignments in their dreams and agendas are contributing to the reproduction and transformation of meanings of culture and place in Central Kiso.[1] I then follow with discussions of three scholarly works that illustrate other ways ethnographies of encounters are useful for writing social geographies.[2] I conclude by reflecting on how ethnographic writing, and ethnographies of encounters in particular, can help us capture both the complexity and the contingency of sociospatial processes.

Stories of Encounter, Place, and Culture in Central Kiso

My work on Filipina–Japanese marriages in Central Kiso considers the ways that recent transnational migration is contributing to the reproduction and transformation of cultural and regional identities in contemporary Japan. Since the mid-1980s, an increasing number of men living in Central Kiso have married Filipina women they met at local hostess bars. By 1999, when I was in the midst of doing 23 months of fieldwork in the region, about 60 Filipina women were married to local Japanese men, many in the towns of Agematsu and Kisofukushima, which had a combined population of around 15,000. Several additional Filipina women had married local men and later left their Japanese husbands and families. Although many Filipina women working in hostess bars do not directly exchange sex for money (Parreñas 2006), they are widely disparaged in Japan as prostitutes and foreigners.[3] However, a number of those who have married local men in Central Kiso have come to be viewed by local community members as *ii oyomesan*: ideal, traditional, rural Japanese brides. Some of these women have come to be described as "more typically Japanese than young Japanese women today." Japan is a nation that has long considered itself ethnically and culturally homogeneous. How has this transformation become possible?

Unraveling this paradox of Japanese cultural identity requires considering how discrepant, but not unrelated, social and political economic processes intersected and shaped the unequal dynamics of everyday relationships among Filipina migrants and rural Japanese residents. That is, it involves ethnographically mapping how Filipina migrants' and Japanese residents' desires and agendas came into transformative encounter. First, consider the ways that many Japanese residents made sense of these relationships. Longtime Japanese residents of Central Kiso often complained that the region was suffering from outmigration and what they call a "bride shortage" (*yome busoku*). Some explained that local men had come to view Filipina women working in bars as appealing spouses out of desperation. Since the

1960s, rural areas in Japan have been socially and politically economically marginalized by national modernization projects that have focused on urbanization and
industrialization. Many local residents in Central Kiso complained of a lack of
viable employment in the region since the decline, first, of the local timber economy
and, later, of the domestic tourism economy that temporarily replaced it. The
national government also has increasingly cut funding for local public works projects
that offered construction work to men and women in the area. As has become
common throughout rural Japan, young Japanese men and women have increasingly
left Central Kiso to find work in urban centers. Some Japanese men, and particularly
eldest sons, who traditionally inherited family homes and land and are responsible
for caring for their aging parents, have remained behind. However, many young
Japanese women do not want to join multigenerational rural families. (Young Japanese
women often describe these families as feudal and patriarchal.) These women have
increasingly opted to leave rural areas, to move to cities and marry white-collar
workers, if they marry at all.[4]

A few Filipina women's husbands owned family businesses in Central Kiso (a
green grocery, a construction company, a soba restaurant); however, most of these
men worked in construction or at *pachinko* parlors or bars.[5] Many of these men
were ten to twenty-five years older than their Filipina wives and past the "suitable
age for marriage" (*tekireiki*) when they met the women they married. Many of these
men had few marriage prospects other than Filipina women working in local bars.

Local residents (including local government administrators) were concerned
about the decreasing numbers of children being born in the region. They worried
not only about the area having enough children to provide a labor force and inherit
family land or local businesses, but also about simply qualifying for enough government funding to keep open local schools. As a result, in recent years Japanese residents in Central Kiso have in qualified ways become open to marriages between
local men and Filipina women. I say "in qualified ways" because Filipina women
were accepted only insofar as they proved that they were, as one local resident put
it, "different from ordinary Filipinos." This meant fulfilling their roles as *oyomesan*
(brides and daughters-in-law) by actively participating in their communities and
raising their children, who are Japanese nationals, as "Japanese."[6] Most Filipina
women married to Japanese men in Central Kiso were instructed to do things "the
Japanese way" (*nihon no yarikata*). They were taught to prepare and eat Japanese
foods (with chopsticks, not a fork and a spoon or one's hands, as food is commonly
eaten in the Philippines) and to speak, read, and write in Japanese. Often, they were
also expected to adopt a range of dispositions that Japanese residents identified as
appropriate to being the bride and daughter-in-law in a "Japanese" family, such as
greeting and assisting neighbors in prescribed manners and regularly participating
in community activities.

Filipina women were also seen to embody certain desirable traits. Many Japanese
community members, including several Filipina women's husbands, told me that
Filipina women were attractive spouses because they came from a poor country and
thus had the "good characteristics of traditional Japanese women."[7] Such comments
rehearsed widespread stereotypes in Japan that Filipina women are dependent,
submissive, and supportive of their husbands (Satake and Da-anoy 2006;
Suzuki 2003). A few men also described Filipina women as "exotic" or "cheerful,"

explaining that this made them more appealing than Japanese women. Some also expressed interest in the Philippines and a cosmopolitan desire to learn about Philippines cultures. However, although one man told me that he felt a spiritual bond with his wife, and another said he was looking for someone he felt would be "a good partner," most of the men were looking for marriages with women whom they believed (not necessarily accurately) they could control or would be dependent on them.

In contrast, many Filipina women married to Japanese men in Central Kiso spoke of quite different desires for their lives in the region. Many of these women found pleasure in their marriages, learned to fulfill their roles in their Japanese households, and even came to identify as *oyomesan*. However, their motivations for doing so, and their desires and agendas for their lives in Japan, often differed dramatically from those of their Japanese husbands and families.

Many Filipina women who went to Japan on entertainer visas during the 1980s and 1990s came from poor urban or rural communities in the Philippines. Many suggested that their families were in dire financial situations when they applied to go to Japan. Most of these women did not have postsecondary educations. Thus, they did not have the option of going abroad as nurses, caregivers, or domestic helpers, which required some level of education and thus tended to attract women from middle-class backgrounds (Constable 1997; Parreñas 2001; Pratt 2004). For most Filipina women married to Japanese men in Central Kiso, spousal visas were one of only two types of visas (six-month entertainer visas being the other) under which they could legally work full-time in Japan, and most of these women continued to support their families by working in local factories and bars after they were married.[8]

Yet for these women, transnational labor migration, and marriage, was more than a means for managing political economic marginality. It was also inspired by personal and affective commitments. Most Filipina women I met in Central Kiso told me that they had gone to Japan in search of "a better life." Many said that they had gone abroad for personal empowerment and adventure, describing desires to travel abroad, live in a "modern" place, and be independent. They spoke of goals of putting family members in the Philippines through school or of building homes for themselves and their parents there.

These women found pleasure and satisfaction in their marriages, which sometimes enabled them to achieve some of their personal goals. Because their job contracts and visas were issued for, at most, six months, every Filipina woman in Central Kiso with whom I spoke said that she had not expected to remain in Japan long term when she first came as a labor migrant. All said that they had never planned to marry a Japanese man. Although many of these women said that their marriages were prompted by either pregnancy or political economic considerations (or both), many also shared romantic stories about their courtship with their future husbands while they worked in local bars, sometimes describing their relationships as "love at first sight."[9] Others relayed how they had fallen in love with their husbands because of the men's persistence and attentiveness in taking the women out to fancy restaurants, driving them to church on their days off, and buying them expensive gifts. They explained that even if they had boyfriends in the Philippines, these men could not provide for them in the ways they imagined that their Japanese husbands could. In this way, political economic inequalities between Japan and the Philippines shaped these women's perceptions of their Japanese husbands as desirable spouses.

After their marriages, many of these women expressed desires to be good, loving wives and mothers. These desires reflected dominant, and often Catholic, ideals of femininity in the Philippines. As Catholics, many Filipina women were self-conscious about having worked in bars, even while about 30 percent of these women continued in such jobs after they were married (usually to send money to the Philippines and in some cases to financially assist their Japanese families as well). By demonstrating that they were good wives, mothers, and daughter-in-laws, Filipina women challenged negative stereotypes attached to them based on their bar work. Moreover, by fulfilling their roles as *oyomesan* in their Japanese families, Filipina migrants were granted a measure of legitimate social status in the region. Caring for elderly in-laws was also an important dimension of Filipina women's roles in their Japanese households in Central Kiso. Many of these women stressed that caring for the elderly was a value instilled in them growing up in the Philippines, and they took pride in the ways that doing so positively reflected on them as "Filipina."

Many Filipina wives of Japanese men in Central Kiso also took pleasure in those elements of their married lives in rural Nagano that resonated with the dream of a modern middle-class life that had initially inspired them to go to Japan. For example, they enjoyed decorating their homes, going out to dinner with their husbands, and shopping in large discount stores in regional cities. Filipina women in Central Kiso sometimes spoke with pride of their husbands' ability to contribute to household expenses, even if the men's salaries were not large by Japanese standards. Even after marriage, these women's commitments to their families in the Philippines remained strong, and many of them described this as a primary concern. Some Filipina women's Japanese husbands supported their desires to send money to their families to the Philippines, which these women greatly appreciated.

In other ways, however, many Filipina women's lives in Central Kiso did not approximate the dreams of a glamorous, "modern," middle-class world that had inspired them to go abroad. These women's goals for their lives often contrasted strikingly with those of their Japanese families. In some cases, their husbands did not support their desires to help their families in the Philippines. Some Filipina women said that their husband's treatment of them changed after their marriage. Others of them faced abuse or other serious and destructive problems in their relationships. Filipina women married to Japanese men in Central Kiso also often became disillusioned with life in the Japanese countryside. All of those who had met their husbands in local hostess bars had first been assigned to work in the region by their promoters in the Philippines, and many expressed surprise and disappointment to find themselves in a rural area. They spoke of desires to live in more lively, exciting, and "modern" places.

To get a sense of the overlaps and tensions that developed among the dreams and expectations Filipina women and Japanese men brought to their relationships, consider, first, what two Filipina women I knew – Ana and Marites – said about their decisions to go to Japan as migrant laborers.

Ana had come to Kisofukushima from the Philippines on an entertainer visa during the 1980s. Like most Filipina wives in the region, she had met her Japanese husband while working in a local Filipina hostess bar. At the time we met she was raising two elementary-school-aged children.

Ana and I were sitting in the house I was renting on the Kiso River when I asked her how she had imagined Japan before she came to live there.

"I thought Japan would be different from the Philippines," she replied in Japanese. She then switched to English, "I thought that, like America, it would be … What is the word? Very 'modern.'"

"You mean like New York or Los Angeles?" I asked for clarification. I knew that she had never been to the United States, and I was trying to get a sense of how she had imagined "America."

"Exactly!" she replied. "I thought that it would be like that. I pictured Japan modern-style. I thought it was like America. When I came to Central Kiso, I was like 'What?! What's this?!' There was nothing but mountains. There was nowhere to go out. I've been to the Philippine countryside. It's like this. But I didn't think there were places like that still in Japan."

When I met Marites in Manila, she similarly drew attention to the ways that dreams of an exciting and "modern" world like "America" had shaped her desires to go to Japan, stressing desires for the beauty and glamor that it promised. Marites was the younger cousin of another Filipina friend married to a Japanese man in Central Kiso. Twenty-four years old at the time, she was soon to return to Japan on her sixth six-month contract as an entertainer, and she had agreed to an audio-recorded interview about her experiences abroad.

"I just wanted to go," Marites told me in a mixture of Tagalog and Japanese when I asked her why she had first decided to work in Japan. We were sitting in the small house that her cousin had built for her mother. Marites told me that she had for some time wanted to go to "America," but after watching many young women from her neighborhood, including her older cousin, go to Japan as entertainers, she had also grown curious about what life was like there. She had wondered, "When they return to the Philippines, why are they so beautiful? Their hair is colored. They bring bracelets, necklaces. So I was curious. I wanted to go to Japan. Maybe I could be beautiful."

Marites suggested that dreams of independence had also prompted her to go abroad. She told me that after watching her cousin return from Japan and purchase a house and business for her family, she decided that she wanted to try to do the same. She explained, "I saw my cousin. She didn't need to work as a prostitute. But she had a brain. So she could build her own house. I thought to myself, why not me? Maybe I'm capable of that too. So I wanted to try, on my own, without help from anybody."

Now compare Ana's and Marites's comments to a conversation I had with the husband of a Filipina woman in Central Kiso, a man I'll call Tanaka-san, about his decision to marry his Filipina wife, whom he had met while she was working in a local hostess bar:

"I didn't marry her because she was beautiful," Tanaka-san explained in Japanese. "In Japan we have a saying: In three days you get sick of a beautiful woman and used to an ugly one."

I tried to mask my discomfort with his comments as he continued. "I dated a lot of women before I married my wife, a lot of Filipinas that I met in local bars."

"How many?" I asked.

Tanaka-san waved his fingers in the air and replied that he couldn't count them all on both hands.

"Why so many?" I pushed.

"I dated them to see if I liked them enough to marry them. You don't know if you want to marry someone until you date them."

"But why Filipinas?"

"After I hit thirty I started to like Filipinas."

"Why?" I pressed.

Tanaka-san paused. "I really couldn't say." He reflected for a moment and then replied, "They have the good characteristics of traditional Japanese women. I like that in a woman. They look up to their husbands. They respect them. They listen to what their husbands say."

"What attracted you to your wife out of all the Filipinas you dated?" I asked.

"I couldn't really say," Tanaka-san responded. "Part of it was that the timing was right. I also liked her because she was poor. Because I wanted to help her."

These ethnographic anecdotes suggest that both Filipina migrants and Japanese men crafted senses of self through their relationships. They also suggest that members of these groups brought desires for certain kinds of futures to their interactions. Yet, while their visions may have in some places overlapped, they were also strikingly different.

Both Marites and Ana told me that they were inspired to go to Japan by dreams of glamor and travel that resonated with their images of "America," a place that, on account of legacies of US colonialism in the Philippines, often figures as a site of wealth and privilege there. Both women also wanted to help their families while asserting their independence. Although many Filipina women in Central Kiso expressed a reluctance to return to the Philippines, citing the lack of job opportunities for women of their educational and class backgrounds, as well as concerns about crime and corruption there, they were generally not that impressed with what they had found in rural Japan. Many of them, as noted above, expressed surprise and disappointment to find themselves living in a mountainous rural area that in few ways approximated the exciting and glamorous visions that had led them to go abroad. Indeed, their lives in rural Nagano were marked by the irony that, as it turned out, the best possible route for them to realize their desires for independence, travel, modernity, and glamor was through performing roles as "traditional brides" in marriages with rural Japanese men.

Yet for Japanese men in Central Kiso, relationships with Filipina women working in hostess bars offered them an opportunity to present themselves as rich, powerful, modern, and desirable "Japanese" men, even while the very presence of these women in the town testified to rural Japanese men's undesirability in an urban Japanese imagination. Through their relationships with Filipina women, Central Kiso men asserted Japan's modernity and advancement vis-à-vis what they maintained was the "poor" and "undeveloped" Philippines. They attempted to overlook the social and political economic marginality of their own lives vis-à-vis those of elite white-collar workers in metropolitan centers like Tokyo by focusing instead on their conviction that life anywhere in Japan was better than life in the Philippines. They crafted a sense of masculinity based on feelings of both benevolence toward and superiority over "poor" Filipina women. These men also expressed nostalgia for traditional times and gender roles, and they believed that Filipina women were more amenable to those views than contemporary Japanese women were likely to be.

The awkward resonances and misunderstandings that emerged in relationships among Filipina migrants and Japanese residents have ironically been contributing to the creation of new senses of regional and cultural identities in Central Kiso (see Faier 2009). As I mentioned, Japan is a country that has long imagined itself as ethnically and culturally homogeneous, and Japaneseness has long been imagined as a biogenetically based identity. Rural Japan, in particular, has often been characterized as the last bastion of an untainted, traditional, and essentially Japanese way of life. However, because of some Filipina women's commitments to challenging negative stereotypes related to their work in bars, to being "good wives and mothers" in Japan, and to supporting their families in the Philippines (which their spousal visas enabled them to do by allowing them to legally remain and work in Japan), since the 1970s, these women have come to be identified and sometimes to identify themselves as ideal *oyomesan* in Central Kiso. These women have learned to prepare traditional Japanese foods and maintain Japanese homes. They participate in local *taiko* drum circles and the PTA at their children's schools, and work in their families' businesses. They are now raising "Japanese" children.

Japanese residents' attitudes toward these women and toward the Philippines have also been changing. In large part because these women have accommodated Japanese residents' expectations (but also occasionally because Japanese community members have shifted their expectations to accommodate these women), some Japanese community members told me that they have come to see Filipina women as more appealing brides and daughters-in-law than young Japanese women – as "more typically Japanese than young Japanese women today." In some cases, Filipina women's Japanese husbands have adopted the women's children from previous relationships in the Philippines and are raising them as their own. Some Filipina women's husbands told me that their wives had convinced them to move to the Philippines when they retired, something these men had never contemplated doing before their marriages. In sum, unequal relationships among Filipina migrants and longtime Japanese residents have been transforming life in the region, reshaping the everyday ways that local Japanese residents and Filipina women craft senses of self and imagine their place in the world. Filipina women are now playing active and material roles in reproducing meanings of Japanese culture and identity in Central Kiso, and through their everyday practice, they are reshaping the ways that Japaneseness and local identities are understood there.[10]

Social Geographies of Encounter Take Many Worldly Forms

In the previous section, I ethnographically mapped the different paths that led Filipina migrants and Japanese residents to their encounters in Central Kiso. I also suggested some of the ways the alignments and misalignments among their desires and agendas are contributing to the transformation of regional and cultural identities in the region. By evoking the dynamics of these encounters through personal narratives, my ethnography of encounters situates these changes in Central Kiso within larger political economic processes, while recognizing the geographical, cultural, and historical specificity of these transformations as they play out in different people's lives.

Transnational migration lends itself particularly well to an ethnography of encounters because it enables new processes of spatial and cultural production by facilitating the coming together of discrepantly situated dreams, agendas, and political-economic histories. An ethnography of encounters illustrates the micro-practices of everyday life through which these processes take shape. However, studies of other kinds of sociospatial processes can also be productively approached in such a manner. Below, I offer three examples of scholars who take, what I am calling, an "encounters approach" to ethnographic discussions of other topics of interest to social geographers – the making of natural-cultural worlds, the emergence of spaces of abjection, and the coproduction of race and place. In the interest of creating a multidisciplinary conversation, I draw here on the work of ethnographers who dialogue with geographers from outside the discipline to illustrate some of the possibilities their approaches offer for writing social geography. I suggest that by narratively mapping the situations they describe through stories of different kinds of encounters, each of these scholars illustrates, in a different way, what Massey describes as "the existence in the lived world of a simultaneous multiplicity of spaces: cross-cutting, intersecting, aligning with one another, or existing in relations of paradox or antagonism" (Massey 1994: 3).

First, Julie Cruikshank's *Do Glaciers Listen?: Local Knowledge, Colonial Encounters, and Social Imagination* focuses on how human encounters with other people and with the natural world figure in the production of local knowledge about place. Drawing on oral histories, travelers' journals, and research by geophysical scientists, she considers how glaciers variously impacted the ways that native populations and European colonialists in the Pacific Northwest understood their relationships with the environment and each other during the late nineteenth and early twentieth centuries. By moving across the perspectives and experiences of members of these different groups, Cruikshank evokes the very different ways that glaciers figured as active, animate beings for them. She demonstrates that while indigenous communities depicted glaciers' behaviors as situated within a moral universe where nature and culture could not be separated, Europeans, applying Enlightenment categories and scientific instruments, wrote about them as natural phenomena distinct from social ones. Illustrating the ways these different understandings of glaciers impacted relationships among members of these groups, Cruikshank shows how natural and cultural histories became intertwined and the imaginative force glaciers exerted on regional histories and social relationships, with lasting effects into the present.

In *Vita: Life in a Zone of Social Abandonment,* João Biehl traces the social, medical, and political economic intersections in contemporary Brazil that have resulted in camps where poor people with HIV, mental illness, and other incapacities are left to die. Focusing on the story of a woman named Catarina who has been left in a camp called Vita, Biehl draws together, as if piecing together a puzzle, the converging forces that create such spaces of social abandonment. He evokes the voices and perspectives of various people and organizations – those of medical and psychiatric institutions, the state, family members, pharmaceutical industries – that in different ways shaped the course of Catarina's life, illustrating how their objectives cut across, reinforced, and disrupted each other, and ultimately led to her neglect. In doing so, he asks us to consider the kinds of lives that different social arrangements support and the institutional and social spaces that correspond with such social formations.

Finally, Jacqueline Nassy Brown shows us that blackness in Liverpool cannot be understood apart from the intertwined histories through which the city has been made as a local place. In *Dropping Anchor, Setting Sail: Geographies of Race in Black Liverpool*, Brown illustrates how differently situated actors – African seamen, British shippers, white English and Irish women, Afro-Caribbean and West Indian migrants, and African American servicemen – figure in the making of Liverpool as a "local place," and she pushes us to consider the ways its "localness" is implicated in the production of race and gender there. Drawing out the different ways that men and women in Liverpool reference Liverpool's cosmopolitan past and "black America" as they craft senses of place, Brown demonstrates that "local" is a cultural category that differentially shapes the ways geographies of race are experienced and lived by differently situated people.

By bringing together multiple perspectives, experiences, and social forces, these three books offer ethnographies as narrative maps that evoke sociospatial configurations through stories about intersecting lives and histories. Moreover, because these narratives focus on the ways that different lives and perspectives come together, they highlight the unequal relations of power, including the displacements and disconnections, through which social geographies emerge. By introducing glaciers into stories of colonial encounters between native populations and explorers and settlers in the Pacific Northwest, Cruikshank shows how European scientific representations of this territory ignored native populations' understandings of the landscape as animate and overlooked possibilities of simultaneously natural-cultural worlds. Biehl pushes us to question why and on what basis some lives are rendered valuable and others unsalvageable, and he invites us to explore the ways such social relationships are spatialized, in part by the erasure of places occupied by those who have become "the ex-human" (2005: 24). Vita, he explains, "is the place where living beings go when they are no longer considered people" (Biehl 2005: 2). Brown draws attention to the ways gender and national differences shape the spatial production of blackness. She thereby illustrates how it is produced through conflicting social geographies that are shaped by power differentials within and across black communities.

By using ethnography to narrate the dialogic coming together of different stories, dreams, and voices, Cruikshank, Biehl, and Brown draw attention to social geographies that resist conventional and straightforward representations of space. Cruikshank's text enables us see the glaciers described by native populations in the Pacific Northwest as animate beings that defy scientific and cartographic representations. Biehl shows us that the official omission of Vita on city maps is a defining feature of such spaces where those who are no longer considered productive members of their communities are abandoned. Explaining that her black Liverpudlian informants felt marginalized and invisible in standard geographic representations, Brown focuses on stories her black informants tell about Liverpool. In doing so, she illustrates how place "is made not merely with bricks and mortar, but with politics, culture, (and) historical narrative" (2007: 378). As these three writers craft their ethnographies as multilayered stories, they evoke cross-sections of the intersecting spaces through which different social worlds take shape. They show us the resonances and tensions that develop among different sociospatial imaginings, and they highlight how these are positioned within larger relations of power.

Writing Worlds of Encounter

Some geographers who conduct ethnographic research have qualified the use of ethnographic methods, explaining that such methods offer deep and intensive analyses at the expense of broad and extensive comparisons (e.g., Herbert 1997). They have suggested extending the reach of an ethnographic analysis by using ethnographic cases as representative of broader trends or combining them with comparative or quantitative methods (Herbert 2000).

Yet the value of ethnographic writing lies in its ability not only to illustrate or complement broader modes of analysis but also to challenge those modes by describing the complexity of social life operating beneath the mantle of what appear to be sweeping political economic trends. By focusing our attention on the intimate details of everyday life, ethnographic writing shows us how sociospatial processes play out on-the-ground, illuminating the nuances and contradictions that studies of large-scale political economic forces often miss. By paying attention to differently situated perspectives, ethnographic writing demonstrates what is lost in translating across geographically distinctive cases and, thus, what cannot be incorporated into a comparative or holistic framework. By relating the stories of individuals, and describing their dreams, goals, and daily practices, it offers a glimpse of who lies behind statistical data, and enables us to relate to "research subjects" as people. Moreover, by focusing on people's accounts of their lives and desires, it teaches us to pay attention not only to the content of what they say but also to the poetics and politics embedded in their statements – the ways they understand the world and the desires that motivate their actions. We can then relate *these* understandings and desires to broader political economic processes, gaining a window into the quotidian and contradictory ways these processes are inhabited, embodied, and enabled.

In the past, ethnographers have focused on the experiences of single populations, exploring their everyday spatial and cultural practices and the ways that they negotiate large-scale political economic processes. However, recent interest in global connections – both colonial and contemporary – has refocused scholarly attention on relationships among members of differently situated groups. Ethnographies of encounters can help geographers grasp these processes. By bringing different voices together, these ethnographies help us see how places are made through the unequal coming together of discrepant genealogies of meaning and desires. They show us the specific ways that larger relations of power are "articulated" (Hart 2004) in distinctive settings and the ways history, culture, and geography shape their manifestations.

Consider my earlier discussion of recent relationships among Filipina women and Japanese men in Central Kiso. I suggested that for both Filipina women and Japanese men in Central Kiso, marriage was, in part, a strategy for negotiating the marginality each group felt within broader nationalist projects of capitalist modernity. However, these projects have played out in geographically and culturally specific ways for both groups. For Japanese men, relationships with Filipina women were a means of negotiating marginality based on their residence in a depopulated rural region of Japan. For Filipina women, relationships with Japanese men were a means of trying to realize dreams of "a better life" and find opportunities for upward mobility unavailable to them in the Philippines – dreams and realities shaped by colonial and neocolonial histories in the Philippines. Moreover, Filipina women and Japanese community members had different investments in the idea of "oyomesan." Japanese community members engaged it as a means for claiming a mode of Japaneseness that privileged their everyday realities; Filipina women did so to maintain not only lives in Japan but also relationships with family in the Philippines and to craft a positive sense of Filipina identity. By juxtaposing the experiences and perspectives of Filipina women and Central Kiso residents, then, I illustrate some of the ways they are differentially situated within larger formations of power while offering a sense of them as people whose desires both aligned and misaligned. I show that meanings of Japaneseness are not simply being reproduced but also transformed through seemingly "traditional" cultural categories. And I suggest that what it means to be an ideal Japanese bride today in rural Japan – a place long considered the last bastion of a pure, traditional Japanese way of life – is the product of the unfulfilled desires and new possibilities that emerged in everyday relationships among Filipina migrants and rural Japanese townsfolk.

For me, then, writing social geography as an ethnography of encounters has meant narratively *remapping* "Japan" not as the site of a homogeneous, traditional, and pure culture but as a contingent product of unequal encounters among Filipina migrants and rural Japanese residents. Maps, we know, are always more than straightforward representations of places; maps are positioned narratives about them (Ghosh 2005; Hanna and Del Casino Jr., eds. 2003; Wood, Kaiser, and Abramms 2006). The kinds of maps/stories we create about the world shape the kinds of actions that become possible in it. Ethnographies of encounters conjure worlds of intersecting, contradictory, and unruly dreams. They show us how these dreams remake places, cultures, and identities as people's desires align and misalign across relations of power. By drawing our attention to these dynamics, ethnographies of encounter offer new ways for mapping the intimate and unequal processes through which social geographies emerge. In doing so, they can also help us imagine geographically and culturally situated strategies for transforming these dynamics to create a more just and equitable world.

Notes

1   My ethnographic discussion is based on fieldwork I conducted in Japan between September 1998 and August 2000, and also during return visits to Japan in 2005, 2006, and 2007. It also draws on four trips I made to the Philippines, three in 2000 and one in 2006. In Japan, I visited women at home and at work, attended events they organized, and also lived with three Filipina-Japanese families. In the Philippines, I stayed with and interviewed women's families there. In all, I conducted more than 100 interviews with Filipina women, their Japanese community members and local government officials, and with women's families and recruiters in the Philippines. I have also worked since 1995

with NGOs in the Tokyo metropolitan area that provides shelter, legal assistance, and emotional support to Filipina migrants throughout Japan. I became acquainted with some Filipina women in Central Kiso through my involvement with these NGOs.

2 "Central Kiso" is not officially recognized as a region in Japan. I use the term to refer to a cluster of mountain towns and villages in Kiso County that included the towns of Kisofukushima and Agematsu and several nearby villages (Kaida, Kisō, Ōtaki, Mitake, Hiyoshi, and Ōkuwa). Almost all Filipina women in Kiso County lived in this region of county.

3 Foreign women who enter Japan as "entertainers" are not legally permitted to have sexual relationships with customers. However, in some cases, such relationships do develop.

4 For discussions of rural Japanese marginality and rural Japanese men's corresponding problems finding brides, see Higurashi 1989; Knight 1994, 1995; Niigata Nippōsha Gakugeibu (ed.) 1989; Sato 1989; and Tamanoi 1998. Bernstein (1983), Jolivet (1997), and Kelsky (2001) discuss Japanese women's desires to marry urban, white-collar men.

5 Pachinko is a gambling game that is something like a combination of pinball and slots.

6 By this, local residents meant teaching children exclusively to speak Japanese, to identify as "Japanese," and to generally do things "the Japanese way" (*nihon no yarikata*). For discussions of similar expectations placed on Filipina brides in other parts of rural Japan, see Kuwayama 1995, Shukuya 1988, and Suzuki 2003.

7 Local Japanese residents had investments in such claims, which enabled them both to negotiate their own marginality within urban-centered discourses of modernity in Japan and to criticize young, urban Japanese women. These claims also resonated with racialized discourses of Philippine underdevelopment. They, thus, enabled rural Japanese residents to identify themselves, in contrast, as full participants of a modern, developed, Japanese nation (see Faier 2009).

8 At least one-third of the Filipina women married to Japanese men in Central Kiso at the time of my primary research continued to work in hostess bars. Many other Filipina women expressed desires to do so. In addition to working in factories, some Filipina women in Central Kiso also did piecework at home.

9 See Constable (2003) and Piper and Roces (eds.) (2003) for explorations of the ways that emotional attachments and political economic considerations are intertwined in the lives of marriage migrants.

10 I use the term "Japaneseness" to refer not to a fixed category of identity but to contingent and relational formations of meaning and practice that are produced in cultural encounters between Filipina migrants and Central Kiso residents. I mean to highlight the ways that categories of geocultural identity such as "Japanese" and "Filipina" are themselves produced in relationships within and among members of these "groups." A more extended discussion of my understanding of Japaneseness can be found in Faier 2009.

References

Abu-Lughod, L. (1993) *Writing Women's Worlds: Bedouin Stories*. Berkeley: University of California Press.

Bernstein, G.L. (1983) *Haruko's World: A Japanese Farm Woman and Her Community*. Stanford, CA: Stanford University Press.

Biehl, J. (2005) *Vita: Life in a Zone of Social Abandonment*. Berkeley: University of California Press.

Brown, J.N. (2005) *Dropping Anchor, Setting Sail: Geographies of Race in Black Liverpool*. Princeton: Princeton University Press.

Brown, J.N. (2007) Response: "In the eye of the beholder: placing race and culture." *Antipode* 39 (2): 376–81.

Clifford, J. (1986) Introduction: partial truths. In J. Clifford and J.E. Marcus (eds.), *Writing Culture: The Poetics and Politics of Ethnography*. Berkeley: University of California Press, pp. 1–26.

Clifford, J. and Marcus, G.E. (eds.) (1986) *Writing Culture: The Poetics and Politics of Ethnography*. Berkeley: University of California Press.

Constable, N. (1997) *Maid to Order in Hong Kong: Stories of Filipina Workers*. Ithaca, NY: Cornell University Press.

Constable, N. (2003) *Romance on a Global Stage: Pen Pals, Virtual Ethnography, and "Mail-Order" Marriages*. Berkeley: University of California Press.

Cruikshank, J. (2005) *Do Glaciers Listen?: Local Knowledge, Colonial Encounters, and Social Imagination*. Vancouver: University of British Columbia Press.

Davis, L.K. (1994) Korean women's groups organize for change. In J. Gelb and M.L. Palley (eds.), *Women of Japan and Korea: Continuity and Change*. Philadephia: Temple University Press, pp. 223–39.

DeLyser, D. (1999) Authenticity on the ground: engaging the past in a California ghost town. *Annals of the Association of American Geographers* 89 (4): 602–32.

Faier, L. (2009) *Intimate Encounters: Filipina Women and the Remaking of Rural Japan*. Berkeley: University of California Press.

Geertz, C. (2001) *Available Light: Anthropological Reflections on Philosophical Topics*. Princeton: Princeton University Press.

Ghosh, A. (2005) *The Shadow Lines*. New York: Mariner Books.

Gilmore, R.W. (2007) *Golden Gulag: Prisons, Surplus, Crisis, and Opposition in Globalizing California*. Berkeley: University of California Press.

Hanna, S.P. and Del Casino, Jr., V.J. (eds.) (2003) *Mapping Tourism*. Minneapolis, University of Minnesota Press.

Hart, G. (2004) Geography and development: critical ethnographies. *Progress in Human Geography* 28 (1): 91–100.

Herbert, S. (1997) *Policing Space: Territoriality and the Los Angeles Police Department*. Minneapolis: University of Minnesota Press.

Herbert, S. (2000) For ethnography. *Progress in Human Geography* 24 (4): 550–68.

Higurashi, T. (1989) *"Mura" to "ore" no kokusai kekkon*. Tokyo: Jōhō Kikaku Shuppan.

Jolivet, M. (1997) *Japan: The Childless Society*. New York: Routledge.

Katz, C. (1994) Playing the field: questions of fieldwork in geography. *Professional Geographer* 46 (1): 67–72.

Katz, C. (2004) *Growing up Global: Economic Restructuring and Children's Everyday Lives*. Minneapolis: University of Minnesota Press.

Kelsky, K. (2001) *Women on the Verge: Japanese Women, Western Dreams*. Durham, NC: Duke University Press.

Knight, J. (1994) Rural revitalization in Japan: spirit of the village and taste of the country. *Asian Survey* 34 (7): 634–46.

Knight, J. (1995) Municipal matchmaking in rural Japan. *Anthropology Today* 11 (2): 9–17.

Kondo, D. (1990) *Crafting Selves: Power, Gender, and Discourses of Identity in a Japanese Workplace*. Chicago: University of Chicago Press.

Kuwayawa, N. (1995) *Sutoresu to kokusai kekkon*. Tokyo: Akashi Shoten.

Marcus, G.E. and Fischer, M.M.J. (1986) *Anthropology as Cultural Critique: An Experimental Moment in the Human Sciences*. Chicago: University of Chicago Press.

Massey, D. (1993) Power-geometry and a progressive sense of place. In J. Bird, B. Curtis, T. Putnam, G. Robertson, and L. Tickner (eds.), *Mapping the Futures*. New York: Routledge, pp. 59–69.

Massey, D. (1994) *Space, Place, and Gender*. Minneapolis: University of Minnesota Press.

McHugh, K. (2000) Inside, outside, upside down, backward, forward, round and round: a case for ethnographic studies in migration. *Progress in Human Geography* 24 (1): 71–89.

Merrill, H. (2006) *An Alliance of Women: Immigration and the Politics of Race*. Minneapolis: University of Minnesota Press.

Mountz, A., Moore, E.B., and Brown, L. (2008) Participatory action research as pedagogy: boundaries in Syracuse. *ACME: An International E-Journal for Critical Geographies* 7 (2): 214–38.

Nagar, R. (2006) *Playing with Fire: Feminist Thought and Activism through Seven Lives in India*. Minneapolis: University of Minnesota Press.

Niigata Nippōsha Gakugeibu (ed.) (1989) *Mura no kokusai kekkon*. Akita City: Mumyōsha Shuppan.

Parreñas, R. (2001) *Servants of Globalization: Women, Migration, and Domestic Work*. Stanford: Stanford University Press.

Parreñas, R. (2006) Trafficked? Filipina migrant hostesses in Tokyo's nightlife industry. *Yale Journal of Law and Feminism* 18 (1): 145–80.

Piper, N. and Roces, M. (eds.) (2003) *Wife or Worker? Asian Women and Migration*. Boulder: Rowman and Littlefield, Inc.

Pratt, G. (2004) *Working Feminism*. Philadelphia: Temple University.

Pulido, L. (2006) *Black, Brown, Yellow, and Left: Radical Activism in Los Angeles*. Berkeley: University of California Press.

Satake, M. and Da-anoy, M.A. (2006) *Firipin-Nihon kokusai kekkon: tabunka kyōsei to ijū (Filipina–Japanese Intermarriages: Migration, Settlement, and Multicultural Coexistence)*. Tokyo: Mekong Publishing.

Sato, T. (1989) *Mura to kokusai kekkon*. Tokyo: Nihon Hyōronsha.

Shukuya, K. (1988) *Ajia kara kita hanayome: mukaeru gawa no ronri*. Tokyo: Akashi Shoten.

Steedly, M.M. (1993) *Hanging without a Rope: Narrative Experience in Colonial and Postcolonial Karoland*. Princeton: Princeton University Press.

Stewart, K. (1996) *A Space on the Side of the Road*. Princeton: Princeton University Press.

Suzuki, N. (2003) Transgressing "victims": reading narratives of "Filipina brides" in Japan. *Critical Asian Studies* 35 (3): 399–420.

Suzuki, N. (2007) Marrying a Marilyn of the tropics: manhood and nationhood in Filipina–Japanese marriages. *Anthropology Quarterly* 80 (2): 427–54.

Tamanoi, M.A. (1998) *Under the Shadow of Nationalism: Politics and Poetics of Rural Japanese Women*. Honolulu: University of Hawaii Press.

Till, K. (2005) *The New Berlin: Memory, Politics, Place*. Minneapolis: University of Minnesota Press.

Tsing, A. (1993) *In the Realm of the Diamond Queen*. Princeton: Princeton University Press.

Visweswaran, K. (1994) *Fictions of Feminist Ethnography*. Minneapolis: University of Minnesota Press.

Wood, D., Kaiser, W.L., and Abramms, B. (2006) *Seeing through Maps: Many Ways to See the World*. Oxford: New Internationalist Publications.

Wright, M. (2006) *Disposable Women and Other Myths of Global Capitalism*. New York: Routledge.

Chapter 13

Participatory Praxis and Social Justice

mrs c kinpaisby-hill[1]

Wednesday January 21, 2009.
Mestizo Arts & Activism Collective.
Salt Lake City, Utah, United States.

*"What! This is not the Esteban I knew! This is not Esteban!!!" Lily was very upset.
"How could they say this about him? Can't they let him die peacefully? … They
don't even know him." Together, we were reading the newspaper to learn more
about the killing of Esteban Saidi, a 16-year-old Latino immigrant who was shot
after school by his classmate Ricky Angilau, a 16-year-old Pacific Islander. Alleged
to have been in retribution for a stabbing that happened the weekend before, the
tragedy pointed to the racial tensions between communities of color. In response,
the city's gang task force and police started an intensive surveillance campaign of
the west side neighborhoods of Salt Lake City where these youth lived. The media
also focused their attention on "the gang problem," "violent teenagers," "dysfunc-
tional families," and young people "without guidance or morals."*

*Young people of color were now on the defensive. Brothers, cousins and class-
mates reported being stopped by the police when "DWL" ("Driving While Latino"),
reflecting the ongoing racial profiling on the West Side. Gang task force units entered
school premises, looking students up and down in the hallways and at lunch.
Schools heightened their own security and "locked down" (prevented entry and exit
to campuses during school hours). The tension was palpable.*

*The heavy-handed negative response of the police, the school, and the media,
contrasted starkly with the tender grief the young people grappled with as they tried*

A Companion to Social Geography, First Edition. Edited by
Vincent J. Del Casino Jr., Mary E. Thomas, Paul Cloke, and Ruth Panelli.
© 2011 Blackwell Publishing Ltd. Published 2011 by Blackwell Publishing Ltd.

to make sense of the loss and tragedy. Several of the young people on our research team knew Esteban very well. Lily went to church camp with him. Jorge lived a block away from Esteban and played with him as a kid. A few went to school with him remembering him as the quiet boy who often sat in the back of the classroom. For those of us who never met Esteben, we listened to stories about him. Together we processed the shock, the loss, and the wider social impacts of his death. We created a space for grieving collectively. And, beyond this, we spent time talking and thinking through the roots of racial divisions within our communities and ways to bring people together. We began a participatory action research project.

Introduction

Throughout this chapter, we draw on the example of the Mestizo Arts and Activism project introduced above – one of several collaborative research projects we are engaged in – as a means to open a window on participatory praxis. Participatory approaches have found increasingly wide support in social geography (see Del Casino 2009; Kindon 2010; Fuller and Kitchin 2004; Pain 2004). Over seventy years of innovation and development in numerous contexts, many in the global south, lie behind these recent developments (see Kindon et al. 2007). Participatory praxis is marked by its thoroughly collaborative approach to research and education, and by its explicit action orientation (McTaggart 1997). It seeks to challenge the hierarchical social relationships that usually characterize academic research (Wadsworth 1998) by changing *how* data are collected, *what sort* of new knowledges and *what impacts* result; and, crucially, *who* directs investigation, and to *whom* any benefits arising from experience, learning, and findings of research accrue.

This chapter will not attempt to provide detailed accounts of participatory methods and techniques. For these you need to look elsewhere (e.g., Kindon et al. 2007; Kindon 2009b). Rather, drawing periodically on the Mestizo case study, we will argue four broad points. First, that engaging a *participatory* frame in research, which recognizes and utilizes the reflexive capacities of participants, makes for more fully *social* geographies. Second, that social geographies informed by participatory praxis offer one very effective means to *address*, not merely explain, socio-spatial injustices which are so often our focus. Third, that in its enactment, participatory praxis helps destabilize the unhelpful dualism between theory and practice that still persists across much of social geography. Fourth, that through an example of emotion/affect we can illustrate the important interplay of theory and practice.

We hope that this chapter will inspire you to explore how participatory praxis might affect how you think and do social geography, and perhaps make a contribution to wider struggles towards a more socially just world. In this regard, as you read the chapter, the strategy behind our playful *nom de guerre*, mrs c. kinpaisby-hill, should become apparent: we seek to disrupt the stifling individualization of knowledge (so prevalent in the Anglo-American context) and the governing effects of citation practices, whilst also drawing attention to forms of academic accountability beyond audit and accountancy (see mrs kinpaisby 2008). Our pseudonym also recognizes the social conditions and relations in and through which this chapter was produced, and manifests the transformations that participatory praxis has wrought in our own work. Our experience of talking, thinking, and writing together is that it often becomes difficult to differentiate our individual contributions in our final output – our collective thoughts become something more than the sum of the parts. And that, as we hope to show, is the value of participatory praxis (for other examples of collective writing see Ian Cook et al. 2008).

Towards More Fully Social Geographies

We want to redefine social geography in light of our experiences with participatory praxis, orienting it away from being a spectator theory of knowledge towards a collaborative, more fully *social* theory of knowledge production.

How "social" is research in social geography?

Unsurprisingly, social geographies are concerned with "the social." What this actually means has often been taken for granted, but, when defined, it is generally viewed as the sphere of life encompassing human relations, identities, and social reproduction (for example Hamnett 1996; Pain et al. 2001; Panelli 2003; Smith et al. 2010). This conceptualization provides a neat boundary around the sub-discipline, although within it there is constant development and change. The boundary between the social and other supposedly discrete domains (the economic/political/cultural/environmental) must be recognized as porous and the domains as relational (Smith et al. 2010). As we have argued elsewhere (Kindon 2010; Pain 2004), there are many points of connection and many synergies between social geographies and participatory approaches to knowledge construction. We begin by interrogating four of these, as we build an argument for "more fully social" social geographies.

First, there is a prominent belief in social geography that *sociality* and *relationality* are central to understanding society and human environments: the idea that life is experienced first and foremost in light of connections with others. A second and related belief, which transcends a variety of philosophical perspectives, is that people are active participants deeply implicated in the production and reproduction of their everyday social worlds. Third, there is the desire (among qualitative social geographers at least) to "enter the life world" of respondents and to take seriously their accounts of their experiences. Fourth, there is growing awareness of the situated nature of knowledge production and a sensitivity to the question of *whose* social geographies are represented in academic accounts (see Panelli 2008).

In many ways then, participatory geographies are pushing at social geography's open door. Indeed, given social geographers' long history of sensitivity to power relations in society and in research processes, they might recognize the paradox they face: that while social geographers study *sociality*, their research practices are frequently *solitary*. While social geographers may value their respondents' perspectives and active production of their social worlds, and are sometimes apprehensive about their own ability to represent those other lives, they rarely entrust respondents with the ability to undertake research themselves. So, the relations that normally produce knowledge in social geography are firmly hierarchical, its theories of society likely to be disconnected from those whose lives they describe. In significant ways then, the praxis of much social geography can end up reinscribing the very power geometries it sets out to problematize.

The critique offered by participatory approaches suggests that despite the best intentions of researchers, the practices of conventional research are still often extractive, even imperial:

1   Researchers are experts who design and implement programs of data collection and analyze and present the findings to other academic and policy professionals;
2   Respondents and communities are sources of raw materials to be mined, the data extracted from the field and transported to universities to be processed along linear lines of analysis and manufactured into expert knowledge;
3   The understandings produced are (increasingly) focused on the creation of marketable products to be circulated among scholastic and policy making consumers; and
4   The benefits of research accrue primarily (in something uncomfortably close to a zero sum game) to the institutions that control the means of research production, and to the elite workers they so unequally reward; not to those who provided the raw materials for research (see Kesby et al. 2005; Louis 2007).

Participatory praxis: socializing the production of social geography

Participatory praxis seeks to engender a more fully "social" model of research in which the segregated geographies and Fordist labor divisions described above are broken down (Heyman, 2007). Participatory praxis involves researchers working together with participants to examine issues of mutual interest, undertaking collective analysis in situ (i.e. in communities, with participants) in order to facilitate co-learning and engender social change. The work and impacts of research are immediately socialized as their results are directly available to participants, and the knowledge resources generated are circulated through participating communities where they may be used to produce strategic effects. Such praxis has many forms, but participatory action research (PAR) is one of the most common, and has a range of characteristics that distinguish it from conventional research (see Box 13.1).

As we have discussed elsewhere (Kesby et al. 2005; Kindon 2009b), participants' levels of engagement with research initiatives vary. We sympathize with the concern that "participation" is now often commodified as a technical fix, used by a bewildering range of international institutions, governmental and non-governmental organizations (and less reflexive researchers) to "deliver" participation in top down ways that contradict the underlying philosophy of the approach (Cooke and Kothari 2001; Greenwood 2002; Hayward et al. 2004). However, we do not agree with more evangelical proponents that anything less than the "gold standard" of complete collaboration, deep participation and community led initiatives is inherently suspect. In our view, such "all or nothing" readings not only mask the messiness of participation as a form of embodied and situated research practice, but also curtail a wider, intelligent, and politically strategic use of participatory praxis at specific

---

Box 13.1   Key characteristics of liberatory participatory action research

Participatory action research:

- Aims to change social practices, structures, and media which maintain social inequality and injustice, and which arbitrarily bound the possibilities for human existence;
- Recognizes that participants are competent, reflexive and capable of participating in all aspects of the research process;
- Facilitates and augments participants' reflexivity and competence through the research process;
- Addresses substantive issues that participants themselves identify as relevant to their lives;
- Involves participants and researchers in collaborative processes for generating knowledge;
- Attempts to engage participants in a diverse range of experiences so that marginalized voices are not excluded from the research findings or co-learning opportunities;
- Pursues an iterative cycle between analysis, learning, action, and further analysis;
- To a significant extent, measures the credibility/validity of knowledge derived from the research process according to whether the resulting action solves problems for the people involved and increases participants' and communities' self-determination; and
- Often works at a "local scale" with "communities," but seeks ways to make connections to broader scales, both in terms of analysis and action, and resists the temptation to imagine that communities are inherently socially cohesive or spatially distinct.

Source: adapted from Brydon-Miller et al. 2003; Fisher and Ball 2003; Greenwood and Levin 1998; Greenwood et al. 1993; McTaggart 1997; Park et al. 1993; Reason and Bradbury 2006. The term "liberatory PAR" comes from Torre and Ayala (2009) to signal a particular intent.

---

stages of otherwise conventional research projects (Kesby et al. 2005; Kindon 2009b). Moreover, while communities have independent capabilities, they may also decide there are benefits from the specialist skills, resources, solidarity, "credibility" and catalytic effect (Stoeker 1999) that external researchers can bring to a project (so long as they work respectfully, develop joint agendas and are prepared to learn from participants and research partners).

Participatory praxis involves, then, a decentering of the lone scholar in the research process. An external researcher may play a vital role, but community partners bring their own expertise and insights to the design and conduct of research in ways that make the production of knowledge about the world more collaborative (see Fine et al. 2001; Smith 1999; Torre and Ayala 2009). This alternative episte-mology is based on an ontology that recognizes that all human beings have consid-erable capacities for creative thinking, analysis, and reflexivity. Elite, official sites of knowledge production such as universities, government departments, think-tanks, and research consultancies do not have a monopoly on intellectual activity. Critical thinking, learning, and theorizing about the world also take place in everyday life and within numerous "grassroots" sites such as community organizations and activ-ist groupings. Humanist social geographers have long viewed such capacities as a substantive element of the social worlds they seek to interpret, but a participatory turn enables these capabilities to be harnessed as a resource for research design, methodology, analysis and action (Kesby and Gwanzura-Ottemoller 2007). As bell hooks (1990) has articulated, those who occupy the social margins often have understandings and perspectives on their lives and experiences that those in privi-leged positions may find more difficult to perceive. Participatory action research is founded upon "the understanding that people – especially those who have experi-enced historic oppression – hold deep knowledge about their lives and experiences, and should help shape the questions, [and] frame the interpretations" of research (Torre and Fine 2006: 458). In so doing, participatory praxis can generate more fully social geographies in which all of those involved share the work of producing knowledge that may inform constructive social change.

The Mestizo Arts & Activism Collective (http://mestizoactivism.blogspot.com) is a social justice think tank in Salt Lake City, Utah that engages young people as cata-lysts of change, through active civic engagement in a model integrating community-based arts, community-based participatory action research, and activism. A university-community partnership, the Mestizo Collective, involves faculty, artists, activists, university, and high school students working collaboratively to address social injustices in our community. Our inquiry is youth led and places emphasis upon the particular contribution and access young people bring to understanding their everyday lives. The opening vignette illustrates our process. Our study was motivated by outrage and the disjuncture between the ways that young people of color were pathologized in the media, and the very different understandings that the young people had of themselves. The tragic loss of Esteben Saidi created an opening for us to witness up close the injustices manifest in media accounts, the police response, and the implications for young people of color. Research became a space for engaging in critical analysis and everyday social theorizing. We collectively designed and carried out an investigation into perceptions of young people of color (working with peers, parents, and "authority figures" – teachers, policy-makers, social workers, and so on) that threw media stereotypes into critical perspective[2].

This project illustrates how a critically engaged, liberatory, participatory praxis involves a particular idea of "the social" enacted: the idea that all knowledge is, to greater or lesser extent, socially produced. So, the epistemology of participatory social geographies is founded upon the assumption of collaborative knowledge production. Socializing geography via participatory praxis is about challenging current assumptions about who is capable of thinking spatially; of seeing the spatial and scale dimensions of injustice and of resistances; and of developing praxis that can reconstitute understandings of space and place.

Engaging Social Justice and Change in Participatory Social Geographies

Action research in communities

These more fully "social" research processes speak to a further concern at the heart of social geography. As we have suggested, the sub-discipline has long focused its scholarship on social and spatial inequality (e.g., Ley 1983; Pain et al. 2001; Smith et al. 2010). Again, connections and synergies are evident with advocates of participatory research who share these convictions. However, for us, a commitment to social justice on paper is a necessary but not sufficient contribution of our intellectual labor. While many social geographers now also seek to take the impacts of their research "beyond the journal article," a participatory turn demands *collective* dissemination, and does not restrict action to offering evidence based advice to policy makers (see Cahill and Torre 2007; Fine and Torre 2008).

Participatory praxis takes seriously the old Marxist adage that the point of intellectual endeavor is not merely to comment on the world, but to change it (Marx, 1845; and see Castree et al. 2010). More than this, in the tradition of the radical educator and writer who influenced early PAR, Paulo Freire (1970), collaborative research is itself a project of social justice that begins the process of socio-spatial change by being, in Gandhi's words, the change it wants to see (Chatterton et al. 2007). Participatory approaches also have strong connections to feminist and Indigenous geographies, which emphasize an ethic of care and relationship over an ethics of risk mitigation (Cahill et al. 2007, 2008; Lawson 2007; Halse and Honey 2005; Tuck and Fine 2007; Kindon and Latham 2002; Smith 1999). So spaces normally reserved for data collection are turned into arenas for learning and action, and the ethic that respondents should be left unaltered is replaced with an ethic that research can directly challenge injustice and facilitate social change. In this sense, participatory geographies have strong ties to the original ethic of radical geography, and have much to offer a new generation seeking "relevance" for radical scholarship (see Heyman 2007; Mitchell 2008; Weis and Fine 2004; Fuller and Kitchin, 2004).

Once again the Mestizo Arts and Activism Collective is instructive. The participatory action research project focused on interrogating the representations of young people that circulated via the media and which also informed the policies and actions of the authorities. Through discussions and analysis a process of conscientization took place (see Freire 1970) whereby participants came to new understandings about their everyday lives. The youth researchers determined that not only were state enforcement agencies in Utah pursuing policies based on inaccurate stereotypes that did not represent their own experiences and understandings of their daily struggles, but it was also allocating significant resources to the "gang task forces" that acted on, and helped to reproduce these stereotypes.

Our research took place alongside an initiative by local activists who organized in response to the racial profiling tactics of the police. Two of the university students/ mentors involved with the Mestizo Collective were also key players with the FACE Movement (www.facethestruggle.blogspot.com) and the Brown Berets, the groups that joined together to demand that city/state authorities match the funding pouring into the police's "gang task force," dollar for dollar, with after school prevention programs (at the time of writing this demand has yet to be met). To model how this might work, they even started a mentorship program at the school where Esteben Saidi had been killed. The explicit goal of the mentoring project was to overcome divisive tensions between multicultural communities and to bring them together around issues of mutual interest. In its praxis, the mentoring sessions parallel our participatory action research process (e.g., "Week 3 deconstructing the situations we live through/ how to manage our struggles, and week 4: strategizing change in our communities") (see FACE Movement 2009).

Participatory praxis "close to home"

Participatory praxis disrupts the unhelpful boundaries between researcher and researched and between community and university. It opens spaces for new knowledge and ways of being in the communities we study, and can also help facilitate transformation of the academy and of our practices as academics (see Chatterton et al. 2007; Kindon 2008; mrs kinpaisby 2008; mrs c kinbaisby-hill 2008). In this regard, participatory geographers are prominent in recent debates urging colleagues to consider the ways in which they facilitate social justice and change in and through their labors within the academy – through teaching (Kindon and Elwood 2009); activism within the University (Chatterton et al. 2007); and inter-professional relations (Kindon 2008; mrs kinpaisby 2008; Heyman 2007) *as well as* through research. By applying modes of participatory praxis to the ways in which we work with students in classrooms and colleagues in board meetings, we can begin to make changes to the hierarchical socio-spatial relations that often govern our own day-to-day lives.

Such an agenda has further implications with regard to how universities orient themselves outwardly, which fold back into the issues raised above about participatory praxis in research. Here again, participatory geographers should be pushing at an open door. In the United Kingdom "public engagement" has become a buzzword in the university sector with universities increasingly keen to expand their links with the localities and communities around them. In the United States, this orientation has a longer history in the form of "civic engagement" oriented through the delivery of service-learning courses. Our own discipline has a long history of such engagement (Bunge 1969) and there is now a reawakened enthusiasm for "public geographies" (Heyman 2007; Hawkins et al. forthcoming; Fuller 2008; Mitchell 2008). Furthermore, the latest iteration of the UK academic research audit process[3] proposes to take account of the "impact" that academic work has beyond the academy (HEFCE 2009). Our concern is to ensure that participatory praxis informs and shapes the politics and practice of such initiatives because the meaning and consequence of "impact" are not given but subject to interpretation. For example, the primary goal of "public engagement" can be the generation of income, student numbers and business opportunities (Demeritt 2005; Castree 2000); "public geographies" may be "research as usual," just disseminated more effectively. Instead, the agenda for the "communiversity" envisages a two-way interaction between scholars and communities, where communities (local or global) may define and conduct geographical research with academics, but based on their own priorities. Here the imperative of research "impact" is inherent in the research process, and is experienced in terms of social justice and socio-spatial change rather than being primarily a metric in a process of neo-liberal academic accountancy (mrs kinpaisby 2008).

Destabilizing Dualisms through Participatory Praxis

Building on the last two sections, we would like to say more about the contribution participatory praxis can make to understandings of theory. We want to do more than simply argue that participation is primarily a theory of knowledge rather than a method for data collection (Reason and Bradbury 2006), or point to its firm epistemological and ontological foundations (see above). The radical theoretical roots of participation call for us to resist an abstract division between contemplation and practice (see Marx, 1845), or between "pure" and "applied" research (Pain 2006). Rather we would emphasize the connectedness between doing and knowing (see Cahill 2007b; Cameron and Gibson 2005; Torre and Ayala 2009; Torre 2009; Tuck 2009). Participatory *praxis* is a form of theoretically informed action, and a practical and grounded form of theorization: "discovery cannot be purely intellectual but must also involve action; nor can it be limited to mere activism, but must include serious reflection: only then will it be praxis" (Freire 1970: 47). Such a perspective helps us to reflect on, and challenge, some of the unexamined assumptions that too often frame the way researchers think about theorizing. We also want to suggest that participatory praxis has an important role to play in understanding the nature of theory more generally. In particular it can help destabilize the dualism between theory and practice that is still persistent across much of social geography.

Participatory praxis: an alternative ontology of theorizing

We think there are at least five important points that participatory praxis demonstrates about theory:

1 Theorizing is not a time–space-limited activity

Too often the practice of theorizing is imagined as necessarily taking place *after the fact*; once empirical data collection has taken place and the researcher has left the field. By comparison, participatory praxis, like other forms of humanistic and qualitative inquiry (particularly grounded theory), explicitly recognizes and encourages iterative cycles of action and reflection within and during the research process, and its theorizing of research issues is inductive and emergent. However, unlike other forms of qualitative research, participatory praxis does not only involve the lone researcher in the process.

2 Theorizing is messy because society is complex and contradictory

Classically, the role of theory is to "bookend" data collection in the field: providing an explanation and framework for research design, a structure for an organized program of fieldwork, and a coherent and structured explanation of findings. In some respects, the meta-theory of participation is no different; however, in its praxis, a participatory approach allows for much more messiness in the process of theorizing. Again, like good qualitative researchers, those pursuing a participatory approach allow themselves to be led and surprised by what participants share; but they go a step further. Participatory praxis involves a commitment to openness, negotiation, and exchange; all theoretically informed preconceptions about the purpose, direction, and outcomes of research are subject to change through interactions between participants. Other field researchers may patiently listen to the complex and often contradictory accounts of respondents, before retiring to make a coherent analysis in private occasionally asking respondents to comment on their interpretations. By comparison, participatory researchers analyze problems and contradictions directly with multiple participants (who may have different, or opposed, interpretations) and together they make sense of (theorize) complex issues in order to formulate joint understandings. In that sense participatory praxis is continually theoretically engaged, and there is a tight fit between its ontology ("what is the world like?") and its epistemology ("what is the best way to build knowledge?"). If the world is social, messy, contradictory, compromised, provisional and structured by unequal power relations, then perhaps a process of isolated contemplation by individual researchers is not the best way to theorize it.

3 Theorizing is a social, embodied, and embedded praxis relevant to the lives of ordinary people

Perhaps one of the strongest myths about theorizing is that it is the stage of research that has the least relevance to the lives of people outside of academia. Moreover, it is the expertise, distance and supposed objectivity of the scholar that enables him or her to perform this function. While Donna Haraway's (1991) "god-trick" critique exposed this imaginary of theoretical praxis, it remains surprisingly prevalent, even in the work of those social geographers who are at pains to stress the situatedness and subjectivity of their work. This is because most scholars still generally reserve for themselves the task of theorizing and the capacity for self-reflexivity, however sensitively and provisionally they undertake this task. Participatory praxis extends a poststructural feminist agenda in a number of ways. It embraces the idea that the situated biographies and positionalities of researchers influence their thinking, but goes beyond the idea of a singular moment of individual self-reflection. Rather, it views theorizing as ontologically distinctly social: it is a citational process, a relational bricolage of one's own and other's ideas, and it is an embedded and constantly ongoing praxis of coming to knowing through iterative cycles, moving between experiences of everyday social life and individual and collective analysis and reflection. Participatory researchers then make three further important moves: first, we recognize that this is also the way in which participants come to know the word. Second, we realize that our own socially embedded theorizations will be enriched by explicitly engaging participants directly in a collective process of sense-making. Third, we realize that *they* [participants] can do it" (Chambers 1997: 131) – in other words, analyze, reflect on, and theorize the socio-spatial issues that are of significance to them (see Cahill 2007b).

4 Theorizing is founded in contextual validity

Theorizing in participatory praxis implicitly involves "jumping scales" – moving from personal experiences to social theorizing, and back again. Elsewhere, we have identified this as contextual validity (Cahill 2007b). This expanded notion of validity involves drawing connections between the personal and broader structural forces in our research. This means that triangulation[4] not only involves a conscious engagement of difference and multiplicity, but also involves the active situation of our analyzes in global/political/social/economic contexts (Fine et al. 2000). In other words, how might our analyzes consciously triangulate micro- and macro-interpretations and attempt to capture "how the intimate and global intertwine" (Pratt and Rosner 2006: 15)? Paying attention to the interdependence of scales in our analyzes, and drawing out these connections, harbors the potential to extend the power and reach of our interpretations.

5 Theorizing can be for contextual action not just general understanding

Finally, and in line with everything we have already said, participatory praxis challenges the notion that the role of theory is only to provide general explanations with broad applicability in order to improve understanding of social phenomena. Participatory praxis shows that when generated in situ with participants, and grounded in the specifics of people and place, theorizing can be directly relevant to lived realities in ways that inform action for social change (see Cameron and Gibson 2005). So, far from being atheoretical, participatory praxis is transformative of what theory is and what it does.

Once again the Mestizo Arts and Activism Collective is illustrative of how theorizing may happen in practice, and how participatory praxis recognizes, plugs into and augments this process. Our youth research team chose to focus upon the relationship between how young people of color were misrepresented in the media and racial profiling. Collaborative analysis within the project enabled the young people to be expert commentators on their own lives and to explore and theorize more fully the ways in which media representations and institutional interventions impacted and played out in their communities. Not only were the theorized understandings that emerged more fully social, they also generated strategies for action that would produce alternative representations and practical social change (see also Cahill 2007a).

We are not suggesting that such collaborative analysis and theorizing progresses without differences, disagreements, conflict and struggle. Participatory researchers do well to consider strategies for conflict resolution to help deal with difficulties when they arise (see Trapese 2007). Nor are we saying that any external researcher's role as facilitator means that they must necessarily lay aside their own expertise, opinions, powers of analysis, or theoretical contributions (Kindon 2009a). Participants often value these and stir them into the mix of their own thinking. As Chatterton et al. (2007) suggest, showing solidarity with participants is not about agreement for its own sake, but rather is about respectful critique and finding productive ways to work through differences (see also Askins and Pain forthcoming; Pratt et al. 2007; Torre 2009). But we *are* saying that the praxis of the co-production of knowledge pushes scholarship to ask new questions and find new answers beyond western and other elite knowledge framings (Panelli 2008; Pain 2009; Sanderson and Kindon 2004; Weis and Fine 2004).

Theorizing participation

Our discussion of theory would be incomplete without reference to recent theoretical critiques of participatory praxis itself. If participatory praxis is to live up to its feminist sensitivities to the situatedness of knowledge (see Reason and Bradbury 2006), and is not to repeat the mistakes associated with the certainties of Marxism, it must recognize that it is itself a form of power (Kesby 2005: 2007). We cannot imagine that our research techniques are neutral technologies, or seek to "change things for the better" (see Wadsworth 1998), without recognizing that our own sense of social justice is inevitably situated and provisional.

Participatory researchers have over the years engaged in a healthy level of internal critique (e.g., Guijt and Shah 1998), despite accusations to the contrary (Cooke and Kothari 2001); but the poststructural critiques of participation that have more recently emerged from development studies (Cooke and Kothari 2001) and geography (Hickey and Mohan 2004; Williams 2004) have been invaluable. We agree that advocates of participation cannot imagine themselves to be wholly benign agents without desires or agendas of their own (Kapoor 2005); when participation is pursued in certain ways, even sometimes when it is pursued "deeply," true to its original spirit, negative power effects may result (see Box 13.2).

Despite these important insights, we are critical of commentaries that seek to demolish the notion of participation altogether, and we prefer to argue for a more constructive interaction between poststructural theory and participatory praxis (Cahill, 2007a; Cameron and Gibson 2005; Kesby 2005, 2007; Kesby et al. 2007). Certainly advocates of participation must recognize that the ground rules, philosophy and social relations established within a participatory arena are forms of governance that will shape people's inherent capacities for reflection and analysis in particular ways, and that will produce particular expressions of efficacious human agency, rather than bypass power or release people's "real selves." Certainly the understandings, theorizing and actions generated will be partial and situated. But still, the praxis of critically informed and reflexive participatory research better enables the limitations of context to be grappled with collectively, *with* participants in communities.

We find it helpful to think about the kind of empowerment that can be produced by projects like Mestizo Arts and Activism along poststructural lines, viewing empowerment as a power effect of participatory praxis rather than as a revelation or release from oppression. This helps us theorize how to sustain those effects: for example, the Mestizo youth researchers are rolling out the social relations of collaboration and discussion to new arenas beyond the small café where the project originally began, inspiring new cohorts of young people to critically examine racial and generational stereotypes.

So, precisely because we take seriously the suggestion that there is no escape from power (Foucault 1983), we are prepared to work with less oppressive and constricting forms of power – such as participatory praxis – because they offer practical and

---

Box 13.2   Some negative power effects of participatory approaches

- De-legitimization of research methods that are *not* participatory;
- Production of participants as subjects *requiring* "research"/"development";
- Production of suitably disciplined subjects as *participants* expected to perform appropriately within participatory processes;
- Retention of researchers' control whilst presenting themselves as benign arbiters of neutral or benevolent processes;
- Re-authorization of researchers as experts *in* participatory approaches;
- Romanticization or marginalization of local knowledge produced through participatory processes;
- Reinforcement of pre-existing power hierarchies among participating communities;
- Legitimization of elite local knowledge simply *because* it is produced through participatory processes; and
- Legitimization of neoliberal programs and institutions (such as the World Bank) that also deploy participatory approaches and/or techniques.

Source: Cooke and Kothari 2001; Cornwall and Brock 2004; Henkel and Stirrat 2001; Kapoor 2002; 2005; Kothari 2001; 2005: Mohan 2001; 2007; Mosse 1994; Sanderson and Kindon 2004.

---

grounded ways for researchers and ordinary people to deconstruct and challenge more oppressive forms of power like racism (see Kesby 2007).

Working with emotion/affect in participatory praxis

In this final section, we turn to an example to illustrate our argument so far about the intersections of theory, practice, and social justice in social geography. We briefly examine emotion/affect[5] in participatory practice, as one way in which participation may be theorized, and one way in which it informs theory. We choose this example for two reasons. Firstly, because advocates of participatory praxis and of activism have long recognized the central role that emotion plays in motivating and sustaining engagement in alternative approaches to research, and are increasingly examining this explicitly (see Askins 2009; Pickerill 2008). Secondly, because while we welcome the insights of those who urge attention to the many events that take place before and after conscious reflexive thought as a means to produce more fully *human* geographies (Anderson 2009; see also Woodward and Lea 2009), we believe that critically informed participatory praxis offers social geographers ways to explore emotion/affect, *with research participants* in ways that work towards social justice (Kesby and Pain 2010).

As the vignette that opened this chapter illustrated, participatory research frequently emerges from strong emotional responses to the existence of social injustice, and certainly requires a major emotional investment (see Cahill 2004; 2006; 2010;

Kindon 2010). While other researchers are also motivated by emotional responses, participatory praxis is marked by "an epistemological shift to *emotion with*, rather than *emotion of* or *compassion for*" those with whom we work (Pain 2009: 480; original emphasis). Kapoor (2005) is right to insist that advocates of participation must inspect the desires that drive them, since their passions may not be as benign as they would like to think. However, desire is apparent in all modes of academic work. Just as is the case with power (Kesby 2005, 2007), emotion/affect cannot be escaped, and so must be worked with. In our opinion, participatory praxis offers productive ways to work with desire (Kindon 2009a) and with emotions/affects (Kesby and Pain 2010; Pain 2009; Wright 2008). The strength of participatory praxis is that it is more able than most research approaches to let the messy and often contradictory issues of motivations, roles, power, and emotion/affect become part of the process of investigation, and in a way that is open to inspection, challenge and renegotiation from participants themselves. Moreover, participatory praxis works with desire, emotion, and affect in ways that are grounded, politically engaged, accessible, and orientated toward social justice, rather than in ways that are impenetrably complex and intellectually elitist.

In the Mestizo Arts and Activism Collective, youth researchers were angered and frustrated by the portrayal of young people in the West Side of Salt Lake City as "gangbangers" (gang members). Reading the news coverage following the death of Esteben Saidi initiated an affective response: at the same time that the youth researchers processed the tragic loss and violence in our community, they were forced to confront blatant misrepresentations of people they knew. The knife edge of stereotypes cut sharply through their vulnerable feelings of grief. Shock shifted quickly to anger and a desire to do something (see Freire 1970; Guishard 2009; hooks 1995). The Mestizo Collective was inspired by these feelings of outrage as they designed and developed their participatory action research project on media stereotypes of youth of color. The research project became a space to process affective responses to injustice: the iterative flow back and forth between affect and cognitive analysis occurred in ways that were empowering for the young people, and gave them the energy to actively pursue social justice and change. Along similar lines, Askins (2009), in her discussion of activism, dismantles any separation in praxis between emotion/affect.

These embodied and cognitive interplays connect to our earlier comments on theory. Ontologically, theorizing is *embodied* as well as social: doing, acting, manipulating, talking, and listening are linked to thinking via affective responses, feelings and emotional insights. In our experience, theorizing can be a passionate, fearful, painful, depressing, and/or thrilling process. The material as well as social praxis of participation (see Askins and Pain in press) greatly facilitates embodied analysis, theorizing, and strategizing among participants and researchers, making for more fully human – and social – social geographies.

Our final point is that emotion/affect can be one of the vital, socially transformative outcomes of a process of participatory praxis. Building on our previous statements about empowerment being a power effect of participatory discourse and practice (Kesby et al. 2007) and our desire to link "affect with effect" (Kindon 2010), we are interested to think more about how emotion and affect might be aroused as a means to distanciate and sustain empowerment beyond immediate arenas of participatory projects (Kesby and Pain 2010). Here again, the creative, visual and performative research techniques associated with participatory praxis have a crucial role to play (see Askins and Pain forthcoming; Boal 1979; Cahill et al. 2008; Kindon 2003; Cieri and McCauley 2007; Hume-Cook et al. 2007; Krieg and Roberts 2007; Rohd 1998; Wright 2008). Clearly, these formats have the potential to capture, express and transmit emotional as well as other forms of knowledge and experience (see Cahill 2004; Cieri and McCauley 2007), and to stimulate affective responses among those who participate in their production and amongst those who witness them, as part of more accessible and social forms of research dissemination.

Towards More Fully Social Geographies

In this chapter, we have argued that the ontology and epistemology of participatory praxis challenge normative approaches to research by offering a collaborative approach which breaks down distinctions between researchers and participants, as well as taking the integration of theory and practice seriously. This approach represents a collective means to understand a shared world and offers social geographers the means to make social geographies more fully social. We have suggested that participatory praxis can help take researchers beyond social geographies that never make it beyond the journal article, offering potential to address as well as explain the socio-spatial injustices which are so often the focus of the sub-discipline.

We have also challenged the common notion that participatory approaches are oppositional to theory, atheoretical or under-theorized. We argued that not only is participation a theoretically informed praxis, but also that it reveals theorizing itself to be a practical, embodied, situated and social practice. Far from being under-theorized we argue that participatory praxis can be transformative of what theory is and does.

We explored some of the emotional dimensions of participatory praxis to demonstrate that academic theory does not have to be remote and impenetrable. Instead, if deployed through a participatory epistemology, intellectual insights can not only be meaningful and accessible to, but created by, people beyond the academy. Following a long lineage of radical geographers (e.g., Bunge 1969; Fuller 2008; Heynen 2006), we suggest that social geographies be explored in more participatory ways that are grounded and orientated toward a politics of social justice.

Notes

mrs c kinpaisby-hill thanks the amazing Mestizo Arts & Activism Collective, the changemakers of Salt Lake City, for their commitment and critical engagement to social justice. Special thanks to Jarred Martinez, Asaeli Matelau, and Yamila Martinez who coordinated the youth media representations project. We are also most grateful to the FACE Movement and the Brown Berets of Salt Lake City for their inspiring activism. Caitlin would also like to thank Matt Bradley and David Quijada for their ongoing collaboration, intellectual engagement, support, and love.
1   mrs c kinpaisby-hill acknowledges a debt of gratitude to Mike Kesby (University of St Andrews), Rachel Pain (Durham University), Sara Kindon (Victoria University of

Wellington), and Caitlin Cahill (City University of New York), and as well as the many research participants s/he has worked with over the years.

2   The youth media representations project team included Eduardo Alfaro, Orion Chacón-Hurst, Laura Cobian, Norma Nuñez, Carmen De Anda, Joanna Alfaro, Alex Mejia-Sosa, Adriana Rodriguez, Isaac Murray, and Pablo Albarca. The team was mentored by Jarred Martinez and Asaeli Matelau.

3   British academic institution are assessed regularly (every four years or so) to measure the quality of their research output. Scores are awarded, and used to unequally distribute a significant proportion of available government funding for research – see the Higher Education Funding Council for England's "Research Excellence Framework": www.hefce.ac.uk/Research/ref/ (accessed 08.02.10).

4   In trigonometry and cartography, researches can determine the exact location of a point in space by measuring angles to it from either end of a known base line (creating a triangle). In social science the term is used to describe the process of comparing findings generated by using different data gathering techniques or approaches in order to come to a fuller understanding of a phenomenon under study. The degree to which a researcher believes that "triangulation" enables them to *definitively* "locate/position" or "know" a phenomenon, depends very much on their epistemological persuasions.

5   In simple terms: affect can be thought of as a raw unmediated feeling or reaction whilst emotion is the social cultural meaning associated with a particular assemblage of affects. We use the term "emotion/affect" here given their close inter-relation in practice (see Askins 2009).

References

Anderson, B. (2009) Affect (and) non-representational theory. In D. Gregory, R. Johnston, G. Pratt, M. Watts, and S. Whatmore (eds.), *The Dictionary of Human Geography*, 5th edn. Chichester: Wiley-Blackwell, pp. 8–9, 503–5.

Askins, K. (2009) "That's just what I do": placing emotion in academic activism. *Emotion, Society and Space*, 2 (1): 4–13.

Askins, K. and Pain, R. (in press). Contact zones: participation, materiality and the messiness of interaction. *Environment and Planning, D.*

Boal, A. (1979) *Theatre of the Oppressed*. London: Pluto Press.

Brydon-Miller, M., Greenwood, D., and Maguire, P. (2003) Why action research? *Action Research* 1 (9): 1–28.

Bunge, W. (1969) *Atlas of Love and Hate*. Detroit: Society for Human Exploration.

Cahill, C. (2004) Defying gravity: raising consciousness through collective research. *Children's Geographies* 2 (2): 273–86.

Cahill, C. (2006) "At risk"? The fed up honeys re-present the gentrification of the Lower East Side. *Women Studies Quarterly* (special issue *The Global & the Intimate*, ed. G. Pratt and V. Rosner) 34 (1–2): 334–63

Cahill, C. (2007a) The personal is political: developing new subjectivities in a participatory action research process. *Gender, Place, and Culture* 14 (3): 267–92.

Cahill, C. (2007b) Participatory data analysis. In S. Kindon, R. Pain, and M. Kesby (eds.), *Participatory Action Research Approaches and Methods: Connecting People, Participation and Place*. London: Routledge, pp. 181–7.

Cahill, C. (2010) Why do they hate us? Young people raise critical questions about the politics of race and immigration. *Area*, DOI 10.1111/j.1475–4762.2009.00929.x.

Cahill, C. and Torre, M. (2007) Beyond the journal article: representations, audience, and the presentation of participatory research. In S. Kindon, R. Pain, and M. Kesby (eds.),

Connecting People, Participation and Place: Participatory Action Research Approaches and Methods. eds. London: Routledge, pp. 196–205.

Cahill, C., Sultana, F., and Pain, R. (2007) Participatory ethics: policies, practices & institutions. *ACME: An International E-Journal for Critical Geographies* 6 (3): 304–18.

Cahill, C., Bradley, M., Castañeda, D., et al. (2008) Represent: reframing risk through participatory video research. In M. Downing and L. Tenney (eds.), *Video Vision: Changing the Culture of Social Science Research.* Cambridge: Scholars Publishing, 207–28.

Cameron, J. and Gibson, K. (2005) Participatory action research in a poststructualist vein. *Geoforum*, 36 (3): 315–31.

Castree, N. (2000) Professionalisation, activism and the university: whither "critical geography?" *Environment and Planning A* 32 (6): 955–70.

Castree, N., Chatterton, P., Heynen, N., et al. (2010) *The Point Is to Change It: Geographies of Hope and Survival in an Age of Crisis.* Oxford: Wiley-Blackwell.

Chambers, R. (1997) *Whose Reality Counts? Putting the First Last.* London: Intermediate Technology Publications.

Chatterton, P., Fuller, D., and Routledge, P. (2007) Relating action to activism: theoretical and methodological reflections. In S. Kindon, R. Pain, and M. Kesby (eds.), *Participatory Action Research Approaches and Methods: Connecting People, Participation and Place.* London: Routledge, pp. 216–22.

Cieri, M. and McCauley, R. (2007) Participatory theatre: creating a source for staging an example in the USA. In S. Kindon, R. Pain, and M. Kesby (eds.), *Participatory Action Research Approaches and Methods: Connecting People, Participation and Place.* London: Routledge, pp. 141–9.

Cook, I. et al. (2008) Writing collaboration, http://writingcollaboration.wordpress.com/1-outline/. Accessed October 1.

Cooke, B. and Kothari, U. (eds.) (2001) *Participation: The New Tyranny?* London: Zed Books.

Cornwall, A. and Brock, K. (2005). What do buzzwords do for development policy? A critical look at "participation," "empowerment" and "poverty reduction." *Third World Quarterly* 26 (7): 1043–60.

Del Casino, V. (2009) Social activism/social movements/social justice. In V. Casino, *Social Geography: A Critical Introduction.* Oxford: Wiley, pp. 154–82.

Demeritt, D. (2005) Commentary: the promises of collaborative research. *Environment and Planning A* 37 (12): 2075–82.

FACE Movement (2009) Home page: empowering the community through unity action and education, www.facethestruggle.blogspot.com. Accessed October 3.

Fine, M. and Torre, M. (2008) Theorizing audience, products and provocation. In P. Reason and H. Bradbury (eds.), *The Sage Handbook of Action Research*, 2nd edn. Thousand Oaks: Sage, pp. 407–19.

Fine, M., Torre, M., Boudin, K., et al. (2001) Participatory action research: from within and beyond prison bars. In P. Camie, J. Rhodes, and L. Yardley (eds.), *Qualitative Research in Psychology: Expanding Perspective in Methodology and Design.* Washington, DC: American Psychology Association, pp. 173–98.

Fine, M., Weis, L., Weseen, S., and Loonmun, W. (2000) For whom? Qualitative research, representations and social responsibilities. In N. Denzin and Y. Lincoln (eds.), *The Handbook of Qualitative Research.* Thousand Oaks: Sage, pp. 107–32.

Fisher, P. and Ball, T. (2003) Tribal participatory research: mechanisms of a collaborative model. *American Journal of Community Psychology* 32 (3–4): 207–16.

Foucault, M. (1983) On the genealogy of ethics: an overview of work in progress. In P. Rabinow (ed.), *The Foucault Reader.* New York: Pantheon, pp. 340–72.

Freire, P. (1970) *Pedagogy of the Oppressed*. New York: Continuum Publishing Company.

Fuller, D. (2008) Public geographies: taking stock. *Progress in Human Geography* 32 (6): 834–44.

Fuller, D. and Kitchin, R. (2004) Radical theo/critical praxis: academic geography beyond the academy? In D. Fuller and R. Kitchin (eds.), *Radical Theory, Critical Praxis: Making a Difference beyond the Academy?* Praxis E-Press, pp. 1–20.

Greenwood, D. (2002) Action research: unfulfilled promises and unmet challenges. *Concepts and Transformation* 7 (2): 117–39.

Greenwood, D. and Levin, M. (eds.) (1998) *Introduction to Action Research: Social Research for Social Change*. Thousand Oaks: Sage.

Greenwood, D., Whyte, W., and Harkavy, I. (1993) Participatory action research as a process and as a goal. *Human Relations* 46: 175–92.

Guijt, I. and Shah, M. (eds.) (1998) *The Myth of Community: Gender Issues in Participatory Development*. London: Intermediate Technology.

Guishard, M. (2009) The false paths, the endless labors, the turns now this way and now that: participatory action research, mutual vulnerability, and the politics of inquiry. *Urban Review*, 41 (1): 85–105.

Halse, C. and Honey, A. (2005) Unraveling ethics: illuminating the moral dilemmas of research ethics. *Signs: Journal of Women in Culture and Society* 30 (4): 2141–62.

Hamnett, C. (1996) *Social Geography: A Reader*. London: Hodder Arnold.

Haraway, D. (1991) Situated knowledges. In D. Haraway (ed.), *Simians, Cyborgs, and Women: The Reinvention of Nature*. New York: Routledge, pp. 183–203.

Hawkins, H., Sacks, S., Cook, I., et al. (forthcoming) Organic public geographies: "making the connection." *Antipode*.

Hayward, C., Simpson, L., and Wood, L. (2004) Still left out in the cold: problematising participatory research and development. *Sociologica Ruralis*, 44 (3): 95–108.

HEFCE (2009) *Research Excellence Framework: Second Consultation on the Assessment and Funding of Research*. HEFCE, London, www.hefce.ac.uk/Pubs/HEFCE/2009/09_38/. Accessed October 3.

Henkel, H. and Stirrat, R. (2001) Participation as spiritual duty: empowerment as secular subjection. In B. Cooke and U. Kothari (eds.), *Participation: The New Tyranny?* London: Zed, pp. 168–84

Heyman, R. (2007) "Who's going to man the factories and be the sexual slaves if we all get PhDs?": democratizing knowledge production, pedagogy, and the Detroit geographical expedition and Institute. *Antipode*, 39 (1): 99–120.

Heynen, N. (2006) "But it's alright, Ma, it's life, and life only": radicalism as survival. *Antipode*, 38 (5): 916–29.

Hickey, S. and Mohan, G. (eds.) (2004) *Participation: From Tyranny to Transformation? Exploring New Approaches to Participation in Development*. New York: Zed.

Hooks, B. (1990) Choosing the margin as a space of radical openness. In B. Hooks, *Yearning: Race, Gender, and Cultural Politics*. Boston: South End Press, pp. 154–53.

Hooks, B. (1995) *Killing Rage: Ending Racism*. New York, Henry Holt.

Hume-Cook, G., Curtis, T., Potaka, J., et al. (2007) Uniting people with place through participatory video: a Ngaati Hauiti journey. In S. Kindon, R. Pain, and M. Kesby (eds.), *Participatory Action Research: Connecting People, Participation and Place*. London, Routledge, pp. 160–9.

Kapoor, I. (2002) The devil's in the theory: a critical assessment of Robert Chambers' work on participatory development. *Third World Quarterly* 23 (1): 101–17.

Kapoor, I. (2005) Participatory development: complicity and desire. *Third World Quarterly*, 26: 1203–20.

Kesby, M. (2005) Re-theorizing empowerment-through-participation as a performance in space: beyond tyranny to transformation. *Signs: Journal of Women in Culture and Society* 30 (4): 2037–65.

Kesby, M. (2007) Spatialising participatory approaches: the contribution of geography to a mature debate. *Environment and Planning A* 39 (12): 2813–31.

Kesby, M. and Gwanzura-Ottemoller, F. (2007) Sexual health: examples from two participatory action research projects in Zimbabwe. In S. Kindon, R. Pain, and M. Kesby (eds.), *Participatory Action Research Approaches and Methods: Connecting People, Participation and Place.* London: Routledge, pp. 71–80.

Kesby, M. and Pain, R. (2010) Emotional and affective geographies of participation. Unpublished conference paper. Association of American Geographers, Washington DC, 2010.

Kesby, M., Kindon, S., and Pain, R. (2005) Participatory approaches and techniques. In R. Flowerdew and D. Martin (eds.), *Methods in Human Geography: A Guide for Students Doing a Research Project*, 2nd edn. London: Longman, pp. 144–66.

Kesby, M., Kindon, S., and Pain, R. (2007) Participation as a form of power: retheorisng empowerment and spatialising participatory action research. In S. Kindon, R. Pain, and M. Kesby (eds.), *Participatory Action Research Approaches and Methods: Connecting People, Participation and Place.* London: Routledge, pp. 19–25.

Kindon, S. (2003) Participatory video in geographic research: a feminist practice of looking? *Area* 35 (2): 142–53.

Kindon, S. (2008) Keynote address to the inaugural International Conference of Participatory Geographies, University of Durham, Durham, January.

Kindon, S. (2009a) Thinking through complicity. Paper presented to the Association of American Geographers Conference, Las Vegas, April.

Kindon, S. (2009b) Participatory action research. In I. Hay (ed.), *Qualitative Methods in Human Geography*, 3rd edn. Melbourne: Oxford University Press, pp. 259–77.

Kindon, S. (2010) Participation. In S. Smith, S. Marston, R. Pain, and J. Jones (eds.), *The Handbook of Social Geography.* London: Sage, pp. 517–45.

Kindon, S. and Elwood, S. (2009) Introduction: more than methods – reflections on participatory action research in geographic teaching, learning and research. *Journal of Geography in Higher Education*, 33 (1): 19–32.

Kindon, S. and Latham, A. (2002) From mitigation to negotiation: ethics and the geographic imagination in Aotearoa New Zealand. *New Zealand Geographer* 58 (1): 14–22.

Kindon, S., Pain, R., and Kesby, M. (eds.) (2007) *Participatory Action Research Approaches and Methods: Connecting People, Participation and Place.* London: Routledge.

Kothari, U. (2001) Power, knowledge and social control in participatory development. In B. Cooke and U. Kothari (eds.), *Participation: the New Tyranny?* London: Zed, pp. 139–52.

Kothari, U. (2005) Authority and expertise: the professionalization of international Development and the ordering of dissent. *Antipode* 37 (3): 402–24.

Krieg, B. and Roberts, L. (2007) Photovoice: insights into marginalisation through a "community lens" in Saskatchewan, Canada. In S. Kindon, R. Pain, and M. Kesby (eds.), *Participatory Action Research: Connecting People, Participation and Place.* London: Routledge, pp. 150–60.

Lawson, V. (2007) Geographies of care and responsibility. *Annals of the Association of American Geographers* 97 (1): 1–11.

Ley, D. (1983) *A Social Geography of the City.* New York: HarperCollins.

Louis, R. (2007) Can you hear us now? Voices from the margin: using indigenous methodologies in geographic research. *Geographical Research* 45: 130–9.

Marx, K. (1969) [1845] Thesis on Feuerbach. In K. Marx and F. Engels, *Selected Works*, vol. 1. Moscow: Progress Publishers, pp. 3–13, www.marxists.org/archive/marx/works/1845/theses/theses.pdf. Accessed November 15.

McTaggart, R. (ed.) (1997) *Participatory Action Research: International Context and Consequences*. Albany: State University of New York Press.

Mitchell, K. (2008) *Practising Public Scholarship: Experiences and Possibilities Beyond the Academy*. Antipode Book Series. Chichester: Wiley-Blackwell.

Mohan, G. (2001) Beyond participation: strategies for deeper empowerment. In B. Cooke and U. Kothari (eds.), *Participation: The New Tyranny?* London: Zed, pp. 153–67.

Mohan, G. (2007) Participatory development: from epistemological reversals to active citizenship, *Geography Compass* 1 (4): 779–96.

Mosse, D. (1994) Authority, gender and knowledge: theoretical reflections on the practice of participatory rural appraisal. *Development and Change* 25: 497–526.

mrs. kinpaisby (2008) Boundary crossings: taking stock of participatory geographies: envisioning the communiversity. *Transactions of the Institute of British Geographers* 33 (3): 292–9.

mrs. c. kinpaisby-hill (2008) Publishing from participatory research. In A. Blunt (ed.), *Publishing in Geography: A Guide for New Researchers*. Oxford: Wiley-Blackwell, pp. 45–7.

Pain, R. (2004) Social geography: participatory research. *Progress in Human Geography* 28 (5): 652–63.

Pain, R. (2006) Social geography: seven deadly myths in policy research. *Progress in Human Geography*, 26 (5): 647–55.

Pain, R. (2009) Globalised fear? Towards an emotional geopolitics. *Progress in Human Geography* 33 (4): 466–86.

Pain, R., Barke, M., Gough, J., et al. (2001) *Introducing Social Geographies*. London: Arnold.

Panelli, R. (2003) *Social Geographies: From Difference to Action*. London: Sage.

Panelli, R. (2008) Social geographies: encounters with indigenous and more-than-white/Anglo geographies. *Progress in Human Geography* 32 (6): 801–11.

Park, P., Brydon-Miller, M., Hall, B., and Jackson, T. (eds.) (1993) *Voices of Change: Participatory Research in US and Canada*. London: Bergin and Harvey.

Pickerill, J. (2008) A surprising sense of hope. *Antipode* 40 (3): 719–23.

Pratt, G. and Philippine Women Center of BC and Ugnayan ng Kabataang Pilipino sa Canada/Filipino-Canadian Youth Alliance (2007) Working with migrant communities: collaborating with the Kalayaan Center in Vancouver, Canada. In S. Kindon, R. Pain, and M. Kesby (eds.), *Participatory Action Research Approaches and Methods: Connecting People, Participation and Place*. London: Routledge, pp. 95–103.

Pratt, G. and Rosner, V. (2006) Introduction: the global and the intimate. *Women's Studies Quarterly* 34 (1–2): 13–24.

Reason, P. and Bradbury, H. (2008) *Handbook of Action Research*. London: Sage.

Rohd, M. (1998) *Theater for Conflict, Community and Dialogue*. Portsmouth, NH: Heinemann.

Sanderson, E. and Kindon, S. (2004) Progress in participatory development: opening up the possibility of knowledge through progressive participation. *Progress in Development Studies* 4 (2): 114–26.

Smith, L. (1999) *Decolonizing Methodologies: Research and Indigenous Peoples*. New York: Zed Books.

Smith, L. (2007) On tricky ground: researching the native in an age of uncertainty. In N. Denzin and Y. Lincoln (eds.), *Handbook of Qualitative Research*. Beverly Hills: Sage, pp. 85–108.

Smith, S., Pain, R., Marsden, S., and Jones, J. (2010) *The Sage Handbook of Social Geographies*. London: Sage.

Stoeker, R. (1999) Are academics irrelevant? Roles for scholars in participatory research. *American Behavioral Scientist* 42 (5): 840–54.

Torre, M. (2009) Participatory action research and critical race theory: fueling spaces for nos-otras to research. *Urban Review* 41 (1): 106–20.

Torre, M. and Ayala, J. (2009) Envisioning participatory action research entremundos. *Feminism and Psychology* 19 (3): 387–93.

Torre, M.E. and Fine, M. (2006) Participatory action research (PAR) by youth. In L. Sherrod (ed.), *Youth Activism: An International Encyclopedia*. Westport, CT: Greenwood, pp. 456–62.

Tuck, E. (2009) Re-visioning action: participatory action research and indigenous theories of change. *Urban Review* 41 (1): 47–65.

Tuck, E. and Fine, M. (2007) Inner angles: a range of ethical responses to/with indigenous and decolonizing theories. In N. Denzin and M. Giardina (eds.), *Ethical Futures in Qualitative Research: Decolonizing the Politics of Knowledge*. Walnut Creek: Left Coast Press, pp. 145–88.

Trapese (2007) *DIY: A Handbook for Changing Our World*. London: Pluto.

Wadsworth, Y. (1998) What is participatory action research? *Action Research International*, November 1998, Paper 2, www.scu.edu.au/schools/gcm/ar/ari/p-ywadsworth98.html. Accessed October 3.

Weis, L. and Fine, M. (2004) *Working Method: Research and Social Justice*. New York: Routledge.

Williams, G. (2004) Evaluating participatory development: tyranny, power and (re)politicisation. *Third World Studies Quarterly* 25 (3): 557–78.

Woodward, J. and Leam J. (2009) Geographies of affect. In S.J. Smith, R. Pain, S. Marston, and J.P. Jones (eds.), *Handbook of Social Geography*. Sage, London, pp. 154–75.

Wright, S. (2008) Practising hope: learning from social movement strategies in the Philippines. In R. Pain and S. Smith (eds.), *Fear: Critical Geopolitics and Everyday Life*. Aldershot: Ashgate, pp. 223–34.

Chapter 14

Using Social Geography

Sarah Johnsen

Much of what is now regarded as front-line research in [human geography] has little practical relevance for policy; in fact one might even say little social relevance at all. (Martin 2001: 191)

A great many geographers think that what they are doing or saying is of relevance when its impact is in fact minimal. (Dorling and Shaw 2002: 630)

Introduction

Social geographers, together with academics working in other subdisciplines, have invested a great deal of energy debating the "relevance" and impact of their work. Their angst about this issue has adorned many pages of academic journals over the past few decades and has, at times, generated intensely heated debate. The statements made by Martin (2001) and Dorling and Shaw (2002) above are illustrative of the assertions made by a number of scholars regarding the failure of geography to address issues that "matter" and/or "make a difference."

Reviewing this literature, and reflecting upon my own experiences conducting policy-oriented research, this chapter examines the main axes around which the so-called "relevance" debate has revolved, most particularly the charge that geography has only had a limited impact on social policy. It concludes that much of the heat generated has been misdirected because contributors often fail to fully appreciate the impact that geographers have *beyond* academe. The chapter also considers some of the actions taken by social geographers to ensure that their work addresses issues of significance and has a tangible positive impact on the communities of interest, highlighting some of the complex implications of taking social geography beyond the academy in so doing.

A Companion to Social Geography, First Edition. Edited by
Vincent J. Del Casino Jr., Mary E. Thomas, Paul Cloke, and Ruth Panelli.
© 2011 Blackwell Publishing Ltd. Published 2011 by Blackwell Publishing Ltd.

Before focusing on these issues in depth, however, I want to step back briefly and contextualize the discussion that follows by providing an introduction to my own stance on, and experiences of conducting, policy-relevant research.

"Confessions" ...

Research relevance and impact are issues that I am confronted with and challenged by on a very regular basis. Most of my work revolves broadly around the issue of homelessness – one of those pernicious social ills that is often used as a yardstick to measure how well society deals with its most vulnerable members (Jones and Johnsen 2009) and is therefore a subject of great fascination for scholars from a range of disciplinary backgrounds. As is the case with many other vulnerable groups (Liamputtong 2006), the difficult and often desperate circumstances of homeless people create particular challenges for researchers. Working in a rather different context, Ward (2005) recalls having been taken aback during the course of his PhD fieldwork when an interviewee – a leading political figure within a UK city – unleashed a verbal volley questioning the process and utility of academic research. I confess that I encounter similar challenges from homeless people and the agencies providing services for them so frequently that I now regard them a standard occupational hazard (so to speak).

In an earlier paper, Paul Cloke, Jon May, and I recorded the response of a homeless man, Don, when we told him about our research exploring the uneven geographies of homelessness. An individual who rarely embellished his viewpoints with social niceties, Don's response was rather blunt: "What the fuck's the point of that?" (Cloke et al. 2003: 1). His conviction of the uselessness of our enterprise was founded on a number of factors, particularly the potential waste of money spent on research when it could have been put to more practical use, and his assumptions regarding our inability to influence anyone important with our findings and thereby actually change things (Cloke et al. 2003). Don's test of utility, which has been echoed in challenges posed by dozens of interviewees in intervening years (albeit usually in less expletive-replete parlance), has become a tough evaluative yardstick by which I continually gauge the relevance and potential impact of my research[1].

Like many of my colleagues working in this field (see for example Blomley 1994; Cloke et al. 2000; Cloke 2002) I harbor an ethical and political commitment to people who experience homelessness. How could I not when witness to the exclusionary barriers they face on such a regular basis? I also share Martin's (2001) belief that as a critical social science, human geography has a duty to engage with public policy issues and debate – to deepen our knowledge and understanding of society in the hope that our work might be used to combat the causes and consequences of social injustice and/or inequality. Many will no doubt believe this to be symptomatic of naive idealism. Perhaps it is; but I remain entirely unrepentant. Whilst I would never want to belittle esoteric or "blue skies"[2] scholarship, I *personally* would not want to stand in a day center full of homeless people and attempt to justify the expenditure of public funds on such work. A desire to devote my work to improving understandings of the causes and consequences of social exclusion, and to provide evidence to inform the development of strategies to combat this, was central to my move from a geography department into a social policy department a few years ago, where I continued, insofar as possible, to follow and contribute to geographical debates.

Given this experience, I inevitably find myself reflecting fairly regularly on broader debates about relevance in social geography. This has brought me to consider the complexities of what is meant by relevance and the ways social geographers can and do engender change.

Intellectual Turns and Cul-de-sacs: The Axes of the Relevance Debate

The concept of relevance can be defined in a number of ways, but most contributions to geography's relevance debate interpret it in terms of "application," that is, the use of research to develop approaches or tools to address specific social problems (Staehli and Mitchell 2005a). Most contributors do in fact evaluate the discipline's *relevance* almost solely in terms of its *impact*, thus conflating the two issues (see for example the Dorling and Shaw (2002) quotation at the opening of this chapter). A desire to affect change has been evident within the discipline's scholarship since the term "social geography" was apparently first coined in the nineteenth century by the French anarchist academic/activist Reclus who, along with other early social geographers such as DuBois, Addams and Abbott, sought to understand the causes of social ills and called for an end to injustice and discriminatory practices that reproduced inequality (Del Casino 2009). Such aspirations have been evident throughout geography's history, and no more so than in the area of welfare geography which emerged in the 1970s. A leading figure in the field, Smith (1974: 289) asserted that:

> The question of who gets what *where* and how provides a framework for the restructuring of human geography in more "socially relevant" terms ... requir[ing] us to identify the desirable or undesirable aspects of human existence, to find out and measure how these are allocated between groups distinguished by place or area, and to examine and if possible model the processes which lead things to be as they are. (emphasis in original)

He went on to argue that geographers should not only work toward better understanding of such social processes, but also respond, as applied geography "requires that this knowledge be put to the service of society, in the design of predictably "better" spatial allocation of the benefits and penalties of modern life" (Smith, 1974: 290). Social geography thus has a long tradition of applying spatial science to questions of social change, in an attempt to address inequalities in the welfare of the "haves" and "have nots" (Del Casino 2009).

Yet there has been, and continues to be, a profound sense of dissatisfaction amongst geographers regarding the impact that the discipline has actually had. For example, in the mid-1970s Coppock (1974: 1) noted that British geographers had "made less of a contribution than [had] members of other disciplines" to the development of public policy. In the United States, Berry (1973) similarly noted that geography was little used by government departments or their policy machinery. Nearly a decade later, Knight's (1986) exasperation was clearly evident in his call for geographers to "take our heads out of our armpits and *do something useful,*" by "addressing some real problems" (p.334, emphasis in original). More recently, Peck (1999: 131) sparked a debate in *Transactions of the Institute in British Geographers* by claiming that "geographers have on the whole been conspicuous by their absence from substantive policy debates" (see the responses from Banks and Mackian 2000; Pollard et al. 2000; also Peck 2000). This was followed by the perhaps most oft-cited, and heated, exchange (containing reference to "gratuitous insults" and "wilful misunderstandings," even) between Massey (2001, 2002), Martin (2001, 2002), and Dorling and Shaw (2002) in *Progress in Human Geography*.

The debate amongst these and other scholars has revolved, primarily, around four interrelated axes: the alleged decline in the attention paid by geographers to "genuine" social problems; the privileging of theoretical approaches over empirical/applied research; increased pressure to publish in prestigious academic journals; and the ineffectual communication of research findings beyond academia (and indeed the geography discipline itself). Each of these is discussed below.

Decline in attention to social problems

Central to the arguments made by many contributors to the relevance debate is the claim that recent developments in the discipline – most notably the cultural and postmodern turns – have diverted geographers' attention away from issues that "matter." Blomley (1994) suggests that a postmodern humility and associated reluctance to speak for "the other" has led to a disciplinary self-silencing on issues affecting marginalized groups. Martin (2001: 195) asserts that the surge of interest in identity, embodiment, the self and other has deflected attention from "practically oriented social investigation," with the consequence that issues such as class, inequality, and poverty have all but disappeared from the human geographer's lexicon.[3] He concludes that "it is difficult to envisage how the vague abstractions and epistemic and ontological relativism of much of human geography research ... can form the basis of critical public policy analysis" (Martin 2001: 196). He thus calls for the discipline to enhance its relevance via recourse to a "policy turn," which would seek to exert a direct influence on policy-making processes. Several other commentators are sympathetic to the principles underpinning such a wish (see for example Peck 1999; Imrie 2004; Beaumont et al. 2005). Dorling and Shaw (2002), however, are pessimistic about the potential utility of any such turn – suggesting that it would run a serious risk of turning into a cul-de-sac given the alleged failure of most geographers to achieve, or indeed be genuinely concerned about, political change:

> many, if not most, "geographers" are focused neither on achieving political change nor on communicating with the world "out there" ... They *are* concerned with thinking about (and understanding and explaining) spatial relationships, not with changing them, and that is precisely why they are geographers. (Dorling and Shaw 2002: 632, emphasis in original)

Whilst acknowledging the positive contributions that the cultural turn has brought to understandings of society and space, as does Martin (2001), Cloke (2002: 588) also alludes to its apparently immobilizing effect:

It almost seems as though as we become theoretically more sophisticated in identifying difference and differentiating identity, so our ability to offer imaginative and practical guidelines for doing something about anything appears to be diminishing.

Pain (2003: 650) also agrees that the cultural turn "has led some away from earlier ideals of a progressive social geography which focuses on social problems and their resolution," but argues that there is no need for social geographers to feel defensive, for "while some of the products of the cultural turn never came near to passing the "so what?" test, others challenged and breathed new life into the traditional interests of social geography." She also emphasizes that many social geographers *do* continue to make a sizeable contribution to policy – but notes that their involvement is not always highly visible in the usual academic outlets (Pain 2003), and that literature on policy relevance tends to exclude already marginalized "development" research and that conducted in poorer regions of the world (Pain 2006; see also Willis 2004). Pain (2006: 256) thus resists any temptation to exhort social geographers to "do more" policy-oriented research, concluding that "hundreds are doing plenty" already.

I agree with Pain's (2003, 2006) assertion that social geographers are having a greater effect on policy than is commonly acknowledged – partly because, as she suggests, the outputs of academics working in this field do not necessarily grace the pages of academic journals and thereby remain unrecognized by the academy. I also concur with Pain's (2003) belief that that approaches inspired by recent theoretical turns can indeed "breathe new life" into the study of policy-relevant issues. My recent work with Paul Cloke and Jon May on the geographies of homelessness is, I believe, illustrative. In this, consideration of fluidity, mobility and performance opened up new understandings of homeless people's identities and uses of space, which led to a conceptual re-mapping of the "homeless city" and their co- (and often contested) constructions of the "spaces of care" offering them support (Cloke et al. 2008). These are especially relevant to ongoing debates in Britain regarding the role of soup runs and day centers, which provide food and other basic services for homeless people – with some quarters claiming that these conflict with the prevailing policy imperative of getting homeless people "off the streets" (Johnsen et al. 2005a, 2005b).

Privileging of theoretical approaches over applied/empirical research

The second axis of the relevance debate relates to the divide between theoretical and "other" (namely applied, empirical, policy-oriented) work, and the intellectual privileging of the former over the latter (Martin 2001; Staehli and Mitchell 2005b; Phelps and Tewdwr-Jones 2008). In discussing the place of "grey" geography, for example, Peck (1999: 131–2) argues that academic practice "privileges abstract and 'scientific' knowledge over practical and policy-orientated knowledge," and that "more often than not, policy work is seen as 'bad science' or at least second-rate research." The privileging of theoretical work is reinforced by research assessment procedures such as the Research Assessment Exercise (RAE) in Britain where greatest value was placed on pure and thus supposedly more "prestigious" research (Martin 2001). Policy research can thus, as Burgess (2005) notes, be "risky" in terms of establishing an academic career.

Prejudice against policy study within human geography has, according to Martin (2001), been underpinned by two key factors: first, a view that policy research can easily be subverted by other organizations or commissioners, thereby threatening the independence of interest, thought and method that is the hallmark of academic research; and second, the charge that policy analysis rarely contributes to theory. There are however many examples of social geographers successfully conducting policy-relevant work without compromising their academic independence or integrity – see Imrie (2004), Martin (2001), and Pain (2006) for overviews of such work. There is also evidence that rigorous empirical and/or applied work can and does make a valuable contribution to the development of theory (Banks and Mackian 2000; Pain 2003). Peck's (1999) differentiation between "shallow" and "deep" policy analysis is helpful in elucidating this. Shallow policy analysis, he explains, is tackled within the confines of stated aims and objectives (i.e. by "geographers-as-impact-evaluators"), whilst deep policy analysis sees the policy process as a contested, politicized domain in which the parameters and exclusions of policy-making are themselves objects of critical investigation (i.e. "geographers-as-state-theorists"). He notes that groups conducting these two different sorts of research may not agree on much – constructing as they do the objects, means and ends of policy research quite differently (Peck 1999).

I would like to propose, however, that projects designed to evaluate the outcomes of specific policies (i.e. apparently "shallow" analysis) need not preclude researchers from reflecting upon the broader parameters, presumptions, and premises of such policies (i.e. conducting "deep" analysis). Let me give an example based on a recent study conducted with Suzanne Fitzpatrick which examined the impact of the increasing use of enforcement initiatives (such as arrests and Anti-social Behavior Orders) on the welfare of people involved in begging, street drinking, and rough sleeping in England. The main report, funded by a voluntary-sector body, was explicitly policy and practice oriented, containing specific recommendations regarding the deployment of coercive measures and associated supportive interventions, and highlighting the circumstances under which such measures were more likely to lead to positive or negative outcomes for the individuals targeted (Johnsen and Fitzpatrick 2007c). Subsequent academic papers based on this work made two key contributions to theoretical debate. One reflected on the implications of the findings for understandings of homelessness and street culture – posing a serious challenge to the revanchist orthodoxy dominating geographers' work on homelessness (Johnsen and Fitzpatrick 2010), while another drew upon moral and political philosophy to interrogate the ethical basis for the use of "control" in social welfare provision for vulnerable groups (Fitzpatrick and Johnsen 2009).

A decade ago Markusen (1999: 872) asserted that:

> The term "theorist" is often applied to those who deal mainly in abstractions and abjure empirical verification, rather than to those who take up knotty problems, hypothesize about their nature and causality and marshal evidence in support of their views.

Indeed, it seems that this generally remains the case today. But surely the "best" theories are those that tackle "knotty problems" and are founded upon and/or hold up under detailed empirical investigation? On this point, I would like to (delicately) suggest that expressions of snobbery within geography are not entirely unidirectional. Whilst applied/empirical researchers sometimes bemoan the alleged conceit of pure theoreticians and blue-skies thinkers, perhaps we should be honest about the fact that we sometimes express our own form of "moral superiority," founded on the fact that we get our hands dirty and consider ourselves to be more effective at engendering change? Such musings aside, there is substantial evidence to confirm that by actively engaging with the messy realities of everyday life, policy research can, and sometimes does, make a very positive contribution to theory without compromising academic integrity.

Pressure to publish in prestigious academic journals

The prioritization of academic publications over other forms of dissemination has also been identified as an institutional barrier constraining research relevance (Staehli and Mitchell 2005b). Major funding bodies typically ask for information about any proposed dissemination to user groups in grant applications (Ward 2005), and some now require impact statements outlining the predicted implications for these groups. But, current assessments of a project's overall quality are still determined largely by its *academic* output. Academic reputations depend to a very significant degree on the perceived theoretical content of publications, which are themselves arranged in strict discipline-specific hierarchies (Phelps and Tewdwr-Jones 2008). Research assessment practices have been directly implicated in this phenomenon, with Ward (2005) arguing that the avenues in which geographers are valued for publishing their work have narrowed in recent years, particularly in the UK.[4] This phenomenon could, arguably, further constrain geographers' limited contribution to interdisciplinary exchange (Massey 2001; Phelps and Tewdwr-Jones 2008). Whilst by no means are all guilty, geographers have often been accused of introspection due, in large part, to an apparent proclivity to talk only to one another (Knight 1986; Gregson 1993).

Controversially, Dorling and Shaw (2002) suggest that geographers with a desire to communicate their messages to a wider audience are actually likely to find a more congenial home in other academic disciplines such as social policy or planning, or outside of academia altogether. My move from a geography department to a social policy department might in many ways be regarded an example of the former. I would however like to suggest that when making such an assertion, Dorling and Shaw (2002) drew attention to a very important, and often overlooked, issue. In the earliest years of my career, during my PhD study and first postdoctoral research post, I befriended many young social geographers who shared my frustration at the great emphasis placed on academic publications and limited value attached to feeding back to and/or impacting the communities in/with which we worked (see below; also Cloke et al. 2000). Of these, several are now employed in interdisciplinary policy-focused academic departments or, alternatively, working in government or charitable sectors – where they (usually) report feeling that their work has a more immediate and tangible impact on the communities with which they operate. Upon reflection, it was the obvious connection to the "real world" and all its fascinating complexities that attracted many of us to geography in the first place, at school and/or as an undergraduate. Yet, paradoxically, it seems that many would-be academic geographers are forsaking the discipline: turning to others or pursuing alternative career paths altogether in order to reconnect with that "real worldness" in an environment where they feel their "grounded" work is accorded greater value.

Moreover, conversations with central government policy-makers and staff in homelessness agencies over the years have indicated that many studied social geography at undergraduate, and sometimes postgraduate, level. Such individuals may not be actively publishing in geography journals – and one might therefore cynically suggest be inadvertently depriving the discipline of opportunities to massage its "impact ego" – but their geographical insights and imaginations will undoubtedly influence how they see the world and construct strategies to combat the social problems they encounter. To assess the relevance of social geography purely according to the influence of its proponents' pontifications in geography journals is to employ a very restricted view of who does geography, how and where (see also Imrie (2004) who makes a similar point). Indeed, it seems that Lee's (2002: 627) assertion that "perhaps the most effective applied – and political – work that geographers might do is to teach and proselytize it well and critically" is particularly apposite. Social geographers, however defined,[5] are not confined to the hallowed halls of university geography departments; nor are their outputs and impacts restricted to the pages of geography journals in the way that many contributors to the relevance debate appear to believe. Indeed, we really should avoid underestimating the long-term impact that high-quality geography teaching has in shaping students' views of the world (Kindon and Elwood 2009; Staehli and Mitchell 2005a), especially regarding issues such as social and spatial injustice/inequality.

Ineffectual communication beyond academia

The final axis of the relevance debate is the argument that geographers are not good at communicating their ideas outside the academic community. Much of the criticism in this regard relates to the language used to convey geographical concepts and research findings. Dorling and Shaw (2002: 634), for example, purport that:

> It is possible that geographers are being ignored because people can neither hear nor understand them ... Regardless of what they have been saying, the language and expression of the "reconceptualising geographers" has often become an elitist jargon.

Similarly, Blomley (1994: 383) argues that much academic work of a self-proclaimed critical, progressive or emancipatory nature is so heavily coded and jargon-ridden that the "battle cries, all too frequently [are] in a language that [makes] sense only to the cognoscenti," that is, can only be understood by those with "superior" specialist knowledge. On this issue, the use of neologisms (new words or meanings of words) has proved a particular bone of contention within the discipline (see for example the exchange between Dear and Flutsy 1998; Jackson 1999; Sui 1999). Phelps and Tewdwr-Jones (2008) quite rightly point out that the non-transferability of terminology constrains interdisciplinary dialogue. I suspect that I am not alone in having found myself on more than one occasion defending (sort of) geographers from charges of writing "incomprehensible nonsense" when colleagues from different disciplinary backgrounds have attempted to engage with their work. I confess that I am not entirely unsympathetic to such critiques; and again, suspect that I am not alone in feeling that way.

On this issue, a number of scholars claim that the use of obscure or esoteric language can sometimes be indicative of attempts to mask "fuzzy concepts" and/or a lack of empirical evidence (Markusen 1999). No doubt we have all read academic manuscripts where we suspect this to be the case, hence my reticence to defend some scholarship from the critiques described above. I believe there is a place for blue-skies work, of course, but like some of my colleagues I am riled by articles that make *empirical* claims in the absence of empirical evidence, especially when these are decreed in unnecessarily complex language. Almost all intellectual communities develop specific languages and use technical terms or jargon to communicate complex phenomena (Staehli and Mitchell 2005a), and it is wholly appropriate that these are used to convey findings or ideas accurately to audiences familiar with that vernacular. Difficulties arise, however, when convoluted language is used to communicate simple messages that could be expressed just as easily, and arguably more effectively, in "plain English."

Languages used within policy circles are not necessarily any less opaque or more inclusive than the esoteric dialects employed by some academics, however (Staehli and Mitchell 2005a). A key challenge in maximizing the impact of academic research on policy and practice, therefore, is to tailor outputs to relevant audiences, most notably by translating findings into "policy/practitioner-speak" when appropriate. Fluency in such "speak" is an essential tool for social geographers working in policy-oriented fields: indeed there really is no excuse for illiteracy in the policy world affecting the social groups and phenomena we study. My colleagues and I generate an array of outputs tailored to different audiences that the academic community will almost certainly be entirely unaware of. I doubt that many (if indeed any) of my articles in practitioner magazines such as *Police Professional* or *ROOF* (see for example Johnsen and Fitzpatrick 2007a, 2007b) will ever cross the desks of those in academic circles – although these publications may arguably have a much wider readership and potential "impact" than any academic outputs from the same projects.

The use of highly theoretical language will *of course* restrict the accessibility and palatability of publications to non-academic audiences. Surely that is so blindingly obvious that it need not be the subject of such extensive debate? Personally, I believe the argument that such language has limited the influence of social geography on policy has been overstated, for two reasons. Firstly, such an argument fails to acknowledge the impact of outputs tailored to non-academic audiences because these are typically disseminated in domains beyond the reach of the academic radar. Secondly, and perhaps more significantly, it is premised on the assumption that people constructing policies actually *read* academic outputs – when the reality of the matter is that they don't (or do so very rarely). Policy-makers *do not* spend hours pouring over the pages of academic journals: information exchange between academic and policy communities occurs in very different forums. If social geographers want to actively contribute to policy debates, they require a nuanced understanding of and engagement with policy-making politics and processes. The remainder of this chapter elaborates upon this argument, highlighting the complex implications of taking geography beyond academe's so-called ivory towers.

Politics and Proselytism: Taking Social Geography beyond the Academy

Academics can have a direct, and powerful, impact on policy outcomes when invited onto government advisory boards because of their known expertise in a particular subject area. It is however far more likely that their influence will result from contributions to a wider stock of knowledge, which may, over time, cumulatively lead to a shift in policy approach via the gradual "sedimentation" of new insights and perspectives (Weiss 1979; Hanney et al. 2002). Influences are rarely instantaneous: rather, academic contributions will be engaged with by policy-makers at varying times, to different degrees, depending upon the political "issues of the day" and priority accorded to these.

Policy-making and policy research are both highly politicized endeavors. The dissemination of high-quality research does not mean it will somehow naturally be listened to or acted upon in the ways researchers hope or anticipate. Beaumont et al. (2005: 118) for example reveal how the lofty ideals of researchers committed to social and political change can sometimes be besmirched by the politicization of the research process, emphasizing that academic arguments are not always judged on merit alone. They recall instances where some of their own work has been reoriented and even discredited by stakeholders when findings proved unpalatable. These resonate with my experiences on a small number of projects, most notably one that highlighted the prevalence of a very politically sensitive issue. The commissioners attempted to undermine the validity of one particular finding by subjecting it to a level of scrutiny far beyond that of those they "approved." Pressure to remove any reference to that particular finding was successfully resisted due to the safeguard of a contractual clause dictating that the funders could not reasonably withhold permission for publication in outlets other than the commissioned report. Such conflicts are clearly diplomatically challenging, but need not lead to the compromise of academic credibility (Beaumont et al. 2005).

It is not just politically contentious messages that are unattractive to policy-makers, but also those that are complex (Peck 1999). As Keynes once famously said: "There is nothing a politician likes so little as to be well informed; it makes decision-making so complex and difficult" (cited in Davies et al. 1999: 3). Some messages are not reducible to the kind of soundbites attractive to policy-makers and politicians. Indeed it is not always possible, or desirable, to provide simple messages and clear guidance for policy-makers, because doing so demands a level of clarity that cannot always be generated out of complex social phenomena (Imrie 2004). There can therefore be a profound tension between the "degree of certainty" desired by policy-makers, and the complex realities that academics seek to understand and convey.

Several commentators argue that to increase the impact of academic work, and the willingness of policy-makers to grapple with more complex ideas, those doing policy-oriented research need to become more proficient in the art of persuasion or proselytism (Martin 2001; Dorling and Shaw 2002; Phelps and Tewdwr-Jones 2008). Some argue that greater fostering of in-depth and sustained relationships with policy-makers, and dissemination in accessible forums, are needed to facilitate this process (Massey 2000; Burgess 2005). Doing so not only generates opportunities for mutual information exchange, but also fosters greater appreciation of the constraints that policy-makers work within (Weiss 1979; Burgess 2005; Pain 2006) and thus theoretically enables researchers to construct more "feasible" policy recommendations. It also potentially allows academics to develop greater credibility in the eyes of policy-makers which may, in turn, make the latter more amenable to "bigger ideas," particularly those which may be unfamiliar, appear counter-intuitive at first, and/or challenge the prevailing political status quo (Massey 2000). Moreover, as Massey (2000) argues, such relationships generate opportunities for one of the more radical contributions of social scientists – that being the _re_formulation of questions that are being asked, or identification of questions that ought to be asked but are not.

One should however avoid taking a purely centrist view on who has the power to bring about change. Beaumont et al. (2005) emphasize that there are a plethora of non-state institutions mediating between individual citizens and the state with which researchers might engage. In the homeless sector, for example, these might include voluntary-sector service providers and campaigning organizations operating at national, regional and local scales. These organizational actors are often in a position to respond to academic research findings even when central government bodies prove either disinterested or resistant. My work on the role of soup runs with Paul Cloke and Jon May (Johnsen et al. 2005b), for example, has recently been used by voluntary sector organizations to counter attempts to criminalize the street-level provision of food to homeless and other vulnerable people in London. In 2007 a government body proposed that the "unfettered distribution" of free food on public land, especially by organizations wishing to assist homeless people, be made an offence subject to a fine of £2,500, on the grounds that it causes public nuisance (London Councils 2007). Our findings were cited in many responses to the proposed ban which emphasized the importance of soup runs for people ineligible for mainstream welfare services (such as destitute asylum seekers), that the demise of soup runs could potentially lead to an increase in survivalist crime such as shoplifting, and that far from being "unfettered" the city's soup runs were being increasingly coordinated via interagency forums. The proposed ban was eventually overturned in the face of extreme opposition (Johnsen and Fitzpatrick 2010).

Some academics have actively involved organizational actors via participatory or action-oriented research which destabilizes traditional relationships between "the researcher" and "the researched" by giving participants an active role in research design, implementation, and/or resultant actions (Kindon et al. 2007). The contributions of participatory methods are relayed in detail in Chapter 13 of this volume. Whilst approaches which attempt to counter the historical tendency for research to be done _to_ or _for_ "othered" groups are extremely attractive on many fronts, I would like to suggest that their utility in my particular field is constrained by the fact that authorities do not automatically accord equal legitimacy to the contributions of all social groups. Klodawsky (2007) reported that the potential involvement of homeless people in a panel study on homelessness in Canada, for example, was viewed unfavorably by other stakeholders given normative assumptions about their chaotic lives and "unreliability." Similarly, some of the views expressed by authorities at research advisory groups I have sat on indicate that they are only assured of the legitimacy of homeless service user views when these accord with (that is, are "verified" by) those of support service staff or other more allegedly "reliable" informants. There is therefore a very real risk that unpalatable findings from participatory research will be disregarded by authorities on the grounds that such studies are not sufficiently "robust," regardless of how well executed they might be[6].

Doing policy-relevant research cannot be, and never is, a politically neutral manoeuvre (Cloke et al. 2003, 2008), but carries with it the very real risk of unanticipated or unintended consequences (Phelps and Tewdwr-Jones 2008). To draw examples from my work on the geographies of homelessness mentioned earlier, reporting the existence of different homeless "scenes" in different cities – in that some are more attractive to homeless people on grounds of greater availability of support services and/or affordability of illicit street drugs – could potentially be used to fuel protests against non-local homelessness and prejudice against "outsider" homeless people (Cloke et al. 2003). Equally, detailed insights into homeless people's uses of different spaces of the city could be used to inform those whose response to homelessness is to argue for a firmly policed purification of space (Cloke et al. 2003). There are no unassailable means of controlling how ideas are used once they have entered the public domain, and the potential for findings to be misrepresented always exists (Imrie 2004), particularly where the tabloid media is involved (see for example Dentith et al. 2009).

Much of this begs the more fundamental question of how much power to influence policy academics *should* have. Should our aim be to actively shape the direction that public policy takes? Or, rather, should we simply see ourselves as contributors to the pool of knowledge and evidence base from which authorities may draw? Academics are often considered somehow more "objective" than other members of the population. Most of us, however, work in the subject areas we do because we are interested in, perhaps even passionate about, them. This being so, perhaps we are not in the best position to advise on the kind of trade-offs that policy-makers must make in the context of restricted resources? There are no simple answers to such questions, but they should not be skirted in debates about the relationship between research and policy.

Conclusions

In reviewing the axes of the relevance debate, this chapter has highlighted concern that social geography and other sub-disciplines should "matter" and "make a difference." The intensity of heat generated by the debate has perhaps been excessive, and some of the critiques involved misdirected. It is true that many geographers have been guilty of talking only to one another, in academic forums, and in highly theoretical (and therefore exclusionary) parlance. That said, many of the contributors to the relevance debate have failed to recognize the impact of social geography *outside* academia – in part because of a lack of awareness of outputs tailored to non-academic audiences (and their impacts), and partly because of their tendency to deploy a narrow view regarding who "does" social geography and in what contexts. When these issues are taken into account, it seems that social geographers might justifiably spend less time lamenting their lack of relevance.

If the subdiscipline's role in public policy formation is to be increased in the future, however, such energies might valuably be redirected to greater engagement with the forums and politics of policy-making, most particularly via development of more sustained relationships with people in decision-making positions. Banks and Mackian (2000) argue that the "water is warm" for British geographers interested in policy research. The policy arena I operate in appears no exception, if the recent publication of a central government strategy entitled "Place matters" is anything to go by. This states that "place is increasingly important as a focus for policy" and signals a wish to "further develop ... thinking about the implications of putting a focus on place at the heart of what the Department does" (Communities and Local Government 2007: 11). What is this, if not an invitation?

Any response to such invitations will require a willingness on the part of social geographers to expose themselves to the sort of political and public criticism and degree of accountability that are routinely faced by our planning colleagues, as well as an acceptance that attempts to intervene in progressive ways on behalf of particular social groups may "fail" or lead to unintended outcomes (Phelps and Tewdwr-Jones 2008). Moreover, if the potential of social geography to affect policy is to be fully realized, those charged with assessing research quality need to pay greater heed to its societal impacts. There are glimmers of such changes on the horizon in the UK (UUK and HEFCE 2008).

The relevance debate will no doubt continue to rumble on in years to come. There are some indications that policy-oriented research may potentially be regarded more highly in the future by the academic community, but in the meantime Don's litmus test – "What the fuck's the point of that?" – will serve as my primary guide in determining what matters and has the potential to make a difference.

Notes

1   It is however very difficult to assess the potential impact of any research prior to its outset, or indeed immediately after completion (Ward 2005). A project might be highly "relevant" to the development of policy, but many factors may mitigate its actual "impact," as noted later in the chapter.

2   Blue skies research is generally understood to be scientific enquiry that is curiosity-driven and holds no expectations regarding how its findings might be applied, or indeed even whether they should be.

3   Martin (2001: 195) is careful to acknowledge that these issues had previously been "woefully neglected" areas in human geography which deserved the greater attention received. He is however critical of the tendency for geographers to recast such issues in narrow, even essentialist, cultural terms whilst relegating or neglecting the real material, political and historical circumstances in which identities and inequalities are forged.

4   This trend may alter in future years, however, given the forthcoming Research Excellence Framework (REF) which will apparently place greater emphasis on "impact" (UUK and HEFCE 2008).

5   This inevitably raises the question of "who is a social geographer?," and does a person stop being one if/when they leave academia? Personally I think not, although others may disagree.

6   On the other hand, it may be equally likely that policy-makers will emphasize the extent of service user involvement – claiming it lends findings greater credibility – should the conclusions accord with political priorities at the time.

References

Banks, M. and Mackian, S. (2000) Jump in! The water's warm: a comment on Peck's "grey geography." *Transactions of the Institute of British Geographers* 25: 249–54.

Beaumont, J., Loopmans, M., and Uitermark, J. (2005) Politicization of research and the relevance of geography: some experiences and reflections for an ongoing debate. *Area* 37: 118–26.

Berry, B.J.L. (1973) A paradigm for modern geography. In R.J. Chorley (ed.), *Directions in Geography*. London: Methuen, pp. 3–22.

Blomley, N.K. (1994) Activism and the academy. *Environment and Planning D: Society and Space* 12: 383–5.

Burgess, J. (2005) Follow the argument where it leads: some personal reflections on "policy-relevant" research. *Transactions of the Institute of British Geographers* 30: 273–81.

Cloke, P. (2002) Deliver us from evil? Prospects for living ethically and acting politically in human geography. *Progress in Human Geography* 26: 587–604.

Cloke, P., Johnsen, S., and May, J. (2003) "What the ****'s the point of that?" The cultural geographies of homelessness. Paper presented at the RGS-IBG Annual Conference, September 5, London.

Cloke, P., May, J., and Johnsen, S. (2008). Performativity and affect in the homeless city. *Environment and Planning D: Society and Space* 26: 241–63.

Communities and Local Government (2007) *Place Matters*. London: Department for Communities and Local Government.

Cloke, P., Cooke, P., Cursons, J., Milbourne, P., and Widdowfield, R. (2000) Ethics, reflexivity and research: encounters with homeless people. *Ethics, Place and Environment* 3: 133–54.

Coppock, J.T. (1974) Geography and public policy: challenges, opportunities and implications. *Transactions of the Institute of British Geographers* 63: 1–16.

Davies, H.T.O., Nuttley, S.M., and Smith, P.C. (1999) Editorial: what works? The role of evidence in public sector policy and practice. *Public Money and Management* 19: 3–5.

Dear, M. and Flutsy, S. (1998) Postmodern urbanism. *Annals of the Association of American Geographers* 88: 50–72.

Del Casino, V. (2009) *Social Geography: A Critical Introduction*. Malden: Wiley-Blackwell.

Dentith, A.M., Measor, L., and O'Malley, M.P. (2009) Stirring dangerous waters: dilemmas for critical participatory research with young people. *Sociology* 43: 158–68.

Dorling, D. and Shaw, M. (2002) Geographies of the agenda: public policy, the discipline and its (re)"turns." *Progress in Human Geography* 26: 629–46.

Fitzpatrick, S. and Johnsen, S. (2009) The use of enforcement to combat "street culture" in England: an ethical approach? *Ethics and Social Welfare* 3: 284–302.

Gregson, N. (1993) "The initiative": delimiting or deconstructing social geography? *Progress in Human Geography* 17: 525–30.

Hanney, S.R., Gonzalez-Block, M.A., Buxton, M.J., and Kogan, M. (2002) *The Utilisation of Health Research in Policy-making: Concepts, Examples, and Methods of Assessment*. Health Economics Research Group Report no. 28. Uxbridge: Brunel University.

Higher Education Funding Council for England (HEFCE) (2007) *Research Excellence Framework: Consultation on the Assessment and Funding of Higher Education Research Post-2008*, Policy Consultation Document 2007/04, HEFCE. London.

Imrie, R. (2004) Urban geography, relevance, and resistance to the "policy turn." *Urban Geography* 25: 697–708.

Jackson, P. (1999) Postmodern urbanism and the ethnographic void. *Urban Geography* 20: 400–2.

Johnsen, S. and Fitzpatrick, S. (2007a) A high risk strategy. *Police Professional* July: 23–4.

Johnsen, S. and Fitzpatrick, S. (2007b). Soft target? *ROOF* November/December: 28–30.

Johnsen, S. and Fitzpatrick, S. (2007c) *The Impact of Enforcement on Street Users in England*. Bristol: Policy Press.

Johnsen, S. and Fitzpatrick, S. (2010) Revanchist sanitisation or coercive care? The use of enforcement to combat begging, street drinking and rough sleeping in England. *Urban Studies* 47: 1703–23.

Johnsen, S., Cloke, P. and May, J. (2005a) Day centres for homeless people: spaces of care or fear? *Social and Cultural Geography* 6: 787–811.

Johnsen, S., Cloke, P., and May, J. (2005b) Transitory spaces of care: serving homeless people on the street. *Health and Place* 11: 323–36.

Jones, A. and Johnsen, S. (2009) Street homelessness. In S. Fitzpatrick, D. Quilgars, and N. Pleace (eds.), *Homelessness in the UK: Problems and Solution*. Coventry: Chartered Institute of Housing, pp. 38–49.

Kindon, S. and Elwood, S. (2009) More than methods: reflections on participatory action research in geographic teaching, learning and research. *Journal of Geography in Higher Education* 33: 19–32.

Kindon, S., Pain, R., and Kesby, M. (2007) *Participatory Action Research Approaches and Methods: Connecting People, Participation and Place*. London: Routledge.

Klodawsky, F. (2007) "Choosing" participatory research: partnerships in space–time. *Environment and Planning A* 39: 2845–60.

Knight, P.G. (1986) Why doesn't geography do something? *Area* 18: 333–4.

Liamputtong, P. (2006) *Researching the Vulnerable*. London: Sage.

Lee, R. (2002) Geography, policy and geographical agendas: a short intervention in a continuing debate. *Progress in Human Geography* 26: 627–8.

London Councils (2007) *London Local Authorities Bill: proposals for a tenth Bill for deposit in November 2007*. London: London Councils

Markusen, A. (1999) Fuzzy concepts, scanty evidence, policy distance: the case for rigour and policy relevance in regional studies. *Regional Studies* 33: 869–84.

Martin, R. (2001) Geography and public policy: the case of the missing agenda. *Progress in Human Geography* 25: 189–210.

Martin, R. (2002) A geography for policy, or a policy for geography? A response to Dorling and Shaw. *Progress in Human Geography* 26: 642–4.

Massey, D. (2000) Editorial: practising political relevance. *Transactions of the Institute of British Geographers* 25: 131–3.

Massey, D. (2001) Geography on the agenda. *Progress in Human Geography* 25: 5–17.

Massey, D. (2002) Geography, policy and politics: a response to Dorling and Shaw. *Progress in Human Geography* 26: 645–6.

Pain, R. (2003) Social geography: on action-orientated research. *Progress in Human Geography* 27: 649–57.

Pain, R. (2006) Social geography: seven deadly myths in policy research. *Progress in Human Geography* 30: 250–9.

Peck, J. (1999) Editorial: grey geography? *Transactions of the Institute of British Geographers* 24: 131–5.

Peck, J. (2000) Jumping in, joining up and getting on. *Transactions of the Institute of British Geographers* 25: 255–8.

Phelps, N.A. and Tewdwr-Jones, M. (2008) If geography is anything, maybe it's planning's alter ego? Reflections on policy relevance in two disciplines concerned with place and space. *Transactions of the Institute of British Geographers* 33: 566–584.

Pollard, J., Henry, N., Bryson, J., and Daniels, P. (2000) Shades of grey? Geographers and policy. *Transactions of the Institute of British Geographers* 25: 243–8.

Smith, D.M. (1974) Who gets what *where*, and how: a welfare focus for human geography. *Geography* 59: 289–97.

Staeheli, L.A. and Mitchell, D. (2005a) Relevant-esoteric. In P. Cloke, P. Crang, and M. Goodwin (eds.), *Introducing Human Geographies*. Abingdon: Hodder Arnold, pp. 123–34.

Staeheli, L.A. and Mitchell, D. (2005b) The complex politics of relevance in geography. *Annals of the Association of American Geographers* 95: 357–72.

Sui, D. (1999) Postmodern urbanism disrobed: or why postmodern urbanism is a dead end for urban geography. *Urban Geography* 20: 403–11.

Universities UK (UUK) and Higher Education Funding Council for England (HEFCE). 2008. User-valued Research in the Research Excellence Framework (REF), HEFCE and UUK workshop, 31 October 2008, available at www.hefce.ac.uk/research/ref/resources/.

Ward, K. (2005) Geography and public policy: a recent history of "policy relevance." *Progress in Human Geography* 29: 310–19.

Weiss, C.H. (1979) The many meanings of research utilization. *Public Administration Review* 39: 426–31.

Willis, K. (2004) "Distant geographies": international understanding and global cooperation. *Geoforum* 35: 399–400.

Part III   Matters and Meanings

A Companion to Social Geography, First Edition. Edited by
Vincent J. Del Casino Jr., Mary E. Thomas, Paul Cloke, and Ruth Panelli.
© 2011 Blackwell Publishing Ltd. Published 2011 by Blackwell Publishing Ltd.

Introduction

Mary E. Thomas

Life. What a fantastically simple and unimaginably complicated word. In this part of the *Companion* our authors were asked to undertake the impossible task of writing about it. Their chapters illustrate the lively production of social geographies, the geographies in turn that give life, and that force lives to intertwine, inter-define, collide, emote. Life is felt, experienced, unconscious, embodied. But what also matters is the materiality of life that cannot be easily narrated or even defined through the social differences and experiences we have learned to see and study in geography. Therefore, the chapters read together offer a range of understanding about what the "social-spatial" even is, and what it means. They offer not merely a state of the field; they push our conceptualization of what the field can be.

The part is framed by an overarching interest in foregrounding the vibrancy of life. Social difference itself guarantees a sort of energy to the everyday, in that differentiation proceeds through affinity, identification, conflict, cooperation, competition – the context, of course, matters for how the meanings of difference are worked out, or not. Vibrancy, therefore, is not an optimistic word necessarily, and the chapters illustrate the range of energies that authors pose for our consideration: from the bio-physical to psychic and other head spaces, these social geographers ask how life becomes understood and just as importantly, what the limits of our theories might be. While some suggest that representations of life stymie the creative tasks of social geography, others stress the ways that words spoken or imagined invade life and thus are intimately woven within the social-spatial.

This part of the *Companion* starts off with Gail Davies' "Molecular Life" (Chapter 15). She covers important ground for the part by showing us that life is lived at the most minute of scales. Is the atom merely the ingredient of life, or vital life itself? Davies challenges such a reductionist reading of the molecular, instead fashioning an intricate tale of biopower, elitist western values of self-determination in biomedicine, globalized capital, and biosecurity. Visualization techniques and biomedicine change the ways people live, procreate, and die; animals suffer in labs; and the geographies of nation state and capital shape molecular life in the smallest of spaces. Yet the biopower of capital and the state, and of the laboratory, must face the capacities of the material. Agency is not just the domain of the human, an intent. She asks us to consider the ontology of the human body, its differences manifested through social terms yet exceeding those. How does the materiality of matter, matter? What are possible methodologies that could tackle questions bridging the expert languages of science, the politics of difference, the vitalist ontologies of molecular life?

In "Psychic Life" (Chapter 16), Hester Parr and Joyce Davidson explore the effects of the unconscious in social geographies, and they ask how psychic geographies can be measured, accessed, and re-interpreted. These methodological questions are especially troubling, given that the unconscious by definition is never directly approachable. But as the authors say, "However and wherever psychic life takes place, we ought to remember that people already live there." One "where" they spend some time considering is the city. For example, urban psychic spaces have served as fodder for the psychogeographies of situationists who agitate for utopian geographies of the city based on desire, play, mobility (the *dérive*), and experimentation. But Parr and Davidson caution that our attentions should not stay solely with radical or hopeful psychic spaces; urban psychic life is also disruptive, especially for those who suffer trauma or mental illness. Parr and Davidson insist that these, often painful, stories show the intimate connections that psychic life forges with many spatialities. While the unconscious might be the most personal of spaces, it exists through profoundly social and spatial relations.

Sexual life has often seemed to be personal space by many geographers, in fact, too personal: Gavin Brown, Kath Browne, and Jason Lim suggest that geography has yet to understand the centrality of sex to social life (Chapter 17). The authors organize their chapter around a series of questions which each answers in turn, weaving their positions into a joined advocacy for theorizing the sexual. This structure highlights their divergences of sexual thinking and the opportunities they mark for future agenda on topics ranging from homo- and heteronormativity, queer politics and subjects, and politicized sexualities. The three authors situate their perspectives on queer geographies and politics, questioning especially the grammar of queer. Should it be solely a verb, indicating a critical stance on normativity, or an adjective describing subjects, or even a noun representing gendered arrays of bodies and practices? Brown, Browne, and Lim raise the political stakes in these debates and insinuate the possibilities for sexual life and lives.

If some geographers consider sexual life too saucy for academia, then Deborah Thien's "Emotional Life" will further perturb those propitious desires to curb social geographies to "scientific" inquiry (Chapter 18). In fact, Thien points to the resistant, feminist advent of emotional interests given pressures exerted by masculinist academia to be rational and dispassionate scholars. What wonderful advent, despite the risks: emotional geographies have shown just how central the felt is to the doing and being of life. Thien turns to love in her chapter to explore the social framings of relationships shaped and delineated by emotions, yet undone by emotions at the same time. Most of all, love maps the contours of intersubjectivity in motion; Thien calls this a movement between. She then turns to her research on the emotional spaces of the Royal Canadian Legion, a veteran's organization. The resonance of these spaces allows men to feel, to be together, and to just be, whatever that might mean to them. Thien considers the emotional dynamics of places, which always matter, even if words might not fully or adequately encapsulate how they feel.

Togetherness takes a whole new meaning with a shift to Keith Woodward's "Affective Life" (Chapter 19). While you might start the chapter wondering what work, ticks, waterboarding, and Volvox have in common, Woodward's treatise illustrates the lessons that each offers the others. Starting with the "waltz" of a many-celled alga called Volvox, Woodward points to the force of context, contact, and vibrant life. Capitalism, he argues, is like the contingency of the relative relations, content, and situatedness of any dynamic system; it does not present a spatial system of determination, but affects from the body of the worker to the controlled space of the factory are emergent and specific. Woodward's aim is to consider the precariousness of affective life and to point to the unfolding of relations and sociality as potential. Affective forces, in other words, do not merely settle into rote and insular spaces, but affect or generate continually interactive contexts beyond the human body or identity. Yet, affect is also a target for control, as Woodward points out in the case of waterboarding: the prisoner's body is tortured to its affective limits.

Chapter 20, "Embodied Life," relates intimately to other Matter and Meanings themes. Bodies, Isabel Dyck tells us, are located between culture and biology, with materiality as crucial to consider as the framing discourses that shape, for example, how we approach issues of disease, bodily specificity, and yes, the emotional ways subjects deal with changing bodies in changing places and spaces. More than a "surface" for discourse, however, Dyck points to the embodied experiences of material life and form, and the possible agency that bodies produce. She draws on two research projects to elucidate her arguments. In the first, an examination of food, bodies, and place, Dyck shows how international migrants and refugees negotiated and formed new contexts of embodiment through self and family diet, shopping, health concerns, cooking, and eating. In the second, Dyck considers "bodies in need of care" and homespace. Both projects illustrate how places and bodies articulate subjectivity as dynamic, contested, and experienced. The social geographies of bodies are framed by discourse, but a concept of embodied life also, Dyck insists, "admits meanings, emotions, materialities, and differentials in power."

With new challenges to discourse analysis in social geography from non-representational theory (NRT), Chris Philo offers a manifesto of sorts for wordy geographies. In "Discursive Life" (Chapter 21) Philo takes seriously the contributions of NRT while resolutely challenging any turn away from words in social geography. Words are generative, he argues, forming lives and materialities within relating layers of discourse, historical and spatial specificity, objects, practice, and subjects. He draws on his on-going engagement with Foucault to show that there is a precision to words that come to matter, and that cause and reproduce powerful meaning. The methodological possibilities stemming from wordy geographies include ethnography, particularly for the consideration of how "trivial" words ground subjects' sense of themselves and their worlds. Philo constructs a parallel project in the chapter with the use of text boxes and figures; his example is a fascinating look at the words and discourses surrounding a "mad" artist and institutionalized patient in a "lunatic asylum" in nineteenth-century Britain, William James Blacklock. Archived words, clinical words, delusional words, lettered words, loving words: discursive life.

In Chapter 22 Julian Holloway turns us to action: the practices and enacting of the "Spiritual Life." Holloway references each religious ritual as a re-presencing; perhaps a repeated act of sameness but with an attending affect of different, new encounters with the sacred or divine. First considering the deconstructive Christian theology of Mark Taylor, and then the Deleuzian theory of immanence, Holloway frames the spiritual life as a constant potentiality, a being faithful that emergent divinity is. Rituals, he claims, are practices of the specific events of this faithfulness in the divine's emergent presence in one's life. They mold space and time, and seek to affect dispositions toward the spiritual. Holloway then turns to the question of whether these dispositions are effective, through the example of the Christian sermon. By examining two Easter sermons, he points to their affective registers and the ways they work through a faithfulness of the sacred and the divine. Sermons are political attempts to mobilize affect toward a non-secular sensibility, Holloway argues, that marks life as fully imbued with and by the divine.

The final chapter in this part offers a view on the social geographies of "Virtual Life" (Chapter 23) that might surprise you. Mike Crang points to the many ways that virtual lives are intensely social in their articulations of the non-virtual. That is, Crang offers examples showing that usage of internet networking or gaming might have more to tell us about the "real" world than virtual worlds. Gaming has remained remarkably situated in recognizable geographies, drawing on maps and urban landscapes rather than inventing new or aspatial visions of interaction. Email, social networking, and new media increase face-to-face encounters, rather than the opposite. And importantly, stratification and global outsourcing of labor are continued and even exacerbated with growing ICT (Information and Communication Technologies) use, rather than ameliorated through new virtual worlds often promising even access to technology and information exchange. Crang also highlights examples from Singapore and South Korea to show the different forms of sociality–and attempts at social control–enacted by online media.

Perhaps more than the other parts of the *Companion*, this part deals directly with the emotive, and the tonalities of life: sometimes torturous, sometimes joyous, and certainly most often not falling into these extremes. Yet like the rest of this book, "Matters and Meanings" also stays focused on the question of why social geographies, well, matter. While their understandings of how power works might be at odds, there is a shared commitment by the authors that attending to life's many essences is a critical endeavor. It means taking seriously even the most mundane spaces, and working toward epistemologies of social geographies that are open to varied experiences, ontologies, bodies, technologies, subjects, matter–all of the stuff of life.

Chapter 15

Molecular Life

Gail Davies

An Invitation to the Molecular

It was caught by the wind, flung down on the earth, lifted ten kilometres high. It was breathed in by a falcon, descending into its precipitous lungs, but did not penetrate its rich blood and was expelled. It dissolved three times in the water of the sea, once in the water of a cascading torrent, and again was expelled. It travelled with the wind for eight years: now high, now low, on the sea and among the clouds, over forests, deserts, and limitless expanses of ice; then it stumbled into capture and the organic adventure. (Levi 1985: 226)

You are sitting in an overly air conditioned room at a geography conference. The chairs are uncomfortable and the lack of natural light tiring, but the talks are topical and important. They encompass a selection of contemporary social geography concerns, touching upon questions of identity and community, of inequality and justice, and of the production and experience of bodily difference. One presentation explores the viral socialities in the unfolding swine flu pandemic: the exclusionary spatial practices of biosecurity, involving school closures and quarantines, alongside the cultivation of new forms of association in the form of flu buddies and interspecies suffering. Another addresses the changing experience of corporeal difference through genetic counseling offered to populations at risk of inherited chromosomal alterations: making visible those individuals with affected recessive genes, at the same time as offering a space, through pre-implantation genetic diagnosis, to decide not to extend this particular community. A third traces the fate of human oocytes on the move: the technological processes which enable this vital entity to transfer from one body to another, and the transactional values – this much for this egg, that much for another – which remind us that not all bodies are marked the same. These

A Companion to Social Geography, First Edition. Edited by
Vincent J. Del Casino Jr., Mary E. Thomas, Paul Cloke, and Ruth Panelli.
© 2011 Blackwell Publishing Ltd. Published 2011 by Blackwell Publishing Ltd.

presentations, recollected here as a composite of cases, underscore that any consideration of social geography has to engage not only with the collective practices of human identity, but also the openness of the human body to other human bodies, and their intersection with a range of lively technological and organic nonhuman others. Such encounters are not determined by these biotechnologies, but offer new opportunities, making new possibilities present, whist at the same time containing new risks and generating new forms of absence.

The identification of links between social geographies and biological agencies is not new, yet something is shifting in these accounts, indicated by their changing modes of illustration. Only a few years ago, the graphics distracting you from your slight physical discomfort, re-engaging your attention, would have been images of the potential and fragility of human bodily experience as seen from the outside. These might be figured by individual suffering or populations at risk, alongside icons of the clinical institutions and scientific professionals we turn to for bodily care. Yet, you look up. On this occasion, the text floats over a field of color, the background is loosely differentiated into shapes both rounded and spiked, objects both distinct and interconnecting. The first is illustrated with the surficial textures of a spherical virus; the second dominated by the complex arrangement of genes upon a human chromosome; the third features the emergence of life from the landscapes of in vitro fertilization. These are images of molecular life.

If you haven't had this experience, you easily can replicate it by going to Google images. Type in any of the above biomedical nouns. The resulting images are often stunning, with vibrant fluorescent colors and intriguingly amorphic shapes. They both offer an epistemic authority to new visualization techniques in science, whilst at the same time being immediately accessible, to those privileged with access to an internet connection. Beyond this electronic encounter, they can be found, temporarily stabilized, in a variety of printed forms circulating outside of the laboratory or clinic: in the glossy brochures of health tourism, in the mass marketing of public health campaigns, in the speculative dossiers of investment funds, as well as the perhaps less ubiquitous literatures of academic writing. At each point, they promise a new space to think about and act in relation to bodies, materiality, and collective life. Yet, for scientists, for publics, and for social geography this move is ambiguous. To talk of molecular life is to shift in a number of potentially contradictory dimensions: towards both complexity and reductionism, towards hope and fear together, towards the expectation of therapeutic interventions as well as undesired risks, towards new understandings of difference and concepts that dissolve boundaries between matter.

I have used such images in my own work. My interests in molecular life are long standing, but often indeterminate, oscillating between encounters with science, policy, and geography. They originate with undergraduate experiences in chemistry. These were inspired, in part, by the writing of Primo Levi who opens the chapter. But they were abandoned after a year of bench chemistry left few points from which to engage questions about the implications of knowledge practices outside the laboratory. What followed was a shift to geography, and a series of experiments with standpoints from which to ask questions about the relations between biology and society. These take from science and technology studies an interest in the performative practices of science, from anthropology a methodology for intervention, and from geography an attitude towards empirical engagements motivated by the production of difference, whether of biological difference or the differential authority to speak for biology in the public sphere. Current work has taken me back to the laboratory, or more precisely the realm of laboratory animal science, and the production of what are varyingly called biological resources, model organisms, or mice (it depends on whom you talk to). The motivation for this research is to understand the role of space and relation in the development of post genomics: the practices involved in translating genomic knowledges into therapeutics through breeding mouse models of human disease, moving research from bench to bedside and facilitating international collaborations in science. To translate is to move information and materials across borders, which are organized around species, disciplines, institutions, and states. One of the strengths of molecular biology is its ability to mobilize bodily materials and extract value from the things it makes mobile. As I hope to demonstrate in this chapter, the questions of post genomics are inherently spatial; they are also questions of geography.[1]

In what follows, I pick up these intersections between biology, society, and space, tracing social geography's emerging engagement with questions of molecular life. Such work is by necessity interdisciplinary, engaging with the philosophy and practice of biology, and in the process encountering other disciplines in the social sciences. Yet I would argue this engagement can, and should, retain a distinctively spatial inflection. I start this review outside of geography, looking to sociology, anthropology, and science studies to foreground the configurations of biotechnology enabling molecularization as a way of mobilizing life. Here geographers have contributed attentiveness to geopolitics to the different inflections of thinking about biopolitics or biocapital; something about which there is still more to say. The following section picks up a different intersection of work on geography and the molecular, one that is more concerned with the properties of matter and the nature of agency. The epistemological and ontological challenge to social geography is perhaps more profound here, though again there are many dimensions yet to be explored. In concluding, I pick up some of these challenges, not seeking a premature synthesis between the two strands of work, but pointing to a selection of projects attentive to the "matter of mattering," in ways which are open to the potential for novelty from these emergent forms of life and attuned to questions of accountability and the promise of justice.

The Mobilization of Molecular Life

The genealogy of molecular life as an object of enquiry within the social sciences is varied, including a complex body of work outside of geography, and a smaller subset within. It is more likely to be encountered as an explicit term within the sub discipline of cultural geography, in reconceptualizing ideas about the matter of more-than-social life. I turn to these in the following section. I start with work in sociology, anthropology, and science and technology studies, for it is here that research attending to the production of cellular or molecular matter is linked most closely to extant geographical questions around the production of community, identity, and difference. In this, I identify and introduce two strands of research. The first is most closely associated with the work of Nikolas Rose and Paul Rabinow

(Rabinow and Rose 2006; Rose 2007), and their rereading of Foucauldian biopower within a landscape composed not only of populations and individuals – the twin poles of Foucault's work (1978) – but also of molecules, genes, cells, and genomes. In the second, perhaps more diffuse set of literature, attention shifts from the governance of the self and citizenship, to the operation and implications of global biological enterprise (Franklin 2005; Sunder Rajan 2006; Cooper 2008). The interdependencies and elaborations emerging from these arguments are imaginatively mapped by Helmreich (2008) as bifurcating species of biocapital. Here it is the time-spaces of biocapital that interest most. Whilst both offer insight into how the biological is made mobile through its intersection with a range of novel biotechnologies, forms of capital, emerging norms, and other institutions, their temporal-spatial imaginaries and thus their implications for thinking about social geography are somewhat different.

For Rose, molecularization is a "style of thought" in contemporary biology, which envisages vital life processes at the molecular level; a scale at which entities such as cells, DNA, and other biological entities can be identified, isolated, manipulated, and recombined. His account of the emergence of molecular life gives primacy to the role of visualization techniques in ushering in a new set of epistemic ground truths for science, which have impacts far beyond the laboratory (Rose 2007: 14; see also Cartwright 1995). In this definition, molecularization is associated with "the reorganization of the gaze of the life sciences: their institutions, procedures, instruments, spaces of operation, and forms of capitalization" (Rose 2007: 44). Developments in contemporary biology are central to this account, but this is not the new genetic determinism, and corresponding challenge to human nature, feared by writers such as Francis Fukuyama (2002) or Jurgen Habermas (2003). The mapping of the human genome at the start of the twenty-first century revealed less than the expected number of genes, opening up an interpretative gap between genetic code and organism, ushering in a post-genomic emphasis on complexities, and the interaction between the genetic sequences, cellular pathways, and epigenetic factors which constitute life at a molecular level. The work of reflecting on the relation between biological inheritance and the management of lived complexity is not just carried out by scientific experts, but is, Rose argues, increasingly part of the practices of governing health and citizenship projects.

What this molecular re-scaling does enable is the relocation of biological entities from their position within specific individuals, bodies, organs, or tissues. It renders them mobile; "to be regarded, in many respects, as manipulable, and transferable elements or units, which can be delocalized – moved from place to place, from organism to organism, from disease to disease, from person to person" (Rose 2007: 15). Once delocalized, biological entities can then be reinserted back into the diverse contexts of pharmaceutical development, reproductive technologies, national biobanks, stem cell science, genomic medicine, and so on. Biology is thus not the only agent, or indeed the determining element, in shaping these new concepts and circulations of vital properties. A convergence of factors outside of the laboratory, including the increasing porosity between university science and corporate biosciences (Sunder Rajan 2006), new rights over the intellectual property in biological materials and knowledge (Parry 2004; Sunder Rajan 2006), an alliance between the speculative logics of biology and international finance (Sunder Rajan 2006; Cooper

2008), and the bureaucratic extension of professional bioethics, financial regimes, and scientific standards (Barry 2001; Rose 2007) all facilitate this increasing circulation of biological materials. The life sciences, economics, and ethics are increasingly interpenetrating and undergoing transformation, with potentially profound implications for our understanding of social geographies.

For Rose, efforts by people to inhabit these new forms of life can be understood through an emerging "somatic ethic" and the centrality of technologies of optimization. The first suggests we increasingly relate to ourselves as fleshy corporeal beings, experiencing, acting, and articulating our identities through the languages of biomedicine. The second locates these practices within technologies of life which are oriented to the pursuit of best possible futures, for optimization. Bodies may be biologically differentiated or susceptible to different kinds of biological risk, but this is just the starting point for life understood through a world of risk, susceptibility, prudence, and foresight for individuals and for those somatic experts – the genetic counselors, patient groups, screening companies, psychiatric practitioners, pharmaceutical companies, insurance brokers, legal representatives, and so forth – who would counsel, guide, or treat them. From this forward looking flow of molecular possibilities, new configurations of community and relations to expertise emerge. These emerging biosocialities connect individuals and experts as they "work on themselves in the name of individual and collective life and health" (Rabinow and Rose 2006, 195; see also Petryna 2002; Rose and Novas 2005; Gibbon and Novas 2008).

What is noteworthy here is that such collectivities are figured as future orientated; molecular life opening up new spaces for both uncertainty and importantly for action. These personalized encounters with biomedicine are not characterized by a fateful acceptance of biology, but rather constitute an opportunity for engagement. As Rose explains, "the idea of susceptibility brings potential futures into the present and tries to make them the subject of calculation and the object of remedial intervention" (Rose 2007: 19). From this, "our somatic, corporeal, neurochemical individuality now becomes a field of choice, prudence, and responsibility. It is opened up to experimentation and to contestation" (Rose 2007: 40). What is traced in this work is a growing capacity to modulate the vital properties of human life, rationalities of government which place increasing emphasis on the responsibilities of individuals to manage their own bodily care, novel experts and authorities guiding and shaping these life choices, and increasing national competitiveness around the potential value circulating in these bioeconomic practices. In this, there is an "elective affinity" between the search for shareholder value and human values, and between contemporary economies of vitality and emerging somatic ethics (Rose 2007).

Rose does acknowledge that such hopes inevitably involve a flip-side, the specter of value judgments based on biology. As he puts it, "every act of choice opened up by the new biomedicine does indeed involve a judgment of value in a field of probabilities shaped by hopes ... The field of choices, judgments, values and hopes about life itself is the territory of our new vital politics" (50–1). Yet this is still framed within the notions of choice: "thanatopolitics" – the politics of death – are seen as exceptional forms of biopower associated with conditions of absolute dictatorship (Rabinow and Rose 2006). Further, these are judgments of value made in a field of probabilities shaped by future expectations; there is less on the values shaped by the inheritance of past circumstances. Yet, in a world in which access to resources, health care, science, and government are *already* unequally distributed, it is pertinent to ask, as does Braun, "for whom is the molecular age an ethopolitical age that is defined and experienced primarily as a matter of choice and the individual management of risk?" (Braun 2007: 7; see also Kearns and Reid-Henry 2009). The individual and collective identities Rose and others distinguish are those subject positions and spatial-temporal imaginaries predominantly enabled under advanced liberal governmental regimes.

These, of course, are not the only subject positions of interest to social geographers. Here further work, exploring the finer textures linking molecularization as a techno-scientific project, globalization as a political-economic venture, and postcolonialism as an ethical imperative is required. The mobilization of molecular life is not only facilitated by bringing potential futures into the present, but also by the historical geographies of expropriation and landscapes of accumulation which pattern these flows of biovalue. Many notable works have begun to map these circulations.[2] The complex temporal and spatial forms of globalizing capital, or more specifically biocapital, are explored through ethnographic work by Sunder Rajan (2007; see also 2006). This traces the global networks of economic, social, and ethical relations around the health industry, which span continents, locating outsourced clinical trials on the Indian subcontinent, whilst simultaneously expanding the market for drug consumption elsewhere. Waldby and Mitchell (2006) articulate the forms of value structuring contemporary human tissue economies – the blood, organs, skin, embryos, and stem cell lines stored and distributed for research and therapeutic purposes (see also Franklin et al. 2000). Other work extends the gaze from human biology to explore the scientific, social, and commercial histories shaping international and postcolonial practices of animal breeding and biocapital accumulation (Franklin 2007; see also Holloway et al. 2009), or plant-based biological extraction and the geopolitics of bioprospecting (Hayden 2003; Castree 2003).

Such work draws attention to the uneven flows of biovalue which are mobilized through the molecularization of life and its ensemble of institutional and other intermediaries. It also reminds us that places, communities, and countries are not abstract spaces; rather, the subject positions accessible to people emerge in their corporeal, political, social, and spatial complexity in distinct material locations. Such work also brings back consideration of territorial imaginations, so returning the nation state, as well as capital, to a key role in shaping these geographies of molecular life. Yet again, this intersection does not predetermine its modality, but it does raise new questions of belonging in relation to the intersections between molecular and human mobility.[3] The work of Catherine Nash on the remaking of ideas of Irish identity and relatedness through geneticized genealogies, shows how molecular technologies can be incorporated into the other practices through which diasporic communities mobilize and make sense of origins and connections, reshaping communities of shared descent (Nash 2008). Conversely, the politicization of emerging disease risks through discourses of protecting homeland security has profound implications for the extension of biosecurity practices into extraterritorial locations to control the potential flow of viral agencies across state borders (Braun 2007). Territorial power retains its importance in structuring many of our imaginative possibilities, despite the proposition of alternative and emerging temporo-spatialities around biosociality (Dillon 2007; Ingram 2008; Schlosser 2008).

Thus, the mobilization of molecular life here does not operate according to singular logics. This is captured in discussion of the potential for geographers to contribute further to understanding the complex scales constituting genetic and health geographies (Nash 2005; Kearns and Reid-Henry 2009). However, more than this, understanding the social geographies of molecular life requires both attentiveness to the calculation of possible futures, and also a careful consideration of past inheritances. Molecular life is patterned in different ways and in different contexts by biological susceptibility, by speculative capital, by territorial imaginaries, by clinical practices, and by past and present human mobilities, challenging and at times exceeding our understanding of what makes human bodily difference and how to pursue the promise of justice. In my own work, I have grappled with this in relation to the unequal geographies and ethical complexities of organ transplantation in London (Davies 2006). In this small case study, a conjunction of organ matching protocols, the institutional organization of transplantation teams, and the location of Southeast Asian communities at risk of hypertension results in the surprising emergence of communities with unequal access to transplanted organs on either side of the River Thames. Thus, I would argue openness to the unexpected social groupings emerging from biotechnological developments is constructive, for it would be extraordinary to predict this patterning and the political questions emerging from it. This is not to suggest emergence flows freely; it is weighted by history. The concept of inheriting well is central to the ethics of present flourishing animated by Donna Haraway (2008). Haraway takes insight from the work of Australian anthropologist Deborah Rose, transferring reflections from the frontier practices of white settler communities to the new landscapes of biotechnology. Taking care of generation (with all the biological and social implications that term embodies) is not only about a somatic ethics inspired by facing the future, but also the ethical imaginations evoked by learning how to face the past.

Matters of More-than-social Life

Haraway's recent concern has been with companion species, as the embodiment of material "past-presents" and as a point from which to reflect on all manner of lively relationalities of becoming with human and non-human others. It is these slightly different questions of community, identity, and difference to which I now turn. Whereas the preceding section explored the mobilization of molecular life, here I open up questions about materiality, relationality, and agency, picking up a slightly different set of theoretical debates over geography, the molecular, and the bio. These have figured more prominently in cultural, rather than social geography, but have implications for considerations of the social too.[4] Here, focus is directed to the philosophical and processual issues in attending to how we think about and intervene in the molecular stuff of life itself, rather than studying "molecular life" as an actor's category in contemporary technological, political, economic, or social practices. In this endeavor, not only are changing concepts of the biological understood through intersections with a range of technical, material, and institutional practices, the abstract category of the social itself is thoroughly disassembled to be refigured as a series of interlocking and interactive material capacities. The preoccupation with questions of representation and epistemology within the linguistic framing of geography's cultural turn is followed by a more thorough-going interrogation of the importance of ontology to geography, to the different ways that matter matters (Jackson 2000; Anderson and Tolia-Kelly 2004; Anderson and Wylie 2008). Adapting an evocative phrase from Sarah Whatmore (2006), this asserts that the geographies of molecular life are always more-than-social.

The narrative of molecularization in the preceding section began with the postwar trajectories of molecular biology from the laboratory to the clinic and beyond. The starting point here is earlier, the definitions emerge from characterization of the atomic scale, and the questions raised are those fundamental to western thinking about matter and representation. The work of Karen Barad (2003, 2007) is a useful guide for locating and summarizing these discussions, though geographical engagement has drawn on a diverse range of writing on the philosophy of ontology, including Jane Bennett, Gilles Deleuze, Joseph Rouse, Baruch Spinoza, and others. Barad takes us back to early Greek philosophy and the development of atomic theory – the theory which suggests matter is composed of discrete units called atoms. It is in the responses to this theory that she locates the philosophical problems associated with representation, notably the separation between representation and ontologically separate entities awaiting representation, the divergence of thinking about ontology and representation in the physical and social sciences, the problem of realism in the former and the idealism of social constructivism in the latter. As she puts it, "Democritus's atomic theory introduces the possibility of a gap between representations and the represented. 'Is the table a solid mass made of wood or an aggregate of discrete entities moving in the void?' Atomism poses the question of which representation is real. The problem of realism in philosophy is a product of the atomistic worldview" (Barad 2003: 806). In its place, she seeks to insert an alternative ontology by replacing the representationalist belief in an ontological distinction between words and the worlds, with a performative understanding of the discursive and other practices involved in the formation of subjects and the production of bodies. Barad calls her approach agential realism; its central concern is the "intra-action" of humans and nonhumans. These intra-actions are "materialist, naturalist and posthumanist" and accord "matter its due as an active participant in the world's becoming, in its ongoing 'interactivity'" (Barad 2003: 803). The "knower" does not stand outside of these relations to study them; there is no such exterior observational point; we are part of the world in its ongoing intra-activity. Barad's engagements with materiality emerge out of philosophical and practical engagements with experimental physics, so it is helpful to follow her arguments into the logics of contemporary nanotechnology to explore them further. This has been a fertile terrain for geographical reflections on matter, emergence, governance, and control (Anderson 2007; Kearnes 2006; Last 2007), tracing debates as scientific researchers explicitly seek to harness the creative potential of molecular life at the smallest scale.

In its simplest and perhaps most potent form, early advocates of nanotechnology offered the vision of building desired properties through the rearrangement of molecular matter. As one early proponent puts it, "Arranged one way, atoms make up soil, air and water; arranged another they make up ripe strawberries. Arranged one way they make up homes and fresh air; arranged another, they make up ash and smoke" (Drexler 1986: 3; cited in Kearnes 2006: 61). This manifesto suggested the physicality and commonality of all matter at the smallest scale, the instrumental rearrangement of atoms by the miniaturization of mechanical processes, and the potential to manufacture commodities, clean up pollution, or produce life from the bottom up. A new industrial revolution is heralded, to be built atom by atom. It is tricky to see the active participation of matter in such reductionist models of materiality. However, this mechanistic version of nanotechnology is increasingly challenged "by the sheer physical possibility of autonomous nanoscale machines and the precision necessary to 'create' them" (Kearnes 2006: 70). Instead, a bio-mimetic model of nanotechnology is gaining currency. In place of a vision of individual atoms directly engineered by researchers, this emerging strand proposes to harness molecular evolution, making use of the creative potential of existing biological systems, which are always already involved in interactions and organization at the nanoscale. Matter has internal creative potential. Its key systems, such as the DNA molecule, are already seen as the functional equivalent of self assembling and self organizing molecular machines. Matter at the smallest scale is reimagined as lively, mobile, organizing, and productive of difference; it is creative, but undirected. The theoretical biologist Brian Goodwin makes a similar point when he suggests evolution is "like a dance – it's not going anywhere, it's exploring a space of possibilities."[5] The hope for bio-nanotechnology is to direct this creativity. The intrigue for geography is the potential dialogue with the vitalist ontologies of Deleuze and Bergson in rethinking and reengaging with matter, emergence, and difference at the molecular scale (Kearnes 2006).

Of course, much of what is traditionally of interest to social geographers is manifest in contexts which appear overwhelmingly complex compared to the rarefied spaces of nanotechnology development. The project of making such vitalist ontologies meaningful in diverse research contexts remains to be explored. As Anderson and Wylie reflect, "If matter names that which is common, it is an odd commonality, always entangled in the heterogeneous logic of difference" (Anderson and Wylie 2009: 2). What such insights do challenge us to reconsider are complex questions around the material connections between life and nonlife, humans and nonhumans, and the ontologies of human bodily difference. Some of the most intriguing, and at times controversial, reflections on where this might engage the questions and methods of social geography have been around understanding different human corporeal forms through the notion of molecular assemblages (Saldanha 2006) or apprehending different human affective states through the molecular scale (McCormack 2007). Without space to do justice to both these strands, I continue with attention to the relations between spatiality, difference, and race in the work of Saldanha. Questions of affective life are picked up later in this volume.

Critical social scientific explanations of human difference, including race, gender, sexuality, and so forth, have been dominated by an emphasis on the powerful role of language and discourse in structuring the experience of human difference since the 1980s. This has been advanced through the theoretical resources of the cultural turn, it has been a robust counter to the crude genetic determinism of work like *The Bell Curve* which linked educational achievement with biological notions race (Herrnstein and Murray 1994), and it has also opened up political spaces to talk about interventions to language and social policies which might promote anti-racist and other anti-discriminatory practices. When critical social theorists did encounter molecular explorations of categories like race, there was an assumption that more research would eventually erase this as a meaningful corporeal category, confirming the place of race as a social construction. Saldanha cites Gilroy's claim that "race-thinking is already fading away in an age of molecular biology" (2006: 14). Yet, debates about the biological basis of race have failed to go away; instead questions of human biological diversity are reinvigorated in the move to personalized medicine (Fullwiley 2007), assisted by the increasingly ease of genome sequencing, the shifting emphasis of the Human Genome Diversity project, and the integrative aims of systems biology. Given this resurgence, there are calls for social scientists to reengage with the materiality of difference, in ways which are open to the complex corporeal differences emerging from more nuanced understandings of the vitality of matter. That the questions raised are too important to leave solely to scientists is part of the argument. Saldanha goes further, suggesting recognition of the materiality of human difference is important to the political and ethical project of building cosmopolitanism attentive to the "real, tangible differences between bodies that matter in face-to-face encounters" (Saldanha 2006: 14). Moving race out of the dualistic classifications of earlier eras, he argues for "a critical biogeography of human phenotype" (Saldanha 2006: 17). Race here is understood as a molecular assemblage, a viscosity in the flow of connections between bodies and properties. This rejects the grid-like dualities and spatialities of older race typologies. "The molecularisation of race would consist in its breaking up into *a thousand tiny races*. It is from here that cosmopolitanism should start: the pleasure, curiosity, and concern in encountering a multiplicity of corporeal fragments outside of common sense taxonomies." Moreover, this project "takes into account the real barriers to mobility and imagination that exist in different places; cosmopolitanism has to be invented, not imposed" (Saldanha 2006: 21, original emphasis). We are back to the questions of spatiality, mobility, and imagination raised by the molecularization of life, but within the framework of a different, livelier understanding of materiality.

This critical biogeography of human phenotypes can, if desired, be taken to explore other forms of human corporeal difference, such as gender, and also further towards the posthuman in tracking the interspecies and microbial ecologies that constitute what we commonly understand as individual human bodies (Hird 2009). At this scale, human bodies turn out to be communities themselves; something which Haraway suggests has a companionable and ethical, if no longer strictly social, dimension:

> I love the fact that human genomes can be found in only about 10 percent of all the cells that occupy the mundane space I call my body; the other 90 percent of the cells are filled with the genomes of bacteria, fungi, protists, and such, some of which play in a symphony necessary to my being alive at all, and some of which are hitching a ride and doing the rest of me, of us, no harm. I am vastly outnumbered by my tiny companions; better put, I become an adult human being in company with these tiny messmates. To be one is always to become with many. (Haraway 2008: 3)

In the contexts of molecular biology, divergent nonhuman capacities not only co-constitute the materiality of our bodies, but also shape the potential trajectories of human becomings. The experimental practices of genetics, and the use of model organisms to identify diagnostics and therapeutics, incorporate molecular pathways from diverse animal capacities, such as the modern laboratory mouse (Haraway 1997; Davies 2010), into the circuits through which human materiality and meaning is composed. Drugs and food, as well, intervene in the affective molecular properties of thought and action (Barry 2005; McCormack 2007). The boundaries between organic and inorganic life, individual, and assemblage are less clear cut in this attention to the diverse ecologies of the human body. Liberal notions of the autonomous subject, central to concerns about biotechnology, are challenged too (Jasanoff 2006). Humans are not the sole authors of life; the coincidence between the individual self and the genetic body found in the biopolitics of Rose is disputed (Braun 2007). To be one is always to become with many. The contours of a posthuman social geography are beginning to be sketched out (Coyle 2006; Panelli 2009), but the full extent to which the existing ontological and epistemological commitments of the social sciences are disturbed through these explorations of the materiality of being, becoming, thinking, and interacting are yet to be explored.

Experimentation and Social Inquiry

In this final section, I turn to some of these methodological challenges. The scope of social enquiry is considerably broadened by attention to the complex contours of molecular life, whether understood through a changing ontology of bodies, nature, and matter; as a changing focus for identity, subjectivity, and community; or as a changing realm of commodification, speculation, and potential control. The specific interests of social geography, encompassing but not limited to considerations of diversity, difference, and inequality, fit somewhat uneasily within these emerging territories. If inequality is a key consideration, how far should we follow the more celebratory potentials of biosociality, itself reliant on access to the already unequally distributed resources of developed democracies? In a discipline with intellectual capital invested in understanding identities through socio-spatial practices, how are we to make sense of definitions of difference organized around the level of molecular assemblages? The limits of our own disciplinary languages are challenged too. How far do we have to understand the detailed and ever changing technoscientific frameworks of genetics, nanotechnology, systems, or synthetic biology? How can our own analysis exceed these framings to open up spaces for considering and shaping alternative trajectories, when already "life is outrunning the pedagogies in which we train" (Fischer 2003: 37)?

Answering all these questions is outside of the scope of a short chapter: much of the research is still to be done. But in concluding, I identify three research tactics, which characterize early empirical experiments in tracing the social geographies of molecular life. These include multi-sited ethnographies, interactive research within scientific and policy-making communities, and innovative forms of public engagement. All three modes of inquiry stage encounters between experimentation as a powerful way of developing knowledge in the physical sciences, experimentation as an emerging form of engagement in the social sciences (Thrift 2008), and the forms of regulation and governance through which experimental enterprise unfolds (Petryna 2008). Notions of the experiment are historically contingent and vary according to scientific, economic, and other regulatory mechanisms. What these experimental practices have in common is some form of projection from the past-present to the future, and thus performative dimensions. For social scientists, the experiment is increasingly conceived less as what can be known through precisely controlled conditions, and more about creative forms of world-making. These are not reductive, but additive, a bringing together of different elements. They are subject to processes of reflection, learning, and repetition, but this iteration is motivated less by the pursuit of a closer approximation between language and its referents, and more by the branching production of meaning and possibility. Although this might seem counterpoised to traditional definitions, or stereotypes, of the nature of scientific experimentation, this is the way experimentation can increasingly be understood in physical sciences too (Barad 2003). Thus, experimentation offers one point at which to open up dialogue between the social and other sciences, as well as attending to the performative dimensions of social scientific research itself.

The importance of ethnography is its attention to sites and to practice, both in research and in writing strategies (Braun 2006). As suggested above, developments in biotechnology are invariably accompanied by both narratives of hope, of improved wellbeing and increased share value and conversely by speculations abut danger, of further genetic reductionism and social and economic marginalization. The biotechnological interventions facilitated by molecular life are added to human–nonhuman collectives that are already dynamic. One cannot know the protagonists in advance – the challenge is how to understand livable co-presence (see for example Bingham 2006; Hinchliffe 2007). A focus on assemblages, irreducible to singular logics or spatialities, requires a flexible approach to both field study and to processes of interpretation, which can be facilitated by ethnographic sensibilities. Tracing this mobility is rarely possible from one site: the materials, metaphors, times, and spaces of molecular life are multiple, and so ethnographies are increasingly multi-sited. This inserts the researcher into the midst of things, folding inside and outside, "here and there," key elements to what Marcus and Fischer (1986) call an "experimental moment in the human sciences." Examples include the collaborative "para-sites" run by Center for Ethnography at the University of California Irvine, which stage encounters on the boundaries between field site and academic seminar, co-producing space outside conventional notions of the field and relations of research central to the conceptual work within projects (Marcus 2000).

Ethnography in this guise is a form of intervention, yet it is not the only form. There are other initiatives too, often coming from those engaged in understanding the extant institutional imperatives of policy formation or public engagement in science and technology, as they meet these unfolding relations between biology, technology, and collective social life. Here, there are further questions of how to intervene, how to reinvent practices of social action research and opportunities for institutional learning with attention to multiple human–nonhuman relations which are themselves unfolding and emergent. In taking a more institutional perspective, these explicitly acknowledge that, whilst assemblages are necessarily emergent and precarious, temporal dimensions are critical to identifying persistencies, path dependencies, and other practices which reinforce inequalities. Here there have been experiments with embedded social science researchers taking reflective opportunities and alternative values into science centers and networks (Doubleday 2007), alternative techniques for the social appraisal of technology opening-up plural framings within policy contexts (Stirling 2008; Burgess et al. 2007), and efforts to bring political questions and scientific framings into closer dialogue by taking engagement activities up-stream (Wilsdon and Willis 2004). The forms and outcomes of these experimental policy engagements differ, but together they seek in some way to hold those authoring the hopeful expectations of technoscience to account (Brown and Michael 2003), against the ever present potential for what Brian Wynne identifies as reductionist returns (Wynne 2005).

This work also demonstrates that the interplay of social imaginaries embedded in new technologies play an important role in authoring and contesting such expectations. In such work, there is discussion about the emergence of new molecular logics and relations, but there is also talk about the existence of the points at which new forms of biology might interact with given categories of ethics, species, and society. Different imaginations of the social are enacted as powerful political rhetorics in both public and scientific discussions of new technology. For diverse publics, the social remains one vocabulary through which claims can be made for responsibility and careful practices from institutional actors charged with the governance of forms of molecular assemblages that may overrun organizational authority, but still require bringing to account. For scientists too, the discursive and material formation of publics is integral to technological development. There is a co-production of science and society (Jasanoff 2004), whose contours are subtle and shifting. New technologies may have democratic potential, but equally technologies already embody particular versions of the public, society, or cultural norms. Ellis et al.'s (2009) work on the Barcoding of Life Initiative, a taxonomic project using DNA segments to provide species level information, provides a nuanced exploration of this tension. The development of a hand-held device enabling nonexperts to identify species is linked, unproblematically by its proponents, to enhanced public understandings of biodiversity and to the democratization of biology. Yet, at the same time, this new technology imagines its publics as inadvertently empty, awaiting this technological tool to access nature. Ellis et al. suggest such representations of the social give life to innovation, but are also necessarily incomplete and contestable. Diverse social capacities both to engage with the existing materialities of nature and cope with future uncertainties may be underplayed in the imaginaries guiding the development and governance of new technologies. This new technology cannot perform its promised democratic potential, for society is always more plural than such innovations presume.

Such competencies can be tricky to elicit within conventional forms of public engagement, and indeed experiments in democratizing science often serve to reinstate the authority of science by erasure of the public knowledges through the terms they are invited to participate (Wynne 2006). As Braun puts it, "Biotechnology's high capitalization and specialization means that immense challenges stand in the way of any sort of informed critique and public debate" (Braun 2007: 13). Yet, to pursue politics only through repertoires of critique and debate is to accept a particular definition of the molecular, of sensing, and of social understanding. At this point, the two broad literatures on molecular life have the potential to inform each other. This is a critique Angela Last (2007) performs through innovative public engagement on nanotechnology. Here talk is supplemented with playful tactile experiences involving magnets, modeling clay, thick liquids, and polystyrene balls, exploring participants' engagements with the nature and agency of matter itself. Through such sensual experimentation, participants are moved to find their own inquiries, freed from narrowly instrumental forms of engagement, posing such questions as "what are we in relation to matter?" (Last 2007: 99; see also http://mutablematter.wordpress.com/, accessed December 4, 2009). This work is still to be fully written up, and I look forward to their answers. It is precisely the complex and multilayered question of "what we are in relation" that geographers and others are asking around the emerging socialities, imaginaries, and materialities of molecular life.

These questions return us, finally, to the opening narrative of a single carbon atom composed by Primo Levi. This offers inspiration, but also caution to the challenge that molecular life offers to social geography. These emerge as we follow the story further. Levi continues with the capture of the carbon atom by photosynthesis, its incorporation into human body as wine, its exhalation into the atmosphere, its uptake by vegetation and return to human form, via milk, to the author's body, where it propels the movements of writing, producing the inscription of the final dot of the last sentence of the chapter and of the book. It is a powerful and moving fiction of beautifully wrought spaces and interwoven temporalities. The intricate and ethical connectivities between materialities, agencies, bodies, and life are vividly written into this short piece. Yet, at the same time as Levi creates this intensity, he simultaneously undercuts its authority. The stories of molecular life are necessarily endless he says. Carbon "says everything to everyone, that is, it is not specific, in the same way that Adam is not specific as an ancestor ... I could tell innumerable other stories, and they would all be true: all literally true, in the nature of their order and data. The number of atoms is so great that one could always be found whose story coincides with any capriciously invented story" (Levi 1985: 225, 232).

The same questions face the encounter between social geography and molecular life. Which stories shall we tell, which relations do we trace, what differences can we describe, what intensities and ethical relations do we try to affect? In making these decisions, social imaginaries do not disappear, they are markings in the milieu in which risks are already written, identities already narrated, resources already assembled. Social life is itself an achievement that includes nonhuman practices and agencies, as well as an abstraction that has a place and a value in the imaginative negotiation of collective life. In this sense, both the molecular and the social are spaces of thought and action shaping the material, ethical, spatial, and temporal imaginations through which we enact the "organic adventures" of molecular life.

Notes

1   I am grateful for the generous contribution of comments and papers from Mary E. Thomas, Kaushik Sunder Rajan, Matthew Kearns, and Angela Last. This chapter has been written with the support of the Economic and Social Research Council Research Fellowship (Award no RES-063-27-0093).

2   Metaphors of mapping dominate attempts to understand the social and spatial transformations wrought by ongoing scientific innovation at the molecular level. The language

of mapping points has resonance in at least three related ways. Firstly, it recognizes the importance of mapping as an epistemic and technical practice in biology through which the spaces of the body are made accessible to study and intervention at the molecular level, for example through gene mapping or functional magnetic resonance imaging (Gaudillière and Rheinberger 2004). Secondly, the cartographic metaphor points to the spatial organization of contemporary biotechnology as an activity, which is not confined to the laboratory, but is also manifest within sites of industrial production; city, state, and regional economies; contemporary governance and public debate; campaigning strategies and clinical spaces (Hall 2003; Parry 2004; Greenhough 2006). Thirdly, though probably not finally, an emphasis on the unfolding cartographies of molecular life gestures towards the ways in which social scientists might understand the relation between social and molecular processes, as patterns of mutual interference which are not deterministic, but have their own distinctive topographies.

3   This leaves aside, for now, the centrality of nonhuman migrations to the management of zoonotic diseases through biosecurity measures (Braun 2007). I turn to points about the more-than-human nature of these intersections of geography and molecular life in the following section.

4   For cultural geography, a critical engagement since the 1990s has been the reconceptualization of the relations between culture and nature, rather than exploration of those between biology and society (Braun and Castree 2001; Braun 2004). This has in common a concern to avoid shades of deterministic and reductionist thinking, but takes a different starting point and has somewhat different aims and outcomes.

5   Quoted in the obituary for Brian Goodwin, *Guardian*, August 9, 2009: www.guardian.co.uk/ theguardian/2009/aug/09/brian-goodwin-obituary. Accessed December 4, 2009.

References

Anderson, B. (2007) Hope for nanotechnology: anticipatory knowledge and the governance of affect. *Area* 39: 156–65.

Anderson, B. and Tolia-Kelly, D. (2004) Matter(s) in social and cultural geography. *Geoforum* 35 (6): 669–74.

Anderson, B. and Wylie, J. (2008) On geography and materiality. *Environment and Planning A*. Online first version. doi:10.1068/a3940.

Barad, K. (2003) Posthumanist performativity: toward an understanding of how matter comes to matter. *Signs: Journal of Women in Culture and Society* 28 (3): 801–31.

Barad, K. (2007) *Meeting the Universe Halfway: Quantum Physics and the Entanglement of Matter and Meaning*. London: Duke University Press.

Barry, A. (2001) *Political Machines: Governing a Technological Society*. London: Athlone Press.

Barry, A. (2005) Pharmaceutical matters: the invention of informed materials. *Theory, Culture and Society* 22 (1): 51–69.

Bingham, N. (2006) Bees, butterflies, and bacteria: biotechnology and the politics of nonhuman friendship. *Environment and Planning A* 38 (3): 483–98.

Braun, B. (2004) Nature and culture: on the career of a false problem. In J. Duncan, N. Johnson, and R. Schein (eds.), *A Companion to Cultural Geography*. Oxford: Blackwell Publishing, pp. 151–80.

Braun, B. (2006) Environmental issues: global natures in the space of assemblage. *Progress in Human Geography* 30 (5): 644–54.

Braun, B. (2007) Biopolitics and the molecularization of life. *Cultural Geographies* 14 (1): 6–28.

Braun, B. and Castree, N. (2001) *Social Nature: Theory, Practice, and Politics*. Oxford: Blackwell.

Brown, N. and Michael, N. (2003) A sociology of expectations: retrospecting prospects and prospecting retrospects. *Technology Analysis and Strategic Management* 15 (1): 3–18.

Burgess, J., Stirling, A., Clark, J., Davies, G., Eames, M., Staley, K., and Williamson, S. (2007) Deliberative mapping: a novel analytic-deliberative methodology to support contested science-policy decisions. *Public Understanding of Science* 16: 299–322.

Cartwright, L. (1995) *Screening the Body: Tracing Medicine's Visual Culture*. London: University of Minnesota Press.

Castree, N. (2003) Bioprospecting: from theory to practice (and back again). *Transactions of the Institute of British Geographers* 28 (1): 1–21.

Cooper, M. (2008) *Life as Surplus: Biotechnology and Capitalism in the Neoliberal Era*. London: University of Washington Press.

Coyle, F. (2006) Posthuman geographies? Biotechnology, nature and the demise of the autonomous human subject. *Social and Cultural Geography* 7(4): 505–23.

Davies, G. (2006) Patterning the geographies of organ transplantation: corporeality, generosity and justice. *Transactions of the Institute of British Geographers* 31 (3): 257–71.

Davies, G. (2010) Captivating behaviour: mouse models, experimental genetics and reductionist returns in the neurosciences. In S. Parry and J. Dupre (eds.), *Nature after the Genome*. London: Sage, pp. 53–72.

Dillon, M. (2007) Governing terror: the state of emergency of biopolitical emergence. *International Political Sociology* 1: 7–28.

Doubleday, R. (2007) Organizing accountability: co-production of technoscientific and social worlds in a nanoscience laboratory. *Area* 39 (2): 166–75.

Drexler, K.E. (1986) *Engines of Creation: The Coming Era of Nanotechnology*. New York: Anchor Books.

Ellis, R., Waterton, C., and Wynne, B. (2009) Taxonomy, biodiversity and their publics in twenty-first-century DNA barcoding. *Public Understanding of Science*. Online first version, doi:10.1177/0963662509335413.

Fischer, M. (2003) *Emergent Forms of Life and the Anthropological Voice*. London: Duke University Press.

Foucault, M. (1978) *The History of Sexuality*, vol.1: *An Introduction*. Tr. Robert Hurley. New York: Random House.

Franklin, S. (2005) Stem cells R us: emergent life forms and the global biological. In A. Ong and S. Collier (eds.), *Global Assemblages: Technology, Politics and Ethics as Anthropological Problems*. Oxford: Blackwell, pp. 59–78.

Franklin, S. (2007) *Dolly Mixtures: The Remaking of Genealogy*. London: Duke University Press.

Franklin, S., Lury, C., and Stacey, J. (2000) *Global Nature, Global Culture*. London: Sage.

Fukuyama, F. (2002) *Our Posthuman Future: Consequences of the Biotechnology Revolution*. New York: Picador.

Fullwiley, D. (2007) The molecularization of race: institutionalizing human difference in pharmacogenetics practice. *Science as Culture* 16 (1): 1–30.

Gaudillière, J.P. and Rheinberger, H.J. (2004) *From Molecular Genetics to Genomics: The Mapping Cultures of Twentieth Century Genetics*. New York: Routledge.

Gibbon, S. and Novas, C. (2008) *Biosocialities, Genetics and the Social Sciences*. London: Routledge.

Greenhough, B. (2006) Decontextualised? Dissociated? Detached? Mapping the networks of bioinformatics exchange. *Environment and Planning A* 38: 445–63.

Habermas, J. (2003) *The Future of Human Nature*. Cambridge: Polity Press.

Hall, E. (2003) Reading maps of the genes: interpreting the spatiality of genetic knowledge. *Health and Place* 9 (2): 151–61.

Haraway, D. (1997) *Modest_Witness@Second_Millenium.FemaleMan©_Meets_ Oncomouse™: Feminism and Technoscience*. London: Routledge.

Haraway, D. (2008) *When Species Meet*. London: University of Minnesota Press.

Hayden, C. (2003) *When Nature Goes Public: The Making and Unmaking of Biosprospecting in Mexico*. Oxford: Princeton University Press.

Helmreich, S. (2008) Species of biocapital. *Science as Culture* 17 (4): 463–78.

Herrnstein, R. and Murray, C. (1994) *The Bell Curve: Intelligence and Class Structure in American Life*. New York: Free Press.

Hinchliffe, S. (2007) *Geographies of Nature: Societies, Environments, Ecologies*. London: Sage.

Hird, M. (2009) *The Origins of Sociable Life: Evolution after Science Studies*. London: Palgrave.

Holloway, L., Morris, C., Gilna, B., and Gibbs, D. (2009) Biopower, genetics and livestock breeding: (re)constituting populations and heterogeneous biosocial collectivities. *Transactions of the Institute of British Geographers* 34: 394–407.

Ingram, A. (2008) Domopolitics and disease: HIV/AIDS, immigration, and asylum in the UK. *Environment and Planning D: Society and Space* 26 (5): 875–94.

Jackson, P. (2000) Rematerializing social and cultural geography. *Social and Cultural Geography* 1 (1): 9–14.

Jasanoff, J. (2004) *States of Knowledge: The Co-production of Science and the Social Order*. London: Routledge.

Jasanoff, J. (2006) Clones and critics in the age of biocapital. *Biosocieties* 1: 266–9.

Kearnes, M. (2006) Chaos and control: nanotechnology and the politics of emergence. *Paragraph* 29 (2): 57–80.

Kearns, G. and Reid-Henry, S. (2009) Vital geographies: life, luck and the human condition. *Annals of the Association of American Geographers* 99 (3): 554–74.

Last, A. (2007) Mutable matter: exploring the physicality of the nanoscale. In D. Ramduny-Ellis, A. Dix, J. Hare, and S. Gill (eds.), *Physicality: Proceedings of the Second International Workshop on Physicality*, pp. 99–101, www.physicality.org/Physicality_2007/ Entries/2008/3/8_Physicality_2007_proceedings_files/Physicality2007Proceedings.pdf. Accessed December 4, 2009.

Levi, P. (1985) [1975] *The Periodic Table*, tr. R. Rosenthal. London: Penguin.

Marcus, G. (2000) *Para-sites: A Casebook against Cynical Reason*. London: University of Chicago Press.

Marcus, G. and Fischer, M. (1986) *Anthropology as Cultural Critique: An Experimental Moment in the Human Sciences*. Chicago: University of Chicago Press.

McCormack, D. (2007) Molecular affects in human geographies. *Environment and Planning A* 39: 359–77.

Nash, C. (2005) Geographies of relatedness. *Transactions of the Institute of British Geographers* 30: 449–62.

Nash, C. (2008) *Of Irish Descent: Origin Stories, Genealogy, and the Politics of Belonging*. London: Syracuse University Press.

Panelli, R. (2009) More-than-human social geographies: posthuman and other possibilities, *Progress in Human Geography*. Online first version, doi:10.1177/0309132509105007.

Parry, B. (2004) *Trading the Genome: Investigating the Commodification of Bioinformation*. New York: Columbia University Press.

Petryna, A. (2002) *Biological Citizenship: Science and the Politics of Health after Chernobyl*. Princeton: Princeton University Press.

Petryna, A. (2008) Experimentality: on the global mobility and regulation of human subjects research. *PoLAR: Political and Legal Anthropology Review* 30 (2): 288–304.

Rabinow, P. and Rose, N. (2006) Biopower today. *Biosocieties* 1: 195–217.

Rose, N. (2007) *The Politics of Life Itself: Biomedicine, Power and Subjectivity in the Twenty-first Century*. Oxford: Princeton University Press.

Rose, N. and Novas, C. (2005) Biological citizenship. In A. Ong and S. Collier (eds.), *Global Assemblages: Technology, Politics and Ethics as Anthropological Problems*. Oxford: Blackwell, pp. 439–63.

Saldanha, A. (2006) Reontologising race: the machinic geography of phenotype. *Environment and Planning D: Society and Space* 24 (1): 9–24.

Schlosser, K. (2008) Bio-political geographies. *Geography Compass* 2 (4): 1621–34.

Stirling, A. (2008) "Opening up" and "closing down": power, participation, and pluralism in the social appraisal of technology. *Science, Technology and Human Values* 33 (2): 262–94.

Sunder Rajan, K. (2006) *Biocapital: The Constitution of Post Genomic Life*. London: Duke University Press.

Sunder Rajan, K. (2007) Experimental values: Indian clinical trials and surplus health. *New Left Review* 45: 76–88.

Thrift, N. (2008) *Non-representational Theory: Space, Politics, Affect*. London: Routledge.

Waldby, C. and Mitchell, R. (2006) *Tissue Economies: Blood, Organs, and Cell Lines in Late Capitalism*. London: Duke University Press.

Whatmore, S. (2006) Materialist returns: practising cultural geography in and for a more-than-human world. *Cultural Geographies* 13: 600–9.

Wilsdon, J. and Willis, R. (2004) *See-through Science: Why Public Engagement Needs to Move Upstream*. London: Demos.

Wynne, B. (2005) Reflexing complexity: post-genomic knowledge and reductionist returns in public science. *Theory, Culture and Society* 22 (5): 67–94.

Wynne, B. (2006) Public engagement as a means of restoring trust in science? Hitting the notes, but missing the music. *Community Genetics* 9 (5): 211–20.

Chapter 16

Psychic Life

Hester Parr and Joyce Davidson

On the Geographies of Psychic Life

What is psychic life, and why should it be of concern to geographers? These questions may be especially pertinent given that there has been a long tradition in geography, successfully challenged by feminist and critical thinkers, to restrict the focus of study to decidedly *non*-intimate matters, such that bodies, emotions, and even homes, were long considered beyond the geographical pale. Thanks to a range of critical approaches that draw on interdisciplinary theoretical resources we are now well aware that the "geographies closest in" (Longhurst 2001; Davidson and Milligan 2004) exert an enormously significant impact on every aspect of our lives, individually and socially. The space of the psyche might be considered "closest" of all, and at least as significant in terms of its impact on social geographical life, at least as complex and intertwined as the body. So, while not traditionally considered a legitimate part of geographers' terrain, some geographers have begun to look at what constitutes psychic life.

In what follows, we first sketch out a broad picture of what an interest in psychic life might involve for geographers, and then go on to explore this in quite specific ways. In using "psychic life" we might immediately begin to enroll a range of ideas and entities in explanation of it (for example, the mind, the unconscious, the psyche, the pre-cognitive), all of which have varied interdisciplinary lineage and which require us to think differently in terms of how we envision attendant spatialities. For our purposes in this chapter, we are using "psyche" and "psychic life" to refer broadly to the unconscious (acknowledging that there are various traditions within psychoanalytic and phenomenological thought which define this differently). We outline a body of work on "psycho-social geographies," showing clearly why geographers have been interested in how psychic life appears to express itself in social

A Companion to Social Geography, First Edition. Edited by
Vincent J. Del Casino Jr., Mary E. Thomas, Paul Cloke, and Ruth Panelli.
© 2011 Blackwell Publishing Ltd. Published 2011 by Blackwell Publishing Ltd.

and spatial forms. We then turn to methodological questions of how geographers have investigated dimensions of psychic life, with a particular emphasis on the relationality of research methods. Here we explicitly address how geographers have struggled with the influence (and promise) of psycho-dynamic and psycho-therapeutic practice and consider methods that move beyond the linguistic in an attempt to trace the contours of psychic life. Turning away from *why* and *how* questions, we then introduce some examples of research which privilege different manifestations of psychic life in particular geographical contexts. We deliberately use research stories of the city to show how different approaches to thinking about the psyche can privilege playful and political experimentation emphasizing voluntary "drifting"; and more problematic and disruptive psychic inhabitations and mobilities of the city. The oppositions we draw through these examples are intended to illustrate the complex nature of unconscious life, and its insistent insurgence into everyday places and spaces.

In discussing psychic life we may immediately elaborate something of why geographers might have avoided or neglected this area until recently. In part, this is related to the "otherness" of the unconscious as research material; otherness that, as Philo and Parr (2003: 285) explain, is derived from the *substance* of what is entailed and related to a "domain, full of 'deeper' drives, passions and repressed psychic materials returning in 'distorted' form," which "inevitably unsettles those geographers who feel more comfortable dealing with the conscious, self-aware and apparently self-directing human being who makes rational decisions on the basis of available information." Holding aside the false dualisms implied here, psychic life is thus depicted as potentially containing disruption and danger. To use psychoanalytic vocabulary, psychic landscapes are often "uncanny" (Wilton 1998). This term (adapted from Freud) refers to something which is familiar, but unsettling and leads us to ask whether analysis of psychic life is all about evoking slightly familiar, deeper, "distorted" aspects of selves which have some kind of distinctive spatial expression.

This question is not easy to answer, and if we look at how geographers have engaged with what we might call aspects of psychic life, we see a history of encounter ranging from the psychological measurement of spatial behaviors (for a summary see Gold 2009) to cognitive studies (Kitchin and Freundschuh 2000) to psychosocial enquiry (Sibley 1995) to emotional geographies (Davidson, Bondi and Smith 2005; also see Thien's chapter, this volume) to "more than representational," "precognitive triggers," and "enduring urges" (Lorimer 2005: 84). Psychoanalytic and post-humanistic theories of self and self-development have contributed a particular dimension to understanding the interior geographies inherent in psychic life (Bondi 1999; Callard 2003; Kingsbury 2003; Pile 2005). Other thinking circulates around the psycho-geographies of cities (Pile 2005; Pinder 2005), which we will discuss later. This partial list begins to convey the very different ways in which geographers have sought to engage with "interior geographies." So, to what does "psychic life" refer?

Although psychic life is impossible to define precisely, we argue it is connected to vaguely articulated (or alternatively, too precisely delineated) notions of mind, spirit, emotion, and (un)consciousness. Our treatment of the geography of psychic life here is gestural, illustrative, and partially related to our interests (which lie in the realm of disrupted psychic lives). Our objective is not to produce an exhaustive literature review, but rather to write *through* aspects of existing work on psychic life, demonstrating that one of the most vibrant examples of human geography research at present is that which drops its disciplinary safeguards and confronts anxiety (Davidson 2003), affect (Anderson 2006), boredom (Anderson 2004b), emotion (Bennet 2004), memory (Thien 2005), love (Wylie 2005), and play (Bingley 2003), and asks "what of anger, disgust, hatred, horror, stress, isolation, alienation, fear, terror, dread, decay, loss, denial?" (Lorimer 2005: 90). In listing these studies, we do not limit psychic life to these bounds, but rather use these references as a starting point in order to ask: what geographical insights can thinking on psychic life expose?

Psycho-social Geographies

Explaining what the geographies of psychic life might entail means initial recourse to some now classic examples of work on the psychic dimensions of social and spatial life by David Sibley (1995: 2003) and Robert Wilton (1998).[1] Sibley and Wilton's research has been particularly influential for social geographers, perhaps especially because of what their work *does*, "on the ground," with psychoanalytic approaches. Sibley and Wilton work with ideas which emerge from Freudian psychoanalysis and psychoanalytic theory in order to explicate how reflecting on the psyche and theories of the self can serve to connect the unconscious, emotions, and the social and material world (Sibley 1995). In other words, these geographers put particular psychoanalytic theories to practical geographical use. By discussing examples and case studies with contemporary geographical relevance – including those dealing with common reactions to the proximity of negatively stereotyped "others" (such as "gypsies" and individuals with HIV/AIDS) – they provide meaningful insights into often neglected (psychic) aspects of, and influences on, geographical experience and relations. Specifically using object relations theory, developed through the work of theorists such as Donald Winnicott and Melanie Klein (see Bondi 2007, Kingsbury 2004 for succinct overviews), Sibley posits that early childhood development structures the constitution of the world into "good" and "bad" objects. These experiences and psychic readings of the world are argued to translate to adult behaviors around particular types of people and places, also culturally codified as "good" and "bad." This is itself related to the socialization of western adults, in particular, to clean, dirt-free environments that are regulated for polluting agents. The complex "purification" processes that we are exposed to as children in connection not only with bio-health issues, but also with respect to a variety of human and non-human "others" (separate entities which may be potentially harmful), is fundamental in this understanding of how psychic development is linked to (later) social and spatial behaviors.

Taking a particular geographical example – the British countryside – Sibley traces exclusionary psychic processes as they manifest in spatial practice, with particular relation to "gypsies" and New Age travelers. Referring to parliamentary debates around proposed legislation to manage such geographically unstable minorities (the Movable Dwellings Bill in 1908 and 1911, and the Criminal Justice and Public Order Bill in 1993), Sibley (1995: 104) demonstrates how members of these groups were constructed as deviant, polluting, and threatening presences that required social and spatial control. Debates were often explicitly couched in terms of perceived problems with sanitation and morality, portraying gypsies and travelers as sources of impurity that must be kept in their place, far apart from the "moral (and land-owning) majority" (p. 104). Arguably, individuals belonging to such groups were felt to be harmful precisely because they did not stay predictably put, so could not be trusted to maintain clear boundaries between, or keep a respectful distance away from, the psychically vulnerable selves and spaces of others. Always potentially on the move, they embody threatening potential to close protective distance, threaten borders, and eliminate sense of control. Returning to Sibley's account of the early origins of such definitive boundaries, "[t]he sense of border in the infant in western society [e.g., border between pure and unpure/dirty] becomes the basis for distances from 'others' ... [this feeling] assumes a much wider cultural significance [in adulthood]" (p. 7).

The distances discussed above therefore indicate a spatial dimension to psychic development. Sibley goes on to argue that social stereotypes, which codify people or objects as pure or dirty, act as a catalyst for these deeply held unconscious feelings which surface as we exercise social and spatial choices in adulthood, for example, who we choose to "keep company" with or exclude, keep close or at a distance. Moreover, this explanatory potential "scales up" in Sibley's account in ways relevant for understanding aspects of xenophobia, social phenomena given spatial expression through, for example, the rigid policing of national (e.g., US) borders to exclude "alien others" (p. 108). In these psycho-social-spatial arguments, then, "separation is a large part of the process of purification – it is a means by which defilement or pollution is avoided" (p. 37).

Wilton's (1998) work offers further insights into psycho-social geographies, elaborating particular understandings of the geographies of psychic life by examining community resistance to the siting of an AIDS hospice in a particular Los Angeles neighborhood. He reveals that engaging with this controversy – an emotionally charged debate firmly grounded in threatened changes to material geographies – allows him to:

> explore the interrelationship between the individual psyche and the morphology of the surrounding social landscape. I argue that a link between psyche and spatiality can help to explain the problematic nature of encounters with difference because moments of proximity represent challenges not only to an established spatial order (when something or someone is out-of-place), but also to the integrity of individual and collective identities. (p. 174).

In a similar vein to Sibley (1995), Wilton is concerned with questions of difference and how particular dimensions to psychic life are spatialized around what is constituted as "other." He emphasizes, however, that the social and the psyche are in dynamic interplay, and that while human psyches strategize in complex ways to maintain their structure, social situations constantly arise that upset that balance (Wilton 1998: 175). In his particular case study, Wilton finds that plans for facilities to house socio-culturally defined "others" disrupt the social order of a community, creating anxieties and resistance among residents who then attempt to reassert control over space felt to be "theirs." While objections were explicitly couched in practical and familiar terms of potential impact on property values and traffic flow (p. 180), Wilton's analysis of interview material suggests that there is something else (something "deeper") involved, perhaps relating to ownership and identification of a different kind. He reveals that anxieties relate to perceived threats to community and – more significantly – individual boundaries in complex socio-spatial and psychically meaningful ways now partially familiar through our consideration of Sibley. That is to say, residents worry about contagion through proximity to bodies considered sufficiently impure and potentially polluting to threaten the integrity, the very basis of embodiment and identity. The undeniable because proximate presence of AIDS, it is argued, symbolizes for residents disturbingly familiar phenomena of vulnerability and mortality: its manifestation brings the threat of illness and death "too close," leading to socially extreme, even "violent" reactions (p. 181). The strong unconscious framing of our human conduct in society and space can, in other words, be upset, disrupted and challenged, usually when questions of identity are involved. Ultimately, however, and "[d]espite all our efforts, difference refuses to keep its distance because it originates within us" (p. 182).

Wilton employs psychoanalytical concepts of the uncanny (or *unhemlich*; Freud 1919) and the abject (Kristeva 1982) to elaborate how and why we often become unsettled and disrupted, especially around social difference as exemplified here by the specter of ill bodies, thus emphasizing the fragility of our psychic life. Here, quite ambiguous divisions between self and society can precipitate anxiety or psychological crisis, with social and spatial consequences. These concepts, Wilton (1998: 180) argues, offer "a valuable means to demonstrate the connections between psyche, body and society, and the way in which these are sustained spatially, both at the level of the individual and within the surrounding social system."

Although Wilton's writing is now over ten years old, it still emerges as distinctive in that it tries to work through these complex psychoanalytical concepts in relation to difficult empirical situations: in this case the siting of an AIDS hospice, but his approach would no doubt prove valuable for investigations of other complex and charged social geographies. Wilton's resulting case study highlights in quite profound ways how processes of individual and collective anxiety surround sites of bodily decay and death, and how this anxiety structures social responses to these places. Such feelings have unconscious and often unrecognized origins, but they have real, material effects on our social geographies. However, as Wilton quite rightly points out, in most geographical studies there tends to be little attention "paid to the complexity of individual psyches and their relationship to the surrounding spatiality" (p. 180).

This lack of attention to psychic complexity is also curious, given that, as Kingsbury (2008: 109, emphasis added) explains, "[t]he unconscious *pulsates from within* everyday consciousness" and so it is uncontainable and influential, always already "out there," making its presence felt in and so involved with the world beyond the individual. "It emerges," he suggests, when "'things go wrong.' It speaks capriciously and stubbornly through distortions or symptoms exemplified by dreams and slips of the tongue" (p. 109). While individual/"internal" dreams may be a source of fascination, they might provide questionable source material for the typically trained geographer (see below). Other and everyday unconscious "external" expressions of the kind attended to by Sibley and Wilton do, however, constitute more accessible matters for research, ones of enormous potential significance for social and spatial life. Exploration of such material does not so much involve getting "inside participants' heads"; rather (and rather more safely and ethically), it entails exploring how psychic content is made manifest in geographical contexts as diverse as Los Angeles neighborhoods, the British countryside, or the US–Mexican border, through affective products or processes. By thus attending to the social and spatial life of the psyche, we can begin to appreciate how it *plays out* in research contexts, and how "unconscious aspects of human life are manifested in the social world" (Bondi 2007: 6).

Such work requires taking seriously the notion that psyches as well as selves are socio-spatial phenomena (also see Butler 1997). Several other critical and cultural geographers have recently contributed to understandings of psychic life in such respects. Kingsbury (2008), Pile (2005), and Pinder (2005) (discussed further below), have stressed the need to extend our vision of unconscious affects beyond the realm of experience of identifiable individuals, whether on the couch in their own homes, at work or at play, on their own, or in small groups, to show how they (and their psyches) have socio-spatial repercussions, shaping wider socio-cultural and material environments. Much can, of course, be learned about psychic life by studying material cultures as represented, for example, in travel magazines (Kingsbury and Brunn 2004), and other popular and/or political imaginings more often associated with cultural studies (Kingsbury 2008; Thomas 2007). Attention paid not just to dreams, then, but "dream factories" of Hollywood (Aitken 2009) and "dream cities" of consumption (Pile 2005), for example, is of relevance here. The psyche has a cultural life too, but given our chapter's focus on *socio*-spatial relations, we mention such work in passing in order to highlight the psyche's formidable power and potential reach in geographies of all kinds. Psychic life has intriguing and important implications for our work at every geographical scale, and we return to consider the particularities of urban psychic life below. We move now to discuss how we might begin to know psychic life (through research methods) in order to understand psychic geographies better.

How We Know Psychic Life

Having established *why* geographers should be concerned to investigate psychic life, we now move to consider questions of *how* they do so. Like any other place of interest to geographers that is at least partially, or in some senses, "unknown" or "unmapped," psychic landscapes present us with myriad challenges when we begin to consider how we might, first, negotiate access in order to gather or generate data (Parr 2001), and second, interpret and re-present our findings to others (Pain 2004). When we think first about how we might "get there," there simply is no clear route that allows us to visit the psychic lives of others (or perhaps even of our*selves*) and should we find ourselves there, whether by accident or design, we have no way of knowing whether the landscape before us would be seen similarly by others. When we come to re-present what we "see" or otherwise find, any form we give to such findings must inevitably say as much if not more about our own prejudices, politics, positioning – our social geographies – than it does about the "place" itself. Access and representation, in other words, are never straightforward or direct. They are always mediated, negotiated, and arguably to a degree apart from those realms more typical of socio-geographical research. However and wherever psychic life takes place, we ought to remember that people already live there. As with seemingly exotic and "undiscovered" landscapes of concern to geographers/explorers of the past, others are already familiar – if not always intimately acquainted – with its contours and ostensibly "hidden" depths.

Qualitative methods of a kind typically used in social geography research are designed to ask questions about the experiences of others. They allow us to conduct in-depth, sometimes formal, often intimate interactions, intended to create opportunities for participants to reveal something of what (a certain part of) the world is like for them, always on the understanding that we can never *really* be where they are or step into their shoes, even briefly. The circumstances of each individual's being-in-the-world are entirely unique, even if they look remarkably similar from the outside. Learning lessons from feminist and critical approaches to such exceptional subjectivities (Moss 2002), social geographers aspire to facilitate sensitive observational and conversational exchanges – in ethnography, individual interviews, and group discussions – where participants can show or most often *say*, how things look and feel from their angle. By soliciting and attending closely to first-hand accounts of a matter of shared interest, we thus hope to gather or "co-construct" meaningful insights that we might then re-present to interested others as we disseminate our "findings." Such methods, as Davies and Dwyer (2007) point out, have been more or less central to the practice of human geography throughout its history. Of course, there are "changes in the way they are being conceived and carried out, and there are transformations in the way these methods are being used to make claims to understanding and intervening in the world" (p. 257). However, and as Davies and Dwyer continue, "it is hard, though perhaps not impossible, to imagine what a radically new form of qualitative research practice might look like" (p. 257). Some would say that when it comes to investigating psychic life, radical new methods are exactly what we need. In what follows, we consider how and why such claims are made, outlining a few commonly cited pros and cons of what we might call "traditional" geographical methods for researching psychic life.

As geographers begin to think and write about unconscious realms, perhaps the primarily *different* dilemma we face is that of handling uncommonly intangible phenomena. It's tricky to get your hands on and head around the stuff of psychic life because it is so often inaccessible even to its "owner." Traditional methods are arguably intended to work with what participants tell us, how they (*say* they) think and feel about any given focus of research. Yet, with psychic life, there is clearly so much more than can be *consciously* thought or felt, and questions continually arise about how we might "get at" this peculiarly unwieldy material. While we would argue that we should be less concerned to access or "take" what is not freely given, at the same time, we stress the need for sensitivity to (even unintended) affects and effects of psychic life as we conduct our research; for psychic phenomena do not always, or perhaps ever, remain entirely out of sight, but rather emerge "out of mind" to bubble up and spill over into shared and social space.

The methodological and epistemological challenges hinted at above are witnessed in various examples of contemporary research. Bennett (2009), for example, refers to research relationships, a key concern for feminists. She states, "I am also intrigued by how far empathetic psychic space of a (research) relationship can stretch as feelings 'grow' or 'develop' beyond meetings through dreams, reflecting, writing and relating experience to others" (p. 248). We can of course never control for such phenomena, but we might at least aspire to better understand what they do and mean. Specifically, Bennett requires – beginning to employ language reminiscent of other professional procedure – "practices that explore how our fantasies and defences affect fieldwork, analysis, and writing" (p. 249). Simultaneously, this prompts uncertainty for Bennett in terms of how to exercise "empathy" and be "emotionally present" in "non-judgemental" ways as this involves "taking" from a therapeutic context into a research situation (p. 249). Clearly, although investigating psychic life or making sense of its manifestations does not mean importing methods from psychoanalysis wholesale, many researchers echo Bennett's sense of the "hurdles that geographers face when they engage with practices developed in a psychotherapeutic setting" (p. 244). Thomas, for example, highlights one important aspect of the challenge facing geographers when she writes: "it has been difficult to juggle the benefits of psychoanalytic theory with the conundrum of how to do so without psychoanalyzing individual interviewees when examining their articulations" (Thomas 2007: 538). Few of us would know – or indeed would want to know – how to begin to psychoanalyze participants, and we should not assume this could ever be straightforward. However, and as Paul Kingsbury (2009: 482) explains, there may be more similarities between procedures in each profession than we are aware:

> much of critical human geographical research in general, is replete with processes that are comparable to free association, transference, and analytic listening. Think, for example, of the free association-like meandering pronunciations and spontaneous ramblings voiced during unstructured interviews. Think, for example, of the powerful transferential bonds that often develop between the researcher(s) and researched during interviews, ethnographies, and focus groups. Think, for example, of the degree to which geographers, especially feminist and poststructuralist geographers, attempt to foster data collection techniques and conditions that are comparable to analytic listening: attendance to self-reflexivity, mindful of multiple viewpoints and meanings, recognition of the situatedness and partiality of knowledge, vigilance towards power dynamics, and appreciative of the ineluctable instability of insider/outside and researched/researcher distinctions.

The similarities Kingsbury refers to here are, however, largely *unacknowledged*, and he insists that "extensive critical assessments of the validity, value, and potentiality of psychoanalytic methods in human geography are rare" (2009: 481). In fact, Kingsbury continues, "geographers have found psychoanalytic concepts much more valuable than psychoanalytic methods" but by attending more closely to the latter, they "may begin to formulate exciting and truly radical research projects wherein psychoanalytic methods are as equally relevant and valuable as the psychoanalytic concepts that they ultimately depend on" (p. 486).

Bingley (2003) is amongst those who have begun to take up this challenge by exploring what (one form of) such research might look like. Without ever coming close to psychoanalyzing participants, Bingley has employed "tactile" methodologies that draw her psychotherapeutic and geographical training together to investigate "sense of place" in non-traditional ways. Working with groups, Bingley organized workshops "where methods drawn from art therapy and humanistic psychotherapy allowed participants to express their experience at a non-verbal, sensory level" (p. 330). In particular, she used "sandplay" to provide a way of "getting beyond" habitual responses to explore meanings ordinarily inaccessible in everyday (conscious) life and thought; the freedom of "playing" with sand facilitates free "expression of elements in the psyche," and it is such non-cognitive material that Bingley aimed to uncover. It is crucial to note, however, that Bingley's innovative approach was never used in isolation from conventional methods, as participants' creations became foci for reflection and discussion: Bingley worked – i.e. talked – with her participants to jointly make sense of what they had made, and so meanings (research findings) are co-constituted in context as "space is made for reflection and the emergence of the unconscious process into conscious awareness" (p. 340). This important aspect of the research process might then reasonably be described in time-honored terms as focused group discussions, albeit ones with a different – therapeutically informed – dynamic and activities-based twist. Talk – then text, via transcripts – is at least as important to Bingley's research as performative act and artifact which, while pivotal to her method, could never stand alone as primary source materials. Artistic activities and their products can be observed, but to make *sense* that is shared, their meanings must be generated with rather than imposed on participants. This leaves feminist and other established methodologies in a place of still crucial importance. For Bingley, then, "using psychotherapeutics" does not constitute a radical methodological departure, but "simply offers a greater scope for opening up existing fields of exploration" (p. 340).

Bingley's methods for investigating aspects of psychic life are clearly effective and potentially valuable, but we would question whether such methods are suitable for all. Any approach capable of stirring up psychic flows and bringing unconscious material to the surface should give us cause for caution, for what resources might those of us untrained in psychotherapeutics have to help smooth over that which we've troubled? While similarly trained and experienced in therapeutic methods, Liz Bondi's approach is perhaps more readily, safely accessible to typically trained geographers. Bondi (2003, 2005, 2006) draws existing connections between certain psycho-analytic/-therapeutic and feminist approaches closer still, and according to Kinsgbury's recent assessment, her "work on empathy and identification is one of the most incisive and extensive theoretical assessments of the potential benefits of using psychoanalytic methods in geography" (Kingsbury 2009: 483). Focusing in particular on Bondi's approach to psychoanalytic understandings of identification and empathy, Kingsbury emphasizes that this approach is intended to theoretically enhance and enrich existing methodologies, those already acutely sensitive to questions "about how positionality and power infuse the dynamics between the researcher and the researched." Using insights from psychoanalysis need not mean abandoning existing methods, but can rather involve continuing to work to refine them. This serves to enrich our research, but also, and importantly, to protect (especially psychically sensitive) research participants.

Turning to one last alternative attempt to investigate unconscious/non-cognitive life, geographers have recently engaged with non- (or more-than-) representational

approaches (Lorimer 2005; Thrift 2007) that provide somewhat different conceptual resources for researching psychic phenomena. Geographers whose work falls
under the convenient umbrella term of Non-Representational Theory (NRT) tend
to favor performance and activity over proclamations and products as foci for study,
arguing that ineffable events provide (or perhaps represent?) more immediate
embodied engagement with the world beyond, beneath, or before language (Nash
2000, and see Philo, this volume). That is to say, NRT develops a focus on what
can be *shown* over what can be said. According to Davies and Dwyer (2007: 259),
NRT practitioners "seek ways of going beyond words, or indeed are suspicious of
words"; in other words, they stress what people do over what they *say* they do,
which leads to "direct" study of such events as dance (McCormack 2005) and music
(Anderson 2004a). Such research emphasizes the "how" and the "now" of performance, the ephemera of immediate experience arguably inaccessible via traditional
"textual methodologies" (Anderson et al. 2005). We would, however, and following
James Hillman (1991: 28), caution against the turn away from language that characterizes much recent and influential work in this field, and suggest that words, too,
are performances:

> As one art and academic field after another falls into the paralyzing coils of obsession
> with language and communication, speech succumbs to a new semantic anxiety. Even
> psychotherapy, which began as a *talking cure* – the rediscovery of the oral tradition of
> telling one's story – is abandoning language for touch, cry and gesture. We dare not
> be eloquent ... Our semantic anxiety has made us forget that words, too, burn and
> become flesh as we speak.

Words are at risk of being taken-for-granted, perhaps stripped of power and promise
because they are so much more familiar in social geography research, but they are
every bit as vital, involved, and potentially out of the ordinary as music and dance
(or sandplay). While we agree that actions do, indeed, often speak louder than
words – think also of tears, obscene gestures or even "inactions" such as demonstra-
tive, *meaningful* silences – such embodied expressions also often require contextual
interpretative frameworks in order to make *sense* and can be just as representational
(and limited) as expressions "straightforwardly" verbal. Unconscious affects will
always be excessive, elusive, and resistant to interpretation and representation –
such limitations are integral to the nature of representation (Smith et al. 2009: 12)
– but to lose sense of the import of language would be to risk losing something of
inimitable value, a cornerstone of communication, which is what researchers *do*.

Psychogeographies and the Psychodynamic City

Having explored something of *why* and *how* geographers understand aspects of
psychic life, we turn now to consider some examples of work which explicitly
explore it, and in very different ways to that discussed above. We begin with geographies of the city. Cities, according to urban social geographers, are concentrations
of dreams, fantasies, memories, (un)conscious journeys, play, and struggle. As an
easily recognizable geographical entity, the city is perhaps a good place to start to
tell some stories about these aspects of psychic life. There are different ways to tell

these stories by focusing on the psychic life of the city itself (Pile 2005), or the psychic dimensions to everyday lives within the city. If we think about how these two are intertwined and interdependent, then a useful starting point is research on psychogeographies[2] (see Pinder 2005).

In some of our comments above about psycho-social geographies, there is a risk that the psyche becomes a deterministic blue-print for related spatial organization, especially with regards to questions of social difference. Research on psychogeography shows how some people quite deliberately try to resist this taken-for-granted imprinting and other organizing structures of social life and seek to "shake up" the psychic dimensions of the city. For example, Pinder (2005) traces the complex history of "the situationists," a grouping of activists and artists intent on forms of urban revolution and circulation of utopian ideals. Although there is much to say about this history and its implications for thinking about geography, our focus is to appropriate some aspects of thinking about utopian geographies of the city in order to relay how some groups voluntarily sought to engage in "an experimental investigation of the free construction of daily life" (Debord 1963, quoted in Pinder 2005: 5). To cast this solely as a project that was all about deliberately disrupting the kinds of (un)conscious psycho-spatial relationships we have described above, would be to deny complex anti-capitalist and radical roots of this movement, but some aspects of this urban experimentation can be argued to do just that.

In the later 1950s, groups of situationists and associated avant-garde activists deliberately sought to investigate and change urban spaces through new kinds of geographical research and action (often focused on Paris and London), including that labeled "psychogeography" (Pinder 2005: 128). Part of the general motivation for this approach was bound up with a will-full reclamation of urban space in terms of "desire, encounter and play" (p. 150) including mobile drifting (called "the *dérive*"). This practice, and others like it, were intended as an "oppositional mode of living in the (capitalist) city" (p. 152), and activists used the term psychogeography to "investigate different ambiences and zones in cities, and to attend to the relationship between social space and mental space and between urbanism and behaviour." A lived psychogeographical approach might engender a new map(ping) of the city whereby cartographies of unfettered desire and fluidity characterize movement. However, as Pinder argues, in fact the maps produced by situationists often indicated "active and conscious" senses of behaviors (p. 155). Indeed, the outworkings of various strands of situationism included thinking and architectural drawings, maps, various forms of art work as well as practices, all deliberately designed to comprise an "irrational embellishment of the city" (p. 155). The contradiction between these *consciously* designed efforts to disrupt the city and also create irrational form stands in stark contrast to *less voluntary* examples of the embodiment of psychic disruption that we detail further below.

The story of psychogeography, and more broadly the situationists and associated movements, is orientated around urban experimentation and new relationships with mental and social space which supposedly allow "spaces and times to be thought and lived otherwise" (p. 265). These themes have also been present in other work which has constructed the psychic life of cities as "real." This term is associated with the work of psychoanalyst Jacques Lacan, and refers to that which falls outside the symbolic order, that which resist or defies symbolization or signification, but which allows us to approach socio-spatial experience in novel and fruitful ways. For Lacan, the real is a powerfully affective and meaningful presence in our lives, but at the same time, it (or at least its *meaning*) is somehow disturbingly absent, unreachable; it is "this something faced with which all words cease and all categories fail, the object of anxiety par excellence" (Lacan 1991: 164). Pile's recent work on the city (1996, 2005) shows how he understands this entity as a "state of mind and body" (Pile 1996: 210), as he illustrates a psychoanalysis of urban space, which he contends, "contains the psychodrama of everyday life" (p. 246). Analysis of city life, he says, should be expanded to include "shadows, irrationalities, feelings, uto-pianisms" (Pile 2005: 2) and other elusive or "unspeakable" and "real" phenomena. Geographies, particularly those associated with (often intensively occupied and emotionally charged) city spaces, are psycho-dynamic. Geographers, Pile argues, can gain by using psychoanalytic theory to help explain this precisely because the spa-tialities of everyday life are so difficult to articulate. It is needed, he argues, to deal adequately with "the complex psychodynamics of place" (p. 100). Pile's work in effect excavates a related, but very different psychogeography of urban space, one that is "phantasmagoric" in orientation, and concerned with "occluded spatialities" which are ghostly and dream-like (p. 3). These descriptors and claims are ones informed by early Freudian psychoanalysis and clearly seek to locate aspects of human psychic life as real, in real cities (p. 3). Pile observes that this work in general focuses on the aspects of human psychic and emotional life that are *positively unset-tling*: "psychogeography was turned towards desire, towards excitement, towards life" (referencing Debord's playful disruptions of even the "tiniest details" of city life, p. 14). Despite this, Pile discusses the "grief work" of cities (using examples such as London and New York; and see Rose 2009) where "dark histories" and the "tradition of dead generations … weighs so heavily on the lives of city dwellers" (Pile 2005: 156). For many, "the city is haunting because it gathers together so many ghosts: there are so many reasons to be haunted" (p. 151). There is so much we cannot know about city spaces – about who lived (and died) before us, or about how they lived (and died) (p. 147) – that they become intensities of uncertainty, associated with "spooky" happenings, legacies of terror, and the uncanny. Nonetheless, Pile notes that cities can also be places of forgetting and "where the ordinary violences of everyday life are simply lived through" (p. 155). It may be that in discussing the psychogeographies of urban life, we are at risk of thinking about psychic life in and of cities as something playful, radical, deliberately unset-tling, and we must remember that there are always other stories to be told.

Disruptive Psychic Life and Traumatic Geographies

To understand geographies, and particularly city geographies, as psychodynamic and phantasmagoric, brings questions of irrationality, emotion, memories, and psychologies to the fore of analysis. Here we highlight less deliberate psychogeog-raphies of mobility, whereby disruptive aspects of psychic life entail tenuous, stressed relationships in and with city spaces. These are also stories of the real city, but ones lived out by marginalized and disenfranchised city dwellers experiencing psychic trauma.

Caroline Knowles (2000) writes of "post-asylum geographies of madness" (see also Wolch and Philo 2000) and charts the lives of deinstitutionalized psychiatric patients in Montreal, Canada. Here lives "are not static, fixed in place, but in the process of many journeys from one place to another" (Knowles 2000: 83). Individuals, often diagnosed with schizophrenia and coping with a variety of psycho-social challenges, find themselves in poverty in an inadequate welfare system and are often forced to wander the city streets daily in search of shelter and warmth. Movements between city spaces and shelters happen because of preferred regimes of care, of different rules about acceptable behavior and policies of exclusion (p. 91). In the enforced, semi-voluntary wanderings in the city, many people "move through, rather than occupy city space" (p. 92), "reading" the city for atmospheres of tolerance. Knowles asks "what kinds of being-in the-world are formed by having no place to spend time?" (p. 97). The semi-permanent "walking exile" of a life in motion challenges the (positive) radical analysis of the potential of the *dérive* (and see Cresswell 1997 on alternative imaginings of "the nomad"). What psychogeog-raphies are at play here? Knowles (2000: 161) is blunt in her answer concerning the psychic disruption we commonly call madness:

> It is the most shifting of personal landscapes and raises questions about the nature of identity and existence itself. It is about the rapid retreat of the certainties of the taken for granted and that formed the basis of being in the world. It is an experience that left people shaken about how and who to be in the world.

Is it not this flux which psychogeographers seek and through which they might revision the city?

Looking closely at accounts of homelessness and mental illness, we see that there are huge populations of people who have lived this flux, involuntarily, and who battle its results, illuminating different dimensions to city life. Desjarlais' (1997: 127) critical phenomenology of street and shelter life for people who are homeless and mentally ill emphasizes how continual momentum, in accordance with the atmospheres of city streets, and felt compulsions, lead to a "blunting" of existence: "For many the sensorium of the street involved a corporal existence in which a person's senses and abilities to make sense soon became dulled in response to exces-sive and brutal demands on these." Part of this story of dislocation and negative mobility is purportedly related to a lack of "integrative narrative," a characteristic of the "dreamy, disconnected" life accounts of people diagnosed with schizophrenia (Desjarlais 1997: 5). For people who already experience challenging forms of psy-chological disorientation, the city arguably offers little solace (although see Parr 1999 on delusional geographies). Certainly, there is a lack of evidence of playful experience or radical politics released by these more-or-less involuntary psychoge-ographies of city life. In accounts of the experience of psychic disruption in the city, the struggle for consensual reality dominates individual stories, though often in very distinctive ways:

> Most of the time, I can be walking down the street and hear thousands of voices and I feel terrible, like I'm not really there. (quoted in Parr 1999: 683)

> You mishear things, out of synchronisation with other people, out of sync. You're outside. Things keep jumping into your mind. You feel like an animal, you feel like you're in touch with the universe, y'know you are outside yourself, in touch with nature. You go along with the momentum of the illness, it's just a different way of thinking—you see and hear the world differently. (quoted in Parr 1999: 683)

Through more or less negative experiential accounts of psycho-social disruption – the second quote suggests a valued sense of connection amongst the "chaos" – we also learn more about the importance of boundaries, and the potential disadvantages of endless atmospheric and sensory mobilities. Arguing for recognition of the need for psychological boundaries for coherent subjectivity is not new (see Glass 1985; Kirby 1996; Hekman 2004), but sensitive work on the geographies of panic shows how our everyday experience of the city is also dependent on these (Davidson 2003). Analysis of agoraphobic life worlds highlights the city as a space of anxious contradiction, a geography of irrational affect. Again, this disrupts a romantic visioning of psychogeographical exploration, with the voices of agoraphobic women speaking of the city as a place where "confusion proliferates on every level" (Crawford 1994: 4, in Davidson 2003: 62):

> Your heart starts thumping, you feel like you're choking in your throat, you can't swallow and you feel all dizzy and giddy, and like if it's in a shop, or the church or what, you feel you want to run. (quoted in Davidson 2003: 58)

Curiously, despite our intention to qualify the frames of references for psychogeographical exploration as one which must include negative spatialities, we also return to emphasize the hopeful possibilities engendered by disruptive psychic life experience. In her discussion of agoraphobic lifeworlds, Davidson reminds us that panic is like excitement, and that "to open oneself up to excitement, to learn to endure and even enjoy the potentiality of panic without giving oneself over to it completely is the phenomenal freedom to which the agoraphobic aspires" (p. 67).

Returning cautiously to a position whereby psychic disruption may in itself be a process through which new and better social and spatial experiences are enabled is not to align ourselves (just) with the playful whims of psychogeographers, but rather to entertain the possibility that somewhere in delusional or anxious psychic experience lies hope and possibility of *connection*. This is not always about a radical collective politics of hope (Harvey 2000) in which new utopian city visions are rescued and redrawn (and see Pile's 2005 comments on this), but that (un)conscious engagements with material spaces may be an important part of a "therapeutic" and psychological search for a sense of boundedness for those who reluctantly occupy transient and distressing psycho-mobilities. For those in delusional states, for example, by physically locating themselves in crowded parks and city malls, the movement, crush, and performances of bodies may be(come) meaningful in transitional realms of dislocation, the flow being reminiscent of both material reality and fantastical states (see Parr 1999). For people seeking consensual reality, then, this reminiscence may be important in trying to locate senses of boundedness. As Pile (2005: 3) warns however, structures of feeling that emerge from and in cities are always "contradictory, mobile, changing and changeable." Mapping and anchoring psychic life in cities is clearly a risky and complicated endeavor.

Conclusion

This chapter explores some dimensions of the geographies of psychic life and the geographical insights that thinking on psychic life might expose. We have considered why geographers might be reticent in investigating these very interior geographies, and yet simultaneously pointed out how psychic interiors are intimately and insistently expressed in wider, external social and spatial worlds. Using work influenced by psychoanalysis, we have shown how concerns about relatedness (e.g., self and other) have dominated geographical accounts of psycho-social forms and structures. "Relating" has also been a key theme of methodological exploration into deeply felt "placing" in the world, and we have shown the radical ways in which some geographers have sought to expand their tool-kit of "how to know" the social psyche. Relating to (and connecting with) the world through sand, dance, music, and so on provide just some examples of the exciting mediums through which geographers are beginning to encounter psychic life. In providing some examples of contemporary research on psychic life we explored different aspects of relating to cities: on the one hand we reviewed particular – psychogeographical – attempts to artistically disrupt perceived constructions of (atmospheric) city life, and on the other, we contrasted writing on such playful mobilities with research that painfully reveals the traumatic material of dislocated psychic lives. The latter strategy reflects both our interest in emotionally challenging or atypical lifeworlds, but also gestures towards the promise inherent to exploration (and indeed experience) of disturbed and distorted psychic landscapes.

Notes

1   See further examples of psychoanalytic thinking in the special themed issue of *Social and Cultural Geography* (2003), 4.
2   The term "psychogeography" is described by Guy Debord as "[t]he study of the specific effects of the geographical environment, consciously organized or not, on the emotions and behavior of individuals" (quoted in Coverley 2006: 10).

References

Aitken, S.C. (2009) Analysis of movies and films. In R. Kitchin and N. Thrift (eds.), *International Encyclopedia of Human Geography*. Oxford: Elsevier, pp. 196–200.

Anderson, B. (2004a) Recorded music and practices of remembering. *Social and Cultural Geography*, 5 (1): 3–20.

Anderson, B. (2004b) Time-stilled space-slowed: how boredom matters. *Geoforum* 35: 739–54.

Anderson, B. (2006) Becoming and being hopeful: towards a theory of affect. *Environment and Planning D: Society and Space* 24: 733–52

Anderson, B., Morton, F., and Revill, G. (2005). Practices of music and sound. *Social and Cultural Geography* 6 (5): 639–44.

Bennett, K. (2004) Emotionally intelligent research. *Area* 36 (4): 414–22.

Bennett, K. (2009) Challenging emotions. *Area* 41(3): 244–51.

Bingley, A.F. (2003) In here and out there: sensations between self and landscape. *Social and Cultural Geography* 4 (3): 329–45.

Bondi, L. (1999) Stages on a journey: some remarks about human geography and psychotherapeutic practice. *Professional Geographer* 51: 11–24.

Bondi, L. (2003) Empathy and identification: conceptual resources for feminist fieldwork. *ACME: International Journal of Critical Geography* 2: 64–76.

Bondi, L. (2005) The place of emotions in research: from partitioning emotion and reason to the emotional dynamics of research relationships. In J. Davidson, L. Bondi, and M. Smith (eds.), *Emotional Geographies*. Aldershot: Ashgate, pp. 231–46.

Bondi, L. (2006) Is counseling a feminist practice? *Geojournal* 65 (4): 339–48.

Bondi, L. (2007) Psychoanalytic theory. Online paper archived by the Institute of Geography, School of Geosciences, University of Edinburgh.

Butler, J. (1997). *The Psychic Life of Power: Theories in Subjection*. Chicago: Stanford University Press.

Callard, F. (2003) The taming of psychoanalysis in geography. *Social and Cultural Geography* 4: 295–312.

Coverley, M. (2006) *Psychogeography*. Harpenden: Pocket Essentials.

Crang, M. (2003) Qualitative methods: touchy, feely, look-see? *Progress in Human Geography* 27 (4): 494–504.

Crawford, M. (1994) The world in a shopping mall. In M. Sorkin (ed.), *Variations on a Theme Park: The New American City and the End of Public Space*. New York: Hill and Wang, pp 3–30.

Cresswell, T. (1997) Imagining the nomad: mobility and the postmodern primitive. In G. Benko and U. Strohmayer (eds.), *Space and Social Theory: Interpreting Modernity and Postmodernity*. Oxford: Blackwell, pp. 360–82.

Davidson, J. (2003) *Phobic Geographies: The Phenomenology and Spatiality of Identity*. Aldershot: Ashgate Press.

Davidson, J. and Milligan, C. (2004) Embodying emotion, sensing space: introducing emotional geographies. *Social and Cultural Geography* 5 (4): 523–32.

Davidson, J., Bondi, L., and Smith, M. (eds.) (2005) *Emotional Geographies*. Aldershot: Ashgate.

Davies, G. and Dwyer, C. (2007) Qualitative methods: are you enchanted or are you alienated? *Progress in Human Geography* 31 (2): 257–66.

Desjarlais, R. (1997) *Shelter Blues: Homelessness and Sanity in a Boston Shelter*. Philadelphia: University of Pennsylvania Press.

Glass, J. (1985) *Delusion: Internal Dimensions of Political Life*. Chicago: University of Chicago Press.

Gold, J.R. (2009) Behavioural geography. In R. Kitchin and N. Thrift (eds.), *International Encyclopedia of Human Geography*. Oxford: Elsevier, pp. 282–93.

Hekman, S. (2004) *Private Selves, Public Identities: Reconsidering Identity Politics*. Pennsylvania: Pennsylvania University Press.

Hillman, J. (1991) *A Blue Fire*. New York: Harper Collins.

Kingsbury, P. (2003) Psychoanalysis, a gay spatial science? *Social and Cultural Geography* 4 (3): 347–67.

Kingsbury, P. (2004) Psychoanalytic approaches. In J.S. Duncan, N.C. Johnson, and R. Schein (eds.), *A Companion to Cultural Geography*. Blackwell: Oxford, pp. 108–20.

Kingsbury, P. (2008) Did somebody say *jouissance*? On Slavoj Zizek, consumption and nationalism. *Emotion, Space and Society* 1 (1): 48–55.

Kingsbury, P. (2009) Psychoanalysis. In R. Kitchin and N. Thrift (eds.), *International Encyclopedia of Human Geography*. Oxford: Elsevier, pp. 480–6.

Kingsbury, P. and Brunn, S. (2004) Freud, tourism, and terror: traversing the fantasies of post-September 11 travel magazines. *Journal of Travel and Tourism Marketing* 15 (2–3): 39–61.

Kirby, K. (1996) *Indifferent Boundaries: Spatial Concepts of Human Subjectivity.* New York: Guilford Press.

Kitchin, R and Freundschuh, S. (eds.) (2000) *Cognitive Mapping: Past, Present and Future.* London: Routledge.

Kristeva, J. (1982) *Powers of Horror: An Essay on Abjection.* New York: Columbia University Press.

Knowles, C. (2000) *Bedlam: on the Streets.* London: Routledge.

Lacan, J. (1991) *The Seminar of Jacques Lacan: The Ego in Freud's Theory and in the Technique of Psychoanalysis, 1954–1955.* London and New York: Norton.

Longhurst, R. (2001) *Bodies: Exploring Fluid Boundaries.* London and New York: Routledge.

Lorimer, H. (2005) Cultural geography: the busyness of being "more-than-representational." *Progress in Human Geography* 29 (1): 83–94.

McCormack, D. (2005) Diagramming practice and performance. *Environment and Planning D: Society and Space* 23: 119–47.

Moss, P. (ed.) (2002) *Feminist Geography in Practice: Research and Methods.* Oxford: Blackwell.

Nash, C. (2000) Performativity in practice: some recent work in cultural geography. *Progress in Human Geography* 24 (4): 653–64.

Pain, R. (2004) Social geography: participatory research. *Progress in Human Geography* 28 (5): 652–63.

Parr, H. (1999) Delusional geographies: the experiential worlds of people during madness and illness. *Environment and Planning D: Society and Space* 17: 673–90.

Parr, H. (2001) Negotiating different ethnographic contexts and building geographical knowledges: empirical examples from mental health research. In M. Limb and C. Dwyer (eds.), *Qualitative Methodologies for Geographers: Issues and Debates.* London: Arnold, pp. 181–97.

Philo, C. and Parr, H. (2003) Introducing psychoanalytic geographies. *Social and Cultural Geographies* 4 (3): 283–93.

Pile, S. (1996) *The Body and the City: Psychoanalysis, Subjectivity and Space.* London: Routledge.

Pile, S. (2005) *Real Cities: Modernity, Space and the Phantasmagorias of City Life.* London: Sage.

Pinder, D. (2005) *Visions of the City.* Edinburgh: Edinburgh University Press.

Rose, G. (2009). Who cares for which dead and how? British newspaper reporting of the bombings in London, July 2005. *Geoforum* 40 (1): pp. 46–54.

Sibley, D. (1995) *Geographies of Exclusion: Society and Difference in the West.* London: Routledge.

Sibley, D. (2003) Psychogeographies of rural space and practices of exclusion. In P. Cloke (ed.), *Country Visions.* Prentice-Hall: Harlow, pp. 218–31.

Smith, M., Davidson, J., Cameron, L., and Bondi, L. (2009) Geography and emotion: emerging constellations. In M. Smith, J. Davidson, L. Cameron, and L. Bondi (eds.), *Emotion, Place and Culture.* Aldershot: Ashgate, pp. 1–20.

Social and Cultural Geography (2003) Special themed issue on "Psychoanalytic geographies," vol. 4.

Thien, D. (2005) Intimate distances: considering questions of "us." In J. Davidson, L. Bondi, and M. Smith (eds.), *Emotional Geographies.* Aldershot: Ashgate, pp. 191–204.

Thomas, M. (2007) The implications of psychoanalysis for qualitative methodology: the case of interviews and narrative data analysis. *Professional Geographer* 59 (4): 537–46.

Thrift, N. (2007) *Non-representational theory: Space, Politics, Affect.* London: Routledge.

Wilton, R. (1998) The constitution of difference: space and psyche in landscapes of exclusion. *Geoforum* 29: 173–85.

Wolch, J., and Philo, C. (2000) From distributions of deviance to definitions of difference: past and future mental health geographies. *Health and Place* 6: 137–57.

Wylie, J. (2005) A single day's walking: narrating self and landscape on the SouthWest Coast Path. *Transactions of the Institute of British Geographers* 30 (2): 234–47.

Chapter 17

Sexual Life

Gavin Brown, Kath Browne, and Jason Lim

When and Where does Sexual Life Matter: Introduction

Sexual life matters in all manner of places and in all manner of situations. This chapter explains why. By thinking about how sexual life matters in different social fields, we think about the diversity of phenomena that might fall under the rubric "sexual life." Within the sub-discipline of "geographies of sexualities" (or "sexual geographies"), much work has focused on the lives, experiences, and identities of sexual minorities, especially those who identify as lesbian, gay, bisexual, and/or queer. Research conducted in geographies of sexualities considers the relationships between the identities and practices of sexual minorities and how they face persecution, discrimination, and (consequently) marginalization from heteronormative society – that is, a society that privileges particular versions of masculinity and femininity and assumes that these are opposites that are meant to come together within heterosexual relationships. Sexual life matters, then, because of these questions of power, rights, exclusion, and marginalization, and geographical research has shown that these questions apply in a wide range of spaces: workplaces and corporate boardrooms (McDowell 1997; Kitchen and Lysaght 2003); bars, nightclubs, and other leisure spaces (Binnie and Skeggs 2004; Casey 2007); voluntary organizations (Andrucki and Elder 2007); streets and other public spaces (Valentine 1993); homes and bedrooms (Johnston and Valentine 1995).

The sex that is considered, celebrated, and debated within geographies of sexualities and queer geographies tends to be the unusual, the resistant, the "abnormal." Although the importance of sexuality in the constitution of gender (and vice versa) has been a central consideration in much foundational feminist theorizing, this relationship of co-constitution has become taken-for-granted in certain recent feminist and/or gender geographies, such that the dynamic reproduction of

A Companion to Social Geography, First Edition. Edited by
Vincent J. Del Casino Jr., Mary E. Thomas, Paul Cloke, and Ruth Panelli.
© 2011 Blackwell Publishing Ltd. Published 2011 by Blackwell Publishing Ltd.

heteronormativity itself at times goes unquestioned. We argue that interrogating the normative – the often invisible, taken-for-granted and "common sense" – and by which we do not only mean gender and sexual norms – is crucial to understanding sexual lives. Queer modes of thinking not only interrogate the constitution and production of the normative, but also question how social, economic, political, and cultural privileges accrue to normative identities and practices. In addition to heteronormative privileges, recent discussions have started to question the privileges enjoyed by those who could be termed "homonormative" – most often affluent, apolitical, white, middle class, gay, and male. This concept, however, has not gone uncontested (Brown 2009; Oswin 2008).

This chapter is structured around four questions: why don't geographers talk about sex; what is so wrong with being normal; who and where can be queer; is queer the only way to "do" and "be" political. Queer thinking has had a significant impact on geographies of sexualities, and *Geographies of Sexualities* (Browne et al. 2007), a book we co-edited, was in many ways a "queer" book. However, we recognize the limits of queer approaches and write this chapter to bring other modes of theorization into conversation with queer approaches. The questions we chose to frame the chapter problematize the centrality of queer modes of thinking, analysis, and politics in recent studies of sexual life. From this starting point, we also question how "normality" is constructed, critiqued, and experienced, and how claims to normality are used and contested in discussions of sexual life. We start with these questions because they are pressing and unresolved, but our answers indicate that they cannot be considered in isolation and that they throw up broader sets of issues about sexual life. There are other questions that could be posed, such as: how we can theoretically and practically explore issues of intersectionality; what are the implications of using queer as a method rather than thinking of it as a mode of subjectivity; how can we deploy identity categories, such as Lesbian, Gay, Bisexual, and Trans (LGBT), to create solidarity and effect change; what are the overlaps and divergences between theorizing sexual lives and living them; and how can we talk about sexual lives outside the Global North without reinscribing Eurocentric concepts or presuming inescapable difference from them. We hope that this chapter creates the space to begin to pose such other questions and to explore further the mutual constitution of sexuality and space.

We have chosen to present the main body of this chapter as a conversation among three voices. We have done so to highlight the partiality of any engagement with social life; there are multiple ways of understanding and engaging with sexual life, as well as of asking questions about it. Each voice alternates in dialogue with the others throughout the chapter. By presenting our thoughts in this manner, we acknowledge our individuality and distinctiveness. Despite our on-going collaboration over a number of years (Browne et al. 2007; Browne and Lim 2008; Lim and Browne 2009), the three of us approach the study of sexual life (and social geography) in different ways. We are conscious that each of us takes distinct positions and trajectories in relation to other aspects of geography – whether they are urban geography, social geography, or cultural geography – and this positionality also shapes our interests and the questions we pose regarding sexual life. Sometimes we end up coming to the same conclusions despite taking different routes through our analyses; sometimes we start in the same political and theoretical place, but end up reaching very different end points; and at times we do not seek to reach consensus, leaving our differences exposed and explicit. The pages that follow hopefully capture something of the multiplicity not only of our perspectives, but also of the fluid and contingent intersections of personal, political, empirical, and theoretical interests that shape explorations of the social geographies of sexual life. By not writing in a linear manner, we question the conventional structures of academic papers and the "normal" formation of arguments, instead allowing our plural and partial arguments and understandings to emerge from the responses and relays among our various contributions to this chapter.

Why Don't Geographers Talk about Sex?

Gavin: A quarter of a century ago, early sexual geographers taunted their colleagues for their "squeamishness" about addressing the spatial aspects of sexuality (McNee 1984). Today, many introductory human geography textbooks include some reference to "gay space" and the geographies of sexual difference, and journal articles abound addressing these themes and more. Despite this proliferation of sexual geographies, the focus of the majority of this work is still concerned with the spatial expression of sexual identity. It seems there is still squeamishness surrounding the geographical study of sex itself, although some geographers have recently called for more attention to be paid to the actual sexual lives of sexual minorities (Bell 2007; although see Brown 2008b).

Queer theory has offered a robust challenge to the neat alignment of sexed bodies, sexual acts, and identities. We now know that we cannot take for granted that there are only two genders – male and female – that these neatly fit onto bodies with particular configurations of organs and orifices, or that they will "naturally" experience sexual desire for each other. This line of thinking has also highlighted that there can be a divergence between the sexual acts in which people engage and the sexual identities that they claim – "gay men" have been known to have sex with "lesbians," and there are plenty of "straight" men who will have sex with other men when the opportunity arises.

All of these phenomena have been well rehearsed by sexual geographers. It is relatively easy to find work that attends to those time-spaces where gender is "misread" (Browne 2004), or to sites where particular sexual possibilities become overlooked (Hemmings 2002) or where the opportunity for sexual encounters across identity categories are fostered (Nash and Bain 2007). And yet, despite increasing attention to issues of embodiment, emotions, and affect in geographical theory, very little work has yet been published that addresses the materiality and more-than-representational qualities of sex as a spatial practice (Brown 2008b). As Binnie (1997) has acknowledged, early in his career he censored more explicit discussions of sexual acts from his writing out of fear for his promotion prospects and hesitancy about what he was comfortable with his students imputing about his own sexual tastes. These are real concerns and not ones that have diminished in urgency over the last decade. However, I would suggest that sexual geography would be greatly enhanced by paying attention to the materiality of desire, theorizing how sex takes place in space, and analyzing how sex is enacted in very different ways (and accorded very different meanings) depending on where it takes place. Such work would have the potential of reconfiguring our understanding of sexual geographies without recourse to *a priori* sexual identity categories and would carry the potential for developing new configurations of sexual affinity.

Jason: Sexual lives are not simply defined by sexual identities and sex acts, but also by all sorts of practices, institutions, and relations – a list that might include but not be restricted to: marriage; mundane acts of affection and being with others; flirting; reproductive acts of all sorts, not just procreative sex; masturbation; pornography. The materiality of sex that Gavin draws our attention to will thus be very different in all of these aspects of sexual life. Of course, talking (literally and metaphorically) about the ins and outs of intimate sex acts has faced and continues to face a number of prohibitions. Questioning the purposes that these prohibitions serve has led to feminist, queer, and poststructuralist critiques of the distinction between private and public spaces, itself a part of the process of regulation of bodies (especially those of women and sexual minorities) and their proper relations and modes of expression. In my answer to the question "what is so wrong with being normal?" (below), I allude to how shifts in sexual norms redraw the boundaries of what is and is not prohibited. To take one of the examples I remark upon, the spread of pornographic tropes and images in mainstream western culture has already prompted analysis (albeit mainly outside of geography) of the changes that this spread of pornographic forms has had upon modes of subjectivity, sexual activity, sexual relationships, and relationships individuals have to their own bodies and their health. It might be argued, somewhat along the lines of Brian McNair's (2002) argument, that the spread of pornography is at least productive in the sense that it allows a space for sexual acts to be discussed and expressed in public, disrupting the forms of repression and prohibition whose main function is to sustain certain forms of patriarchy (and, one might add, heteronormativity). However, while I would agree that the "sexualization" of everyday cultures affords an opportunity for geographers to also talk meaningfully about sex, I do not think that the very fact of being able to talk more openly about sex is, in itself, liberatory. Rather, sexual acts themselves have effects, are put to uses, and become connected in assemblages of action and relation. For me, this recalls Foucault's argument with Deleuze (Deleuze 1994) about whether pleasure or desire – de Sade or von Sacher-Masoch – offered a greater promise of sexual freedom and innovation. In retrospect, it might appear that both Foucault and Deleuze were trying to do the same thing, but with different concepts, and for both, I would guess, the dangers of the current pornography-driven "sexualization" is that it enslaves bodies within organized and repetitive relations, within an already laid out set of pleasures and within a broader economy of representation. Much better, perhaps, if geographers are to talk about sex, for them to encourage Grosz's (2005) invocation to find sexuality in the *next* sexual encounter, untrammelled not only by prohibition, but by other kinds of enslavement.

Kath: I would disagree with the question: geographers do talk about sex, perhaps not in the ways that Gavin outlines, but in the implicit assumption of what sex is, what cruising is, what counts as sexual practice, morals, deviancy, and so on. So I would extend Gavin's call and build on Jason's appeal to the "next sexual encounter," and ask for a challenge to what is sex, how and where does sex come to be recognized, whose sex are we discussing and what counts as "sex." Why don't geographers deconstruct the phallic, climatic emphasis that is often assumed to be sex? I am thinking here of discussions – such as those found in studies of sex tourism – that often relate solely to male-female sex, usually assumed to be penetrative, always assumed to involve a penis and rarely addressing emotions, intimacy, and other forms of sexual and sexualized contact that is not necessarily genital. What of "lesbian" intimacy (and this is not simply restricted to lesbians, women, etc.), which could be used to deconstruct not just penetrative genital and phallic sex but also ethical considerations of how sexual affinities can and should occur? Such discussions can draw on the extensive range of feminist learning regarding experiments with (non-)monogamy, separatisms, and sex. Recognizing sex in kinships, we could also move beyond such ties, not only to explore other forms of kinships, but also other forms of sex. This can move beyond that which is queer, erotic, and even perhaps sexy. The question then could be, is sexual life necessarily premised on sex? Sexual lives can be formed of intimacies, kinship, and diverse relationship forms that are not premised on erotics, desire, and (sexual) attraction. In these contexts, when sexual life matters may vary not only spatially but also temporally over life-courses, for example stable relationships that move through (and beyond) phases of sex-ual lives.

The question could also be slightly rephrased to consider the sexual lives of sexualities geographers, the discussion of which could be anti-normative, challenging and indeed dangerous for all of the well rehearsed reasons of "revealing" personal sexual details (Binnie 1997; Vanderbeck 2005). Yet what if these "exciting" new ventures were really quite "normal" and "vanilla"? Just plain "boring"? Revelations of (homo)normative sexual practices in a field such as this would perhaps be more damning than discussing queer practices that don't just fuck, but fuck with kinkiness, gender, sexuality, taboos, the regulation of spaces, etc. Could one of the reasons we don't include the sexual lives of geographers be that they are not as interesting as the sexual lives, transgressions, and otherness of alternative, erotic, or kinky sexual practices and spaces? What are we doing by focusing on – even exoticizing – "other" "better" sexual lives? What more could we discuss about lesbian intimacies or reveal about ourselves if we started with our own sex(ual) lives?

What Is So Wrong with Being Normal?

Gavin: Since the inception of queer theory as a distinct body of work, queer theorists have attempted to identify and critique expressions of heteronormativity (Warner 1999) – those social practices by which heterosexuality is expressed as the dominant, ideal, and "normal" way of life. More recently, geographers and other scholars have turned their critical gaze to the "normalization" of lesbian and gay lives in Europe, North America, and elsewhere (Nast 2002; Richardson 2005). The targeting of lesbian and gay consumers by large corporations, the spread of gay marriage and civil partnerships, and the development of other "equalities" legislation have all been critiqued as expressions of "homonormativity," what Duggan (2002) has termed "the sexual politics of neoliberalism." For many critical queer scholars, the fear seems to be that the incorporation of lesbians and gay men into "the mainstream" in these countries has diluted the potential of sexual dissidence to challenge and transgress societal norms in a way that might inspire more progressive and egalitarian social relations. The question remains, are there any "authentically" queer spaces, or are even (self-declared) "radical" queer cultures complicit in the reproduction of a host of normative social relations (for example around class, ethnicity, and gender), as some have asserted (Oswin 2005; Puar 2007)?

My own response to these debates is complex and changing. I am concerned that critical scholarship about sexualities is increasingly disconnected from the lives of many (perhaps most) lesbians and gay men – as Weeks (2007: 9) has cautioned, we should "never underestimate the importance of being ordinary." Always living one's life on the margins of society and bearing the brunt of social disapproval carries a heavy emotional and psychological cost (just as it can also bring unexpected pleasures and benefits). And yet, I remain committed to developing a sexual politics that stresses points of connection, interdependence, and solidarity. What enthuses me more than "paranoid thinking" about (homo)normative complicity is an engagement with those spaces that engage with sexual and gender diversity as a means of prefiguratively experimenting with alternative, post-capitalist social and affective relations (Brown 2007).

I am also conscious that much of the scholarship that critiques "homonormativity" originates from a place of privilege in the metropolitan centers of the Global North. I want to see geography brought back into these studies to address the geographies of (homo)sexualities (and sexual normativities) in a far larger range of (local and national) settings than has been the case in recent years. What does it mean to be "normal" (or not) in small-town America, Krakow, Porto Alegre, or Tehran? How can social geographers engage with the "ordinariness" of socio-sexual relations in these places without imposing analytical models and political debates from the metropolitan centers of the Global North on these studies? (Brown 2008a).

Jason: It is important to ask, as Gavin has done, what uses the concepts of "normal" and "normativity" are put to. We might also ask, alongside these questions, other questions such as "who is normal?" and "what is normal?" In geography, writers such as Phil Hubbard (2000, 2007) have problematized the equation of heterosexuality and normality. After all, if certain heterosexual subjects (such as sex workers, absent fathers, and lone mothers) are sometimes popularly deemed to be "bad" and "other," and if certain practices (such as anal sex and bondage/ domination/sadism/mashochism: BDSM) deemed by some to be risqué are engaged in by heterosexuals as well as by lesbian, gay, and bisexual people, then heterosexuality cannot be considered to be monolithically normal. Moreover, popular understandings of what is "normal" (and who are "normal") change over time and space in line with broader changes in social and cultural practice. The transformation of mainstream sexual representations and practices in many western countries throughout the twentieth century has continued into the twenty-first century, more latterly involving shifts such as the influence of pornographic tropes on mainstream culture, among many others. Feminist critiques – both academic (e.g., Gill 2007) and popular (Levy 2005) – have pointed to some of the implications of such changes upon aspirations, practices, sexual subjectivities, and the relation of individuals to their own bodies.

One of the tendencies that have marked both the rise of so-called "postfeminist" forms of heterosexual culture and the rise of purportedly homonormative institutions and practices is the neoliberal appeal to the empowerment of the individual.

Gay marriage, civil partnerships, equalities legislation protecting workers and consumers on the basis of sexual orientation are understood as "rights gains," but it
is the individual who has gained these rights. Yet, many academics on the political
left in Europe remain critical of a political model based on struggles for rights,
questioning how such a model entails an implicit liberal conception of the subject,
the impoverishment of conceptions of the dynamic, social and embodied nature of
politics and ethics, and a false universalism that tends to project western terms onto
others (see Ahmed 1998). Recently, however, in the face of the erosion of certain
other civil rights that has taken place especially in the United Kingdom and the
United States under the auspices of the fight against terrorism, the importance of
struggling for and defending human rights is starting to be reassessed. The challenge
for geographers of sexualities, especially queer-influenced geographers, is to gauge
whether such a reassessment translates into the field of sexualities, whether rights
can become pragmatically deployed as a means for effecting certain progressive
ends, and, if so on these two fronts, what the implications are for critiques of
homonormativity.

Of course, this is not to belie the necessity for more strategies than a reliance on
rights discourse. However, if critiques of heteronormativity play a significant role
in political struggle, then the distinction Gavin draws between ordinariness, on the
one hand, and domination, idealization, and normativity, on the other hand, allows
the concept of heteronormativity to be used with more analytical precision. This
distinction between normativity and "being normal" allows the interrogation of
two sets of problems: firstly, those related to the *institutions* that sustain the domination, ideal, and normativity of certain forms of heterosexuality; and, secondly,
those pertaining to individuals and their relation to such institutions.

The problems associated with the institutionalization of heteronormativity
include the suffering and marginalization of sexual minorities that arise from the
force, violence, and persecution involved in policing sexual norms. Systems of sexual
normativity also restrict the freedom – of heterosexuals as well as of sexual minorities – to practice a diversity of sexual acts and relationships. Moreover, both heteronormative and homonormative institutions encourage consumerist lifestyles and
thus have economic effects in terms of the (over)exploitation of workers, farmers,
and resources. They also tend to perpetuate unequal and gendered divisions of labor,
and investments in individual prosperity as against the common good. In questioning heteronormativity and homonormativity, queer discourse has often observed
how the structuring of sexual lives is tied to the structuring of other aspects of social,
economic, and political life, and such observations remind us that although individualistic rights discourses may be important politically, they are by themselves
inadequate.

There is, nevertheless, value in thinking about the set of problems pertaining to
the relationship between individuals and these institutions, not least because it
allows us both to interrogate the costs of complicity with heteronormativity and
homonormativity, and to think about conditions that might encourage individuals
to enroll into anti-normative projects. Distinguishing the problem of heteronormativity from the problem of "being normal" allows for the possibility of individuals
both leading relatively ordinary sexual lives and simultaneously withholding support
from – or actively acting against – the institutions and practices that reinforce the

normativity of certain forms of heterosexuality. "Being normal" in the sense of leading an ordinary sexual life, then, does not necessarily imply complicity with the problematic policing of sexual norms or an investment in economic privilege and underprivilege. One corollary of this line of reasoning is that if queer practices and politics are understood to involve innovative sexual and gender practices and relations, a broader understanding of "sexual" and "gender" practices and relations than those pertaining to interpersonal, embodied, sexual acts and relationships must be maintained.

It is paradoxical that an acknowledgement of the social and economic dimensions of heteronormativity and homonormativity has the effect of focusing attention upon individuals who are enrolled into complicity. Should individuals who are part of the "mainstream" be asked to interrogate and challenge the bases for their own privilege? One only has to look at certain sections of the popular media to see that attempts to do so couched only in terms of rights discourses prompts those who think of themselves as among the "moral majority" to perceive of a revanchist attack upon them. The effect is a reactive and equally revanchist backlash. How might an anti-normative politics operate differently if it is to also challenge institutions at the level of the social and economic? It is with respect to this question that Gavin's invocation of Sedgwick's (2003) warning against "paranoid thinking" is perhaps most apt.

Kath: Jason elucidates the key distinction between the ordinariness of the individual and the force of normative systems, institutions, etc. He identifies the ultimate paradox of this distinction – namely, that institutions, systems, and so on are constituted through performative relations: we are all complicit! Yet even in this complicity, what is wrong with being normal is that it equates with a hierarchy of privilege that should be challenged. The search for individually and collectively abnormal, unusual, different, and non-normative practices, lives, and structures shows that these can occur in the DIY context of queer anarchism, or in the boardroom where lesbian, gay, bisexual, and trans is for the first time on the agenda. Conversely, there are many places where societal "normality" is the opposite of the dominant categories of that particular space. Thus, we can assert that what is normal is spatially, temporally, and socially contingent and specific. For those who are marginalized, excluded, and othered, finding a place in which to be and feel "normal" can be a lifelong goal. There is, as Gavin pointed out, power in the ordinary. Concurrently, being attentive to one's privilege, whatever this may look and feel like, is more than admitting one's place within and outside of "normal." However, such moves can be disempowering, as one can constantly feel the need to be apologetic for being and feeling happy in being "normal," or to prove in some way the contingency of the normal.

Whilst I enjoy reading and hearing about queer anarchist activities, I would never fit in or feel comfortable there (too normal – some might say prudish!). This could be a reflection of a certain sex radical chic that pervades those spaces, which can be off-putting if you aren't a happy pervert! Yet, I move into and out of categories, feelings, and moments of normality. In an Irish supermarket, when my mother introduces my female partner to someone who has just reminded me of how she pushed me around in my buggy I do not feel "normal." Striving to celebrate and discuss these moments once again moves the discussion (and attention) away, hides my embarrassment and legitimates me as at times queer, out of place and not normal. What is so wrong with being normal that I feel more comfortable writing about moments of abnormality, not the moment when my parents accepted my relationship or when we held hands in a supermarket or the thousands of other normal or normalizing moments?

Who and Where Can Be Queer?

Jason: This is a question already loaded with presuppositions. Primarily, the question implies that a person and a place can be identified as "queer." This, however, is not a neutral presupposition. Rather, it presumes a certain definition of "queer." If "queer" suggests the instability of the relationships among sexual and gender identities, sexual desire, and sexual practice (Browne et al. 2007: 8), then the question asks how the challenge to heteronormativity that a realization of this instability might pose is embodied in particular individuals and particular spaces, rather than in, for example, particular practices. Natalie Oswin (2008) argues that attempts to specify "queer spaces" confers on those spaces a homogeneity that cannot exist. All spaces, no matter what kinds of non-normative practices characterize them, are marked by diversities of practices (so that not all are non-normative) and by striations of difference, power, and exclusion (so that even the seemingly queerest spaces can be marked by classed, racial, and gendered differences and exclusions). Queer space, Oswin argues, is contested and disciplinary, just as any other space is. Similarly, attempts to specify who might count as a queer individual quickly run into trouble. Who is it that is "queer enough"? What is the minimum requirement that one has to meet to be counted as "queer"? Probe deeply enough and it will turn out that everybody has to make at least some compromises with power and the everyday norms of social and economic life just in order to get by. Asking who is queer ends up as a kind of competition where individuals are asked to punish themselves for not being "queer enough." It also runs the danger of letting heterosexual individuals off the hook in terms of challenging sexual and gendered norms and power relations: could a "straight" individual ever be "queer enough"? As Oswin (2008: 92) points out, attempts to specify queer individuals and queer spaces recuperates a fictitious resistance/oppression binary, but does so by shifting the burden of resistance from lesbians and gay men onto "queers"; yet, Oswin reminds us, as Foucault suggests, resistance is never in a position of exteriority to power.

By proposing instead "a usage of queer theory as an approach that critiques the class, race, and gender specific dimensions of homonormativities and heteronormativities" (2008: 96), Oswin invokes a mode of deploying queer theory advanced by, among others, Judith Butler (2004). Queer, in this view, becomes less a property to be attributed and more an approach to critiquing the uses of sexuality and the deployment of sexual and gendered norms. These norms have not only been used to marginalize lesbian, gay, and bisexual people. There are many heterosexual individuals and practices that remain non-normative (Hubbard 2002, 2007); sexual norms have been used to create the stereotype of the "terrorist" out of Arab and South Asian men and women (Puar 2007); and racialized stereotypes have been used to construct the very notions of "heterosexuality" and "homosexuality" (Somerville 2000; Eng 2001; Cohen 1997).

The argument made by Oswin, Butler, and others is undoubtedly a powerful one, and I have much sympathy for it. However, in deploying this understanding of "queer," certain questions need to be asked. For instance, does it imply a distinction between, on the one hand, academics and other "critical thinkers" and, on the other hand, "ordinary" people? If "queer" is to be deployed as a critical and interrogative approach, how can we guard against slipping into a situation where those with greater access to the relevant theoretical knowledges have a privileged position from which to apply such queer approaches? How will the practices and organizations of non-academics that also challenge the uses of sexual and gendered norms and that also call themselves "queer" sit in relation to academic practice? Queer activism, alliances, and organizations often seek a locus of commonality among their stakeholders – a point of identification – but they don't necessarily think that this identity exhausts what they do as a collective or as an organization. It is important that queer academic analyses acknowledge this *deployment* of a "queer identity," even as they interrogate what it does and how it also carries out the action of undermining sexual and gendered norms.

Kath: When we were writing this question, I was acutely conscious, as Jason has pointed out, that this question itself could be seen as inherently flawed. To argue that queer can be claimed as an identity and that someone or something can be "queer" – that "queer" can be located – is impossible if queer is anti-normative, critical, fluid, and spatially and temporally specific. I have been rightly charged in policing queer in a way that only sees its challenging non-normativity in theoretical ways (see Browne 2006). Queer is more than theoretical, and it is, as Gavin argues, creative. Queer, then, can be a lived experience and an empowering (if potentially limited, exclusionary, etc.) label. Who are we (as academics) to deny this usage or to critique any deployment of the term?

I moved away from understanding myself as "queer" with the advent of "homonormativity." Fitting in perhaps with Oswin's arguments, I didn't feel "good enough" to "be" queer. The places I went to and socialized in were not "queer"; rather they were apparently steeped in "homonormative privilege." Recognizing my "not queer" identity was for me about recognizing my power and privilege. For me, I could still use queer to open spaces for understanding and discussing manifestations of privilege and power among "us," rather than who or where inherently "is" queer. Yet I have found limits to this, and for years I wrote about "non-heterosexual women," seeing but challenging the term "lesbian" (see Browne 2004, 2007), until it became all too obvious that women and lesbians are continually lost, subsumed, and unnamed within queer, LGBT, and other critical moves (see Browne 2007). Using the term "lesbian" personally and in my writing is about challenging gendered norms, but often these moves are not seen as queer enough. After all, they can reiterate gendered binaries. I see that there is a need to recognize and discuss the power of lesbians, LGBT, queer, and academia, whilst retaining recognition of the often simultaneous disempowerment within these groupings. Currently, in reiterating the binaries that separate powerful from disempowered and that set homo- or hetero-normative in opposition to queer, we may omit the ongoing marginalizations that can coalesce around gendered and/or sexual lives. Such marginalization has been long documented, yet continues to be perpetrated. Recently, the need for collectivities has become increasingly apparent to me. The need to work together could augment ways of addressing (rather than just acknowledging and questioning) privilege and marginalization among us. This is not about finding a "them" to argue against, but rather about positive solidarities. Such a move is problematic, as it will never fully "succeed": it will always be ripe for critical analysis, particularly for any theory that finds "normativity" at every turn. Yet we need to question not only what deploying queer identity does, but also what such queer thinking does. What strongholds, exclusions, and otherings, and, conversely, solidarities and collectivities, are we missing in the searching for and challenging of "new" normativities?

Gavin: Having spent a lot of time over recent years hanging out, playing (and, occasionally, researching) in spaces that describe themselves as being for "queers of *all* sexualities and genders" (see Brown 2007 for a discussion of Queeruption gatherings and related events), and witnessing the variety of people who use those spaces (including several married and monogamous heterosexuals), Kath's comments have forced me to think about who is "not 'queer' enough." Certainly, these spaces (re)produce their own normativities and are complicit in the replication of complex exclusions around age, class, ethnicity, and (dis)ability. But, in other ways, they create a space where a diverse group of people can come together to find new expressions of commonality in the encounters facilitated by those "queer spaces."

In these times and places, people might identify as "queer," but in other arenas they may find that other identities are more acceptable or pertinent. In most aspects of our lives, people tend to read me and my partner as "gay men," even though we both have more affinity for the term "queer" than "gay." For my part, describing myself as queer acknowledges that my sexuality and my masculinity are more multifaceted than most people assume when they find out I have a long-term male partner. In (self-defined) queer spaces, I have no need to explain or justify that I am attracted to masculinity, not (biological) men. Flirtation happens in the encounter. And yet, it also occurs to me that one might need to be socialized in these types of queer spaces in order to fully understand and appreciate the fluid possibilities that can be acknowledged or explored within them. For me, queer is not so much an identity as an ethical stance of openness towards sexual and gendered difference. But, I wonder if this ethical stance can have much social effect unless it is reciprocated by others taking a similar stance? If one's ontological understanding of sexuality and gender appreciates fluidity and immanence, then "queer" (at least in these terms) might make sense. If, however, the ontological basis of one's understanding of sexuality is fixed by and within the hetero/homo binary, then it can be easy to mistake queer refusals to unambiguously identify (as "lesbian," "gay," or "straight") as the last resort of the "closeted" individual. Anecdotally, a gay male colleague from a local university recently told me that he interpreted some of his male students' choice to identify as "queer" as simply a trendy way for "straight boys" to have sex with other men when they were drunk. Perhaps it is. But, perhaps it also reveals that to claim "queer" is not an "either/or" choice made in opposition to other identity categories; instead (depending on the situation), it can be either a "both/and" choice or a "neither/nor" one.

Is Queer the Only Way to "Do" and "Be" Political (about or in Sexual Life)?

Kath: Sex shapes the socio-spatial world in uneven ways. The examination of the political effects of the practices and regulation of sex and sexual lives can involve both critiquing dominant power relationships, surveillances, and governances and examining processes of resistance, transgression, and "queering" of everyday lives. There can be no doubt that in the contemporary United Kingdom and United States the sexual lives of some lesbians and gay men are increasingly privileged whether through legislative change or through commercialization. Contesting such power relations is an important academic endeavor and one that often takes a queer perspective (see Browne 2006). It is also one that regularly engages with anarchy and the contesting of state and capitalist processes as problematic and normalizing.

What I often feel in international "queer" academic contexts is that the ubiquity of queer deconstructions, anti-normalizations, and/or anarchies can mean that other ways of "doing," in particular lesbian and gay, sexual politics are challenged or downplayed. The imperative to challenge all that is normal, generally, and often "the state" and all its normalizations, specifically, can mean that the work I do, for example, to improve the provision of services for lesbian, gay, bisexual, and transgender (LGBT) people (see www.countmeintoo.co.uk), is seen to facilitate state regulation and reiterate problematic identity categories. Similarly, certain forms of collectivities, such as Pride events, can be read (again, in international contexts of journal articles regarding international Pride events) as normative (with the implication that seeking to become "normal" is "wrong"). The local specificity of organizations and events such as Pride in Brighton & Hove, which engages with identity politics and uses large commercial organizations to celebrate LGBT lives, can be vilified as "selling out," lacking in critical or theoretical insights and "real" political effects (but see Browne 2007). Such critiques often ignore the nuanced engagements with a range of communities and the contestation of heteronormativity that organizations like Pride in Brighton & Hove undertake. Engaging with powerful institutions and organizations to effect change that involves compromise is often seen as less preferable to smashing institutions and critiques made from the perspective of the "bigger (global) picture." Conceptualizing the local contingent and spatialized effects of "colluding" with institutional powers including Pride organizations, local councils, and community groups cannot only take place within an anti-statist frame. Broader social geographies that recognize the place of identity politics, the importance of social policy research, and the spatialization of politics offer valuable insights for geographies of sexualities and can help facilitate nuanced engagements with sexual lives that continue to need and offer diverse political engagements.

Gavin: Before addressing some of the issues raised in the previous contribution, it seems important to ensure I stay focused on the main question: Is queer the only way to "do" and "be" political? The answer is clearly, "no." Quite apart from the numerous different interpretations of what a "queer" politics might be (that are shaped by local and national circumstances, as well as translated through numerous actor-networks), the dominant modes of "progressive" political engagements around sexuality in most of the world are liberal ones, which make an appeal for equality between men and women, and between people of different sexual subject positions.

Of course, I am not opposed to formal legal equality for lesbians and gay men; but I question the limits of this equality (Richardson 2005) and don't limit my political aspirations simply to negotiating the best possible terms of citizenship within the political status quo of liberal democracy.

But, let me return to the earlier points about the ubiquity of queer and anarchist critiques of liberal sexual politics. I think there *are* potentially important connections between queer deconstructive methodologies and some aspects of anarchist thinking. I also think some very important contributions to theorizations of sexual life have been written from an anarcha-queer perspective (Heckert 2004; Mattilda 2004; Wilkinson 2009). Conversely, queer theory has informed some very imaginative reconsiderations of how capitalism operates and the potential for post-capitalist politics (Gibson-Graham 2006). But, I contest the notion that anarchist-inspired queer analyses are ubiquitous, or that queer deconstruction and anarchism are the same thing. To say that seems to run the risk of overlooking the breadth of intersections between queer theory and political thought.

While I believe it is important to question and critique the numerous ways in which state institutions and corporations shape contemporary sexual lives, I agree that it is important to recognize ways in which same-sex marriage, civil partnerships, and state-provided welfare provision have given comfort and sustenance to many people in their sexual lives. An anti-statist perspective is not the only political insight that queer scholarship and activism can glean from anarchism. Many strands of contemporary anarchism have broadened that older tradition of anti-statism to encompass an opposition to all forms of domination (Gordon 2007) – a perspective that might owe something to an engagement with queer anti-essentialisms. But, I think if we are to talk about the coincidence of anarchism and some forms of radical queer activism, then it is important to consider not just what anarchism is *against* but what it proposes, what excites it. Anarchism has traditionally been critical of strong divisions of labor – through which some are presented as experts in a particular field and others are not. Politically, this approach can be applied in terms of not looking to the leaders of Stonewall or the Human Rights Commission for solutions, but finding these through grassroots experimentation and negotiation. It can also mean not waiting for local club promoters and gay entrepreneurs to put on a club night that suits your tastes (musical or sexual), but getting together with your friends, engaging in constructive direct action, and "doing it yourselves" – autonomously creating a social space that is not restricted by the sexual and cultural mores of the mainstream lesbian and gay scene. In the field of identity politics, this approach can be extended to resisting clearly demarcated identity categories and celebrating the fluid, contextual possibilities of human sexuality. This might also mean rethinking recent debates about homo- and heteronormativity, and recognizing that the sexual and social lives of "lesbians," "gay men," and "heterosexuals" are more interdependent and entwined in practice than most critical queer analyses can countenance.

Jason: Such is the current popularity of "queer" approaches within geographical studies of sexualities that it is sometimes easy to overlook the longer histories of other intellectual traditions within, more broadly, lesbian and gay studies and feminism that have been and continue to be used to think about contemporary sexual life. Not only do these other traditions provide frameworks for thinking intellectually about sexual life, but they also provide frameworks for political thought and action in the field of sexuality. Certain forms of feminist thought, for example, has been addressed to a wide range of struggles, among them contestations surrounding patriarchal and heterosexist institutions such as family, marriage, and prostitution, and the norms that sustain these institutions by differentiating among both women and men on the basis of a putative sexual morality. Some feminist thinking has also contested sexual oppression, repression, and violence that centers on the body, and has sought to increase bodies' capacities for pleasure. In these ways, not only has feminism shared many of the same concerns as queer theory and activism, but it is probably fair to say that many of the most prominent writers on queer politics – such as Judith Butler, Elizabeth Grosz, Sara Ahmed, or Gloria Anzaldúa – are simultaneously writers on feminist politics.

In the sense that there are other frameworks, such as feminism, for thinking about the politics of sexuality, queer is clearly not the only way in which to conduct such a politics. There is also another sense in which queer is not the only way in which to conduct a politics regarding sexual life. Queer theory and, to a somewhat lesser extent, queer activism has developed as an anti-normative mode of thinking and acting in a mainly Anglo-American context. Its arguments and its presuppositions arise from (and are pitted against) western historical forms – primarily, the binary opposition between heterosexuality and homosexuality, and the alignment among sex, gender, sexual desire, and sexual practice that coagulates into identities that define individuals. While it is the case that these institutions of heteronormativity and homophobia have been and continue to be spread beyond Europe and North America by colonialism and globalization (Binnie 2004), the complexity that arises from how these institutions become modified as they travel and from how they interact with local institutions and formations of sexuality can confound Anglocentric and Eurocentric understandings of queer. Just as critiques of the universalization of the idea of the "global gay" (Altman 2001) direct our attention towards how political struggles around sexuality involve the adoption, adaptation, and transformation of "gay" and "lesbian" identities in different locales (Binnie 2004), so our understanding of queer politics needs to acknowledge how activists outside of Europe and North America will adapt and transform the very bases of queer. It is not merely that, in specific places, activist groups and individuals reconstruct identities and ideas so that it becomes impossible to judge whose identities and ideas are "queer" or something else. Rather, such complexity precludes the possibility of defining a specifically queer way of living. Not only does it not make sense to invoke a queer subject (Oswin 2008), but neither can one appeal to a queer motif that can be applied to everyone or everywhere (see Binnie 2004: 80).

References

Ahmed, S. (1998) *Differences That Matter: Feminist Theory and Postmodernism*. Cambridge: Cambridge University Press.

Altman, D. (2001). *Global Sex*. Chicago: University of Chicago Press.

Andrucki, M. and Elder, G. (2007) Locating the state in queer space: GLBT non-profit organizations in Vermont, USA. *Social and Cultural Geography* 8 (1): 89–104.

Bell, D. (2007) Fucking geography, again. In K. Browne, J. Lim, and G. Brown (eds.), *Geographies of Sexualities: Theory, Practices and Politics*. Aldershot: Ashgate, pp. 81–6.

Binnie, J. (1997) Coming out of geography: towards a queer epistemology? *Environment and Planning D: Society and Space* 15: 223–37.

Binnie, J. (2004) *The Globalization of Sexuality*. London: Sage.

Binnie, J. and Skeggs, B. (2004) Cosmopolitan knowledge and the production and consumption of sexualized space: Manchester's gay village. *Sociological Review* 52: 39–61.

Brown, G. (2007) Mutinous eruptions: autonomous spaces of radical queer activism. *Environment and Planning A* 39 (11): 2685–98.

Brown, G. (2008a) Urban (homo)sexualities: ordinary cities, ordinary sexualities. *Geography Compass* 2 (4): 1215–31.

Brown, G. (2008b) Ceramics, clothing and other bodies: affective geographies of homoerotic cruising encounters. *Social and Cultural Geography* 9 (8): 915–32.

Brown, G. (2009) Thinking beyond homonormativity: performative explorations of diverse gay economies. *Environment and Planning A* 41: 1496–1510.

Browne, K. (2004) Genderism and the bathroom problem: (re)materialising sexed sites, (re)creating sexed bodies. *Gender, Place and Culture* 11: 331–46.

Browne, K. (2006) Challenging "queer" geographies. *Antipode* 38 (5): 885–93.

Browne, K. (2007) A party with politics?: (re)making LGBTQ pride spaces in Dublin and Brighton. *Social and Cultural Geography* 8 (1): 63–87.

Browne, K. and Lim, J. (2008) *Community Safety: Count Me in Too Additional Analysis Report*: Spectrum and the University of Brighton, www.spectrum-lgbt.org/cmiToo/downloads/CMIT_Safety_Report_Final_Feb08.pdf.

Browne, K., Lim, J., and Brown, G. (eds.) (2007) *Geographies of Sexualities: Theory, Practices and Politics*. Aldershot: Ashgate.

Butler, J. (2004) *Undoing Gender*. London: Routledge.

Casey, M. (2007) The queer unwanted and their undesirable otherness. In K. Browne, J. Lim, and G. Brown (eds.), *Geographies of Sexualities: Theory, Practices and Politics*. Aldershot: Ashgate, pp. 125–35.

Cohen, C. (1997) Punks, bulldaggers, and welfare queens: the radical potential of queer politics? *GLQ* 3: 437–65.

Deleuze, G. (1994) Désir et plaisir. *Magazine littéraire* 325 (October): 59–65.

Duggan, L. (2002) The new homonormativity: the sexual politics of Neoliberalism. In R. Castronovo and D.D. Nelson (eds.), *Materialising Democracy: Towards a Revitalized Cultural Politics*. Durham, NC: Duke University Press, pp. 175–94.

Eng, D. (2001) *Racial Castration: Managing Masculinity in Asian America*. Durham, NC: Duke University Press.

Gibson-Graham, J.K. (2006) *A Post-Capitalist Politics*. Minneapolis: University of Minnesota Press.

Gill, R. (2007) *Gender and the Media*. Cambridge: Polity.

Gordon, U. (2007) *Anarchy Alive! Anti-authoritarian Politics from Practice to Theory*. London: Pluto Press.

Grosz, E. (2005) *Time Travels: Feminism, Nature, Power*. Durham, NC: Duke University Press.

Heckert, J. (2004) Sexuality/identity/politics. In J. Purkis and J. Bowen (eds.), *Changing Anarchism: Anarchist Theory and Practice in a Global Age*. Manchester: Manchester University Press, pp. 101–16.

Hemmings, C. (2002) *Bisexual Spaces: A Geography of Sexuality and Gender*. London: Routledge.

Hubbard, P. (2000) Desire/disgust: mapping the moral contours of heterosexuality. *Progress in Human Geography* 24: 191–217.

Hubbard, P. (2002) Sexing the self: geographies of engagement and encounter. *Social and Cultural Geographies* 4 (3): 365–81.

Hubbard, P. (2007) Between transgression and complicity (or: can the straight guy have a queer eye). In K. Browne, J. Lim, and G. Brown (eds.), *Geographies of Sexualities: Theory, Practices and Politics*. Aldershot: Ashgate, pp. 151–6.

Johnston, L. and Valentine, G. (1995) Wherever I lay my girlfriend, that's my home: the performance and surveillance of lesbian identities in domestic environments. In D. Bell and G. Valentine (eds.), *Mapping Desire: Geographies of Sexualities*. London: Routledge, pp. 99–113.

Kitchin, R. and Lysaght, K. (2003) Heterosexism and the geographies of everyday life in Belfast, Northern Ireland. *Environment and Planning A* 35: 489–510.

Levy, A. (2005) *Female Chauvinist Pigs: Women and the Rise of Raunch Culture*. London: Pocket Books.

Lim, J. and Browne, K. (2009) Senses of gender. *Sociological Research Online* 14 (1): www.socresonline.org.uk/14/1/6.html.

Mattilda (ed.) (2004) *That's Revolting! Queer Strategies for Resisting Assimilation*. New York: Soft Skull Press.

McDowell, L. (1997) *Capital Culture: Gender at Work in the City*. Oxford: Blackwell.

McNair, B. (2002) *Striptease Culture: Sex, Media and the Democratization of Desire*. London: Routledge.

McNee, B. (1984) If you are squeamish. *East Lakes Geographer* 19: 16–27.

Nash, C.J. and Bain, A. (2007) "Reclaiming raunch"? Spatializing queer identities at Toronto women's bathhouse events. *Social and Cultural Geography* 8 (1): 47–62.

Nast, H.J. (2002) Queer patriarchies, queer racisms, international. *Antipode* 34 (5): 877–909.

Oswin, N. (2005) Towards radical geographies of complicit queer futures. *ACME: An International E-journal for Critical Geographies* 3 (2): 79–86.

Oswin, N. (2008) Critical geographies and the uses of sexuality: deconstructing queer space. *Progress in Human Geography* 32 (1): 89–103.

Puar, J.K. (2007) *Terrorist Assemblages: Homonationalism in Queer Times*. Durham, NC: Duke University Press.

Richardson, D. (2005) Desiring sameness? The rise of a neoliberal politics of normalization. *Antipode*, 37 (3): 515–35.

Sedgwick, E.K. (2003) *Touching Feeling: Affect, Pedagogy, Performativity*. Durham, NC: Duke University Press.

Somerville, S. (2000) *Queering the Color Line: Race and the Invention of Homosexuality in American Culture*. Durham, NC: Duke University Press.

Valentine, G. (1993) (Hetero)sexing space: lesbian perceptions and experiences of everyday spaces. *Environment and Planning D: Society and Space* 11: 395–413.

Vanderbeck, R. (2005) Masculinities and fieldwork: widening the discussion. *Gender, Place and Culture*, 12 (4): 387–402.

Warner, M. (1999) *The Trouble with Normal: Sex, Politics and the Ethics of Queer Life*. New York: Free Press.

Weeks, J. (2007) *The World We Have Won*. London: Routledge.

Wilkinson, E. (2009) The emotions least relevant to activism? Queering autonomous activism. *Emotion, Space and Society* 2 (1): 36–43.

Chapter 18

Emotional Life

Deborah Thien

1

> Any given moment – no matter how casual, how ordinary – is poised, full of gaping life. (Michaels 1996: 19)

I am 150 meters up the rock face and with a further 150 meters to go, I am not thinking about emotional life; I am feeling it. In this dramatic landscape of Yosemite National Park, California, United States, Lembert Dome climbs to 2,870.91 meters (9,419 feet) above sea level. Clinging to resistant granite, I feel terrified, I feel small, I am about to fall; I *am* panic. Immersed in bodily awareness, my heart lurches inside an unstable chamber. I cannot breathe. I sob without premeditation or conscious reflection, totally abandoned in my fear. All of me wants off this cliff face, yet I am pressed against this sun-warmed stone wall with all the passion of a lover. I am immobilized by "movement-feeling" (Adey 2008: 446). My companion looms above, offers his hand, reassuring words, suggestions for a foothold, then a handhold. I continue to exist in small, precise movements. Safely on top, to a sunset view of seemingly infinite mountain ranges, spectacular and rewarding, I am not moved by the splendid vista and refuse to look to the camera for the celebratory shot (see Figure 18.1). Instead, I contemplate my emotional ascent and its various implications: what is this emotional life? How to untangle its many elements? What role gender, and identity: feminized, feminist? What place the emotional body, when the "exploring body" (John and Metzo 2008) is required? What understandings are needed of relations with others: people, places, things? In succumbing to this terror on the rock face, and to subsequent rescue at the hands of my companion, am I simply reiterating a stereotype I would like to refute? "Of course!" I will reply later, indignant, when I share these thoughts and my companion mildly notes, "there are

A Companion to Social Geography, First Edition. Edited by
Vincent J. Del Casino Jr., Mary E. Thomas, Paul Cloke, and Ruth Panelli.
© 2011 Blackwell Publishing Ltd. Published 2011 by Blackwell Publishing Ltd.

Figure 18.1 On top of Lembert Dome, Yosemite National Park, California, United States. Photo by Gregory Ziolkowski.

plenty of women climbers." As I look unseeing at the red gold peaks, I wonder, am I rejecting this (masculinist) mountain or is it rejecting me?

To extend the "merely emotive" possibility of such an account (Mansvelt and Berg 2005: 257) requires the expansion of this narrative to the broader contours of emotional geography, wherein emotions are "understood as events that take-place in, and reverberate through, the real world and real beings" (Smith et al. 2009b: 2). My experience of touring Yosemite, for example, is more than the "inhale-exhale of landscape" (Lorimer 2007: 94); this is rarefied air, infused with the trade winds of asymmetrical social and historical forces. National park users tend to be a privileged bunch, and not due to the grandeur of a cinematic setting. The officious ranger who delivers the "wilderness ethics" lecture before dispensing our wilderness permits tells us that users of Yosemite are a typically educated cohort – in fact, that this land-use cohort is one of the most highly educated in the United States. To my chagrin, it appears I am one of many academics lurking in the so-called wild. I get his point that the wilderness area users tend to reflect a certain elite demographic (see Byrne and Wolch 2009), who, because of education, class (born into or acquired), and other "accidents of geography" (Kearns and Reid-Henry 2009) are predisposed to feel that "Nature," writ large, is compelling, an appropriate place to feel for. Hailing back to the Romantics, this Nature is a place to feel within, even when without, as for Wordsworth whose English countryside offered him "tranquil restoration" even in the city ("Lines: Composed a Few Miles Above Tintern Abbey, on Revisiting the Banks of the Wye During a Tour, July 13, 1798"). Such "middle- and upper-class sensibilities" together with a history of "eugenicist ideologies about pristine wilderness" are foundational to the American national park system, Jason Byrne and Jennifer Wolch note, and arguably work to exclude many on the basis of income and racialized identity, beginning from the earliest exclusions of native Americans by European settlers (Byrne and Wolch 2009: 5). Underpinned by these inequities, "specific spatial structures" – in this case, the dome and the park itself – "work to organize affect to have certain effects upon motion and emotion" (Adey 2008: 440), not least by positioning specific bodies in very particular social(ized) ways, whether helpless on a cliff, or in commanding possession of it. These social and emotional geographies are elemental to the feeling environment I describe.

The recent development of geography as more explicitly emotional invites a deeper consideration of emotional life, specifically in relation to "the social" and "the geographic." Such a consideration demands an engagement with interpretive and methodological frameworks, both in and outside of geography. We can attend to how and why we shape investigations of the uneven terrain of emotional life; that is, to question what kinds of (social, geographic) arguments we are making, what ontological and epistemological approaches we are turning towards or away from. We can look to "life itself" by way of empirical social geographic accounts; for example, how emotional life presents and what it represents; what emotional life reveals or conceals of social(ized) differences; culturally norma- tive emotional reflexes or gestures; the sociality of emotion; even, when and where emotion lacks.

In what follows, I offer an incursion – in its usual sense of "an undertaking," but also, with the emotion of the rock face lingering, inviting intimations of a rescue party of sorts – into the sometimes inexplicable, always tenuously occupied spaces, social and geographic, of emotional life. The aim: to give emotion some kind of matter, to bear witness to its matter(ing) for a social geographical project.

2

> what hubris to think that the body's reactions to another's emotions and affects are strictly within the realm of the personal and therefore devoid of academic/scientific interest. (Probyn 2005: 135)

Emotion shapes, impresses, moves – we are (differently) vulnerable to the pull of ordinary, emotion-filled life. The etymologically powerful intensity and energetics of this otherwise abstract concept incorporate a telling sweep of meanings: a moving out, migration, transference from one place to another; a moving, stirring, agitation, perturbation (in physical sense); a political or social agitation; a tumult, popular disturbance; any agitation or disturbance of mind, feeling, passion; any vehement or excited mental state; a mental "feeling" or "affection" (e.g., of pleasure or pain, desire or aversion, surprise, hope, or fear, etc.), as distinguished from cognitive or volitional states of consciousness (Oxford English Dictionary Online, 2004).

No wonder then, that emotion is something we all have fiercely held feelings about, and no less so because we are scientists of "the social" or "the geographic" (e.g., Anderson and Harrison 2006; McCormack 2006; Thien 2005a; Powell 2009). For the most part, writing about emotional life has been left to the lyrical skills of poets and prose writers, perhaps wisely. Scholarly treatises, dry almost by definition, can surely only reduce what excites about emotion to banal details and dull distinctions. Literary examples succeed, I suspect, because they convey emotions through making space for readers to feel, without over-telling; they evoke feeling by inviting readers into suggestive moments and unexpectedly supple forms of human experience where emotion bests sentiment, where emotionality takes privately felt if socially held shape. The "distinctiveness of literary expression" (Sharp 2000: 329) depends upon deftly drawn characters, finely rendered narratives, the details, but also I would argue, on our ability to recognize emotional depth and texture. Emotion *works* in story because it *is* story, whether imagined, actual, felt, sensed, visual, or virtual. Emotions grip because emotions *tell* us.

Emotions have been described as "seemingly instinctual, yet clearly a cultural achievement" (Sheller 2004: 225). Emotion, by definition, is not a static thing-in-itself, but relationally[1] constituted, dynamic, and so subject to shifts in position and relative power. R. Darren Gobert, for example, argues that we are all "theorists of emotion" as "the views of emotion that predominate in any given moment require and are constituted by their cultural capitulations" (2009: 73). As such, certain historical and cultural versions of emotion come to have primacy in academic and social terms. Historically notable, the ethos of the Enlightenment project (itself influenced by the Platonic tradition[2]) places emotion as the messy "other" half of ideal rationality. As such, emotions are rather absently-mindedly assigned to a list of "lesser" states such as being female (to its ideal counterpart, male), embodied (as opposed to the pristine world of mind), and subjective (a chaotic other to the ordered realm of objectivity). In this Enlightenment ontic of one and its shadow other, emotions are understood precisely as imprecise, uncontained, without rational edges or as overstepping or overflowing the limits of rational conduct – instinctive not deliberative.

Consider, for example, the chaos implied in each of the following expressions: "boundless grief," "wild with anger," "uncontrollable fury," and "overjoyed." Scientific inquiry has distinguished itself epistemologically in the practices of rationality, intellectual rigour, and (self)discipline; as a set of laws, science in this tradition dispenses with that which oversteps the bounds. Consequently, emotions have been both feminized and privatized within contemporary academies, and this "private/feminine [operates] as a category for containing emotions" (Harding and Pribram 2002: 419). Yet, Alison Jaggar has argued, contra western tradition, that emotion is not "epistemologically subversive"; instead, "the myth of dispassionate investigation has functioned historically to undermine the epistemic authority of women as well as other social groups associated culturally with emotion" (Jaggar 1989: 151). The Kantian subject of this investigative mythology has been epistemologically oppressive, valorizing the construction of some knowledges, while vaporizing others. Safely contained and effectively breathless, emotions can be set aside while the "real" academic labor goes on. Emotion in this reading is necessarily furtive, limited, paradoxically both out of control, and tightly controlled.

If this were the end of the story, this chapter would have little left to say. But not all academic disciplines have cut off or cut out emotion. Emotion as a topic of study has proliferated, if in cautious or stealthy ways. Williams argues that such issues have a "secret history" within sociology, which "just needs re-reading in a new more emotionally informed, corporeal light" (Williams 2001: 3). Why this emotional exposure now? Williams is not alone in attributing much of the outing of emotionality to steady feminist critiques, which have ably deconstructed the binary opposition of emotion to reason (e.g., Haraway 1988; Harding 1986). A consequence of such critiques is that "the disembodied illusions and (masculine) ideals of a rationally controllable world, "untainted" by emotions, are being exposed for what they are" (Williams 2001: 8). That is, to paraphrase Haraway's (1988) famous analysis, a God-trick is undone. Cultural theorist Kathleen Woodward (1996) goes deeper into the psyche, suggesting that the increasing circulation of emotion within scholarly discourses has a particular purpose: to "serve as compensation for the anesthetization of the emotions in everyday life – especially in academic life" (1996: 760). Consider how the very designation "social scientist" insists upon a distanced and distancing investigation of the social. Researchers are instructed to maintain a judicious distance between self and research "subject" to avoid obscuring thought-full judgment and instead to promote an emotion-less evaluative intelligence. What are the effects of such sundering of emotion from encounter, or for that matter, from intelligence? We are encouraged to formulate our theories with an objective eye and to avoid the subjective "I." We condone our own absence. While this Enlightenment legacy is stamped across the social sciences, within geography it has its own territorial flavour (Rose 1993), where "maps and chaps" (Eagleton 1997) are presumed to be natural, if stoic companions. At a glance, emotions might seem, therefore, expectedly absent from geography as a disciplinary practice – but, to paraphrase Davidson et al., we might better say that geography has historically had trouble expressing its feelings (Davidson, Smith, and Bondi 2005). Geography too has had its "secret life."

Collectively, if not too homogenously, the current emotional trend in social and critical thought defies the commonplace notion that emotion is out of place within academic research and practice. This burgeoning research area advances a (poststructuralist) challenge to a strictly rational science by addressing explicitly, academically, emotionally the spatialities of emotions. Following Woodward, perhaps academics are striving simply to enliven our working spaces, seeking a scholarship with affect(ion). Perhaps, in a "cool" atmosphere "saturated with stringent rules of emotionless rationality" (Woodward 1996: 760) wherein too much feeling is uncomfortably warm, repressed emotional matter necessarily finds a way to seep in, or out. Amidst bureaucratically constrained encounters, scholars of emotion can only attempt to transform their boredom, irritation, or anomie by tuning into something more sustaining. However we explain it, despite the risks, everybody's doing it, from social, cultural, and feminist geographers (e.g., Bondi 1999; Smith et al. 2009a; Wood 2002; Airey 2003; Bondi and Fewell 2003; Callard 2003; Thrift 2004); theorists of gender and culture (e.g., Chodorow 1999; Probyn 2005; Ahmed 2002; Harding and Pribram 2002; Sedgwick 2003); philosophers (e.g., Nussbaum 2001); and sociologists (e.g., Jamieson 1998; Hochschild 2000; Williams 2001); to the psychological disciplines (e.g., Matthis 2000; Coburn 2001; Blackman 2004), neuroscientists (e.g., Damasio 2000), and artists (e.g., Nold 2009). In this press of bodies, the prohibitions surrounding the "realm of the personal" (Probyn 2005: 135) are dissolving. Emotional geographies now encompass a steadily expanding and satisfyingly multi-disciplinary scholarship that combines the insights of geography, gender studies, cultural studies, sociology, anthropology, and other disciplines to understand how the world is mediated by feeling; that is, to make sense of this emotional life.

3

No one should be afraid to be theoretical. (Del Casino 2009: 16)

Social geographer Vincent Del Casino's statement is subtly provocative, combining as it does the emotional valences of fear with the contrasting sense of mastery evoked by "being theoretical." To be skillful with knowledge is to be confident, not fearful. Del Casino's intent is to encourage social geographers to explore and to think, yet the suggestive nature of the comment is also a reminder that the terrain of knowledge production is fraught indeed. Being theoretical is a risky business, subject to strict disciplining for the unruly. A feminist theoretical take on emotion is doubly fraught because of two distinct if related issues. First, feminist theorizings have deliberately shifted away from "expert" stances, allowing for tentative, partial understandings and thus requiring a level of acceptance of at least ambivalence with, if not fear about occupying theoretical positions (i.e. positions of epistemological privilege). Second, if the expert account is deliberately value-free and impersonal, there is a certain inevitability of emotionality in its countering: a non-expert, more personal account; yet feminist thinkers have maintained a careful distance from such notions of emotion (but see Jaggar 1989; Irigaray 1996) due to the degree to which it has been negatively feminized. The result, Elspeth Probyn suggests, is a "feminist suspicion of emotion" (2005: 9, 11), just a short pathological step away from paranoia.

A notable response to feminist anxiety about/critique of scholarship as distanced authority is an increasing intimacy of the subjects situated within (and telling the story of) academic texts (e.g., Ellis 1999, 2007). "Feminism, the critique of mastery, the contingency of self-presence and of language," writes Lauren Berlant in a trans-atlantic email interview, "all of that made questions of 'situated knowledge' ever more urgent" (Tyler and Loizidou 2000: 505). As a consequence, feminist scholar-ship in geography (e.g., Rose 1993; Wolf 1996) and elsewhere (e.g., Waring 1996) has intentionally incorporated personalized narratives as an element of scholarly texts. Much of this expression has detailed uneasiness about being the producers of knowledge. Feminist researchers have sought to manage this ambivalence by "naming" or "locating" a researcher's various subject positions. This practice of reflexivity aims, earnestly if not always perfectly, to offer "the view from a body, always a complex, contradictory, structuring, and structured body, versus the view from above, from nowhere, from simplicity" (Haraway 1988: 589). Such intentional embodied first persons have enlivened scholarship in complicated ways, not least by provoking (self) criticism of such personalized texts (see Rose 1997). Berlant (whose interview is itself an example of this personalized scholarship, wherein two other scholars engage her in "conversation" not "theory") comments on her suspi-cion of critiques of this autobiographical bent. She notes insightfully that such critiques are often directed at "the shamed classes, whose voices always seem to take up too much space and take it up so badly, in such an ungainly way" (Tyler and Loizidou 2000: 506). Scholarship rewritten for proximity, not distance, may be uncomfortable, and render its readers uncomfortable too because it does not fit the space accorded to a disembodied authorial voice: a miss-fit.

To further complicate matters, thinking reflexively, engaging with emotion as a feminist scholar/geographer requires a coming to grips with the overdetermined link between the feminine subject position and emotion, as if women would self-evidently be immersed in or even expert at emotion (see Berlant 2008). This association, reflecting long-established and well-documented dichotomies between the realm of the mind and the weight of the body, between masculinized reason and feminized feeling, has far-reaching effects for how and where people feel they can feel. Feminist thinkers have struggled with the social and cultural weight of the body, wrestling with sexual politics and gender in order to come to grips with and make changes within the complexities of a patriarchal world. The long-standing fusion of women with emotion, dismissively cementing emotion as the weak feminine opposite to a masculine intellect, has rendered emotion too risky a topic for serious feminist study. Mona Domosh, considering sexualized spaces notes "a long and often unquestioned tradition of identifying women ... as bodily disruptions to the "normal" course of rational knowledge-making" (Domosh 1999: 429). As a result, she argues, feminist geographers studiously avoided anything that might "reinforce our cultural assumptions about women as always denoting the sexual," and might delegitimize the authority of what until recently, she argues, has been a fairly precarious "body" of knowledge (Domosh 1999: 429). But this feminist aversion to emotion seems to be fragmenting. To research emotions has been, and possibly continues to be, to risk the dismissive gesture, the de-legitimization, but importantly it is also an effort to refuse it. And certainly, the proliferation of feminist scholarship on both sexualities and emotions shows that shifting ground. This reframing, releasing, reconsidering of emotion is an exciting moment for (twenty-first century, western) feminist scholarship.

4

> Everyone knows what the female complaint is: women live for love, and love is the gift that keeps on taking. (Berlant 2008: 1)[3]

The most cursory look at western popular culture demonstrates powerful associations of femininity with emotion. This is by no means to say that women are "naturally" creatures of emotion, nor that men are unfeeling, but rather, that powerful subject positions and social geographies characterize emotional life: women are figured as experts of emotion; emotion is women's terrain. As with any such positionalities, the Womanly figure lives (or dies) in relation to the social norms and spatialized practices that buoy Her up – that is, at the heart of "women is emotion" is an emptiness which is filled in, in different ways, by different means to generate meaningful, recognizable subjects. Berlant argues (elaborating on Spivak 1987): "Any concept that is considered a universal must take on the form of this emptiness, this indeterminate and proliferative abstraction, while appearing to assume a formal continuity that surpasses any particular context" (Berlant 2001: 43). Love, of all the emotions, is the one most strongly coded as feminine – as the epitaph above suggests, for women, love and its various labors are central to life itself. Love fills the emptiness. This universalizing over-association of women with love, love as embodied by and giving meaning to women, is precisely why love's spatialities deserve critical consideration.

The concept of ontology refers to that by which we "make sense," a theory of being. No wonder then, that a search for ontological meaning is an overly popular scholarly pursuit (Barnett 2008). If love can be cast as a theory of emotional knowing by which feelings are made sensible, by which feelings make us sensible, than love is itself an ontology: "The connection between to know and to love is also intimate etymologically" (Arman and Rehnsfeldt 2006: 8). Love makes intimate knowledge of particular places, is comprised of specific spatial practices, and attaches one to an other, whether in affection or despair. Getting love right is at once straight-forward and fraught. Love, arguably, is a "socially approved emotion" (Woodward 1996: 770) with recognizable parts to play in specific places. When love is out of place, for example in the case of the (oxymoronic) caring professional, the nurse, love is precisely not professional and instead may be "threatening," "weak," or "going beyond one's duty" (Arman and Rehnsfeldt 2006: 6, 7).

Conventions of romantic love delimit where love takes place and with whom, from trysts for two to material and virtual spaces for "washed-out love" (Vištica and Grubišic 2009), such as the Museum of Broken Relationships. Lauren Berlant's insightful exploration of the "female complaint" suggests, to value romantic love is at once to be womanly and to know loss, a familiar script of "woman as lack." Romantic love offers meager space for such feminized subjects passively subjected; "intimate publics," in Berlant's terms, are the result, wherein members are believed to share a common emotional story, leaving those outside the normative plot per-sonally adrift (Berlant 2008: viii). While law is the father's domain, the normative story of familial love stems from mothering. To not know this form of love, or to be unable to correctly perform it, is to be similarly emotionally stranded. In his affecting analysis of fathership, Stuart Aitken has argued: "The crisis of fathering [...] is that men do not recognize themselves in contemporary cultures of fatherhood that draw on motherhood as a benchmark and patriarchy as a citational practice" (Aitken 2000: 597). Not only, then, are men normatively prevented from mother-love, but also, where father-law reigns, father-love is contradictory and therefore impossible. Aitken worries about the effects of such a lack of recognizable parenting models for men, coupled with an ongoing performance of patriarchal fictions which manifest as "common sense" (e.g., men as incidental "helper" around the house and not invaluable helpmeet within).

As an aspect of emotional life, love is al/most ordinary, certainly banal, and yet (also) exudes a deep visceral pull: the inhale-exhale of emotion (to paraphrase Lorimer 2007). More than this, love is arguably "the ontological basis for caring and ethical acts" (Arman and Rehnsfeldt 2006: 11). Love is paradoxical; as Arman and Rehnsfeldt note, the concept is traditionally separated into two oppositional elements: "*Eros*, demanding and pleasure-seeking love, and agape, an unselfish love towards one's fellow human being and humanity" (Arman and Rehnsfeldt 2006: 8). And yet, love, feminist philosopher Luce Irigaray argues in *I Love to You*, is a question we need to answer.[4] In her analysis, a statement of subjugation is embedded within contemporary love. Irigaray performs a grammatical exposé, re-figuring the conventional phrase, "I love you," as that which seeks "you" as an object. Love, in this paradigm, signals the intent to possess within an "affective economy" (Ahmed 2002):

> Whether it is a question of our bodies or our words, we remain subject to the power or hierarchy of the one who possesses, of the one who has more or less – knowledge or sex as well as wealth – of the one who can give or make some thing, in an economy of relations (especially amorous ones) subordinate to the object, to objects, to having. (Irigaray 1996: 129–30)

Irigaray recasts "I love you" as "I love to you" in order to discourage this element of subjugation and to encourage a respectful offering, a move towards a "syntax of communication" (p. 113). By incorporating "to" and transforming the transitive (a verb that requires an object) to the intransitive (a verb that does not), she argues that an intersubjective relationship can be maintained between "I" and "you" such that neither party is possessing or possessed. This is significant for the predominantly patriarchal language of loving: "The being is thus never the whole and is always separate (from) inasmuch as it is a function of gender. It cannot, therefore, be in a state of fusion, either in childhood or in love" (p. 107). This maintenance of *autonomy in relation to* is central to Irigaray's theorizing of gender, self, and love (see also Irigaray 2000).

Does this then, solve the painful conundrums of father-love, the hot mess of intimate publics, or the cool temperatures of the workplace? Relationality, itself coded female (see Bondi's 2009 arguments about reflexivity, relationality's methodological doppelganger), is a much vaunted model for feminist theorising utilised precisely because it opens up the idea of fixed identities and fixed relations and instead suggests "we" and the parameters of "we-ness" are fluid, dynamic, and continually (re)situated. In this theoretical frame, love is about more than predetermined social roles and spaces, and instead encompasses the relationships between those subject positions and the dynamic subjectivities that may be expressed by individual women and men. Love trans-formed does not simply fill the emptiness. Women have complex and non-innocent relationships with love made manifest in a multitude of diverse ways – witness the woman clinging to the cliff face.

As an ontological position(ing), such intersubjectivity specifically acknowledges the relational nature of subjectivity, and so admits this diversity. This admittance of difference is integral to the "spatial imperative" of subjectivity:

> Instead of plastering over those differences, we need to stop and address them. Sometimes that stopping will result in silence. And that slash between dis/connections should indicate a pause – a moment of non-recognition that may be expressed as simply as "wow, you really are different from me." (Probyn 2003: 298)

To take one's own measure by moving through others' spaces. As Probyn points out, we see our selves as private projects, but in fact our subjectivities are intensely communal, "a public affair" (2003: 290). We cannot have it otherwise. Love offers one such expression of this publicly private state: "Love may perhaps require secrecy, but it also needs culture and a social context ... Such progress is necessary for the development of the human order" (Irigaray 1993: 104). Love, unbound from its binary roots, is a complex moving target, a collaborative social sculpture. Love, simultaneously of the mind and of the body, offers an everyday metaphysics that can work to sideline easy binaries in favor of more complex (differently mobile and gendered) array of always spatial subjectivities (see Probyn 2003). Herein lies the potential for ontological shakedown (but compare Callard's (2003) cautionary notes on the impossibility of infinitely malleable subjectivities). Thinking through the affective shadings, social configurations, and geographical spacings of love offers a productive if paradoxical place for shaking up and reconfiguring ontological tensions in social geographic thought. As an emotion in motion, love travels. Love traces a movement between; it is a means to find our way through the intersubjective, shared spaces of ordinary, extraordinary, emotional life.

5

> Emotions do things, [...] they align individuals with communities – or bodily space with social space – through the very intensity of their attachments. (Ahmed 2001: 11)

To consider the specificity and diversity of emotional life – living/doing/feeling – requires a breadth of methodological approaches that social geographers already use routinely and expertly in their research, for example, close observation, reflexivity, ethnographic-style interviews, and so on. I have used such techniques for gaining understandings of well-being as expressed in interviews and for examining how people narrate an emotional self via a process of placing the self, whether in material or symbolic ways. I have touched on therapeutic tactics (broadly understood to include, for example, empathy and understanding, following Bondi 2003) for eliciting narratives, whether about intimacy or mental health care (Thien 2005b, 2005c). As I have continued to examine emotions as a feminist, a geographer but not as a therapeutic practitioner, I have also wondered what else might be necessary to trace what is "emotional," to explain how or why various emotional geographies take shape in the everyday, those practices of social spaces and relations.

   In my research about the emotional spaces of the Royal Canadian Legion, a longstanding veteran's organization, I have searched for ways to more adequately convey the depths of emotional texture which the Legion branches hold, particularly in small towns in rural or remote Canadian places. Legion branches are places where Legion members and guests gather for a drink, conversation, a game of pool or a game of Bingo, and places specifically designated to remember lost lives. They are also places where people *feel something*, and furthermore, they are places that even in the telling, are affecting places. In the terms of Donald Winnicott, the quietly influential[5] pediatrician and child analyst, a "facilitative environment" is made, wherein one can be alone but in shared presence with others. In trying to understand what this means for those who variously inhabit the Legions, I have found I also need to consider what it means for a research process.

   I came to dwell on the notion of resonance as a means to consider the emotional environment such as the Legion. Indeed, doing this kind of ethnographic research is itself always an emotional experience (Bosco 2007: 547). I came to this out of being there, or rather, feeling there. The Legions *feel* like places of memory. Remembrance as a national (and nationalist) project is the *raison d'être* of the Legion; their motto "*Lest we forget*" intentionally reminds members of the pain and loss of war, to take care not to forget sacrifices made by untold soldiers, to be thankful for life. As remembrance specialists, the Legion takes centre stage each year at the November 11, Remembrance Day Ceremonies which take place at Legions and cenotaphs around the country. Undoubtedly, this is in part why the Legion branches are deeply resonant places – there are shades here, in the Branches' displays of war memorabilia, and family histories, of death and dying, of, as Audre Lorde writes in her Cancer Journals, "the war we are all waging with the forces of death subtle and otherwise, conscious or not" (Lorde, 1997: 19). Feelings about dying are perhaps the most ordinary of affects (to paraphrase Stewart 2007). All this in a context where the Branches themselves are struggling for survival as the numbers of World War II and Korean War veterans dwindle (as of last year the average age of World War II vets was 82) and some Legions have closed their doors for lack of members.

Resonance is something of an attempt to further qualify empathy, deftly discussed by Liz Bondi in particular (Bondi 2003), by giving it (a) feeling. The notion of resonance encompasses the potential for intensity, for the felt amplification of emotional material. This is particularly salient in the case of the Legion's legitimating of distress and loss for a mainly male clientele. "People think of the Legion as bunch of old men sitting around drinking and telling war stories" a Legion member noted in 2007; while this is not strictly "true," the latent possibility, the potential space (to use another of Winnicot's terms) for men to tell stories painful and sad, is powerful: "In many cultures, men are not expected to speak of grievances. The traumas from war often follow men silently, resulting in unattended physical and/or mental disorder and difficulties in adjusting to the post-war life. Alcoholism, domestic violence, and criminality among men are common phenomena in the aftermath of war." (Brounéus 2003: 10). The normatively militaristic and masculinist Legion spaces popularly associated with heteronormatively "manly" attitudes[6], paradoxically offer a significant space for male feeling in a manner readily accorded to a feminized repertoire of emotionality (that of caring, sharing of emotions, emotionality). Masculinity is privileged as a productive, normative discourse of Legion business, but strong emotional responses not culturally associated with militarism or masculinity, such as ready tears, and the willingness to recount emotionally difficult stories take place here. And much takes place outside of speech, inside fulsome silences.

Writing in a post-9/11 America, Judith Butler argues for feeling as a necessary political goal. Specifically, she writes "without the capacity to mourn, we lose that keener sense of life we need in order to oppose violence" (Butler 2004, xviii). To regain our sense(s), she suggests we ask: "Who counts as human? Whose lives count as lives? And finally, What *makes for a grievable life?*" (Butler 2004, 20, emphasis in the original). Philosopher Martha Nussbaum describes emotions as precisely about love and loss, loving and grieving: "Being able to have attachments to things and people outside ourselves; to love those who love and care for us, to grieve at their absence; in general, to love, to grieve, to experience longing, gratitude, and justified anger" (2007: 23). The legion's resonant spaces are not pure spaces for shared love or shared loss, in the sense of harmonious, or even harmless, reminiscence; "authentic" memories. Among the "effects of spatial proximity," Ash Amin lists "anger, dissonance, distance, visibility, encounter, evasion, mobility, non-correspondence" (Amin 2004: 43); such is the friction inherent in "zones of awkward engagement, creative and unstable places" (Tsing 2005: xi). But, what is clear is that the Legions are places where people *feel something*. An emotional resonance reverberates across varying intensities and differentiated emotional responses. While times, and Legions, have changed, emotional resonances linger on in the material spaces, feelings and practices of differentiated Legion users in local, place-specific branches. Such spatial and emotional dynamism reveals the energetic processes of place-making and their role in emergent, uneven, and sometimes unexpected feeling subjects. Resonance, with its intimations of simultaneous tactility and ineffability is a means of touching upon what is both substantial and shadowy, about feeling in touch, in place, about place resonance. Is this, then, a call to measure "resonance" as methodological strategy? I do not wish to suggest this is something easily measured, or neatly contained; rather this is precisely why I find the idea useful, because it foregrounds the very difficulty of assessing emotionality.

This specific set of (non-identical) places, Legion branches, are not simply aging sets for predictable activity. Considering them as resonant places makes more explicit the ways in which they are "meeting points or conjunctions of experiencing life" (Coburn 2001: 308) in affective ways fashioned in often highly emotive moments, vis-à-vis a multi-sensory, psychic, and material becoming. Explicitly considering resonance offers a means to examine emotional geographies in a way that moves us away from the "inner/outer quandaries" (Beatty 2005: 34) that have beleaguered studies of emotion; that is, away from emotion as a purely internal state (Davidson, Smith, and Bondi 2005) and places us instead in the tangle of the here-and-now. Such a framing captures something of the dynamism of emotion, the active constitution of psychic, social, and spatial relations, and acknowledges the important role of place in subjectivity. Where we are feeling and feeling where we are – the resonance of place.

6

> It is as though the details do not, after all, (have) matter, like ghosts ... the details are at best asterisks or epitaphs to fantasies of will in the privatized world. (Berlant 2001: 55)

On the most American of holidays, July 4, I awake in our verdant woodland campsite in this most iconic of American destinations, Yosemite National Park. It is wonderful and strange. My companion and I, neither of us American, find ourselves in a compelling emotional discussion with a Persian family (latterly of Orange County, California) camped alongside us. In the ordinary manner that conversations wend their way, we are all the while intermittently packing up the tent (us) and cooking breakfast (them). We pull up stakes as we respond to their curious inquiry ("philosophy?") about our previous evening's campfire reading (unbeknownst to us, all three tents were listening as we took turns reading aloud Alain de Botton's *The Art of Travel* into the night). A woman prepares coffee, another sets the table for breakfast, we talk. A passionate dialogue ensues about the Iranian revolution currently taking place in exciting, momentous, epoch-defining ways, about communism, socialism, capitalism. We ruminate on the sometimes disappointing margins of professorial radicalism. The Persian family patriarch recommends we read Harvey's *The Limits of Capital*, an unexpectedly geographical citation for two geographers in the so-called wild. When we conclude with best wishes all around for the American Independence Day, I feel there are no shortage of sharp ironies to be found in our various manifestations of foreign-ness; my official status as an "alien" work visa holder, for example. We farewell one another with the kind of genuine fondness that arises out of the emotional immediacy of such brief connections.

Ordinary life is full of such social encounters, always illuminated by geographic contexts and subtexts, and resonating and reverberating with emotional associations. Full of assumptions, suppositions, and questions too ...

> Emotions are ... intimately and inescapably caught up in the current re-writing of the earth, the production of new, transformed, geographies, and New World Orders, that affect us all ... And yet the question of how we might *feel* as well as think about these transformations is seldom addressed and this is (only one reason) why the world needs emotional geographies and why geography needs to take emotions seriously. (Smith et al. 2009b: 3)

Emotions affectively shade the formerly sharp line between mental and visceral experience (Laurier and Parr 2000; Chodorow 1999; Ahmed 2001), allowing us to trouble[7] the everyday metaphysics of mind versus body, emotion versus reason, which may have begun with Descartes, but which we owe to centuries of studious repetition. Emotions, whether happiness, despair, or grief may be realized in the body: "the internal milieu, the viscera and the musculoskeletal system" and the brain (Damasio 2001: 781), but bodies are always and differently placed, porous (Davidson 2000). Such distinctions matter. Emotional bodies are culturally intelligible bodies – a head tipped down may indicate not only fear or amusement, but also the gesture may be read as one of shame or respect, depending upon the context. Indeed, the notion that knowledge is power is only true for some; for others knowledge may be painfully born: "Dissident knowledge by women (and subjects conventionally overidentified with the body) always bears the burden of its apparent failure to be impersonal *enough*" (Berlant 2001: 48). Knowledge *is* (always, too) emotion.

Coming down from Lembert Dome proved much easier than the terror-filled ascent. The trail, a series of wide and sharp switchbacks allowed for a generous stride and quiet reflection as we made our way down the dusky trail. Lodgepole pine and mountain hemlock, gave way to California red fir, Jeffrey pines, and the Sierra juniper, illuminated by last rays of the day's sun. Feelings fading, traces lingering, telling us. An attention to emotional life, its social geographies, ensures (to paraphrase Berlant above) that the details are not ghosts, that they (have) matter. This is not to end the story, whether of social geography, of love, Legions, museums, or mountain ranges. I will (continue to) do geography as if emotion matters, because, so simply and resolutely, it does.

Mammoth Lakes, California
July 2009

Notes

1   In the context of a psychoanalytically inspired theory of human subjectivity, "relational" means that that which is in relation is not fixed or stable, and as such the relation is a contingent process instead of an assumed, factual reality.
2   See Songe-Møller (2002) for a discussion of "masculine democracy" within the Athenian city state and women's corresponding construction as outsiders. She addresses the gendered dichotomy of unity and plurality within which the masculine principle was unified and the feminine, disharmonious.
3   Berlant goes on to note: "of course that's a simplifying phrase: but it's not false, just partial."
4   The following comments on Irigaray draw from a previous essay on Irigaray and love (Thien 2004).
5   As Martha Nussbaum puts it, "There is no school of Winnicott" (2006: 375).
6   Thus, while militarism is seemingly inseparable from the ideals of masculinity (Enloe 2000), there is room to maneuver.
7   Many have noted the impossibility of outright rejecting dualisms given their entrenched position in our ways of thinking and living (e.g., Irigaray 2000). Davidson (2002) has argued further that dualistic frameworks can assist people who are experiencing identity difficulties while noting that this does not mean we should refrain from questioning and problematizing such binary worldviews.

References

Adey, P. (2008) Airports, mobility and the calculative architecture of affective control. *Geoforum* 39 (1): 438–51.

Ahmed, S. (2001) Communities that feel: intensity, difference and attachment. Paper read at Affective Encounters: Rethinking Embodiment in Feminist Media Studies at University of Turku, School of Art, Literature and Music, Media Studies.

Ahmed, S. (2002) Affective economies, keynote presentation. Paper read at Emotional Geographies, at Lancaster, UK.

Airey, L. (2003) "Nae as nice a scheme as it used to be": lay accounts of neighbourhood incivilities and well-being. *Health and Place* 9 (2): 129–37.

Aitken, S.C. (2000) Fathering and faltering: "Sorry, but you don't have the necessary accoutrements." *Environment and Planning A* 32 (4): 581–98.

Amin, A. (2004) Regions unbound: towards a new politics of place. *Geografiska Annaler Series B* 86 (1): 33–44.

Anderson, B. and Harrison, P. (2006) Questioning affect and emotion. *Area* 38 (3): 333–5.

Arman, M. and Rehnsfeldt, A. (2006) The presence of love in ethical caring. *Nursing Forum* 41 (1): 4–12.

Barnett, C. (2008) Political affects in public space: normative blind-spots in non-representational ontologies. *Transactions of the Institute of British Geographers* 33 (2): 186–200.

Beatty, A. (2005) Emotions in the field: what are we talking about? *Journal of the Royal Anthropological Institute* 11 (1): 17–37.

Berlant, L. (2001) Trauma and ineloquence. *Cultural Values* 5 (1): 41–58.

Berlant, L. (2008) *The Female Complaint: The Unfinished Business of Sentimentality in American Culture*. Durham, NC: Duke University Press.

Blackman, L. (2004) Self-help, media cultures and the production of female psychopathology. *European Journal of Cultural Studies* 7 (2): 219–36.

Bondi, L. (1999) Stages on journeys: some remarks about human geography and psycho-therapeutic practice. *Professional Geographer* 51 (1): 11–24.

Bondi, L. (2003) Empathy and identification: conceptual resources for feminist fieldwork. *ACME* 2 (1): 64–76.

Bondi, L. (2009) Teaching reflexivity: undoing or reinscribing habits of gender? *Journal of Geography in Higher Education* 33 (3): 327–37.

Bondi, L. and Fewell, J. (2003) "Unlocking the cage door": the spatiality of counselling. *Social and Cultural Geography* 4 (4): 527–48.

Bosco, F.J. (2007) Emotions that build networks: geographies of human rights movements in Argentina and beyond. *Tijdschrift Voor Economische En Sociale Geografie* 98 (5): 545–63.

Brounéus, K. (2003) *Reconiliation: Theory and Practice for Development Cooperation*. Uppsala: Swedish International Development Cooperation Agency, Department of Peace and Conflict Research.

Butler, J. (2004) *Precarious Life: The Powers of Mourning and Violence*. London and New York: Verso.

Byrne, J. and Wolch, J. (2009) Nature, race, and parks: past research and future directions for geographic research. *Progress in Human Geography* 33.

Callard, F. (2003) The taming of psychoanalysis in geography. *Social and Cultural Geography* 4 (3): 295–312.

Chodorow, N. (1999) *The Power of Feelings: Personal Meaning in Psychoanalysis, Gender, and Culture*. New Haven and London: Yale University Press.

Coburn, W.J. (2001) Subjectivity, emotional resonance, and the sense of the real. *Psychoanalytic Psychology* 18 (2): 303–319.

Damasio, A. (2000) *The Feeling of What Happens: Body and Emotion in the Making of Consciousness*. London: Vintage.

Damasio, A. (2001) Fundamental feelings. *Nature* 413 (6858): 781.

Davidson, J. (2000) "… the world was getting smaller": women, agoraphobia and bodily boundaries. *Area* 32 (1): 31–40.

Davidson, J. (2002) Women, agoraphobia, and self-help. In L. Bondi (ed.), *Subjectivities, Knowledges, and Feminist Geographies*. Lanham and Oxford: Rowan & Littlefield, pp. 15–33.

Davidson, J., Smith, M., and Bondi, L. (eds.) (2005) *Emotional Geographies*. Aldershot: Ashgate.

Del Casino, V.J. (2009) *Social Geography: A Critical Introduction*. Oxford and Malden: Wiley-Blackwell.

Domosh, M. (1999) Sexing feminist geography. *Progress in Human Geography* 23 (3): 429–36.

Eagleton, T. (1997) International books of the year. *Times Literary Supplement* December 5.

Ellis, C. (1999) Heartful autoethnography. *Qualitative Health Research* 9 (5): 669.

Ellis, C. (2007) Telling secrets, revealing lives: relational ethics in research with intimate others. *Qualitative Inquiry* 13 (1): 3–29.

Gobert, R.D. (2009) Historicizing emotion: the case of Freudian hysteria and Aristotelian "purgation." In M. Smith, J. Davidson, L. Cameron, and L. Bondi (eds.), *Emotion, Place and Culture*. Farnham: Ashgate, pp. 59–76.

Haraway, D.J. (1988) Situated knowledges: the science question in feminism and the privilege of partial perspective. *Feminist Studies* 3: 575–99.

Harding, J. and Pribram, E.D. (2002) The power of feeling: locating emotions in culture. *European Journal of Cultural Studies* 5 (4): 407–26.

Harding, S.G. (1986) *The Science Question in Feminism*. Ithaca: Cornell University Press.

Hochschild, A.R. (2000) Global care chains and emotional surplus value. In A. Giddens and W. Hutton (eds.), *On the Edge: Living with Global Capitalism*. London: Jonathan Cape, pp. 130–46.

Irigaray, L. (1993) *Je, tu, nous : toward a culture of difference*. London and New York: Routledge.

Irigaray, L. (1996) *I Love to You: Sketch of a Possible Felicity in History*. New York and London: Routledge.

Irigaray, L. (2000) *To Be Two*. London: Athlone.

Jaggar, A.M. (1989) Love and knowledge: emotion in feminist epistemology. *Inquiry: An Interdisciplinary Journal of Philosophy* 32 (2): 151–76.

Jamieson, L. (1998. *Intimacy : personal relationships in modern societies*. Oxford: Polity Press.

John, G.E. and Metzo, C.R. (2008) Yellowstone embodied: Truman Everts' "thirty-seven days of peril." *Gender, Place & Culture: A Journal of Feminist Geography* 15 (3): 221–42.

Kearns, G. and Reid-Henry, S. (2009) Vital geographies: life, luck, and the human condition. *Annals of the Association of American Geographers* 99 (3): 554–74.

Laurier, E. and Parr, H. (2000) Emotions and interviewing in health and disability research. *Ethics, Place and Environment* 3 (1): 98–102.

Lorde, A. (1997) *The Cancer Journals*. San Francisco: Aunt Lute Books.

Lorimer, H. (2007) Cultural geography: worldly shapes, differently arranged. *Progress in Human Geography* 31 (1): 89–100.

Mansvelt, J.R. and Berg, L.D. (2005) Writing qualitative geographies; constructing geographical knowledges. In I. Hay (ed.), *Qualitative Research Methods in Human Geography*. Melbourne: Oxford University Press, pp. 248–65.

Matthis, I. (2000) Sketch for a metapsychology of affect. *International Journal of Psychoanalysis* 81 (2): 215–28.

McCormack, D. (2006) For the love of pipes and cables: a response to Deborah Thien. *Area* 38 (3): 330–2.

Michaels, A. (1996) *Fugitive Pieces*. Toronto: M & S.

Nold, C. (2009) www.softhook.com/, March 27, 2008. Accessed February 1.

Nussbaum, M.C. (2001) *Upheavals of Thought: The Intelligence of Emotions*. Cambridge and New York: Cambridge University Press.

Nussbaum, M.C. (2006) Winnicott on the surprises of the self. *Massachusetts Review* 47 (2): 375–93.

Nussbaum, M.C. (2007) Human rights and human capabilities. *Harvard Human Rights Journal* 20: 21–4.

Powell, R.C. (2009) Learning from spaces of play: recording emotional practices in high Arctic environmental sciences. In M. Smith, J. Davidson, L. Cameron, and L. Bondi (eds.), *Emotion, Place and Culture*. Farnham: Ashgate, pp. 114–32.

Probyn, E. (2003) The spatial imperative of subjectivity. In K. Anderson (ed.), *Handbook of cultural geography*. London: Sage, pp. 290–9.

Probyn, E. (2005) *Blush: Faces of Shame*. Minneapolis: University of Minnesota Press.

Rose, G. (1993) *Feminism and Geography: The Limits of Geographical Knowledge*. Oxford: Polity.

Rose, G. (1997) Situating knowledges: positionality, reflexivities and other tactics. *Progress in Human Geography* 21 (3): 305–20.

Sedgwick, E.K. (2003) *Touching Feeling: Affect, Pedagogy, Performativity*. Durham, NC, and London: Duke University Press.

Sharp, J. (2000) Towards a critical analysis of fictive geographies. *Area* 32 (3): 327–34.

Sheller, M. (2004) Automotive emotions: feeling the car. *Theory Culture & Society* 21 (4–5): 221–42.

Smith, M., Davidson, J., Cameron, L., and Bondi, L. (2009a) *Emotion, Place and Culture*. Farnham: Ashgate.

Smith, M., Davidson, J., Cameron, L., and Bondi, L. (2009b) Introduction: geography and emotion – emerging constellations. In M. Smith, J. Davidson, L. Cameron and L. Bondi (eds.), *Emotion, Place and Culture*. Farnham: Ashgate, pp. 1–18.

Songe-Møller, V. (2002) *Philosophy without Women: The Birth of Sexism in Western Thought*. London: Continuum.

Spivak, G.C. (1987) *In Other Worlds: Essays in Cultural Politics*. New York: Methuen.

Stewart, K. (2007) *Ordinary Affects*. Durham, NC: Duke University Press.

Thien, D. (2004) Love's travels and traces: the "impossible" politics of Luce Irigaray. In K. Browne, J.P. Sharp, and D. Thien (eds.), *"Geography and Gender" Reconsidered: Women and Geography Study Group*. London: WGSG, pp. 43–8.

Thien, D. (2005a) After or beyond feeling? A consideration of emotion and affect in geography. *Area* 37 (4): 450–6.

Thien, D. (2005b) Intimate distances: considering questions of "Us." In J. Davidson, M. Smith, and L. Bondi (eds.), *Emotional Geographies*. Aldershot: Ashgate, pp. 191–204.

Thien, D. (2005c) Intimate distances: geographies of gender and emotion in Shetland. Unpublished dissertation. Department of Geography, University of Edinburgh.

Thrift, N. (2004) Intensities of feeling: towards a spatial politics of affect. *Geografiska Annaler Series B* 86 (1): 57–78.

Tsing, A.L. (2005) *Friction: An Ethnography of Global Connection*. Princeton: Princeton University Press.

Tyler, I. and Loizidou, E. (2000) The promise of Lauren Berlant: an interview. *Cultural Values* 4 (3): 497–511.

Vištica, O. and Grubišic, D. (2009) *The Museum of Broken Relationships*, July 14, 2009, www.brokenships.com.

Waring, M. (1996) *Three Masquerades: Essays on Equality, Work and Hu(man) Rights*. New Zealand: Auckland University Press, Bridget Williams Books.

Williams, S.J. (2001) *Emotion and Social Theory: Corporeal Reflections on the (Ir)rational*. London: Sage.

Wolf, D.L. (1996) *Feminist Dilemmas in Fieldwork*. Boulder: WestviewPress.

Wood, N. (2002) "Once more with feeling": putting emotion into geographies of music. In L. Bondi (ed.), *Subjectivities, Knowledges, and Feminist Geographies: The Subjects and Ethics of Social Research*. Lanham and Oxford: Rowan & Littlefield, pp. 57–71.

Woodward, K. (1996) Global cooling and academic warming: long-term shifts in emotional weather. *American Literary History* 8 (4): 759–79.

Chapter 19

Affective Life

Keith Woodward

For Kathleen Powers (1972–2007)

Consider *Volvox* (Figure 19.1). A favorite of highschool and college science classes, *Volvox* is a water-dwelling green alga consisting of a colony of "thousands of biflagellated cells anchored in a transparent, extracellular matrix (ECM) and daughter colonies inside the ECM" (Drescher et al. 2009: 1). The life of this spherical, multi-celled organism involves a continuous interplay of forces and tensions unfolding between its internal, bodily composition and the external environments it negotiates. Within the hollow, transparent ECM, daughter and granddaughter colonies generate in Russian doll-like fashion. Across this sphere's outer surface, the colony of *Volvox* cells work like synchronized swimmers, whipping their tail-like flagella so as to spin the entire body of the sphere in a clockwise motion through the surrounding water. While it is not impossible to conceptualize each cell as a discrete entity, it would be a mistake to abstract it from its colony. For although each has, for instance, its own light-sensitive "eyespot" that aids navigation relative to light sources (Braun and Hegemann 1999; Drescher et al. 2010), it is only by virtue of the cooperative inter-actions amongst its community of cells that *Volvox* life is created, sustained, and empowered to explore its universe. With its multi-generational style of self-organization and its dynamic, aggregated movements through fluid environments, entities such as *Volvox* can help shed light upon the collective nature of the ubiquitous bodily force relations known as "affects."

For example, the complexity of these relations quickly multiplies where one colony swims too closely within proximity of another. In such occurrences, the

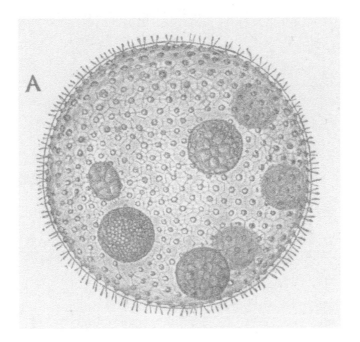

Figure 19.1 *Volvox.* Image from Jordan and Kellogg 1907: 29.

hydrodynamic motion of the surrounding water – the wake caused by the movements of swimming *Volvox* – pulls passing colonies into each other's orbit, *"bind-
ing"* them to one another in rotating "dances" (Figure 19.2). Revolving around
each other like twin stars, "*Volvox* colonies are drawn together, nearly touching while spinning, and they 'waltz' about each other clockwise" (Drescher
et al. 2009: 1). These waltzes are the product of a continuum of forces running
between two swimming colonies and *through* the surrounding water, which is
likewise set in motion by their swimming movements (Figures 19.3 and 19.4). The
relative heaviness or lightness of the body (an effect of the weight of its daughter
and granddaughter colonies) also contributes a dimension of force that impacts
its movement. As the body grows heavier, its motion becomes increasingly polarized,
impacting, in turn, its relative depth and capacities for movement: "When *Volvox*
have become too heavy to maintain upswimming, two colonies hover above one
another near the chamber bottom, oscillating laterally out of phase in a 'minuet'
dance" (p. 1). Orientation and velocity for relating *Volvex* colonies are thus also
affective processes: that is, each is the emergent product of its dynamic, *forceful*
set of relations between itself, other colonies, and its environment. Thus, not only
is the internal coherence of a *Volvox* colony expressed as a complex negotiation
of "outsides," but the character of its orientations *between* bodies is likewise substantially impacted by the forces produced by its "internal" composition (mass,
weight).

This chapter explores how such affective relations play a central role in constituting and expressing social and biological life. In keeping with earlier thinkers

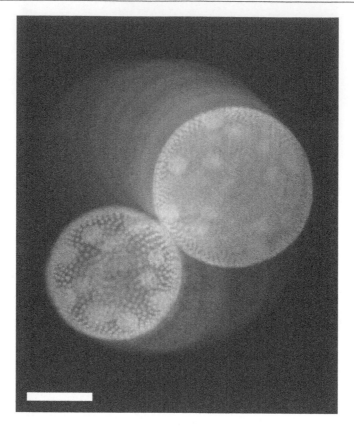

Figure 19.2 Waltzing *Volvox*. Top view: "superimposed images taken 4 seconds apart, graded in intensity." Source: Drescher, et al. 2009: 1; image courtesy of Drescher et al. 2009. Used with permission.

(Tuan 1984), geographers have recently made efforts to connect affect to several key themes in social theory – Anderson (2005, 2006), Saldanha (2006a, 2006b), and Lim (2007), for example, focus upon questions of identity and social difference in cultural theory (race, sexuality, and class, respectively). The central concern here will be to *extend* these discussions to socio-political areas that geography has yet to address in detail, and, in so doing, to meditate upon future sites and trajectories for social geographies of affect. It begins by considering forceful bodies and their relations with their own insides, with others, and with their environments so as to conceive the context of affect as something much different than (though not disassociated from) the human as an enclosed identity/entity. Taking affect as a general condition for material life, it begins this account from the perspective of non-human life. Following this, it goes on to explore how affect lends important critical, political, and material nuances to social research and theory concerning questions of the intensification of work and the precariousness of life. These latter themes illuminate the ways that affect can simultaneously be a crucial resource for the production of social life and the site of exploitation and political differentiation.

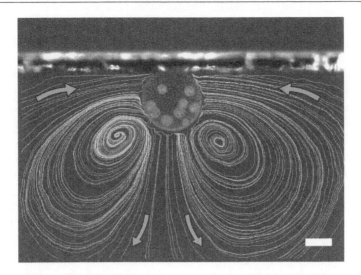

Figure 19.3 *Volvox* flow field. Side view of "a colony swimming against a cover slip, with fluid streamlines. Scales are 200 μm." Source: Drescher et al. 2009: 1; images courtesy of Drescher et al. 2009. Used with permission.

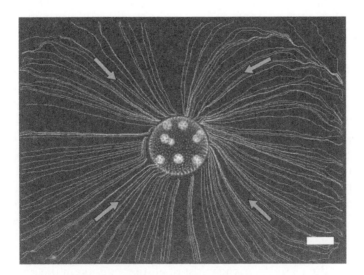

Figures 19.4 *Volvox* flow field. Top view of "a colony swimming against a cover slip, with fluid streamlines. Scales are 200 μm." Source: Drescher et al. 2009: 1; images courtesy of Drescher et al. 2009. Used with permission.

Affect

> In a philosophy of life, the fundamental determinations of existence are foregrounded: body, action, affectivity. (Henry 1993: 4)

To *affect*, to be *affected*: both notions gesture to the dynamic interplay of forces, to the work of bodies upon one another, entangling every life with complex series of material processes, moving both inward and outward, shaping materials and localizing their environments. Recent examinations of the ontology of life identify the differential, "active forces" known as *affects* with sites of transitory, impermanent, and generative transformation (Henry 1973; Deleuze 1988; Simondon 1989). Critical of viewpoints that contain bodies within individual, *a priori* borders, Deleuze introduced affect to explain bodies as *sites:* they are simultaneously *materials* connected by affects (something is affecting, something is being affected) and *surfaces* across which such forces play out. For *Volvox* colonies, these relations become a factor in the composition of life: "The important thing is to understand life, each living individuality, not as a form, or a development of a form, but as a complex relation between differential velocities, between deceleration and acceleration of particles. A composition of speeds and slowness on a plane of immanence" (Deleuze 1988, 123; see also Deleuze 2001). *Volvox* consist of many bodies, each of which exerts a slightly variable and different amount of force, but which contributes to the aggregate movements and dances and which marks a singular point in an emergent continuum of forces between bodies and environments. Drawing upon Jakob von Uexküll's ethology, Deleuze relates such affectivities to regimes of organization (rather than categorical body typologies) wherein individuals take form and dissolve *as* specific, situated engagements (Deleuze 1988; 1994; Deleuze and Guattari 1987: 257; Uexküll 1992; see also: Pearson 1999; Ingold 2000; Agamben 2004; Buchanan 2008; Harrison 2009). Uexküll (1992: 327) shares this vision of externalities as continua with insides: "As the spider spins its threads, every subject spins his relations to certain characters of the things around him, and weaves them into a firm web which carries his existence." The complex site of such relations – what he will call an individual's *Umwelt,* or "environment" – is the unique intersection of a body's specific perceptual and affective capacities with the emergent particularities of its environment. Thus, for example, an important dimension of the world of the fly arises through its capacity to taste with its feet.

Keen to explore the possibility that we may live in a world of many worlds, all formed from different affective engagements Deleuze invokes Uexküll to highlight the role of emergent capacities of forcefully interacting bodies in *composing* environments. Drawing upon one of Uexküll's more famous examples, Deleuze defines "the tick by three affects: the first has to do with light (climb to the top of a branch); the second is olfactive (let yourself fall onto the mammal that passes beneath the branch); and the third is thermal (seek the area without fur, the warmest spot). A world with only three affects, in the midst of all that goes on in the immense forest" (Deleuze 1988: 124). Deleuze's explanation of the tick's actions links percepts with affects, such that parts of its body are affected and mobilized in particular ways by variations in illumination, smell, heat, and so on. And while these percepts – sensing light, for example – sound familiar to normative conceptions of human sensation, the world built from tick affects is relationally, practically, and experientially distinct from that of human life. Though we may encounter, say, a similar distribution of light rays as the tick, these will not be "for" the tick in the same way that they are "for" the human (the tick senses light by way of its skin, which is photo sensitive). Nor are we affected in the same way: sunlight may cause me to squint, to feel warm, to reflect upon skin cancer, or, indeed, to grow melanoma, but its appearance does not attract me towards food, as it does for the tick. Save the last, all of these – from cell growth to thought itself – are ways that I can be affected, due, in part, to the particular way that my body has composed itself (its potential to affect and be affected). They are unavailable to the tick for the same bodily reasons.

Though a "world with only three affects," the tick's universe is the site of substantial complexity when considered in connection with the dynamic environments it must negotiate: the variable movements of the wind that carries animal scents, the contingent arrival of a mammal beneath its branch, the accuracy required to drop upon its moving target, and so on. Likewise, the *Volvox* is subject to the changing environmental specificities of its situation. Consider once again the affects of the swimming cells on its outer surface (see also Uexküll's [1992: 342–3] discussion of the *Paramecium)*, a complicated system whose movement entangles the "work" of the cell flagella, the changing dimensions of the aggregate body, and the hydrodynamics of its watery environment. Further, note the resonances and dissonances with Deleuze's list of tick affects in Drescher et al.'s description of *Volvex* "bound" – dancing – states:

> Four aspects of *Volvex* swimming are important in the formation of bound states ...: (i) negative buoyancy ..., (ii) self-propulsion ..., (iii) bottom-heaviness ..., and spinning ... During the 48 h life cycle, the number of somatic cells is constant; only their spacing increases as new ECM is added to increase the colony radius. This slowly changes the speeds of sinking, swimming, self-righting, and spinning, allowing exploration of a range of behaviors. (Drescher et al. 2009: 2)

Again, everything references the specificity of environments as an interactive condition for life, but here perceptions give way almost entirely to bodily-environmental affects.

While affect is crucial for interactions between material beings and their environments, this relation is something considerably more than a simple means through which individuals and social organizations extend their intentions or realize their goals in an otherwise passive universe. Nor does it describe a relation of cause and *effect*. Rather, affect is better characterized as an "onto-genetic" process whereby disparate materials pull together to organize as a cohering individual: "affectivity indicates and comprises [the] relation between the individualized being and pre-individual reality" (Simondon 1989: 108; my translation). The force of affect *is* this process of material organization: it is not only the product *of* individuals, but more importantly, it *produces* individuals (whether human or non-human, lively or dead matter). For example, *Volvox* movements are coordinated without being intentional, collective without being strategic. It is by virtue of its cells' capacity for working together – their relation of *affective sociality* – that a colony both navigates its world and spirals into bound states with other colonies. The mobility of an individual colony arises from the aggregate force of its cell flagella, just as *Volvox* dancing is the culmination of multiple moving, proximate colonies. In both cases, the communication of complex affects – movements, speeds – between individuated bodies gives rise to their rotating collectivity. Further, these multi-*Volvox* dances are both the producers and the products of dynamic environments, exposing the intimacy between affect and its site of articulation. Rarely are environments passive media for interacting bodies; more often they *actively participate* in the composition of affective sociality. Accordingly, affect expresses not only relations between formal or typological bodies, but also – like the dynamic waters where *Volvox* dance – the forceful proximities and involvements of environments. The materialities attached to such situations, it seems, are something much more than simple spaces *of* affect (as though they were affect's mere production). Rather, they are *affective spaces* – or "sites" (Woodward et al. 2010) – participating fully in immediate, forceful engagements.

So far, I have avoided direct connections between affects and humans and have instead concentrated upon non-human perspectives on affect. I have done this for a number of reasons. First, it illustrates that affect is fundamentally important not only for human life, but for life in general. Second, and following from the first, it reveals that affective life is forged upon a continuum running between beings and their environments – or, to echo Simondon, between an individualized being and pre-individual reality – rather than inner egos divorced from their outsides. Third, it understands affect from the viewpoint of its ontological relevance, an assessment that is considerably more complex than a residuum of human agency or emotion. That is, affect is not solely a humanistic concept. The debates to which these three points form a response – particularly those concerning the relation between affects and emotions (Thien 2005; Anderson and Harrison 2006; McCormack 2006; Tolia-Kelly 2006) – have been well rehearsed within geography and without (Woodward and Lea 2010). It is the philosophical tension at play, for example, in Stengers' (2002) decision to translate Whitehead's notion of "feeling" as the French *"affecter"* ("to affect") rather than *"sentir"* ("to feel"), reasoning that the latter term, "in French, is ... associated with psychological life" (Stengers 2002: 329: my translation). Affect does indeed shed light upon the character of human emotional and psychological life. However, the regime of affect comprises a much larger universe, which allows us to conceptualize our relations with the world in ways that are not restricted to privileging or prioritizing – as was so long the case after Kant – the ego, the self, or Reason as the control center of the cognition and embodiment. Thinkers such as Whitehead and Spinoza, for example, argue that affect plays a significant generative role in the formation of cognitive processes and pathways, though much of this work may be only vaguely or unclearly discerned – if at all – by the affected thinker. This is implicit in Whitehead's socio-metaphysical account of "feeling" that so concerns Stengers: "A feeling ... is essentially a transition effecting a concrescence" (Whitehead 1978: 221). That is, affection plays a key role in human life as a force that engenders thought by bringing cognition into material connection with "outsides" or environments (Nietzsche 2003: 60).

While *Volvox* might appear to be a comparatively simple organism following this description of human thought and feeling, the connections it illuminates between affect, life, and environments is helpful for diagramming the ways that human social life grows and changes by virtue of its situatedness. Consider, for example, the tendency to select the local environments of economically, socially, and politically disenfranchised communities as the sites for dumping toxic industrial waste (Bullard 2000). Here, spatial logics of difference reveal themselves both in the policies of local decision makers and in the impacted community's uneven access to political power.

These differences are materialized in the environmental changes that impact the everyday lives of community members. Eventually, differences in identity, location, and access, become manifest as illnesses and birth defects – affects – through the community's localized exposure to toxic chemicals. Such stories have become a refrain amongst many communities whose local environments have been destroyed thanks to global corporations' neoliberal marriage of political-economic privilege and managerial negligence: Shell Oil's "petro-violence" in the Niger Delta, where oil drilling and gas flaring devastated the farming and fisheries of indigenous Ogoni (Watts 2001, 196; Boele et al. 2001); the Union Carbide toxic gas leak that has killed thousands in Bhopal (Dhara et al. 2002); Coca-Cola's draining of the water table while cynically passing off industrial sludge as fertilizer to Indian farmers (Aiyer 2007: 644); and British Petroleum's oil spill in the Gulf of Mexico that not only wiped out the livelihoods of fishermen, but forced them to seek employment as members of the oil company's cleanup crews (Cohen 2010) doing hazardous work that involved exposure to the highly toxic oil dispersant Corexit (Quinlan 2010).

Now, Braun has rightly noted that theorizing along the lines of affective *life* can lead to a "neo-vitalism" that ultimately lacks much explanatory depth beyond the observation that such systems are emergent: "once the vitalist point has been made, it very quickly becomes uninteresting" (Braun 2008: 675). For many areas of social science – particularly in consideration of what counts as a *solution* for much of social science – I agree wholeheartedly with his objection. However, the *specific* roles that affects play in the production of social bodies and their environments is a considerable exception. What is perhaps most interesting in this connection to the question of vitalism – though it is something that many geographers of affect tend *not* to discuss (but see Harrison 2009) – is the light that it sheds on the *diminishment* of affective life. For example, because affect plays a part in material transformations from the pre-individual to the individual, it offers a language for describing the subtle impacts that environments can have on localized lives from the perspectives of illness and disease, exploitation and oppression. In what remains, I turn directly to these questions, so as to illustrate several of the varied social dimensions of affective human life.

Work

> The way in which men produce their means of subsistence depends first of all on the nature of the actual means of subsistence they find in existence and have to reproduce. This mode of production must not be considered simply as being the production of the physical existence of the individuals. Rather, it is a definite form of activity of these individuals, a definite form of expressing their life, a definite *mode of life* on their part. As individuals express their life, so they are. What they are, therefore, coincides with their production, both with *what* they produce and with *how* they produce. The nature of individuals thus depends on the material conditions determining their production. (Marx and Engels 1970: 42)

Marx was perhaps the first to develop a "political anatomy" of the affects from the perspective of working life, recognizing "labor power" as a bodily force – affect – subjected to the organized control of factory-style production and the abstracted value of capitalist accumulation. One consequence of the historic development of the capitalist mode of production was that labor power became a social relation that paradoxically expressed, supported, and extended and intertwined the processes of life and capitalism. Here accumulation came to be realized through the appropriation of bodily force *as* labor power, whereby affect became simultaneously the source of material *production* and the site of bodily *exploitation*. It is in light of this relation that Read observes, "In discussing the struggle over the working day Marx foregrounds the fact that the relation between the capitalist class and worker is initially, and at its core, a relation of force" (Read 2003: 96). The centrality of affect becomes even more pronounced with the rise of the Fordist factory in the twentieth century, where the speed-up of the production process is accomplished through the introduction of task-specific mechanized technologies distributed along the assembly line. Famously, this innovation had the double effect of "de-skilling" workers while simultaneously intensifying the speed of their exertions in the production process. Nor were these two phenomena unconnected: mechanization made it possible to gradually erase technical knowledge as a human capacity within mass production, thus making raw affect – in the form of simple, potential labor power – the fundamental component of factory labor and worker employment. The assembly line was reduced to a series of fast, specific, repetitive exertions, and consequently, the entire scene of production became a factory of boredom.

Foucault expands upon Marx's analysis of labor, explaining that the institutionalization of modern modes of production subjected the affectivities, spatialities, and temporalites of workers' movements on the factory floor to increasingly specific forms of segmentation and management (Foucault 1977). The segmented organization of labor subjects affect to a "micro-physics" of power, or a political anatomy of bio-political production, through which ever-more acute circuits of force and movement are simultaneously mapped onto and managed through work *spaces* and production *times*. The locations and speeds of the realization of labor power become the sites within which the worker is invited to self-exploit by micro-managing his/her own movements. Neither the individual worker nor a completed commodity are encompassed by this affective relation, but rather a specific set of worker actions exerted upon a specific set of parts. The value of working life, as Marx noted, is not measured against the completed commodity, but is given its own abstract measure on the basis of an aggregate value of production times. Surplus value occurs precisely in the inexactitude of such a measure, through the minor increases in the intensity of application of forces on the line. As Read (2003) notes, one consequence of this situation is the increasing exhaustion of the worker:

> Skilled artisans are transformed into hobbled fragments of a labor process beyond their control. Marx often describes the transformation with a sort of Dickensian attention to the crippling effects of labor reduced to pure repetitive activity, producing what he refers to as an "industrial pathology" ... It is with the development of machines, the transfer of mental skills and physical exertions, into the body of the machine that abstract labor is effectively produced. Capital at this point is at least potentially indifferent to the skill, background, gender, and age of the bodies put to work. Moreover, this transformation of abstract labor, its production by and incorporation into machinery, makes possible a different strategy for the production of surplus value: Surplus value is no longer produced simply by extending the working day beyond the necessary

time to reproduce labor power but through an intensification of labor, speeding up the machines, making labor more productive. (Read 2003: 95)

This assessment highlights two explicitly affective dimensions that capitalist production simultaneously engenders and exploits. On the one hand, the factory becomes a specific affective environment within which tasks are segmented according to circuits of localized force (i.e., one's place, role, or task upon the assembly line). On the other hand, recalling the cooperative movement of *Volvox* cell flagella, the factory is itself the assemblage of specific forms of affective sociality that create the conditions for the accumulation of surplus value, relatively coordinating the speed-up of *all* the segments along the assembly line, causing bodies to move in concert like a human centipede. Transformations in one dimension of socio-affective life will thus play off of those in the affective environment and *vice versa*. In this regard, affective relations in the factory bear further material similarity to those of waltzing *Volvox*. Biopolitical production as the production of a life, draws upon and segments the affective relations that enlist bodies in relations of production. Accordingly, within factory work and mass production, the commodity becomes the aggregate-result of the forceful relations of any number of different workers enlisted by the assembly line. The complex production line finds the commodity immersed in a network of approximate value relative to an entire system of production whose key variable is an averaged intensity of speed and exertion. Where other forms of production, such as piecework, individuate the exertion of labor power by each worker (made manifest in the quantity of commodities produced), the factory line populates the production process with a multiplicity of individuals whose combined, differing exertions – and the mediations of various machines – contribute to an aggregate accumulation of labor power via the production of an aggregate quantity of commodities.

If affective life had become exploitable and subsequently alienated as "labor power" under Fordism, post-Fordism's employment of outsourcing and just-in-time production, along with the emergence of new sectors and styles of consumption, introduced increasing degrees of uncertainty and precariousness into both labor and life. The standard condition of life under the Fordist factory system was the continuous exhaustion of bodily affects, whereas, under post-Fordism, the body becomes subject to further tensions and anxieties by virtue of the insecurity of stable work. I will explore some of the extensions of this precariousness to other areas of contemporary life in the following section, but first I turn to the transformations that this notion presents to the specific question of working life. The growing proliferation of precariousness is particularly pronounced in the growth of sectors employing "affective labor." This notion does not refer to the exertions by workers in a factory that result in the production of material commodities, but rather to forms of labor that engage in "immaterial activity" (Virno 1996: 191; see also Fortunati 2007) resulting in no end-product other than the manifestation of affect or affection within the consumer. It is characteristic of a variety of forms of "reproductive labor," such as service and care work, where "the *object* of production is really a *subject*, defined, for example, by a social relationship or a form of life" (Hardt and Negri 2009: 133; see also Hardt and Negri 2004: 110–11). Not surprisingly, such work tends to blur distinctions between labor and life for both the consumer/employer and the worker. Recalling the repetitive affective segments that characterized the factory, consider, for example, the resonances and dissonances that arise in work involving live-in child care by women immigrant workers in Los Angeles. Hondagneu-Sotelo notes, unsurprisingly, "After what may be years of watching, feeding, playing with, and reprimanding the same child from birth to elementary school, day in and day out, some nanny/housekeepers grow very fond of their charges and look back nostalgically, remembering, say, when a child took her first steps or first learned nursery rhymes in Spanish" (Hondagneu-Sotelo 2007: 39). Here, the affects share factory labor's characteristic tendency toward routine repetition, but also differ in that, rather than engendering surplus values, affective labor contributes directly to the production of a life, affection, and subjectivity.

At the same time, this work constitutes its own regimes of risk associated with "the work of keeping the children clean, happy, well nourished, and above all safe" and Hondagneu-Sotelo observes that, often, "Nanny/housekeepers fear they will be sent to jail if anything happens to the children" (Hondagneu-Sotelo 2007: 39). The manifestation of anxieties in connection with care work and reproductive labor reveals tensions in the amount of precariousness – the "non-self-determined insecurity of all areas of life and work" (Raunig 2004) – to which the worker is subject. Here, affective insecurity appears not only *as a general condition* of a laboring body, but *with the specific construction* of a body as different: raced, gendered, nationalized, classed, othered. The construction and valuation of identities choreographs a variety of encounters between affect and difference: "certain kinds of bodies will appear more precariously than others, depending on which versions of the body, or of morphology in general, support or underwrite the idea of the human life that is worth protecting, sheltering, living, mourning" (2009: 53; see also 2004). Amongst the more under-represented affective modes of work in economic life, this tenuousness and uncertainty becomes a unifying characteristic:

> A freelance designer and a sex worker have certain things in common – the unpredictability and exposure of work, the continuity of work and life, and the deployment of a whole range of unquantifiable skills and knowledges [carried out in] the important spaces of daily life of women ... working in precarious and highly feminized sectors: language work (translations and teaching), domestic work, call-shops, sex work, food service, social assistance, media production. (Precarias a la deriva 2004: 158)

Here a telling component of precariousness, according to Precarias a la deriva, is its coincidence with the feminization and devaluation of certain forms of labor: "In garment factories throughout the world, women make up the majority of the workforce. The feminized shop floor is both a new phenomenon and, especially in the garment industry, the way things have always worked" (Brooks 2007: xviii; see also Wright 2006). In the case of sweat shops, these sites have been known to amplify insecurity and health risks through the manipulation of workers, using "oppressiveness as emotional degradation, including rigorous regulation of toilet access and miserable sanitary conditions" (Ross 2004: 21) – that is, the production of precarious affective environments – as a strategy for increasing the speed of production.

Key questions of social geography thus intersect affect along vectors of the relative construction and valuation of bodies and the relations of force to which they subsequently become subject. Thus, as Wright illustrates, the sexist myths circulated amongst maquila owners that women are untrainable and unreliable serve to both rationalize and perpetuate the deskilling of women's labor through the practice of "pigeonholing ... women into the lowest waged and dead-end jobs throughout the maquilas" (Wright 2006: 82). Such relations amplify the precariousness of work through both the lowered bar on the skills required to complete a task (thus increasing the threat of job loss to the surplus labor force and simultaneously the conditions for self-exploitation) and through the increased repetitiveness of the task performed (thus increasing the likelihood of a worker injury due to the strain of repetition and worker turnover due to boredom [Wright 2006, 56]). While these conditions pertain to the precariousness that gendered bodies in maquilas must work with, they also represent the amplification of insecurity of life. For here, bodily affects such as the endless anxiety over the possibility of job-loss and the decrease in bodily capacity due to repetitive work strain of specific body parts both bear immediately upon life insofar as they increase the possibility illness. It is with this in mind that I now turn to the consideration of life itself.

Life

> It is *at bottom false to say* that living labour consumes capital; capital (objectified labour) consumes the living in the production process. (Marx 1973: 349)

> The owner of labour power is mortal. (Marx 1976: 275)

The affective character of work under capitalist conditions of production calls upon a broader set of concerns regarding the precarious relation between affect and life. Affect theory's attention to the variability of forceful bodies serves to remind us that each body consists of a variety of specific capacities and limitations that distinguish it from other bodies. Here, difference is the embodied, material reality of every affective being. Echoing the spirit of Deleuze and Guattari's (1983) first collaboration in the wake of the student revolts of May 1968, this understanding of difference has frequently been invoked as though it were a passageway to liberation of one form or another. While not to be dismissed out of hand, such notions are too often taken up at the cost of glossing over an equally crucial point: affect not only highlights the potential openness to the world of *every* body, it also reminds us of the real specificity of the limitations of *each* body. Carel's discussion of the body subjected to "chronic illness and disability" is illuminating in this regard. She considers the detailed and daunting affective transformations that occur "when the body loses some of its capacities and becomes unable to engage freely with its environment" (Carel 2008: 25). Taking her understanding of the body from the phenomenologists – Merleau-Ponty, in particular – and their understanding of cognition as something always already bound up in embodiment (see also Gallagher 2005), she explains: "On the phenomenological view, disease cannot be taken as a mere biological dysfunction, because there is nothing in human existence that is purely biological. We *are* embodied consciousness, so consciousness is inseparable, both conceptually and empirically, from the body. Therefore the concept of illness must be reconceived to take this unity into account" (Carel 2008: 14). Complicating the notion of affective environments discussed throughout this chapter, she situates the body amidst a "social architecture" and a "geography of illness": "In the same way that distances increase, hills become impossible and simple tasks become titanic, the freedom to go out into the social world and improvise, to act and interact, is compromised. A new world is created, a world without spontaneity, a world of limitation and fear: a slow, encumbered world to which the ill person must adapt" (Carel 2008: 52; see also Gleeson 1999; Siebers 2001; 2006; Sullivan 2005). Crucially, a motivating dimension of future work in affective social geography must be the introduction of critical wariness toward assuming an abstract, normative ("healthy") body as its site of theorization (and generalization).

Beyond those modes where affects are diminished as a side-effect of capitalist accumulation or a symptom of illness, many of the technical and policing apparatuses that have appeared as consequences and conspirators of the last decade's culture of political securitization concern themselves directly with the precariousness of affective life. In one sense, this has often involved the semiotics of nationalism, such as the representations of nationalist desire that surfaced in the wake of September 11, 2001 (Ó Tuathail 2003; Kingsbury 2007). More recently, it has arisen through the production of morale, disciplinarity, and securitization in a period of international conflict and insecurity (Anderson 2010a; 2010b). However, affects and precariousness have most recently come to be treated as objects in themselves by policing agents who are trained to read bodily affect as signs of *potential* security threats. For example, Martin has described the US Transportation Security Administration's (TSA) stationing of security officers in American airports who monitor passengers for signs of suspicious behaviors:

> "Behavior Detection Officers," or BDOs, [are] TSA employees specially trained to read physiological evidence of fear in the unintentional movements of passengers' bodies. According to TSA, "Behavior analysis is based on the fear of being discovered. People who are trying to get away with something display signs of stress through involuntary physical and physiological behaviors." (Martin 2010: 25; see also Addy 2008)

Governmentality has always been concerned with affect, but generally the emphasis has been upon the production of subjects and practices. It is a considerably different kettle of fish when the intelligibility of affect is subjected to a representational logic that exceeds it. That is, policing for affect, it seems, begs several important questions regarding the possibility of successfully reading and masking the affects, not to mention the fact that the awareness of figures of power or authority often engender any number of involuntary affective responses on the part of others. Given the uncertainty of affect and the fact that individual encounters with policing are far from uniform, the attempt to make it a site of governance risks repeating numerous subterranean modes of identitarian profiling (racist, classist, sexist, heterocentric, and so on). This was long a point of internal conflict, for example, in the management of colonial sexual desires (Stoler 1995). Affect, rather than helping to single out the other, is just as likely to produce waves of nonthreatening-yet-unreadable others, and in this way, a certain uncertainty implicit in the effort to wage a "war on terror" becomes manifest in the indiscernibility of the other. Where this strategy hinges upon an effort to read a moment of involuntary behavior according to representational logics, the use of torture – one of the central controversies of the US wars in the Middle East – reveals the state's appropriation of the precariousness of affect as an apparatus of control.

Gregory recognizes that spaces of torture such as Abu Ghraib differ substantially "from the modern carceral regime described by ... Foucault" (Gregory 2007: 206) and the many leaks and news reports that have appeared since 2005 suggest that these spaces often function more like sites for the expression of whimsical violence and control. However, recently released documents also suggest a kind of "science" of torture whereby military interrogations employed a rigid bio-politics that involved segmenting the capacities of detainee bodies according to different affects so as to better control their reactions. Thus, on the one hand, the Abu Ghraib abuse scandal illustrated how random practices of torture are often informed by the racist imaginaries, colonialist desires, uneven power relations, and perhaps even the occupational boredom of the torturers. On the other hand, the institutionalization of "interrogation techniques" such as waterboarding are rationalized by way of a pragmatics of the management of affective life. Recalling the repetitiveness and segmentarity of Foucault's factory, consider the accounts of the techniques used by the US interrogators found in recently declassified documents:

> Interrogators pumped detainees full of so much water that the CIA turned to a special saline solution to minimize the risk of death, the documents show. The agency used a gurney "specially designed" to tilt backwards at a perfect angle to maximize the water entering the prisoner's nose and mouth, intensifying the sense of choking – and to be lifted upright quickly in the event that a prisoner stopped breathing.
>
> The documents also lay out, in chilling detail, exactly what should occur in each two-hour waterboarding "session." Interrogators were instructed to start pouring water right after a detainee exhaled, to ensure he inhaled water, not air, in his next breath. They could use their hands to "dam the runoff" and prevent water from spilling out of a detainee's mouth. They were allowed six separate 40-second "applications" of liquid in each two-hour session – and could dump water over a detainee's nose and mouth for a total of 12 minutes a day. Finally, to keep detainees alive even if they inhaled their own vomit during a session – a not-uncommon side effect of waterboarding – the prisoners were kept on a liquid diet. The agency recommended Ensure Plus. (Benjamin 2010: npn)

On the one hand, the institutionalization of torture depends upon deeply and specifically segmented set of practices: six 40-second applications of liquid in each two-hour session for a total of 12 minutes per day; the immediate application of water to ensure that it is inhaled by the detainee; even dietary instructions that go so far as to recommend a specific *brand* of liquid diet. These bio-medico-legal prescriptions for interrogation "techniques" can be characterized by two regimes of affect that have concerned us throughout this chapter: (1) they describe the practical segments for affecting the body of the detainee (the timing, duration, and frequency of water application; the use of saline solution), and (2) they provide a recipe for creating the conditions whereby the body of the detainee can be affected (providing a liquid diet; tilting the gurney; damming the water runoff). In total, what they describe is a diagram for the management of life *at its affective limits*. The body of the detainee becomes a surface that can be made subject to a finite number of measured affections.

On the other hand, the effectiveness of this set of techniques, like the factory, depends upon the specific capacity of the detainee's bodily affects to vary in intensity. The regulations thus operate by way of an economy of torture, whereby interrogation techniques simultaneously attempt to minimize the "risk of death" while intensifying the detainee's "sense of choking." That is, the technique appears to exploit the tendency for affect to work outside of representative cognition, for it to operate upon registers that are other than rational. This is something entirely distinct from "the folk cognitivism of media reports that describe waterboarding as the production of the belief in the mind of the victim that he or she is drowning. We should rather describe waterboarding as triggering an evolutionarily preserved panic module that acts by means of a traumatizing biochemical cascade" (Protevi 2009: 148). The logic behind these techniques hinges upon exploring the possibility of utilizing the variability of affective biological responses as a means of overwhelming consciousness. Thus, the techniques capitalize upon the sense of the increasing precariousness of the detainee's life, upon his or her dependency upon a set of affective relations that extend beyond the individuated body of the subject, and upon the biochemical "cascade" that gets triggered during the sudden panic of the experience.

Human life, much like the dances of the waltzing *Volvox*, has a circularity to it through which affects interact, grow, and diminish at the same time that they materially produce the very *ground* for sociality. As a consequence, the social production of affective states reveals something like a complex "structuring" or perhaps "ontogenetic" effect upon, for example, the production of desires (consumption), habits (normativity), or common sense (ideology) that bear immediately upon political life. This, in a broad sense, is what Marx, Lenin (1988), and Althusser (1969) all seemed to have in mind in their criticisms of the notion of spontaneity in the political act (Woodward 2010). For, if it is to be truly spontaneous, then something like ideology must not only play at registers "beyond" or "beneath" the domain rational choice, but in so doing, it must be pieced together (however erratically), segmented, and carried along by the forces and encounters that play out in those subterranean zones. With regard to such affective dimensions concerning the political costs of securitization, the construction of terrorism as both inevitable and indiscernible, and the complicity of (democratic) citizenry in the institutionalization of state torture, Hannah is informative:

> Administration policy since 11 September 2001 ... and the American public's continued willingness to live with it can be explained to a significant degree by a particular discursive construction: the ticking-bomb scenario. To the extent that this scenario frames official and public understandings of the threat of terrorism, it tends to make torture appear more reasonable as a response. The ticking-bomb scenario prompts a reimagining of the landscapes of everyday life as suffused with an unacceptably high level of risk. If unacceptable risk is extrapolated to cover the entire national territory, the imperative to eliminate such risk is intensified. The imagined imperative to eliminate this risk at all costs constitutes an opening for the contemplation of torture. (Hannah 2006: 623)

Hannah's attention to discursive constructions calls upon affective registers that combine accounts of affect as both a site of *concern* for the state and a site of *control*. In particular, the ticking-bomb scenario that he identifies is an imaginary that links the legitimation of state power and, in particular, the power to torture, to the uncertainty of potential security threats. The implied "them-or-me" logic that drives the scenario is obviously deeply flawed, but in its combinatory appeal to affects and xenophobic nationalism it completely achieves the desired effect: mobilizing fear as a strategy for suspending the legitimation crisis introduced by the practice of state torture. In this way, the social dimension of affect falls not only on the body of the detainee, but also circulates well beyond it in the anxieties of the state that tortures. And while they are not of the same *kind,* both modes make precariousness the dominant condition in the constitution of the political field.

Conclusion

Because the majority of this chapter has been devoted to perspectives on affect that highlight its constraint, control, or diminishment in the context of social relations, I conclude by reflecting briefly on the entangled topics of resistance and empowerment. Surin (1996: 212) reminds us that "capitalism, like other social and economic formations, is subject ... to the action of a postcapitalist form which it has to stave off as a condition of remaining in existence." This, as can be judged from the earlier discussion of working life, is accomplished at least in part through the appropriation and exhaustion of affect, the "anticipated potential" (p. 212) of which capitalism captures, manipulates as an abstract value (affect-as-labor power), and finally transforms into a commodity capable of realizing surplus value. Thus, one of the ways that capitalism appears to stave off postcapitalist futures is by accumulating, consolidating, and controlling human affects that might otherwise be "spent" inventing and organizing alternative practices of consumption and exchange. The reemergence of piecework in neoliberal production chains is a recent, obvious example of the attempt to squeeze affect by effectively throwing the responsibility for managing the precariousness of working life back upon the exhausted body of the worker. Further, certain patterns of affect can even be complicit in exploitation insofar as they enable the spread of production relations and environments that serve as conditions for capitalist futures. With regard to the exhaustion of bodily capacities, such "capitalocentric" affects are, for the worker, self-destructive forms of accumulation. This was, for example, the working experience of managerial class in Japan during the economic bubble in the 1970s, 1980s, and 1990s, when the emerging "lean" production system colluded with self-exploitation to cause an enormous rise in cases of *karoshi* (death from overwork) (Nishiyama and Johnson 1997).

Thinking as much from the perspective affect as from the possibility of postcapitalist futures, Starhawk suggests that "a movement is like an ecosystem" (Starhawk 2002: 257). Like the *Volvox* or the factory worker, the sociality of affect (a movement) becomes wedded to the question of environment (an ecosystem). However, while such environments are often the sites of routines that are easily appropriated by forms of capitalist accumulation, more recently affect has been enlisted for the purposes of envisioning, amongst other things, resistance movements, alternative economies (Gibson-Graham 2006), and foundations for organizing alternatives to capitalism (Sitrin 2006). Describing the efforts to develop horizontal styles of political and economic organization in Argentina following the 2001 economic collapse, Sitrin explains:

> One way people in movements describe the territory they are creating is through the idea of *política afectiva*, or affective politics. They are affective in the sense that of creating affection, creating a base that is loving and supportive, the only base from which one can create politics. It is a politics of social relationships and love. To translate this term as "love-based politics" would miss many of the social relationships it implies. (Sitrin 2006: vii)

Sitrin's account locates a continuity between the circulation of affects with the production of environments. In particular, moments in Argentina arose out of the organization of neighborhood assemblies and the recuperation of workplaces and factories. North and Huber (2004) have observed that these spaces – like most bottom-up forms of organization – are hardly immune to conflict and internal dissonance. However, these affective dimensions, as well, figure in the production of a dynamic environment of complex and collaborative social relationships that become sedimented through collective politics. Such sites constitute something akin to what Anderson, following Marx, has called "atmospheres": "atmospheres are generated by bodies – multiple types – affecting one another as some form of 'envelopment' is produced" (Anderson 2009, 80; see also Lazzarato 2004). What is fundamental, then, is not so much affective homogeneity as a certain consistency – even in difference – that serves to give rise to alternative environments or atmospheres. This, it seems, implies studies of affects that emphasize the role of experimentation. This work may mean exploring social organization with an eye toward the articulations of affect that different arrangements enable and curtail. Social life is, after all, a condition for the circulation of affect: it makes sense that we seek out arrangements that ward off the translation of affects into profits (that is, exploitative assemblages). For recent social movements such as those detailed by Sitrin, such relations appear to constitute sites whereby the collective participation unfolds, as Gordon (2010) puts it, in the form of an affective "power-with" rather than a "power-over."

References

Addy, P. (2008) Airports, mobility and the calculative architecture of affective control. *Geoforum* 39: 438–51.

Agamben, G. (2004) *The Open: Man and Animal*. Tr. K. Attell. Stanford: Stanford University Press.

Aiyer, A. (2007) The allure of the transnational: notes on some aspects of the political economy of water in India. *Cultural Anthropology* 22: 640–58.

Althusser, L. (1969) *For Marx*. Tr. B. Brewster. New York: Verso.

Anderson, B. (2005) Practices of judgment and domestic geographies of affect. *Social & Cultural Geography* 6: 645–60.

Anderson, B. (2006) Becoming and being hopeful: towards a theory of affect. *Environment and Planning D: Society and Space* 24: 733–52.

Anderson, B. (2009) Affective atmospheres. *Emotion, Space and Society* 2: 77–81.

Anderson, B. (2010a) Morale and the affective geographies of the "war on terror." *Cultural Geographies* 17(2): 1–18.

Anderson, B. (2010b) Security and the future: anticipating the event of terror. *Geoforum* 41: 227–35.

Anderson, B. and Harrison, P. (2006) Questioning affect and emotion. *Area* 38: 333–5.

Benjamin, M. (2010) Waterboarding for dummies. *Salon* March 9, www.salon.com/news/feature/2010/03/09/waterboarding_for_dummies.

Boele, R., Fabig, H., and Wheeler, D. (2001) Shell, Nigeria and the Ogoni: a study in unsustainable development I. The story of Shell, Nigeria and the Ogoni People – environment, economy, relationships: conflict and prospects for resolution. *Sustainable Development* 9: 74–86.

Braun, B. (2008) Environmental issues: inventive life. *Progress in Human Geography* 32: 667–79.

Braun, F.J. and Hegeman, P. (1999) Two light-activated conductances in the eye of the green alga *Volvox Carteri*. *Biophysical Journal* 76 (3): 1668–78.

Brooks, E.C. (2007) *Unraveling the Garment Industry: Transnational Organizing and Women's Work*. Minneapolis: University of Minnesota Press.

Buchanan, B. (2008) *Onto-Ethologies: The Animal Environments of Uexküll, Heidegger, Merleau-Ponty, and Deleuze*. Albany: State University of New York Press.

Bullard, R.D. (2000) *Dumping in Dixie: Race, Class, and Environmental Quality*. Boulder: Westview Press.

Butler, J. (2004) *Precarious Life: The Powers of Mourning and Violence*. New York: Verso.

Butler, J. (2009) *Frames of War: When Is Life Grievable?* New York: Verso.

Carel, H. (2008) *Illness: The Cry of the Flesh*. Stocksfield: Acumen.

Cohen, E. (2010) Fisherman files restraining order against BP. *CNN* May 31, www.cnn.com/2010/HEALTH/05/31/oil.spill.order.

Deleuze, G. (1988) *Spinoza: Practical Philosophy*, tr. R. Hurley. San Francisco: City Lights.

Deleuze, G. (1994) *Difference and Repetition*. Tr. P. Patton. New York: Columbia University Press.

Deleuze, G. (2001) *Pure Immanence: A Life*. Tr. A. Boyman. New York: Zone Books.

Deleuze, G. and Guattari, F. (1983) *Anti-Oedipus: Capitalism and Schizophrenia*, vol. 1. Tr. R. Hurley, M. Seem, and H.R, Lane. Minneapolis: Minnesota University Press.

Deleuze, G. and Guattari, F. (1987) *A Thousand Plateaus: Capitalism and Schizophrenia*, vol. 2. Tr. B. Massumi. Minneapolis: Minnesota University Press.

Dhara, V.R., Dhara, R., Acquilla, S.D., and Cullinan, P. (2002) Personal exposure and long-term health effects in survivors of the Union Carbide disaster at Bhopol. *Environmental Health Perspectives* 110: 487–500.

Drescher, K, Goldstein, R.E., and Tuval, I. (2010) Fidelity of adaptive phototaxis. *Proceedings of the National Academy of Sciences of the United States of America* 107 (25): 11171–6.

Drescher, K., Kyriacos, C.L., Tuval, I., Ishikawa, T., Pedley, T.J., and Goldstein, R.E. (2009) Dancing *Volvox*: hydrodynamic bound states of swimming algae. *Physical Review Letters* 102 (168101): 1–4.

Fortunati, L. (2007) Immaterial labour and its mechinization. *Ephemera* 7 (1): 139–57.

Foucault, M. (1977) *Discipline & Punish*. Tr. A. Sheridan. New York: Vintage.

Gallagher, S. (2005) *How the Body Shapes the Mind*. New York: Oxford University Press.

Gibson-Graham, J.K. (2006) *A Postcapitalist Politics*. Minneapolis: University of Minnesota Press.

Gleeson, B. (1999) *Geographies of Disability*. New York: Routledge.

Gordon, U. (2010) Power and anarchy: in/equality + in/visibility in autonomous politics. In N.J. Jun and S. Wahl (eds.), *New Perspectives on Anarchism*. New York: Lexington Books, pp. 39–66.

Gregory, D. (2007) Vanishing points: law, violence and exception in the global war prison. In D. Gregory and A. Pred (eds.), *Violent Geographies: Fear, Terror, and Political Violence*. New York: Routledge, pp. 205–36.

Hannah, M. (2006) Torture and the ticking bomb: the war on terrorism as a geographical imagination of power/knowledge. *Annals of the Association of American Geographers* 96: 622–40.

Hardt, M. and Negri, A. (2004) *Multitude*. New York: Penguin.

Hardt, M. and Negri, A. (2009) *Commonwealth*. Cambridge, MA: Belknap/Harvard.

Harrison, P. (2009) In the absence of practice. *Environment and Planning D: Society and Space* 27: 987–1009.

Henry, M. (1973) *The Essence of Manifestation*. Tr. G. Etzkorn. The Hague: Martinus Nijhoff.

Henry, M. (1993) *The Genealogy of Psychoanalysis*. Tr. D. Brick. Stanford: Stanford University Press.

Hondagneu-Sotelo, P. (2007) *Doméstica: Immigrant Workers Cleaning and Caring in the Shadows of Affluence*. Berkeley: University of California Press.

Ingold, T. (2000) *The Perception of Environment*. New York: Routledge.

Jordan, D.S. and Kellogg, V.L. (1907) *Animal Life*. New York: D. Appleton and Co.

Kingsbury, P. (2007) The extimacy of space. *Social & Cultural Geography* 8 (2): 235–58.

Lazzarato, M. (2004) The political form of coordination. *Transversal*, http://eipcp.net/transversal/0707/lazzarato/en.

Lenin, V.I. (1988) *What Is to Be Done?* Tr. J. Fineberg and G. Hanna. New York: Penguin.

Lim, J. (2007) Queer critique and the politics of affect. In K. Browne, J. Lim, and G. Brown (eds.), *Geographies of Sexualities: Theory, Practices and Politics*. Aldershot: Ashgate, pp. 53–68.

Martin, L.L. (2010) Bombs, bodies, and biopolitics: securitizing the subject at the airport security checkpoint. *Social & Cultural Geography* 11 (1): 17–34.

Marx, K. (1973). *Grundrisse*. Tr. M. Nicolaus. New York: Penguin.

Marx, K. (1976) *Capital*, vol. 1. Tr. B. Fowkes. New York: Penguin.

Marx, K. and Engels, F. (1970) *The German Ideology*, In C.J. Arthur. (ed.), New York: International Publishers.

McCormack, D.P. (2006) For the love of pipes and cables: a response to Deborah Thien. *Area* 38: 330–2.

Nietzsche, F. (2003) *Writings from the Late Notebooks*. Tr. K. Sturge. Cambridge: Cambridge University Press.

Nishiyama, K. and Johnson, J.V. (1997) *Karoshi* – death from overwork: occupational health consequences of Japanese production management. *International Journal of Health Services* 27: 625–41.

North, P. and Huber, U. (2004) Alternative spaces of Argentinazo. *Antipode* 36: 963–84.

Ó Tuathail, G. (2003) "Just out looking for a fight": American affect and the invasion of Iraq. *Antipode* 35: 857–70.

Pearson, K.A. (1999) *Germinal Life: The Difference and Repetition of Deleuze*. New York: Routledge.

Precarias a la deriva (2004) Adrift through the circuits of feminized precarious work. *Feminist Review* 77: 157–61.

Protevi, J. (2009) *Political Affect*. Minneapolis: University of Minnesota Press.

Quinlan, P. (2010) Less toxic dispersants lose out in BP oil spill cleanup. *The New York Times* May 13, www.nytimes.com/2010/05/13/business/energy-environment/13greenwire-less-toxic-dispersants-lose-out-in-bp-oil-spil-81183.html.

Raunig, G. (2004) *La inseguridad vencerá. Anti-Precariousness Activism and Mayday Parades. Transversal*, tr. A. Dereig, http://eipcp.net/transversal/0704/raunig.

Read, J. (2003) *The Micro-Politics of Capital: Marx and the Prehistory of the Present*. Albany: State University of New York Press.

Ross, R.J.S. (2004) *Slaves to Fashion*. Ann Arbor: University of Michigan Press.

Saldanha, A. (2006a) Reontologising race: the machinic geography of phenotype. *Environment and Planning D: Society and Space* 24: 9–24.

Saldanha, A. (2006b) Vision and viscosity in Goa's psychedelic trance scene. *ACME* 4 (2): 172–93.

Siebers, T. (2001) Disability in theory: from social constructionism to the new realism of the body. *American Literary History* 13: 737–54.

Siebers, T. (2006) Disability aesthetics. *Journal for Cultural and Religious Theory* 7 (2): 63–73.

Simondon, G. (1989) *L'Individuation Psychique et Collective*. Breteuil-sur-Iton: Aubier.

Sitrin, M. (2006) *Horizontalism: Voices of Popular Power in Argentina*. Oakland: AK Press.

Starhawk (2002) *Webs of Power*. Gabriola Island: New Society.

Stengers, I. (2002) *Penser Avec Whitehead: Une Libre et Sauvage Création de Concepts*. Paris: Éditions du Seuil.

Stoler, A.L. (1995) *Race and the Education of Desire*. Durham, NC: Duke University Press.

Surin, K. (1996) "The continued relevance of Marxism" as a question: some propositions. In S. Makdisi, C. Casarino, and R.E. Karl (eds.), *Marxism Beyond Marism*. New York: Routledge, pp. 181–213.

Sullivan, M. (2005) Subjected bodies: paraplegia, rehabilitation, and the politics of movement. In S. Tremain (ed.), *Foucault and the Government of Disability*. Ann Arbor: University of Michigan Press, pp. 27–44.

Thien, D. (2005) After or beyond feeling? A consideration of affect and emotion in geography. *Area* 37: 450–6.

Tolia-Kelly, D.P. (2006) Affect – an ethnocentric encounter? Exploring the "universalist" imperative of emotional/affectual geographies. *Area* 38: 213–17.

Tuan, Y.-F. (1984) *Dominance and Affection: The Making of Pets*. New Haven: Yale University Press.

Uexküll, J. von (1992) A stroll through the worlds of animals and men: a picture book of invisible worlds. *Semiotica* 89 (4): 319–91.

Virno, P. (1996) Notes on the "general intellect." In S. Makdisi, C. Casarino, and R.E. Karl (eds.), *Marxism Beyond Marxism*. New York: Routledge, pp. 165–72.

Watts, M. (2001) Petro-violence: community, extraction, and political ecology of a mythic commodity. In N.L. Peluso and M. Watts (eds.), *Violent Environments*. Ithaca: Cornell University Press, pp. 189–212.

Whitehead, A.N. (1978) *Process and Reality*. New York: Free Press.

Woodward, K. (2010) Events, spontaneity and abrupt conditions. In B. Anderson and P. Harrison (eds.), *Taking-Place: Non-Representational Theories and Human Geography*. Surrey: Ashgate, pp. 321–39.

Woodward, K. and Lea, J. (2010) Geographies of affect. In S. Smith, R. Pain, S.A. Marston, and J.P. Jones III (eds.), *The Sage Handbook of Social Geographies*. London: Sage, pp. 145–75.

Woodward, K., Marston, S.A., and Jones III, J.P. (2010) Of eagles and flies: orientations toward the site. *Area* 42: 271–80.

Wright, M.W. (2006) *Disposable Women and Other Myths of Global Capitalism*. New York: Routledge.

Chapter 20

Embodied Life

Isabel Dyck

[T]he more the body is studied and written about the more elusive it becomes: a fleshy organic entity and a natural symbol of society; the primordial basis of our being-in-the-world and the discursive product of disciplinary technologies of power/knowledge; an ongoing structure of lived experience and the foundational basis of rational consciousness; the well-spring of emotionality and the site of numerous "cyborg" couplings; a physical vehicle for personhood and identity and the basis from which social institutions, organisations and structures are forged. (Williams and Bendelow 1998: 2)

…[T]he body, or rather, bodies cannot be adequately understood as ahistorical, precultural, or natural objects in any simple way; they are not only inscribed, marked, engraved, by social pressures external to them but are the products, the direct effects, of the very social constitution of nature itself. It is not simply that the body is represented in a variety of ways according to historical, social, and cultural exigencies while it remains basically the same; these factors actively produce the body as a body of a determinate type. (Grosz 1994: x)

As we walk through the spaces of the cities where many of the readers of this chapter may live, we are bombarded with images and messages about bodies. Images that promote goods intended to improve our looks – whether we are men or women – or to ameliorate our illnesses or imperfections and that remind us that we inhabit a particular body that may or may not match normative ways of being. Newspapers and magazines, through articles or advertisements, keep us up to date with science and health news persuading us to look after our bodies or, as new genetic knowledge unfolds, to see them in new ways. We are reminded that the health and appearance of our bodies are our responsibility. It seems our very being is at stake in this forest of information and ideals, where our body is promoted as the vehicle through which we can achieve a valued place in the world. There is also growing awareness that the body is not as natural as commonly thought, especially as we learn how bodies

A Companion to Social Geography, First Edition. Edited by
Vincent J. Del Casino Jr., Mary E. Thomas, Paul Cloke, and Ruth Panelli.
© 2011 Blackwell Publishing Ltd. Published 2011 by Blackwell Publishing Ltd.

can be modified in a variety of self-imposed and biotechnological ways. Although the focus on improvement and desire dominates much advertising and media imagery, images of other bodies impinge our lives, often from other parts of the world, such as those of the starving or those who are subjected to violence under repressive regimes and war torn countries. Closer to home we hear of the problem of "failing" bodies requiring care in old age and ambiguous images of "other" bodies – migrants, refugees, ethnic minorities – which may sit uneasily in national imaginaries.

But in what ways, if any, are such representations of bodies entangled with the materiality of *our* bodies, our everyday experiences and our life chances? What, indeed, *is* the body? How are we to understand the body and its relationship to our identities and lives? The quotes above suggest understanding the body and our bodily experiences is far from simple. In this chapter I aim to provoke a way of thinking about the body that acknowledges the complex intertwining of our physicality, ideas, ideals, and our sense of self, through the notion of "embodied life." In doing this I signal a complex web of relationships, the examination of which both denaturalizes the body and sets it firmly in society and space. I draw on research examples that focus on migrant bodies and bodies defined as in need of care, with a particular focus on the home as a means to elicit the spatiality of embodied life. But let's begin with the quotes above.

The authors suggest the extreme complexity of bodies, the "things" we as social beings "are" and experience, perform and ultimately conduct our lives through. As both quotes suggest, how the body is conceptualized is deeply embedded in theory and methodological practice eschewing any attempt to bring a comprehensive definition of the body – or, as Grosz notes, the plurality of bod*ies*. The singular form denies the social differentiation of bodies according to, for example, gender, race, age, sexuality, and class, and its significance in the ordering and experience of everyday life. The study of bodies in contemporary social science and humanities research has much to do with tracing the processes whereby such differentiation is achieved and takes effect, with a corresponding interest in the meanings of bodies in particular historical, cultural, and geographical contexts. While different theoretical approaches are adopted, there is broad consensus that the body is located at the interface of culture and biology. It is at this interface that analytical focus frequently rests. How, then, can we understand ourselves as both biological and cultural? How do the two intertwine? The notion of the body as a site where meanings specific to time and place are both produced and reflected is well rehearsed; its corporeality cannot be understood outside such historical, social, and geographical context.

The idea of body practices being integrally involved in the constitution of social and geographical worlds is present in a wide range of theorizations of this "most peculiar thing," as Grosz puts it (1994: xi). A strong sense of agency and the creative capacity of the human subject is signalled in such work. Grosz (xi) states, "[B]odies are not inert; they function interactively and productively. They act and react. They generate what is new, surprising, unpredictable." It is this sense of agency that I aim to explore in this chapter, in the course of which I take up the enduring tension in the literature between the body as both a material "thing" and a surface on which powerful cultural discourses are etched or inscribed. As well as political economy and other wider social processes and structures framing local materialities, it is through everyday bodily practices in local contexts that agency is enacted and able to forge change. However, bodily practices themselves are interpreted and put into action in the context of a field of meanings about bodies and spaces, as poststructural work on the discursive construction of bodies tells us. In the rest of the chapter I aim to explore embodied life as active process, constituted by and constitutive of social relations, embedded in place (understood as a point of articulation as social, cultural, economic, and political processes interweave). In the next section I briefly lay out a terrain through which to approach embodiment as a concept; I follow this with research examples to explicate this conceptual groundwork.

Bodies, Embodiment, and Geography

As noted above, ways of understanding embodiment are characterized by an ongoing tension rooted in the relationship between the corporeality of the body – its materiality composed of flesh, bones, muscles, nerves, vital liquids, genes, and other palpable or observable attributes – and discourses about the body, or bodies, which inscribe it with meaning, and through which it is represented visually, in writing or in "talk." The idea of the discursive marking of the body leads to the metaphorical conceptualization of the body as a "text to be read." Understanding the body in terms of its corporeality is familiar to most of us, as we rely not only on body sensations but also on dominant forms of knowledge about the body that circulate in society. In the global north, this is knowledge developed in the context of medical science with its taken-for-granted expertise rooted in scientific method. A long legacy of observational and experimental techniques, with more precise ways of looking and describing the body continually emerging, has established a powerful construct of the body as a knowable object of medical science. This knowledge is used in explaining and treating illness and disease, not only through establishing cause and effect relationships but also through classificatory categories that differentiate bodies (and minds) against statistical and social norms.[1] Measurement has been a key method in such classificatory exercises, and is the basis of the power of scientific medicine as a dominant paradigm of body knowledge (Petersen and Bunton 1997).

As Grosz (1994: xi) stresses, however, animate bodies cannot be reduced to mere objects, and available knowledge created through particular disciplinary methods will "have tangible effects on the bodies studied." Recent advances, for instance, such as the genome project and sophisticated biotechnology, have produced further definitive knowledge of how the body works, with this expanding surety of knowledge about the body influencing how we think of our own and others' bodies in society. Take, for example, the impact of new body knowledge on identifying and addressing deviations from the norm in the form of genetic screening. New responsibilities arise for parents of unborn children who are found to be at risk of degenerative conditions or congenital disabilities. Having to act on such knowledge is not simply a matter of assessing physical outcome, but one that is fraught with less tangible cultural, religious, and moral imperatives that shape "life" (Rose 2007). Normalizing discourses of "appropriate" bodies, ranging from political narratives concerning national identity to public health discourse, serve to form a context within which subjectivities are forged and played out. Amongst these are discourses valuing all life, whatever its limitations and constraints. So while the biomedical model of the body may be used to explore, diagnose, and treat illness, disease, and other forms of variations from established norms, how we understand bodies must necessarily go beyond its physicality or materiality to questions of identity, morality, and the emotional dimensions of being in the world (Howson 2004).

Sociological and anthropological work in particular has drawn on the notion of embodiment to trace the processes through which ways of understanding the inter-connections between different ways of understanding and talking about the body frame experience (see Howson 2004; Petersen 2007; and Williams and Bendelow 1998 for useful overviews). The "lived body" is central to such analysis. Although with roots in phenomenology, which is critiqued for a tendency to neglect power relations, a focus on everyday experience is useful in that it can be employed without discarding either the materiality of the body or its representation through discourse. The concept of embodiment, indeed, can be central to the exploration of how dis-courses are implicated in the constitution of particular kinds of bodies and the social effects of such differentiation. As with the body, there is no single definition, but the following are resonant with the notion of embodied life discussed in this chapter. Stanley and Wise (1993: 196–7) cited in Coffey (2004: 81) describe embodiment as "a cultural process by which the physical body becomes a site of culturally ascribed and disputed meanings, experiences and feelings," signaling the body as a site of resistance as well as the receiver of powerful social and cultural discourses. Turner (1996: xiii) focuses on action with embodiment defined as "making and doing the work of bodies – of becoming a body in social space."

Geographers are particularly sensitive to the spatiality of embodiment and, indeed, cast the body as a space itself. Pile and Thrift (1995: 11), for example, discuss subjectivity as "rooted in the spatial home of the body." The inseparability of bodyspace, subjectivity, and the social and material spaces comprising the every-day environments of routinized activity is reflected in Moss and Dyck's (2002: 50) discussion of the spatiality of embodiment as "lived spaces," where bodies engage "in material practices that produce and reproduce both the meanings of bodies, and the circumstances within which bodies exist." Such engagement, like Turner's idea of bodies "becoming" through their practices, suggests that spaces too come into being through routinized bodily practices.[2] This theme is taken up extensively by geographers, who have long recognized the recursive constitution of people and places. Nast and Pile (1998) look at the relationship between bodies and places in an edited collection of chapters that aims to highlight this co-constitution: as they comment, bodies "make, and are made through, the practices and geographies of places" (1998: 5). Embodied life, therefore, is not simply experienced and con-structed in and through the specificities of spatial and social practices, but incorpo-rates agency – actively involved in constitutive processes of place. Neither bodies nor places are fixed, or final.

Casey (2001), drawing on Bourdieu's notion of habitus and influenced by a phenomenological focus on being-in-the-world, is particularly concerned with the relationship between what he terms the "geographical self" and place. His discus-sion seeks to conceptualize the body as the vehicle between self and place, with bodies engaging with place through spatial practices. He takes into account the differentiation of bodies according to, for example, gender, class, and racial identity, with people bringing different dispositions for ways of self-presentation and being-in-the-world. What is interesting in his analysis, particularly in thinking through diasporic space and migrant populations, is the idea that bodies are carriers of "traces of other places," so that ways of engaging with place is infused with memories, lived through bodily comportment and practices. The lens of embodied memory is helpful in integrating bodyspace in analysis of landscape or place change. What is missing in Casey's account is an understanding of bodyspace as an inscriptive site whereby hegemonic discourses are written on the body through practices of power. In the specificities of place, the corporeality of bodies is read through multiple discourses that position the self in relation to local resources and social relations, as well as within wider processes of power that frame social and economic inclusions. For example, the migrant body, racialized in a context of white hegemonic identities of belonging at a national scale, will be read differently according to other axes of difference, such as class and location. The financial worker's "brown" body in the City of London or Wall Street, New York, or in an upper income neighborhood of either of these cities, will be open to a different interpretation than that of the poor migrant in a deprived area of the same cities – not only due to location but also to comportment and body style.

Tapping into the geographical imagination in pursuing the notion of embodied life, we can ask the following questions. How do people live and experience – do the work of – their bodies in particular places? How do individuals and groups engage and negotiate prevailing discourses about bodies through their own bodily practices? How is space implicated in such discursive construction and negotiation? How are bodily practices engaged in the constitution of the meanings of use of particular spaces and neighborhoods? In the rest of the chapter I explore these questions.

Food, Bodies, Place

The first theme I find useful in exploring the integration of geography into thinking about embodied lives is that of food. Embodiment, as geographers have emphasized, has a spatiality and is closely related to identity. Just as Bondi and Davidson (2003: 338) comment, "to *be* is to be *somewhere*," Bell and Valentine (1997) elaborate ways in which "[W]e are where we eat." The relationship, however, between where everyday food consumption takes place and identity, as performed in "styles of flesh" (cf. Butler 1990), is complex. Pratt in her work on live-in Filipina domestic workers in Canada, mainly employed as nannies, indicates some of this complexity. She discusses the difficulty that food consumption posed for migrant young women living in middle-class homes. Eating with the family Pratt likens to "literally ingesting another culture" (1998: 295) as the domestic worker prepares and partakes of the employer's household meals. Commenting on food as the "standard marker of culture," Pratt notes its place in the women's struggles in the shaping of their subjectivity through processes that racially inscribe their bodies. Notions of nation and home mediate such inscription, Pratt suggests. Research with immigrant and refugee women in Canada and with families living in an East London neighborhood of the UK similarly indicates that discourses of home and nation were at play in food practices (Dyck 2006; Dyck and Dossa 2007; Dyck 2008). In exploring the relationship between routine practices and wider social and political processes, we found food consumption practices to be a significant part of identity performance and the production of certain types of bodies – bodies that were interpreted in the context of both local communities and national narratives focusing on identity orientation. The dynamics of place identity were also shown to be negotiated through everyday practices in the home and neighborhood.

Narratives of food and home

Our work in Canada was with international migrant women from India and refugee women from Afghanistan. The research was concerned with how international migrants and refugees understood and practiced health as they re-made their lives following migration. While the two groups of women had entered Canada through different immigration policy categories – the women from India as "family class" immigrants joining or accompanying relatives through a family reunification scheme and the Afghan women through a humanitarian refugee category – both were socially and politically positioned as "other." Such groups are marginalized in social and political discourse as immigrants less valuable to Canada's nation-building goals than those entering through a points-based category delineating preferred features of human capital (Abu-Laban 1998).

The interview narratives of the women reflected a collective self-understanding of such subordinate positioning in Canadian society, reinforced by particular patterns of access to material resources such as housing and jobs. The Indian women's experience of housing, neighborhood life, domestic work, and paid employment was situated in and crafted through dense, local networks largely based on relatives and other migrants from the same region of India. Following Casey (2001), "traces of other places" were marked on both bodies and landscape through dress, bodily comportment, and a clustering of landscape features expressing "Indianness" such as Sikh temples, a faith school, and shops and markets selling produce and goods from or echoing India. The gendering of activity, reaffirmed by the women's entry into Canada as dependants, was evident as women spoke of their primary role as mothers and wives, despite also often being engaged in paid employment (low skilled jobs primarily filled through migrant labor). The Afghan women, although more highly educated, had not been able to find commensurate work in Canada and most were financially dependent on the State. Some living in households without men – their husbands killed in conflict in Afghanistan – found life financially precarious. With only a fledgeling Afghan community in the area where they had been settled, they were less well endowed with existing social networks to provide informational, emotional, and material support. Unlike the Indian women's experience of ethnic clustering, the Afghan women's bodily practices in experiencing place and re-making home were mediated and shaped through more ethnically diverse city spaces. Similarly, however, they defined their current lives in terms of a primary responsibility for their families and bringing up happy and healthy children, an orientation that had shifted to some degree from earlier lives in Afghanistan where they had been pursuing careers.

Food, health, and home were central themes in the interviews with both sets of women. Food articulated a tension between a remembered, valued past – although also kept alive in the case of the Indian women through transnational links and activity (see Dyck 2006; Dyck and Dossa 2007) – and a revised orientation to current ways of belonging in Canada. We observed continuities of food preparation techniques and consumption practices, brought from and referencing the remembered homeland and drawing on a family repertoire of familiar and valued dishes. Nevertheless, public health discourse on healthy eating had led to the modification of ingredients (less fat might be used) and permeated body ideals, with "fatness" now decried rather than framed as a desirable plumpness. Consumption of the "other" was also prominent in the women's narratives, forming part of a set of strategies used in the re-making home and producing healthy bodies in a context where "a web of meanings of bodies and remembered place ... are constantly reworked" (Dyck 2006: 7). Women wanted their children to neither "lose the taste of home" as one mother put in, nor be outside "Canadian culture." The women saw partaking of Canadian food (although much of this was the internationalized fast food of burger, french fries, pizza, and cola) as well as that of the cultural home as strategic. It positioned children as hybrid citizens embodying cultural capital that would enable them to "get on" in a multicultural nation, both in the private realm of family and home and the public life of the state and its opportunities for employment and status.

Through food practices, which might include the use of particular foods and herbs in keeping the body physically healthy, women were active in bodily re-inscription as different, cultural knowledge claims about health and healing were negotiated at the site of the body. In this process they claimed and came to embody a particular citizen script within the nation state – one that is accompanied by a particular corporeal style. In Turner's terms we can see the body "becoming" in social space. However, this becoming is not simply in reference to social space, but also material space, which, however, is layered with meanings created and reproduced through bodily practices over time. Much of the body work related here takes place in the home, a significant site in the negotiation of meanings and material practices. But its effects carry on into neighborhood and other city spaces as people go about their everyday lives. Notions and practices of the healthy body are reworked in spaces of diaspora and fueled by the particularities of place that are themselves being re-formed and brought into being through bodily practices. In the next section I turn to research that focuses directly on the relationship between place change and bodily practices.

Changing places: identities in flux

Ray (2004) notes that the food practices of migrant cuisine may, in time, effect a reorientation of hegemonic conceptions of place. Further, as Bacchi and Beasley (2002) note, how body materiality is produced is always in relation to other bodies. Through everyday bodily practices, understandings of place, and so the formation of place identity, may reproduce or contest hegemonic conceptions. In a study of family food practices in an East London borough in the UK, similar processes of place change to those reflected on the Greater Vancouver landscape are at play. The study indicated complex interactions as demographic and economic changes intertwine with the everyday realities of the people living there – reconstituting the meaning of place as well as the sense of belonging of its inhabitants (Dyck 2008).

Once a predominantly white, working-class neighborhood, the study area now houses a population reflecting the historical layering of its demographic shifts, including those associated with international migration. Halal butchers, ethnically diverse food stores, and a thriving market carrying produce catering to the cuisines of several parts of the world are amongst material changes that signify on the landscape the demographic shifts and economic polarization of a global city. The collective story of the participants of this neighborhood, a landscape primarily associated with deprivation, although with some internal socio-economic differentiation, was one of change – place change, the change of positionality of some groups within it, and for some individuals, a new area of residence. Interviews with members of 17 local family households, in two cases incorporating three generations, showed the varied relationship to place that individuals and groups have with this modest neighborhood. The voices of white working-class family members who have lived the whole of their life in the neighborhood (several now in their 70s and 80s) comingle with those of more recent residents, either immigrants or children born in the area of immigrant parents.

The study was set up specifically to investigate how households varying in composition, socioeconomic position, length of residence, and ethnicity constructed homes and neighborhoods as healthy spaces. The study's aim was to explore how everyday food practices related to the negotiation of both subjectivity and collective identity in shared neighborhood space. The participants' "food stories" showed food practices were closely associated with the material, social, and symbolic dimensions of an imagined "home." The participants were knowledgeable agents aware of the healthy eating messages contained in public health discourse, but this discourse was mediated by and interpreted through both material constraints shaping routinized behavior and, importantly, notions of a collective ethnicized identity. Food was a potent identity marker as bodily routines, discursive constructions of the body (for example that of the ethnic minority "other" against the white "center"), and space intertwined in the living of identity.

The food practices of the study participants articulated the intersection of food and subjectivity (cf. Gunew 2000), an intersection that can be seen in terms of the working out of subject position in relation to challenges to the "whiteness" of the nation's imaginary. This however was complicated by generational differences in allegiance to and experience of place. Indeed, interviews indicated the ability to construct the home and neighborhood as healthy space through the medium of food practices was embedded in participation in local social networks and institutions, household norms, and access to a changing "foodscape." For example, access to desired foods was interpreted quite differently. The once homogenous "English" foodscape remembered by the older white women living in the area and characterized by traditional butchers, bakers, and greengrocers, has been superseded by local shops on the core shopping streets and a thriving local outdoor market, which provide a wide variety of products directed at local minority ethnic populations. For some area residents this is a situation described as "you can find everything you want" here, but is countered by others in sentiments such as "there are no good shops here" and "there are no butchers."

This contrast in views, reflecting the experiences of the "new" South Asian East Enders and other migrant minority ethnic groups as against older white British East Enders, signals not only place change but a renegotiation of subjectivities through the body. Bodily routines in the home and neighborhood – acts of domestic citizenship whereby home is maintained and secured[3] – express the normative values and home-making work that "support personal and collective identity" (Young 1997: 164, cited in Blunt and Dowling 2006: 5). Older white women in the study claimed for themselves a valued "respectable" working-class identity, associated with an earlier period of this neighborhood, within a discourse of decline and neighborhood instability. They participate in a thriving lunch club, which serves familiar, traditional English food, and they shop in neighborhoods where national supermarket chains ensure a shopping environment where goods and experience are more in tune with their sense of self. Younger people, such as British born Asians and recent migrants from African countries, occupy and traverse different spaces in the neighborhood, embodying and performing identities indicating a different relationship to place.[4] Those who had grown up in the area were "at home" and felt they belonged in the multicultural environment, with its Asian and other ethnic food outlets, lively market, and a collective bodily ethnic presence on the landscape. In a similar way to the women in the Canadian studies discussed above, they are creating new meanings of place as they negotiate their subjectivities in the course of the bodily routines of their everyday life – shopping for everyday needs, taking children to school, partaking of family life, preparing meals, and making friends. While subjugated in a national imaginary with its narrative of a (thinly veiled white) "British identity," homes and the public spaces of this neighborhood are constructed spaces of alternate identities of national belonging, with bodies performing a counter narrative and a collective sense of belonging.[5]

The above studies suggest that the materialities of everyday life, socially and spatially produced and experienced as a dynamic physical environment, intertwine with and are constituted through bodily routines of active human subjects as they "do" identity, identities that are marked through multiple scripts related to gender, race, class and age. These embodied lives reflect and are circumscribed by differential positioning within relations of power; the prominence of international migration is highlighted here. Dossa (2009: 28) notes that it is through story-telling that study participants can "claim some sense of agency and purpose" within these wider workings of power. It is also through the personal stories gained through interviews that we hear of the emotional dimension of embodied life. The "emotional turn" in geography draws attention to ways of understanding the "socio-spatial mediation and articulation of emotion" and how emotions "coalesce around and within certain places" (Bondi et al. 2005: 3). In the following section I draw on work that has emotion at its core.

Vulnerable Bodies, Homespace, and "Expert" Knowledge

Research concerned with the experiences and care of the chronically ill, disabled, or frail elderly brings to light the emotional ladenness of embodiment. Again, in this second theme in exploring the notion of embodied life, I focus on the space of the home, as a material, discursive, and symbolic site closely intertwined with cultural processes that constitute the conceptualization and experiences of vulnerable bodies.

While we tend to take our bodies for granted, if we feel ill, are diagnosed with disease, suffer an accident, or begin to be less able in everyday functioning our body comes to the fore. In these cases, a changed experience of the physical body and the observation by others of such change often is mediated through some form of expert knowledge. This may be, for example, that of the biomedical sciences, such as in a clinical diagnosis, or the codification of contrasts to "normal" functioning in forms to be completed in assessments and in formulating care plans. Studies on long-term care provision in the home and on the experiences of women diagnosed with chronic illness show how expert-knowledge, embedded in policy discourse and medical diagnoses, inflects bodily experience, interaction with others, and how home is reconstituted through bodily encounters. The emotional dimension of embodiment is apparent in the narrative accounts of women with multiple sclerosis (MS) and myalgic encephalomyelitis (ME) (Dyck 1998; Moss and Dyck 2001). While their experiences of what had become an unreliable body was interpreted by themselves and others through a diagnosis that legitimized a particular bodily state, fluctuations of experience – as women had "better" and "worse" days – needed to be accommodated within the particular locales of everyday life. As women realigned themselves to the workplace and home, their stories indicated struggles, frustration, and sorrow as they navigated everyday spaces. Identities were at stake as the diagnosis, as a cultural marker, both assaulted a previously lived able and well identity with one that devalued women as employees and, in many cases, also as wives and mothers in a normative script of gender. In challenging a disabled identity, the home took on a new significance. For some it became a place out of sight where work tasks could be completed or where rest could be taken, as women attempted to conceal the stigmatizing symptoms of a less than able body in the particular context of the workspace. At the same time, as some relinquished everyday tasks of care – of the home and family members – to other family members or paid employees, their changed relationship to homespace was a constant reminder of their reduced bodily capacity.

Turning to research on the provision of long-term care in the home, which included interviews with care recipients and providers and observations of the homespaces within which care was provided,[6] we see particular bodies and homespaces emerge through entwined processes.[7] Jacobs and Smith (2008) discuss the binary between the material dwelling and the home that has characterized work in social and cultural geography, suggesting a predominant focus on the meaning of home and its significance to selfhood has neglected its materialization through "a whole range of human and nonhuman relations that bring the structures and meanings of properties into being" (p. 528). While their focus is primarily on the enlivening of "house as property," we can take from their approach the assemblage of persons, discourses, and inanimate objects that are negotiated through bodily practices as a home becomes a space of care.

Home spaces and care

A trend to locate long-term care of the frail elderly, the chronically ill, and disabled in the home has occasioned considerable discussion and debate, particularly from the vantage point of policy critique and practice issues. Nettleton and Burrows (1994) put forward the view that the movement of much care from hospitals to the community (commonly the home) represents a re-spatialization of disciplinary power, in the form of state management of bodies through spatial practices and cultural inscription. Bodies in need of care are constituted through their etching by discourses about need and eligibility. This may include the social practices of professionals, such as physicians' diagnoses, or assessments codifying experience and/or observed function against set criteria. Shildrick and Price (1996) use the example of forms that welfare claimants must fill in when seeking financial aid for personal care; these require a self-monitoring of bodily functioning throughout the day. Such forms, they claim, exemplify the proliferation of sites where Foucault's notion of disciplinary power is exercised. As individuals self-assess their need, they are exercising self-surveillance and through such disciplining emerge as new objects of knowledge to be acted upon.

One aspect of the study on the provision of long-term care in the home examined how both the meanings of "bodies in need of care" and the homespaces where such care was delivered were interpreted, managed, and negotiated by both care receivers and paid care givers (Angus et al. 2005; Dyck et al. 2005; Dyck and England 2009). Just as bodies in need of care were constructed, so too homes as caregiving sites were materialized through embodied care practices. Homes can be understood as material, discursive, and symbolic sites where a "fit" between an individual's biography and the materiality of the home may be achieved over time (see Blunt and Dowling 2006, and Chapman and Hockey 1999, for elaboration on the diverse meanings and experience of home). Over the life course, and particularly in the face of frailty, chronic illness, or disability, the meanings of home may change as social and geographical worlds shrink through constrained mobility. The centrality of home to a way of being in the world therefore becomes of greater emphasis. This coherence may be threatened when paid for care services are provided in the home. The home's discursive construction as a family dwelling is challenged. Furthermore, the meaning of a home for care recipient and care-giver will differ: for one it is a repository of meaning related to an individual's identity and place in the world (Twigg 1999), while to the other it is a workplace. The case studies showed variation in how homes emerged as caregiving spaces, highlighting the wide variety of homes within which care is provided as well as different configurations of care according to client needs. The case of "Bob" indicates the negotiation of a formal script of care (codified needs and tasks) in relation to a particular setting and care needs.

Bob, suffering from a chronic, terminal illness, received a care package of services provided by a registered nurse (RN), a registered practical nurse (RPN), and a homemaker. He had received care over a period of time, with his condition deteriorating to the extent that he was completely dependent on others for all aspects of personal body care. The care provided to him consisted of bathing, toileting, feeding, changing his clothes, and administering medication. It involved the use of bulky equipment, including a hoist to lift him from the bed. He was confined to his bedroom but Bob was able to stay at home, despite heavy care needs, through the provided nursing and homemaker services, with supplementary care from a male relative living in the house. All involved felt the care provision worked well and respected Bob's wish to remain at home. Yet the working conditions for the caregivers were difficult: they described the house as dirty and cluttered, one of the worse they had to work in. The home was situated in a rural setting, which also added demands to the caregivers' work as the well-based water supply could be threatened in summer. Navigating these different aspects of the material dimensions of care was integral to the process of constructing the home as a care space, a process which involved the mediation and negotiation of policy discourse.

According to practice guidelines, the care-giving space for Bob consisted only of the bedroom, with sharply defined tasks performed there, with the rest of the house excluded from mandated tasks. The homemaker kept the bedroom clean and did Bob's laundry. This oasis of cleanliness was surrounded by other spaces in the home the nurses needed to use, for example the bathroom to wash their hands, and a space to write up care notes and communications with other caregivers. The relative did little in the way of housekeeping and the workers needed to navigate cluttered, unsanitary spaces in other parts of the home in the course of providing care to Bob. It was only through going beyond the legitimate, codified bounds of practice and through mutual cooperation between the care-givers that a care-giving space was constructed that met their criteria for minimal standards under which quality care could be given. The homemaker would wash the dirty towels left for long periods of time in the bathroom, and the nurses brought in water when the well became dry. The care-workers would sometimes pick up Bob's medications on the way to the house to help the relative out and made other cooperative gestures between themselves to overcome the constraints of the care situation.

In these different ways the care-workers used practices of discretion as they constructed the home as a care-giving space. While the practices of care-giving were informed by institutional discourse and care policy, they were negotiated according to other discourses derived from other knowledge, knowledge beyond that of practice guidelines. The emotional dimension of care was apparent in the care-workers' narrative accounts. This included expressions of disgust at the condition of the client's home, although tempered by acknowledgement of the burden of care on the relative. They talked of their attachment to Bob and the rewards of keeping him in his home rather than within the cold walls of an institution. They spoke, too, of efforts to bring some pleasure to Bob, such as teasing and getting him to smile, and of little extras, such as small gifts brought in, so marking an attachment to him. Practices of care actively shape homespace in its ambiguous positioning as both home and workplace.

A large literature on care notes the complex interactions that go into producing a care giving situation (Mol 2008). The bodily practices of care, caught up with the emotional dimension of client-caregiver relations, are at the heart of constructing care and the bodies of the cared-for in the spaces where routine, everyday practices are carried out. As an embodied approach suggests, such bodily practices involve encounters with others and the physical environment and its objects, with discursive formations permeating enactments. Never far away are the emotional dimensions of encounters, emerging as Hubbard states (2005: 121), albeit in a different context, "in the midst of the (inter)corporal exchange between self and world." Care situations are clearly ones where emotions are likely to play a heightened part in the social and physical encounters of embodied life, but increasingly our attention is drawn to the fact that emotions do *matter* in geographical analysis of a wide range of concerns (Bondi et al. 2005: 1). The scrutiny of the emotionality of embodiment is beyond the scope of this chapter, but clearly there is room to further this line of investigation.

Conclusions

This chapter aimed to bring a geographical imagination to notions of embodiment and embodied life. Geographers have emphasized the co-constitution of place and subjectivities, with the spatiality of everyday life providing a potential entry-point in the analysis of how broader processes are articulated in the specific materialities, social interactions, and cultural meanings of everyday life. The everyday, or local, as a point of entry to examining multi-scalar processes and the potential to situate local action in wider processes and relations, is well recognized. The notion of embodiment centers the body, understood as constituted at the interface of biology and culture, in the analysis of complex processes as social and geographical worlds are made and experienced. I used a focus on the home, and by extension the neighborhood, to explore how bodily practices intertwine with meanings of everyday spaces and dominant discourses about bodies in day-to-day life.

Qualitative research strategies provide an array of tools through which to explore the shaping and regulation of subjectivities within the very spaces of everyday life. Investigating narratives of embodied life is an important route in discovering such knowledge, knowledge that emerges from bodily experience in particular sites where bodily practices and social interactions occur. The study of embodied life, therefore, has a radical potential. As Ong (1999: 4) comments, we can understand "everyday actions as a form of cultural politics embedded in specific power contexts." In sum, a focus on embodiment as active process enables the reconciliation of agency and structure in a way that admits meanings, emotions, materialities, and differentials in power.

So when we walk along a city street and engage with a range of media, whether print or visual, we can keep in mind that such body representations constitute one form of discourse through which our fleshy corporeality is mediated, negotiated, and socially constituted. But there are others, many of which we take for granted, such as the expert knowledge of biomedicine which permeates public health messages about what we should eat, how we should conduct our bodies, and the responsibility we have for maintaining them in certain ways. Or categories of difference, such as those based on perceived ethnicity or gender, which frame our understanding of appropriate bodily practices in particular places as well as notions of belonging. Our social geographies are inextricably entangled with our embodied lives.

Notes

The research projects discussed in this chapter were funded with grants from the Social Science Research Council of Canada and The British Academy. I am grateful for their support.
1 The biomedical paradigm does not go uncontested. There is considerable evidence that it is resisted or negotiated through other, but subordinated, knowledge that creates alter-

native ways of understanding the links between health, illness, body, and mind such as various forms of complementary and alternative medicine (CAM) and Indigenous healing systems.

2   See also Gregson and Rose (2000) for a useful explication of the notion of places being brought into being through routinized bodily practices.

3   I am indebted to Jane Jacobs for the notion of domestic citizenship.

4   Only one household (of three generations) included a white British young person. She was in her teens, attended a local school, and had ambitions to make her career in central London. It is not possible therefore to interrogate the meaning of place for this generation from the study data.

5   The instability of collective belonging within a national imaginary that continues to construct ethnic minorities in terms of "the other" was highlighted in 2006 when police raided a house in the study neighborhood where Asians suspected of terrorist activities were living. The considerable media coverage afforded the event signaled a racialized politics of difference, fueled further when the suspects were released without evidence against them found.

6   The Hitting Home project was led by Dr. Patricia McKeever, University of Toronto, and consisted of a multidisciplinary team investigating home care from the perspective of different players, including paid care givers, family care givers, and care recipients. The materiality of the home and its meanings provided a central focus for the project.

7   Foucault's ideas are significant here, with the inscription of policy discourse and biomedical knowledge on the body through institutional and professional practices, exemplifying the workings of "capillary" power. However, the body as text cannot be understood just as a passive surface; rather individuals negotiate and resist inscriptions through everyday practices, "re-writing" both bodies and daily life spaces. Embodied experience, then, encompasses the interpretation and potential re-working of discourses.

References

Abu-Laban, Y. (1998) Welcome/STAY OUT: the contradictions of Canadian integration and immigration policies at the millennium. *Canadian Ethnic Studies* 30: 190–211.

Angus, J., Kontos, P., Dyck, I., and McKeever, P. (2005) The personal significance of home: habitus and the experience of long term home care. *Sociology of Health and Illness* 27: 161–87.

Bacchi, C.L., and Beasley, C. (2002) Citizen bodies: is embodied citizenship a contradiction in terms? *Critical Social Policy* 22: 324–52.

Bell, D. and Valentine, G. (1997) *Consuming Geographies: We Are Where We Eat*. London and New York: Routledge.

Blunt, A. and Dowling, R. (2006) *Home*. London and New York: Routledge.

Bondi, L. and Davidson, J. (2003) Troubling the place of gender. In K. Anderson, K. Domosh, S. Pile, and N. Thrift (eds.), *Handbook of Cultural Geography*. London: Sage, pp. 325–43.

Bondi, L., Davidson, J., and Smith, M. (2005) Introduction: geography's emotional turn. In, J. Davidson, L. Bondi and M. Smith (eds.), *Emotional Geographies*. Aldershot: Ashgate, pp. 1–16.

Butler, J. (1990) *Gender Trouble: Feminism and the Subversion of Identity*. London and New York: Routledge.

Casey, E. (2001) Between geography and philosophy: what does it mean to be in the place-world? *Annals of the Association of American Geographers* 91: 683–93.

Chapman, T. and Hockey, J. (eds.) (1999) *Ideal Homes: Social Change and Domestic Life*. London and New York: Routledge.

Coffey, A. (2004) *Reconceptualising Social Policy: Sociological Perspectives on Contemporary Social Policy*. Maidenhead: Open University and Price.

Dossa, P. (2009) *Racialized Bodies, Disabling Worlds: Storied Lives of Immigrant Muslim Women*. Toronto: Toronto University Press.

Dyck, I. (1998). Women with disabilities and everyday geographies: home space and the contested body. In R.A. Kearns and W.M. Gesler (eds.), *Putting Health into Place: Landscape, Identity and Wellbeing*. Syracuse: Syracuse University Press, pp. 102–9.

Dyck, I. (2006) Travelling tales and migratory meanings: South Asian migrant women talk of place, health and healing. *Social and Cultural Geography* 7: 1–18.

Dyck, I. (2008) Changing places: food, health and belonging in East London. Unpublished paper, presented to the Annual Meeting of the Association of American Geographers, April 15–21, Boston.

Dyck, I. and Dossa, P. (2007) Place, health and home: gender and migration in the constitution of healthy space. *Health & Place* 13: 691–701.

Dyck, I. and England, K. (2009) Homes for care: negotiating the intimate spaces of body work in home-based care. Unpublished paper presented at Homecare Pre-conference Workshop, In Sickness and in Health, April 14–15, Victoria, BC.

Dyck, I., Kontos, P., Angus, J., and McKeever, P. (2005) The home as a site for long-term care: meanings and management of bodies and spaces. *Health & Place* 11: 173–85.

Gregson, N. and Rose, G. (2000) Taking Butler elsewhere: performativities, spatialities and subjectivities. *Environment and Planning D: Society and Space* 18: 433–52.

Grosz, E. (1994) *Volatile Bodies: Toward a Corporeal Feminism*. Bloomington and Indianapolis: Indiana University Press.

Gunew, S. (2000) Introduction: multicultural translations of food, bodies, language. *Journal of Intercultural Studies* 21: 227–37.

Howson, A. (2004) *The Body in Society: An Introduction*. Cambridge: Polity Press.

Hubbard, P. (2005) The geographies of "going out": emotion and embodiment in the evening economy. In J. Davidson, L. Bondi, and M. Smith (eds.), *Emotional Geographies*. Aldershot: Ashgate, pp. 117–34.

Jacobs, J.M. and Smith, S.J. (2008). Guest Editorial, *Environment and Planning A*: 515–519.

Lawson, V.A. (2000) Arguments within geographies of movement: the theoretical potential of migrants' stories. *Progress in Human Geography* 24: 173–89.

Mol, A. (2008) *The Logic of Care: Health and the Problem of Patient Choice*. London: Routledge.

Moss, P. and Dyck, I. (2001) Material bodies precariously positioned: women embodying chronic illness in the workplace. In I. Dyck, N.D. Lewis, N.D., and S. McLafferty (eds.), *Geographies of Women's Health*. London, Routledge, pp. 231–47.

Moss, P. and Dyck, I. (2002) *Women, Body, Illness: Space and Identity in the Everyday Lives of Women with Chronic Illness*. Lanham: Rowman and Littlefield.

Nast, H. J and Pile, S. (eds.) (1998) *Places through the Body*. London and New York: Routledge.

Nettleton, S. and Burrows, R. (1994) From bodies in hospitals to people in the community: a theoretical analysis of the relocation of health care. *Care in Place* 1: 93–103.

Ong, A. (1999) Making the biopolitical subject: Cambodian immigrants, refugee medicine and cultural citizenship in California. *Social Science and Medicine* 40: 1243–57.

Petersen, A. (2007) *The Body in Question: A Sociocultural Approach*. London and New York: Routledge.

Petersen, A. and Bunton, R. (eds.) (1997) *Foucault, Health and Medicine*. London and New York: Routledge.

Pile, S. and Thrift, N. (1995) Introduction. In S. Pile and N. Thrift (eds.), *Mapping the Subject: Geographies of Cultural Transformation*. London and New York: Routledge, pp. 1–2.

Pratt, G. (1998) Inscribing domestic work on Filipina bodies. In H.J. Nast and S. Pile (eds.), *Places through the Body*. London and New York: Routledge, pp. 283–304.

Ray, K. (2004) *The Migrant's Table: Meals and Memories in Bengali–American Households*. Philadelphia: Temple University Press.

Rose, N. (2007) *The Politics of Life Itself: Biomedicine, Power, and Subjectivity in the Twenty-first Century*. Princeton and Oxford: Princeton University Press.

Shildrick, M. and Price, J. (1996) Breaking the boundaries of the broken body. *Body & Society* 2: 93–113.

Stanley, L. and Wise, S. (1993) *Breaking Out Again: Feminist Ontology and Epistemology*. London: Routledge.

Turner, B.S. (1996) *The Body and Society*. London: Sage.

Twigg, L. (1999) The spatial ordering of care: public and private in bathing support in the home. *Sociology of Health and Illness* 21: 381–400.

Williams, S. and Bendelow, G. (1998) *The Lived Body: Sociological Themes, Embodied Issues*. London: Routledge.

Young, I.M. (1997) House and home: feminist variations on a theme. In *Intersecting Voices: Dilemmas of Gender, Political Philosophy, and Policy*. Princeton: Princeton University Press, pp. 134–64.

Chapter 21

Discursive Life

Chris Philo

Introduction: Words, Geographies, and Clearing Some Ground

We live amongst words. Words spoken – uttered, muttered, shouted, whispered; words silently mouthed. Words written – printed, typed, hand-written, scribbled; words written to be widely read or to be kept in secret places. Words on paper, on screens, in books, in memoranda, in diaries, in letters, in text messages, on billboards, on backs of envelopes. Words kept and words lost. Words in sentences, words alone, words making sense, words seemingly nonsensical. Words as information, instructions, intimations; words repeating "facts" and words telling "stories"; words as worldly reflections or words as other-worldly imaginings. Words addressed and circulated to many people; words voiced to a handful of people or just one other; words said to oneself, in one's head and heart, with no material trace at all. Words in recognizable languages, words as symbolic notations, words as strangely "private" vehicles for communicating with oneself. And these words about words, the many kinds of words, the diverse forms that they take and numberless contexts of their production and consumption, are themselves hopelessly inadequate for capturing this multiplicity and variability of words.

Difficult as it is to find the right words to capture all that I mean here, we, human beings, live amongst words. Notwithstanding certain lines of argument to the contrary, it remains widely accepted that the use of words or, more grandly, language, is *the* key distinction to be drawn between the human animal and all of the other animals (life-forms) with whom we share our planet – but this is another point to which I will return. Excepting a few humans with very particular intellectual or somatic impairments for whom *wording* (as in speaking, writing, hearing, reading) is deeply challenging, the rest of us humans do rub along with words almost all of the time, even likely in our sleep and reveries. Try to empty your awake mind of

A Companion to Social Geography, First Edition. Edited by
Vincent J. Del Casino Jr., Mary E. Thomas, Paul Cloke, and Ruth Panelli.
© 2011 Blackwell Publishing Ltd. Published 2011 by Blackwell Publishing Ltd.

all words, and you will quickly appreciate the validity of this statement. Stilling the trains of words that otherwise constantly rattle through your mind might be possible in meditative or drugged states, but for most of us getting the words to cease is practically impossible (and for some it can even become an "illness" when way too many competing words crowd into their wakeful hours). Such a claim is to make no judgment about the precise neurological and physiological processes involved in uttering and scripting words, and all that I have written so far inevitably begs complex questions about what we might mean by "mind" and "brain" (or even "head" and "heart") and, indeed, by the very word "word" itself. Such a claim is also to be agnostic about the exact lengths, connectedness and route-edness of (the trains of) words involved, and I am in no doubt that the variations here – mapped on to myriad bodily and socio-cultural dimensions (age, gender, ethnicity, class, place of origin, etc.) – are so numerous as immediately to demand attention to an immensely complicated historical geography of wording at a range of spatial scales.

This is not the occasion to begin debating such matters at length, and for the remainder of this chapter I will operate with a relatively common-sense understanding of words that I anticipate sharing with most of my readers. This understanding supposes words spoken and written to be a crucial domain of human life, not just in that words allow us to tell each other "what is going on" but also in how words cannot but be central to "making" much that goes on. In other words, by this token words are not merely *reflective* of the world around us, a means for responding to it and sharing these responses in some manner, but are also *generative* of the world, not wholly, to be sure, but in all kinds of ways entering fundamentally into how human facets of the world are planned, enacted, maintained, and transformed.

Here, specifically in the latter claim about the generative as opposed to merely reflective force of words, is the focus for my chapter. I will begin with a mini-example of "words and geography" as discerned from my city doorway. I will then discuss a very recent departure within the conceptualizing of human geography, an orientation termed non-representational theory (NRT), that is highly *critical* of such attention to words. This discussion of NRT will be a spur to two interlocking arguments for *restating* the importance of words, albeit now alert to the hesitations of NRT before words. The first will borrow from the "archaeology" of Michel Foucault, the celebrated French intellectual, to rehearse claims about taking seriously the words of "serious" human discourses. The second will borrow from ethnography, including ethnomethodology and its interest in everyday conversation, to rehearse claims about taking seriously the words of ostensibly "trivial" human chatter (spoken *and* written). The result will be sketchy, given the complexity of the terrain covered, but a small measure of empirical grounding for the overall argument will be provided in three set-apart boxes (see e.g., Box 21.1) of text-and-illustrative-material derived from my own substantive research (Philo 2006).

Words and Geography

As I stand at my doorway this morning, seeing a typical western city street full of buildings and cars, animated by people doing things and moving in all directions, I cannot but also intuit a rich tapestry of words. First, there are the words spoken

---

Box 21.1   The case of William Blacklock

William James Blacklock (1816–58), the son of an established landowning family from Cumbria, north-west England, became an artist in the 1830s, and spent time in London pursuing a professional career. His health deteriorated over time, his eyesight included, but the main problems were what we might now term "neurological," with manifestations taken at the time as impending "insanity." He returned to Cumbria, and on November 28, 1855, he was admitted by his brother to the Crichton Royal Institute (CRI), Dumfries, south-west Scotland, a notable lunatic asylum forever associated with the pioneering "medico-moral" treatments of its long-time superintendent Dr. W.A.F. Browne.

Upon admission Blacklock was described in the CRI *Casebook* (vol. 11, 1854–9) as "egotistical, self-willed, passionate," enduring "elevation of spirits" and with warrant for "suspecting the presence of monomania of ambition and general paralysis." Immediately, it can be seen that the words in the *Casebook* gave a shape – a definition bound up in contemporary nosologies of mental disease – to Blacklock's condition, framing it in a manner suggestive of appropriate treatments and, in effect, setting a template for how Blacklock would be regarded and acted upon for the remainder of his life. The written words here were absolutely crucial. "General paralysis" was a term often used to characterize individuals suffering the physical and mental afflictions of developed syphilis, and in a later case note Blacklock's syphilis was explicitly referenced. Subsequent case notes also reported "epileptiform convulsions" or "attacks," reinforcing the likely somatic basis of his condition, and the final entry for him stated that, on March 12, 1858, Blacklock "[d]ied of convulsions."

There is much that could be said about Blacklock as an artist, and since publishing my 2006 paper I have uncovered further evidence of his relative prominence on the British (London) art scene, exhibiting at the premises of the Royal Academy, the British Institution and the Royal Society of British Artists (head-quartered at the Suffolk Street Galleries) (Johnson 1975: 42; Pavière 1968: 19; Various 1995: 47). This "geography" of exhibiting acquires a sad significance for Blacklock, as can be seen from Box 21.3. Germane to what follows is also the fact that he continued to draw and paint while at the CRI, and examples of his art were included in folders of work by so-called "mad artists" assembled by Browne, itself the basis for a paper that Browne (1880) published anonymously in which Blacklock's pieces were briefly (but without attribution) discussed. Items of Blacklock's art, together with specimens of his letters, found their way into the collection of another nineteenth-century psychological physician, Thomas Laycock, becoming key evidence in a written account where Laycock conceptualized mental disease (see Box 21.2).

---

by the people in the scene – a father to his young daughter in a pushchair, the two furniture removers sitting in their van opposite, the smart businesswoman on her mobile phone, the student murmuring to himself about his revision notes as he races to school. But, secondly, there are also the layerings of words unavoidably "written" into the fabric of countless decisions made by different authorities, public and private, at different times about the location, structure, utilization, and appearance of the road itself, the houses and gardens, the shops and businesses, the school and church, the services to be provided, the local bylaws, and so on.

To this way of thinking, the street is a scene *of* words, tumbling incessantly from all of its occupants and passers-by. True, many of these words have nothing to do with the street – perhaps being about a TV program, yesterday's match, a future meeting, a Geography examination – although some of them may be very directly about the street ("Can we back our van up that drive?") or at least more directly influenced by it ("Speak up, I can't hear you, this is a busy street"). The street is a scene of words, then, but it is also a scene partially constituted *by* words. Not necessarily ones spoken or written *in situ*, although some of them will have been pronounced locally, but certainly ones spoken or written (and heard and read) elsewhere: in all manner of City Council planning meetings from the grandest of chambers to the meanest of back offices, in the studios of architects and designers, in corporate head offices, in city educational departments or religious assemblies, and so on. "Words spoken *in* places, words spoken *about* places," to paraphrase the geographer Allan Pred (e.g., 1989, 1990). What this tiny example conveys is the densely overlapping, networked, and often contested discursive web that is woven around the likes of any one city street, full of words *in* this street as well as *about* this street. The material geography of the street meets, and in many respects is made by, the largely immaterial geography of the words in, of, and about this street.

Here, to me, is a glimpse of the relevance of discursive life for the study of human, and more specifically social geography. Few accounts to date have framed the importance of a discursively attuned human geography in quite the manner implied above, it is true, but it would be possible to reframe various maneuvers taken in the history of the discipline through such a lens. Different strands of perceptual, behavioral, and humanistic geography have taken seriously the "thoughts" (and sometimes "feelings") of people making decisions with geographical consequences – where to site a factory, where to relocate a residence, where to route a railway, where to place a hospital, etc. – and have sometimes began to reflect more deeply on the character of the "language" or "discourse," whether individual or collective, seemingly implicated in such decision-making (e.g., Philo 1989). Contributions to the so-called "cultural turn" (Cook et al. 2000) in human geography have engaged with claims about the routinized conversational encounters central to the formation of everyday "life-worlds," as the meaning-filled horizons for human socio-spatial activity (e.g., Cloke et al. 2004: ch. 10). Closely related has been a "textual turn," in which a focus on "writing worlds" (e.g., Barnes and Duncan 1992; Duncan 1990) has elaborated how discursive constructions ("scripts") of, for instance, nationalism, colonialism, geopolitics, racism, patriarchy, and diverse other power-laden bases of group identity become imprinted on landscapes of all kinds (intimate and expansive, urban and rural, built and "natural"). In another vein, moreover, some other geographers have engaged directly with literature, meaning novels, poems, and other worded artistic creations, with the ambition of appreciating how "imaginative geographies" (after Said 1978) or "senses of place" (e.g., Pocock 1981) may be fashioned in such media, but also sometimes seeking inspiration for other ways of writing academic human geography that arguably dispense with the too-rigid protocols of natural and even social science (e.g., Olsson 1980; Cloke et al. 2004: ch. 11; Dewsbury 2009). Accepting that so much more *ought* to be said about such meetings between academic geography and the realm of words, then, I nonetheless wish to concentrate instead on the provocations of NRT, wherein *all* such meetings, however exactly they are conceived, become open to dispute.

"After Words": The Challenge of Non-representational Geographies

NRT in geography, or non-representational geographies, stems from Nigel Thrift, a British geographer, and his sustained battle against an envisaged tyranny of words in *what* we elect to study, *how* we study, and *how* we report our findings (as academics). A short early piece named the dangers attending to our obsession with "over-wordy worlds" (Thrift 1991), with becoming overly hung up on what people say rather than with what they do. Here he essayed a critique of the "cultural" or "textual turn" – "the ascendancy of texts and writing" – for "becoming biased towards what I will loosely call 'middle-class' concerns with their own myths, modes of representation and 'transnational' worlds" (Thrift 1991: 144). Revealingly, in this connection he also suggested that: "[i]t all starts to smack of something Walkerdine ... and other feminists have constantly warned about ...; namely, the desire to position working-class women's discourses in middle-class ones and so in a sense to regulate them, actually to obliterate difference by patronising it" (Thrift 1991: 145, citing Walkerdine and Lucey 1990). Such strains of critique echo Stephen Gale's (1977) warning against academics projecting their own modes of intellectualizing the world – their own words, concepts, logics, representational preferences, and the like – on to "ordinary" people who probably tend to think, feel, and do things somewhat differently. It is to dispute accounts by human geographers, social scientists, or whoever that depict "us" as *all* mini-intellectuals, critically identifying, assessing, debating, and deciding as we strive to cope with our lives.

Much the same line of critique was pursued by Thrift (2000b) when reflecting on a further decade or so of the "cultural turn" in human geography, but now stirring in an emerging perspective that more overtly emphasized the realm of *practices*, many of which stand as skilled accomplishments that are, in their execution, little-reflected upon, barely registered at the levels of consciousness or cognition (whatever precisely these might entail), and nothing but – or nothing *more* than – "ways of proceeding," of going on, of being-through-doing as a line of flight across the moment-by-moment exigencies of the here-and-now. Through a suite of publications, Thrift (from 1997 to 2008) has assembled an impressive array of arguments demanding that we engage with the "still point" when representation stops or in respect of which representation just does not matter, when thought and hence words are stilled, which arguably lies at the beating heart of most (all?) human activities. Thus, when we dance, paint, craft a sculpture, play a sport, make love or even, more mundanely, walk, wave, put on a coat, turn off a light-switch, etc., the claim is that we are precisely *not* functioning reflexively, knowingly, or self-consciously with an implied chronology whereby, first, we "think" (or speak to ourselves) the nature of the act to be undertaken before, second, we perform the act. If anything, the chronological gap is the reverse, so Paul Harrison (2000) stresses, such that the act occurs and our apprehension of the act – our ability to word it to ourselves – is always racing to catch up, maybe only a matter of nano-seconds behind but always late. Far from being generative, then, thoughts-and-words are here doomed to be secondary in the vast majority of practices occurring from one minute to the next, whether kissing, kicking, or killing, which immediately prompts huge questions as to why academics have spent so much time worrying about what we say as opposed to investigating what we do.

Perhaps the most famous example supportive of this proposition of the chronological gap comes as an aside in an 1884 paper by the US pragmatist philosopher William James. If we see (sense) a bear in the woods, we run away as fast as possible and, if lucky, later think/say "that was a bear"; but what we do not do is stop and think/say "that is a bear" and then run, since that would be curtains for us. This example is instructive, not least because, by implication, it positions us as very much *part* of nature, as just another organism responding by flight from its sensed source of danger, seemingly as an outcome of instinct. It is fairly easy to identify other activities (eating, sleeping, procreating, getting sick, etc., as well as many other acts already mentioned) where such continuities with *all* life, with the vital processes of *all* living, are perhaps paramount – quite possibly determinant of what we do, how we do it, when and where, irrespective of *post hoc* intellectualizations – and infer a wafer-thin divide between "us" and countless other life-forms. The *human* geographies of such activities are arguably much less distinctively human than we might previously have imagined, and, by extension, the claim becomes that even many more ostensibly "human" of our activities – notably skilled ones – still retain within them a substantial *non-human* embodied foundation of practices that mocks their assumed distinctiveness from non-human animal worlds. It is for these reasons that Thrift's advocacy of NRT has been closely allied with a "vitalism" (esp. Thrift 2005; also Greenough 2010), meaning the dynamics and processes of life-in-the-round, humans included but not privileged, and with provocative attention to various non-human "multitudes" or "swarms" (organic [biotic, animal, viral] and technological [machinic, electronic, biotechnic]) that are intimately folded into the co-mingled ecologies of places where humans may (or may not) be dwelling.

Instructively too, Thrift and others following in the tracks of NRT – a broad church, it might be added (see esp. Anderson and Harrison 2010; also Anderson 2009; Cadman 2009) – have started to contemplate what it *is* that tends to impel human actions, if it is accepted that the crucial elements are not conscious or cognitive thought-and-words. At the borders of NRT and the subfield designated as "emotional geographies" focus alights upon such constructions as "emotions" and "feelings" or, with a more psychoanalytic flavor, "passions" and "drives" that arise in or are triggered by, initially at least, the embodied exchange between a human being and their environing world (or possibly in the embodied recall of prior such exchanges). To over-simplify, by such an account, this world – meaning its occupants, human and non-human, and the events transpiring therein – creates sensations to be felt, through eyes, ears, nose, and touch, which are then circulated as "feelings" traversing the body and providing the raw material of an emotional state (happiness, sadness, elation, terror, being gripped, being bored, etc.) that may then be worded in some way. The writers here would largely agree that the sensation is primary, its naming secondary, but that there may be a profound power in the emotional states so released, possibly impelling a person to dangerous courses of action, perhaps violent, sexual, "deviant," corrupt. To the psychoanalyst, these emotional states may encompass previously repressed psychic material – repressed as in ejected from conscious reflection – lurking in the body, buried in wherever physically the *un*conscious (the largely *un*word-able) is reckoned to be located, but still pregnant with the potential for "making things happen," for erupting into conduct (see Kingsbury 2009). To the more "strict" or vitalist non-representationalist, the desires or drives so released are manifestations of a broader, trans-personal surface of *affect* – possibly envisaged as resonating fields of "stuff" present in the ambience of a place, about whose ecologies we are only starting to inquire – that may localize in particular humans but which precedes them, so to speak, constituting a (back)ground out of which the likes of thoughts-and-words may (but need not) take shape.

Throughout the above thumbnail description of a vitalist, NRT-inspired human (or *post*-human) geography, it should be obvious that Thrift and his co-workers have been *battling against words*. Although this is possibly to over-state matters, the impression given is that the enemy for them is words, even if it takes many words to articulate this dislike of words, which may, ironically, prove their point about the inadequacy of words. Indeed, Thrift's compelling 2000a agenda for NRT in human geography is entitled "Afterwords," and it begins with a note – with the printed text deliberately gradually fading as it progresses – about the death of Thrift's father which is at once also a strong auto-critique of the need to put the deceased "into words," the argument being that his father was so much more than, and hence cannot remotely be done justice by, mere words (see also Harrison 2007). Continuing the questioning of "over-wordy worlds," this paper rehearses at length, from diverse theoretical positions, the need to prioritize practices and to resist the seductions of representation, chiefly in what we research (always look for evidence *other* than mere words in the human-geographical situations under study) but also in being more experimental in how we, as academics, might eventually strive to get "after words" in our reporting of ideas, findings, and recommendations. The message is that so little of what "we" humans are, be, and do is worded, at least not firstly or generatively, so that our efforts as geographers must look elsewhere for what matters, is decisive, might make a difference. Moreover, in the editorial to a 2000 collection of papers on performative geographies, Thrift and J.D. Dewsbury effectively position much that has recently passed for human geography and social science, with its cultural and textual obsessions, as "dead," somehow imposing too many alien thoughts-and-words on subject-matters that ought to be allowed to "breathe" more easily. The inference is obvious: words are deadening, and in effect words or the discursive are opposed to life, which hence appears to rule out of court the couplet chosen for the title of this chapter, "discursive life."

"For Words": The Restating of Wordy Geographies

I want to make clear that I regard NRT in contemporary human geography as a fine thing: a fantastic set of provocations, at one level starkly simple and direct, at another forged through some of the most sophisticated "philosophical" engagements yet attempted by any academic geographer. NRT-inspired geographies demand attention, respect, and response (e.g., Lorimer 2005), and to my mind they embody as profound a challenge to the discipline as, say, feminist geographies did from the 1980s. But these geographies are certainly not beyond critique, on all sorts of grounds (theoretical, methodological, substantive, ethico-political) and likely from many different standpoints, although such critique will need to be open-handed in its assessment if it is not to miss the import of what NRT *has* brought to the table. The rest of this chapter aims to proceed in such a spirit, seeking to restate the importance of human geographies attuned to words, especially to the *work* that words do across a range of different registers, but always appreciating that words are *not* everything but rather are always, unavoidably, entirely in and amongst everything else, influencing and being influenced, "having their say" (even if this might not always be decisive).

Perhaps my most general response to NRT, however, is as follows. I *do* agree about so much of being human having nothing to do with words – performed *without* direction from some chattering little "inner controller" drawing up, evaluating, and making decisions on the basis of "mental" representations – and I *do* accept that much of what "we" are, be, and do divulges continuities with *un*worded life-in-the-round. Thus, I can concur with Thrift's (2000b: 37) proposal that 95 percent or even more of our life as humans is non-cognitive and hence non-representational, entailing practices and their impulsions that elude, evade, or simply never-had-anything-to-do-with representations (including the words, spoken aloud or to oneself, integral to [most] representations). But, echoing Bob Dodgshon's (2008) rumination on NRT's sense of time, I harbor the suspicion that this 5 percent remainder of representation-dominated human life is exactly where we need to look if we wish to find the majority of "things going on" which *are* the substance of most worldly human geographies. Dodgshon's point is slightly different, since he stresses how it is in this 5 percent that we humans create the words (the books, ledgers, reports, archives) enabling us to stretch our temporal comprehension of ourselves – our cultures, societies, economies, and polities – across far more than just the nano-second gap between practice and its cognition.

My (simple, even naïve) supplementary claim is that it is this 5 percent of humanity's representational (and hence worded) activity that surely must be the generative ground out of which decisions are made, plans formulated, strategies devised, and tactics agreed, leading to such outcomes as: the laying out of settlements and the building of cities; the creation of agricultural and industrial production systems; the founding, defending, and policing of (nation-)states; the enacting of both social policies and social resistance; etc. (and this list can only be suggestive and evocative). Yes, the words involved may be very different from one another in form, content, and reach – from a detailed paragraph in an international treaty to a shouted invocation on a street corner – and, yes, NRT and other practice-based theories rightly warn us not to privilege any of these words as somehow "pure" originating moments. They will all have come from somewhere, been subject to all manner of shapings dependent on the precise spaces and places involved, perhaps lying in the folded veils of affect, full of pre-worded drives, desires, and passions, and so on. But, to my mind, a serious mismatch occurs if, as it were, we simplistically read the plans from the drives, the policies from the affects. We then miss the processes by and through which worded deliberations intervene, with more-or-less intellectual

sophistication, to formulate problems and solutions, to divine intentions and projec-
tions, to "tell us" (quite literally) how "to go on." It is therefore into the 5 percent
that I now must dive.

Taking seriously "serious" words: Foucauldian excavations

In his introduction to the English-language edition of *The Archaeology of Knowledge*
(Foucault 1972), Foucault expresses his impatience with all forms of what he termed
"totalizing" historical explanation that, in effect, draw together and simplify all of
the diverse components of a given period and place under study, postulating a core
logic – whether it be the forces of nature, some spirit of humanism, the dynamics of
capitalism, etc. – that somehow permeates everything, "capturing" everything in its
jack-booted embrace (also Elden 2001; Philo 1992). More specifically, he declares
his resistance to all ways of dealing with the written "archive" of the past, the repositories of words scripted on pages, that aim to delve *beneath* or *behind* them for the
"real truth" of what they say. He thereby opposes, or at least hesitates before, any
attempt to read them for underlying (psychological) motivations on the part of
authors, individual or collective, or for wider cultural-symbolic systems ("ideologies") or for expressions of, say, a deeper humanist consciousness or capitalist strategizing. All such attempts, and many more, are seen by Foucault as forms of
reductionism that risk avoiding taking seriously the words themselves, not just their
form but also, crucially, their *content* – what they actually, on their immediate surfaces, say about the worlds out of which they emerge and in relation to which they
are frequently supposed to make an intervention. From this perspective, an NRT
position, seeing words as secondary and hence to be looked "past," is also guilty of
a certain reductionism. Instead, Foucault urges us to treat these repositories of words
as "monuments," themselves shallow-buried under soil and sediment, for which
read, *a priori* assumptions. They need to be carefully excavated, the obscuring matter
dug away, their original shapes and lines surveyed, their fragments discovered and
reassembled: in short, their "archaeology" disinterred and laid bare for sensitive
inspection. Foucault wants us to pay close attention to the words themselves, suspending the temptation for our eyes (and concepts) to keep flicking elsewhere, since
for him, notwithstanding all manner of qualifications and elaborations on this theme
throughout his *oeuvre*, it is here – in the words – that the evidential "clues," not just
for human history but for all human social life, past and present, are to be found.

It is true that Foucault's primary standing as an *historian* does color his approach
to words, since he cannot but think of them as *marks*, the scratchings of a quill
pen, the indentations of a printing press, whatever, distributed over the material
remains of yellowing parchments and paper deposited in archives and libraries of
all kinds. It is also true that Foucault faces the particular dilemma of the historian:
namely, that there are so few sources for recovering the past which are *not* entirely
or principally collections of words on pages (cf. Lorimer 2009). Yet, Foucault's
approach to the words arguably spirals way beyond the specialist province of the
historian, and indeed has lessons for any scholar of any period or place, not least
because he repeatedly makes clear that his understanding extends from written
words/statements to spoken words/utterances (which, by this logic, should also
be treated archaeologically as fragile but identifiable "monuments"). In the main

chapters of *The Archaeology of Knowledge*, Foucault develops an ontological vision of how knowledge is pieced together from words; of how an "archive" – which can be as much a body of knowledge available for mobilization in the here-and-now as a dusty cellar of "old" documents – becomes gathered, acquires some measure of order, system, and coherence, is used and re-used and, crucially, may have effects that ramify out into a wider (non-written) world beyond. It is possible to anatomize the self-styled "bizarre machinery" (Foucault 1972, 135; also Philo 1989) that Foucault devises for probing this making of knowledge or "the archive," but it must be acknowledged, with Foucault, that the upshot is not a "method" for straightforward exportation into other substantive studies (e.g., of madness, sickness, deviancy, etc.). Rather, it is a sensitizing device, prompting us to contemplate precisely how it is that words (sometimes) coalesce into statements (written or spoken, often both) which may then circulate, interleave, and even fall into (more or less stable and enduring) discourses.

What is often lost about Foucault's take on words, statements, and discourses, however, is that there is a definite *precision* about exactly what sorts of words, statements, and discourses "matter," in the sense of being the ones worth his sustained attention as an archaeologist-historian social researcher. For the most part, he is *not* interested in everything that gets said, the overall sweep of popular writing, journalism, gossiping, tittle-tattle, whatever form it might take. Or, rather, he is unconcerned with such words except insofar as they create a "horizon" of writing-and-speaking over and against which it may be possible to ascertain *which* (sets of) words and statements – probably located within a discourse amenable to some general description (e.g., a "medico-moral discourse about insanity") – have the particular quality of being "serious speech acts" (Dreyfus and Rabinow 1982: 48) with ontological significance. The objective is to figure out which statements, by dint of who speaks them from what location (social and spatial) and with what degree of authority lent to their utterances as both information and instructions, comprise "serious" contributions to a given body of "serious" discoursing on a given subject-matter whose gravity, at least to the period and place concerned, cannot be doubted.

The further task is to seek for patterns in the "dispersion" of identified words, statements, and discourses, as well as for lines of combat *between* discourses and perhaps also agonistic discords *within* discourses. Increasingly in his later writings, Foucault portrays the land of discourse as fractured, conflictual, itself the site of strategic combats, the theater for disputes between dominant and counter-discourses (e.g., Philo 2007). Another task is to spot moments when something "new" gets said, when the discourses shift, possibly even entailing revolutionary breaks. Two commentators (Lemert and Gillan 1982: 42) memorably discuss how "[t]he historian uncovers reversals in force relations by discovering, in documents, moments when what previously was not said is said and what was said is no longer said." It can be admitted that Foucault's criteria for selecting words and statements for his critical scrutiny ultimately rest on the (largely subjective) craft of the historian-detective in the archive, and his strong assertions about excavating "real" discourses – ones that actually exist(ed), being ontological candidates for existence – cannot be completely divorced from the methodological necessity of choosing only certain documents and archives to consult.

Studiously avoiding the vocabularies associated with discursive constructionism, which risk supposing there to be "nothing outside the text" or attributing a too-simplistic hegemony of text over world, Foucault offers a compelling ontological picture of "real" discourses operative-in-the-world. For him – influenced by the "ABC" of Althusser, Braudel, and Canguilheim – the world is envisaged as a series of overlapping layers, each with its own temporal and spatial specificities, some deemed material (the ecologies of valleys and uplands; the dynamics of capital accumulation) and others more immaterial (the customs and traditions of culture; the teachings and institutions of religion). Sinuously inserted into these layers is then a layer of discourse, "a specific order of historical reality" (Gordon 1979: 34), prompting a vision of a blanket or web of words, books, orators, writers, listeners, and readers stretching unevenly – thicker in some places, where more words circulate, thinner in others – over other layers such as those just mentioned. Foucault appreciates that this discursive layer comprises his chief entry-point for viewing other layers, but he does *not* accord it ontological primacy. Rather, it is taken as but one level *in amongst* the other levels, influenced by them and influencing them in return, even if his main problematic (in his ongoing scholarship) is to spy what drips of influence do leak from the level of the discursive to stain, maybe more forcefully on occasion to channel, the "goings-on" within these other levels. To put it another way, echoing the French title of his book *The Order of Things* (in French, *Les mots et les choses*: Foucault 1970), it is absolutely a case for Foucault of "words *and* things" or even "words *as* things," definitely not "words *instead* of things" or even "things *as* words."

For me, therefore, Foucault advances a compelling perspective that warrants the geographer's prolonged consideration. In practice, it *is* possible to identify great swathes of the very best (notably historical-) geographical scholarship that operate with a deeply Foucauldian sensibility when tracking between discourses of all stripes – economic, social, cultural, political, scientific, medical, fiscal, literary, sexual, etc. – and the vexed spaces of their production, consumption, circulation, and application (e.g., see essays in Crampton and Elden 2007). In Box 21.2, I furnish a small vignette of a Foucauldian probing of words, statements, and discourses entraining a nineteenth-century "mad" patient. More should be said to secure this argument, but I hope that there is sufficient intimation here of a "wordy" Foucauldian geography which foregrounds words (and statements and discourses) as a rejoinder to the NRT critique.

Taking seriously "trivial" words: ethnographic observations

As explained, Foucault's approach does *not* routinely encompass what may, at first glance, be regarded as the more "trivial" words encountered in everyday settings, meaning the chatter of face-to-face interactions (spoken) but also the ephemera of much popular media as well as diverse scribblings (notes-to-self) or jottings (maybe diary entries) – to say nothing of most text messages, voice mails, or e-mails. Such media may of course witness the most profound of statements whose seriousness would be widely accepted by most audiences: that memoranda declaring, yes, we have the "intelligence" proving the existence of a dictator's chemical weapons; or that text message giving the go-ahead to some multi-million pound business

Box 21.2   "Ancestral times and spaces"

My first brush with Blacklock was through a lengthy paper published by Thomas Laycock in an 1875 issue of the *Journal of Mental Science* (see Figure 21.1). What initially drew me to this paper was the presence of a small landscape drawing, a wholly unexpected illustration in such a journal, together with both a doodle and specimens of handwriting from an unnamed mental patient (see Figure 21.2). Contrasting these images, Laycock concluded that the latter indicated "a reversion to the rude comic ideas and execution of boyhood" (Laycock 1875: 166). Contrasting different handwriting specimens – concentrating largely on the form of the writing, *not* on the content of what was actually written (a revealing omission) – Laycock identified "a reversion in thought to birth and to the style of hand-writing of boyhood" (p. 166). As can now be guessed, these drawn and written fragments were originally by Blacklock (and I owe a debt here to historian-archivist Mike Barfoot for making this connection). Blacklock was almost certainly never a patient of Laycock, who practiced in Edinburgh, but Laycock and Dr. W.A.F. Browne were known to one another. And Laycock did visit the CRI, albeit after Browne's death, and by some means obtained Blacklock's art and letters, adding them to his own collection of patient materials upon which he drew extensively in both his clinical instruction and his own writings on mental disease. Laycock's surviving "archive" (held by the Royal College of Physicians, Edinburgh) includes the Blacklock fragments, and it is instructive to compare the originals with what Laycock elected to reproduce in his 1875 paper, noting in particular both what Laycock did *not* use and how he edited lines *out* of the letters (which must have required some artifice on his part; for more discussion of this case, see Philo 2006: esp. 908–12).

It is telling to realize the complete absence of any contextualization of Blacklock's case in the 1875 paper. Indeed, Blacklock is just one evidential peg, among many in Laycock's overall discursive project. The art and letters were mobilized in support of a complex, wide-ranging argument in which Laycock rehearsed his understanding of mental disease as an individual's "reversion" to childhood states of being *or*, even in some cases, to *ancestral* states of being associated with an individual's parents, grandparents, or more distant forebears. For the geographer, it is particularly fascinating that Laycock sought instances of mentally unwell individuals apparently reminiscing about remote times and places of which they had no personal experience. Laycock's overall ambition was to demonstrate his "reversion" thesis; and from unpublished writings, it appears that he thought his thesis to parallel, even to be the equal of, Darwin's theory of evolution. A crucial statement ran as follows: "I accordingly seek in the rude arts of childhood and dements for the analogues of those of uncultured men, and observe that in brain-diseases of men of high culture there is a reversion to the substrata of childhood, ancestors and uncultured man" (Laycock 1875: 165). Such a claim prompted me to argue that, in effect, Laycock was seeking to position "madness" on a grand time-space grid of evolutionary development. In the process, the "otherness" of mental difference (compared to the norms of

(Continued)

"cultured men") was being both captured *and* positioned as inferior, problematic, and in dire need of correction or even eradication (Philo 2006).

When conjoined with a critical reading of other texts by Laycock, notably his mammoth two-volume *Mind and Brain: Or, the Correlations of Consciousness and Organisation* (Laycock [1860] 1869, 2nd edn), it is possible to discern an overall discourse about phenomena such as "germination cells," inheritance, the physio-chemistry of brain, the philosophical psychology of mind, and much more besides. From a Foucauldian stand-point, I have no hesitation in detecting a plethora of "serious" words and statements tumbling together, if not all that elegantly, into a sustained "self-serious" discourse about mind and brain. (And Laycock never appeared to doubt the significance of his own discoursing.) These words entered a much more expansive landscape of medical reasoning about brain, mind, and mental disease tracking across the pages of learned journals from the mid-1800s onwards. Indeed, journals such as the *Journal of Mental Science* were increasingly arising as vehicles for the "authoritative" wording of such matters, complete with myriad recommendations about "real world" treatments, institutions, spaces, and places.

transaction. In principle, Foucault would likely have no hesitation in considering such words, at least insofar as they would sit within broader prevailing discourses of war and capital passing between certain key "powerful" actors. But, there is perhaps another line of thinking to pursue here about the importance of words whose only significance (which is not remotely to decry such significance) lies in a highly personal realm – as when someone says "I love you" – or even of words whose ostensible content is plainly simple, bland, instantly forgettable, barely worth saying – as in "Pass the butter," "I like peanuts," "Green is nice," etc. – and hence light-years from the seriousness for which Foucault is (usually) seeking. As it happens, the vast majority of words that we say, academics included, almost certainly fall into the latter category of things barely worth saying; and seen in this light, it is easy to sympathize with the objection that we live in "over-wordy worlds." For all that, though, there may still be much to be claimed about the place and role of "trivial" words in the performance of everyday human geographies. Recalling my opening mini-example, I start to think less about the prior words of all of those decision-makers who have shaped the physical and social spaces of the street upon which I gaze from my doorway, and more about the words spoken in the street, here-and-now, by the father addressing his daughter, the furniture-removers in their van, the women on the mobile phone, and the student muttering to himself.

There are various conceptual materials available for considering this quite other order of words, and one claim may be that they perform a crucial function for us in routinely anchoring our*selves* in the world, confirming to our*selves* that we are "here," affirming our own presence-in-the-world to both ourselves and to others, from whom we may often hope to hear a response – "I love you too," "Yes, here's the butter." Careful ethnographic inquiry into certain of the words that people say, verbally or on paper, may therefore reveal much about the status of an individual's

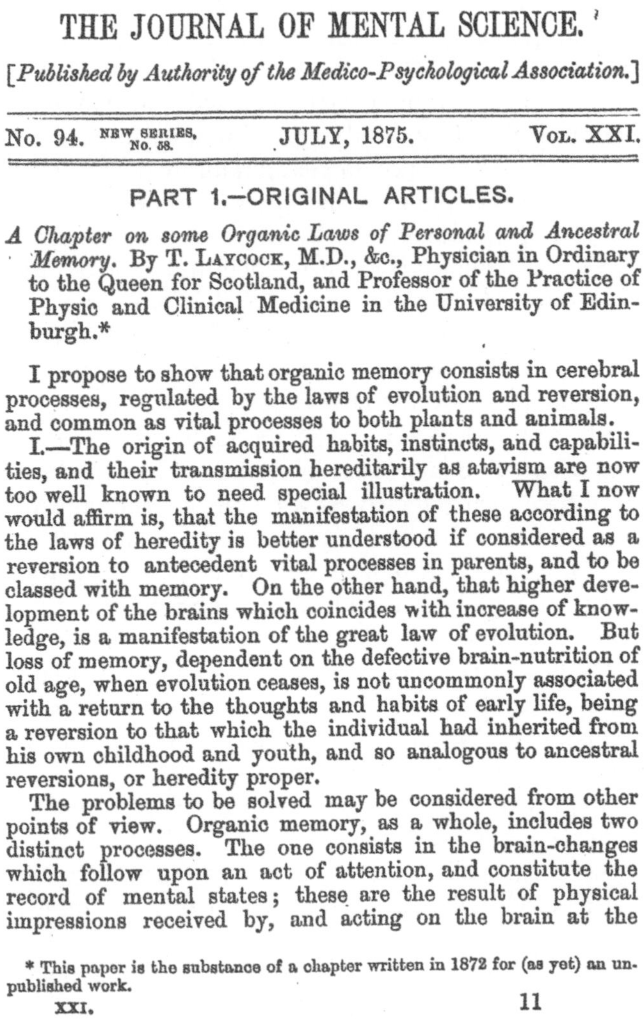

THE JOURNAL OF MENTAL SCIENCE.

[*Published by Authority of the Medico-Psychological Association.*]

No. 94. NEW SERIES, No. 58. JULY, 1875. VOL. XXI.

PART 1.—ORIGINAL ARTICLES.

A Chapter on some Organic Laws of Personal and Ancestral Memory. By T. LAYCOCK, M.D., &c., Physician in Ordinary to the Queen for Scotland, and Professor of the Practice of Physic and Clinical Medicine in the University of Edinburgh.*

I propose to show that organic memory consists in cerebral processes, regulated by the laws of evolution and reversion, and common as vital processes to both plants and animals.

I.—The origin of acquired habits, instincts, and capabilities, and their transmission hereditarily as atavism are now too well known to need special illustration. What I now would affirm is, that the manifestation of these according to the laws of heredity is better understood if considered as a reversion to antecedent vital processes in parents, and to be classed with memory. On the other hand, that higher development of the brains which coincides with increase of knowledge, is a manifestation of the great law of evolution. But loss of memory, dependent on the defective brain-nutrition of old age, when evolution ceases, is not uncommonly associated with a return to the thoughts and habits of early life, being a reversion to that which the individual had inherited from his own childhood and youth, and so analogous to ancestral reversions, or heredity proper.

The problems to be solved may be considered from other points of view. Organic memory, as a whole, includes two distinct processes. The one consists in the brain-changes which follow upon an act of attention, and constitute the record of mental states; these are the result of physical impressions received by, and acting on the brain at the

* This paper is the substance of a chapter written in 1872 for (as yet) an unpublished work.

XXI. 11

Figure 21.1 The title page of Laycock's 1875 paper in the *Journal of Mental Science*.

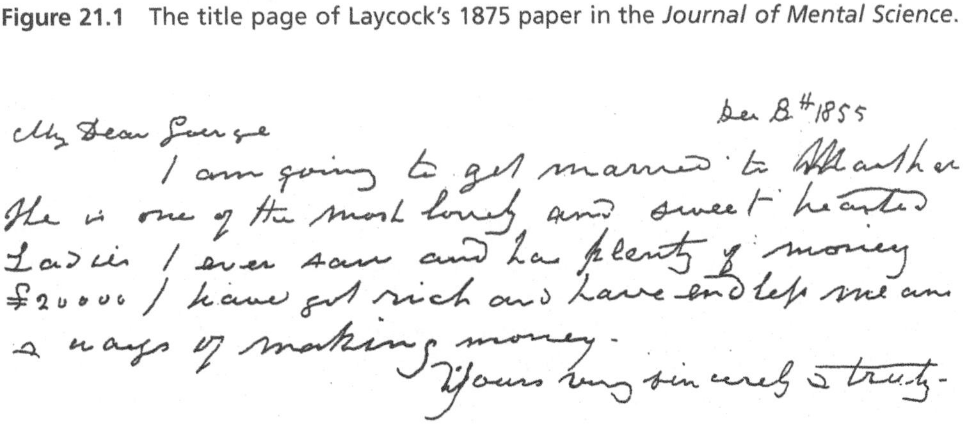

Dec 8th 1855

My Dear George

I am going to get married to Martha She is one of the most lovely and sweet hearted Ladies I ever saw and has plenty of money £20000 I have got rich and have endless means a ways of making money.

Yours very sincerely & truly-

Figure 21.2 Extract from a Blacklock letter of December 8, 1855, as reproduced in Laycock's 1875 paper in the *Journal of Mental Science*. Source: courtesy of the Royal College of Physicians of Edinburgh.

identity: about how strong, fragile, enduring, or mutable it might be; about how someone might be struggling to keep a hold or a check on a given self-identity in the face of external pressures or internal unwellness; about the "masks" that someone might create and wear, performances of self that may be dislocated from their actual or habitual self, and so on. The possibilities are near-limitless, although, with the exception of some health geographers exploring the testimonies of sufferers from mental health problems, phobias, and other conditions (e.g., Davidson 2007; Davidson and Smith 2009), little attempt has yet been made to think more systematically about the sinews connecting words, identities, spaces, and places in the highly personal narrative-biographies of given individuals or social groupings. A small vignette of what might be possible, here through historico-ethnographical research, is given in Box 21.3, where we encounter the words of the same "mad" patient met with in the previous text boxes.

More mundanely still perhaps, it can be speculated that streams of ostensibly trivial words, ones where the contents hardly seem to matter at all, are actually vital in how people negotiate their interactions with one another and with the objects, spaces, and places of the world. The inspiration for such an orientation is "ethnomethodology," the close-grained scholarly study of the micro-scale practices, skills even, of people going about their business, a prime objective of which is to tease out the exact "methods" by which people accomplish their goals, notably insofar as these entail ensuring an ongoing flow of conduct that largely avoids – or at least incorporates all kinds of mini-maneuvers to compensate for – delays, confusions, and embarrassments. Neighboring, catching a bus, walking a dog, buying a coffee: these are emblematic of the "low-level" activities, happily acknowledged as such by the researchers, that attract the attention of ethnomethodologists, who deploy a raft of observational techniques – increasingly, videoing – to record the activities, rendering them available in their completeness for subsequent ethno-analysis (e.g., Laurier 2001, 2004, 2009). Salient for this chapter is that a subset of ethnomethodology is "conversation analysis," which entails subjecting everyday conversations, face-to-face or over the telephone, to the same obsessive pursuit of detail to do with matters of initiation, participation, turn-taking, talking over one another, hesitations and deviations, and simply "keeping the conversation going."

The principal proponent of an ethnomethodological geography, Eric Laurier, cut his teeth on conversation analysis (Laurier 1998), and it is actually from him, less NRT, that I personally learned about the non-representational assault on the notion of a "little person" in "my" head producing cognitive maps for how I should proceed. For Laurier, conversation is taken as just another on-going embodied practice, itself "made up" as people "go along" in their interactions, occurring in the moment, in the flow of conduct, with no necessary break whereby participants think first and then speak. Instead, the supposition is of people here *just* talking, the thought and the talk happening simultaneously or even, if this is possible, with the talk sometimes proceeding the thought – and how often do we accuse someone of failing to "engage their brain" *before* speaking? I wonder if there may also be an abundance of writing "out there" that it is not so different from this model as well, and the scribblings of our "mad" patient in his Box 21.3 letters, especially the twelfth one, could perhaps be understood in this way. To make such an assertion is to raise a questionmark against the NRT insistence that words must be cognitive,

Box 21.3   "Half-mad with love"

A remarkable find subsequent to my main research on the Blacklock case was a collection of twelve letters penned by Blacklock between November 18, 1853, and November 6, 1855. These letters – full of commonplace information and comment, and as such hardly "grand" statements – are included in the family "papers" of James Leathart (1820–95), a Newcastle lead manufacturer, who became an art collector and commissioner with a special interest in the pre-Raphaelites (Various 1989; Morgan 1989: 39). These papers ended up deposited in the archives of the University of British Columbia, Vancouver, Canada. The letters were written by Blacklock to Leathart, primarily seeking to solicit further commissions of his artwork but more generally trying to reassure Leathart that he *was* actively painting, as well as still endeavoring to exhibit in the London galleries. There is an obvious note of desperation in his words, since the period covered, late 1853 to late 1855, was precisely when he was getting sicker. The last letter was dated just 22 days before his committal to the CRI. The letters were mostly addressed from the family home in the village of Cumwhitton, Cumbria, near Carlisle, although some came from other addresses in Carlisle itself, one being expressly noted as "near to Mr Elliot, my medical attendant," who was having "to employ very severe means" to save his employee's sight, and one being from Wilson's Boarding House, Allanby, at the "seaside," where Mr. Elliot had sent his employee for recuperative purposes. There is thus an intriguing micro-geography of *where* Blacklock was living at this time, shaped in part by his deteriorating health, and a prevailing theme of the letters was Blacklock's failing eyesight and other maladies.

Despite writing at tedious length about his maladies, Blacklock always sought to spin a positive prognosis, presumably in an effort to convince Leathart not to give up on him as a competent professional artist able to deliver on commissions. The words make sad reading, however, as he offered up excuse after excuse for *not* completing and sending on pictures for "Mr A[rmstrong]" and "Mr L[eatheart himself?]." A nadir was reached when he insisted that pictures *had* been sent before expressing surprise and anger that they had not arrived, in effect blaming his brother who had been away "from home" for this oversight. It might be wondered, of course, if the paintings had ever been completed. At the same time, Blacklock repeatedly discussed exhibiting in London, talking about pictures that he was preparing for the RA (Royal Academy); seemingly, he did continue exhibiting until 1855 (Various 1995: 55). In one letter he hoped to be in London to meet with Leathart in person, as well as also musing about visiting Leathart in Newcastle or even inviting him over to Cumwhitton. Transparent here were his efforts to retain his place within a network of patronage and acquaintance – itself closely bound into "real" spaces of artistic display, validation, and face-to-face encounter – crucial to maintaining his status as a working, and respected, artist.

Most haunting is the final letter, which is markedly different in its appearance to the previous ones (compare Figures 21.3 and 21.4). Displaying a muddled format, spread over two sides but with the final lines appearing at the top of the

(Continued)

first side, and with lines of tiny scrawled words down the margins at right-angles to the main text, it contained much more grammatical incoherence than was ever the case previously. After a brief comment about his grief at the death of Leathart's brother, Blacklock quickly spiraled into a rambling disquisition of landscape and love, apparently connecting his "love" for a picturesque Lakeland scene (Shap Fell) with his "love" for a woman, "my Bessie," who "has complete power over me." This "Bessie" had not been mentioned before in any letter, and it is uncertain whether she was a real person or a figure of Blacklock's imagining. Somewhat oddly, he went on to say that "she has a good fortune of £28,000 but I would wed her if she had not a farthing." Possibly the most revealing statement was one in the margins, however, which reads: "You must excuse the mistakes and scribbling manner in which I write – I am half mad with love." The extent to which this remark demonstrated Blacklock having some insight into his chang-ing mental health state is hard to gauge, but this letter clearly did *embody* what cannot really be assessed as anything other than a "descent into madness," or at least a slippage into an alternative mental state commonly if problematically designated as "mental illness." Committal to the asylum followed soon after this letter.

The words of these twelve letters, especially the twelfth, clearly illuminate a deeply personal history and geography of change, of an identity-in-crisis, shot through with references to places, spaces, and landscapes vital to a professional artist of the time or, in the case of the Lakeland fells, possibly more of a "spir-itual" home where Blacklock's increasingly troubled emotional state compelled him to seek solace and even "love." Far from the ontological significance of Foucault's "serious" discourses, but tugging at us, the readers, in quite other ways, these letters hence stand as an evocative counterpart to both the authorita-tive case notes and Laycock's 1875 paper (as discussed in Box 21.2).

---

secondary, *post hoc* expressions always running along after the practice, although Thrift himself (e.g., Thrift 2000a: 223; emphasis in original) does acknowledge that "talk" might be treated in this manner: "talk is responsive and rhetorical, *not* rep-resentational; it is there to do things." Indeed, in such an envisioning, the practice-words gap is decisively closed and, in effect, a certain order of words, chatty everyday ones if not the "serious" ones pursued in a Foucauldian mode, becomes potentially retrievable even by a strict NRT-inspired geographer (see also Laurier and Philo 2004, 2006).

Conclusions

The recurrent references to "madness" in this chapter have a particular Foucauldian resonance, since his researching and writing during the 1960s (especially Foucault 1967) implied the notion that we might be able to *escape* from language (e.g., Foucault 1994c). Or, more accurately, he drew parallels – sometimes quite explicitly (e.g., Foucault 1994c) – between the condition of "madness," which for a while he conceptualized as a quite "other" state of human being-in-the-world expressed

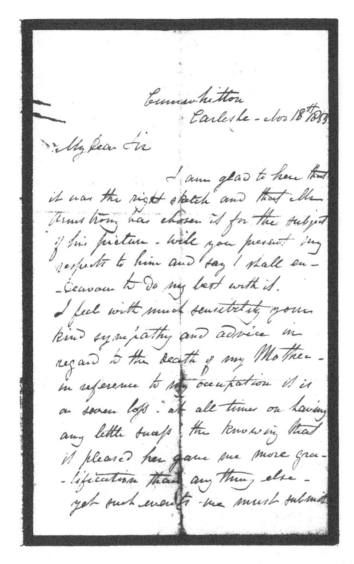

Figure 21.3 Letter from Blacklock to Leathart, November 18, 1853, side 1. Source: courtesy of the University of British Columbia Archive (Rare Books and Special Collections).

in its own distinctive vocabularies, and the striving after quite other, novel, non-establishment forms of literary expression by *avant garde* (and surrealist) authors such as Blanchot, Breton, and others (e.g., Foucault 1994a, 1994b). Indeed, it is clear that, for Foucault at this time, "dreaming of an outside" meant seeking simultaneously for *both* the "outside" of reason *and* the "outside" of conventional language. He was subsequently to abandon this stance, however, coming to see his questing for such "outsideness" as a naïvely "romantic" gesture, sunk within a misguided phenomenology of essences (in the sense of supposing there to be some essential *otherness* somewhere, somehow engrained in madness or the provocations

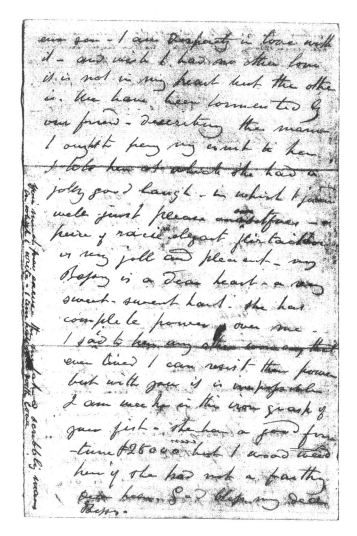

Figure 21.4 Letter from Blacklock to Leathart, November 6, 1855, side 2. Source: courtesy of the University of British Columbia Archive (Rare Books and Special Collections).

of the *avant garde*). This is not to say that he ever entirely dropped his utopian dreaming, but arguably it ceased to be such a motivating force in his work, being replaced by a more sustained critical battle *with* the "realities" of life in its struggle to be freed from oppressions in the historical present, albeit by drawing upon resources – linguistic, conceptual, practical – indelibly available in this present. Such a remark arguably requires much more unpacking, but the immediate point for this chapter is that Foucault, possibly unlike at least some of the non-representational geographers discussed earlier, eventually elected – not, as it were, to set his face against words – but rather to work on, with, and through words (the words available to us, whoever, whenever, and wherever we are).

This, finally, is my proposal when concluding a chapter on "discursive life." I have sought to frame an understanding of discursive life that makes plain the unavoidable importance of words (statements and discourses) in the everyday living and, crucially, making of worldly human geographies. Curiously, though, the *need* to portray matters in this way, to labor the importance of words and geographies, arises precisely because the status of words has been so forcefully critiqued by the remarkably provocative recent strains of NRT in human geography. I have therefore outlined – and, apologies, I have inevitably been led in parts to caricature – the challenge of non-representational geographies, chiefly so that I could then restate the value of wordy geographies, now fully alert to the NRT critique, but still insisting that there *are* diverse registers of words deeply implicated in both the living of everyday "life-worlds" and the making of, say, cities, regions, nations, states, and the countless convulsions – from violence to financial melt-down – always shaking their foundations. Whether proceeding as a Foucauldian excavating "serious" words or an ethnographer-ethnomethodologist spotting "trivial" words, or indeed quite possibly as some other kind of scholar thinking about words in quite some other fashion, the argument must be that words do still matter: that discursive life in all of its wordings, textual and spoken, noisy and silent, continues to be a candidate for having the "last word."

Notes

Big thanks to Mary E. Thomas for her thoughtful and patient editorial guidance. Thanks as well to Eric Laurier and Hayden Lorimer for their typically close and acute readings of an earlier draft, although I have not been adequately able to meet all of their suggestions. I wish to note that this chapter is based on a rather more extensive body of reading, over many years, than is immediately obvious here. There are many other pieces that should ideally be referenced in the main text; and I also owe more explicit debts to some of the pieces listed in the bibliography for specific claims in my argument than may be apparent from my (limited) "words" alone! More formally, I acknowledge the Royal College of Physicians of Edinburgh for permission to reproduce Figure 21.2, and the University of British Columbia Archive (Rare Books and Special Collections) for permission to reproduce Figures 21.3 and 21.4. Thanks as well to Mike Shand for technical assistance with the scanning of the figures.

References

Anderson, B. (2009) Non-representational theory. In D. Gregory, R.J. Johnston, G. Pratt, M.J. Watts, and S. Whatmore (eds.), *The Dictionary of Human Geography*, 5th edn. Chichester: Wiley-Blackwell, pp. 503–5.

Anderson, B. and Harrison, P. (eds.) (2010) *Taking-Place: Non-Representational Theories and Human Geography*. Aldershot: Ashgate.

Barnes, T.J. and Duncan, J.S. (eds.) (1992) *Writing Worlds: Discourse, Text and Metaphor in the Representation of Landscape*. London: Routledge.

Browne, W.A.F. (1880) (Actually published anonymously but the authorship by Browne is undoubted.) Art II: mad artists. *Journal of Psychological Medicine* 6: 33–75.

Cadman, L. (2009) Non-representational theory/non-representational geographies. In R. Kitchin and N. Thrift (eds. along with co-eds.), *International Encyclopedia of Human Geography*, www.elsevierdirect.com/brochures/hugy/index.html.

Cloke, P., Cook, I., Crang, P., Goodwin, M., Philo, C., and Painter, J. (2004) *Practising Human Geography: The Construction and Interpretation of Geographical Knowledge*. London: Sage.

Cook, I., Crouch, D., Naylor, S., and Ryan, J.T. (eds.) (2000) *Cultural Turns/Geographical Turns: Perspectives on Cultural Geography*. London: Longman Higher Education.

Crampton, J.W. and Elden, S. (eds.) (2007) *Space, Knowledge, Power: Foucault and Geography*. Aldershot: Ashgate.

Davidson, J. (2007) "In a world of her own." Re-presentations of alienation in the lives and writings of women with autism. *Gender, Place and Culture* 14: 659–77.

Davidson, J. and Smith, M. (2009) Autistic autobiographies and more-than-human emotional geographies. *Environment and Planning D: Society and Space* 27: 898–916.

Dewsbury, J.D. (2009) Avant garde/avant garde geographies. In *International Encyclopedia of Human Geography*, www.elsevierdirect.com/brochures/hugy/index.html.

Dodgshon, R.A. (2008) In what way is the world really flat? Debates over geographies of the moment. *Environment and Planning D: Society and Space* 26: 300–14.

Dreyfus, H.L. and Rabinow, P. (1982) *Michel Foucault: Beyond Structuralism and Hermeneutics*. Brighton: Harvester Press.

Duncan, J.S. (1990) *The City as Text: The Politics of Landscape Interpretation in the Kandyan Kingdom*. Cambridge: Cambridge University Press.

Elden, S. (2001) *Mapping the Present: Heidegger, Foucault and the Project of a Spatial History*. London: Continuum.

Foucault, M. (1967) *Madness and Civilization: A History of Insanity in the Age of Reason*. London: Tavistock.

Foucault, M. (1970) *The Order of Things: An Archaeology of the Human Sciences*. London: Tavistock.

Foucault, M. (1972) *The Archaeology of Knowledge*. London: Tavistock.

Foucault, M. (1994a) The thought of the outside. In J. Faubion (ed.), *Michel Foucault: Aesthetics, Method and Epistemology* (being vol. 2 of *Essential Works of Michel Foucault, 1954–1984*). London: Penguin, pp. 147–69.

Foucault, M. (1994b) A swimmer between two words. In J. Faubion (ed.), *Michel Foucault: Aesthetics, Method and Epistemology* (being vol. 2 of *Essential Works of Michel Foucault, 1954–1984*). London: Penguin, pp. 171–4.

Foucault, M. (1994c) Madness and society. In J. Faubion (ed.), *Michel Foucault: Aesthetics, Method and Epistemology* (being vol. 2 of *Essential Works of Michel Foucault, 1954–1984*). London: Penguin, pp. 335–42.

Gale, S. (1977) Ideological man in a non-ideological society. *Annals of the Association of Americal Geographers* 67: 267–72.

Gordon, C. (1979) Other inquisitions. *Ideology and Consciousness* 6: 25–48.

Greenough, B. (2010) Vitalist geographies: life and the more-than-human. In B. Anderson and P. Harrison (eds.), *Taking-Place: Non-Representational Theories and Human Geography*. Aldershot: Ashgate, pp. 37–54.

Harrison, P. (2000) Making sense: embodiment and the sensibilities of the everyday. *Environment and Planning D: Society and Space* 18: 497–517.

Harrison, P. (2007) "How shall I say it …?" Relating the nonrelationational. *Environment and Planning D: Society and Space* 33: 487–503.

Johnson, J. (1975) *Works Exhibited at the Royal Society of British Artists, 1824–1893, and the New English Art Club, 1888–1917*. Woodbridge, UK: Antique Collectors' Club, Baron Publishing.

Kingsbury, P. (2009) Psychoanalysis/psychoanalytic geographies. In R. Kitchin and N. Thrift (eds. along with co-eds.) *International Encyclopedia of Human Geography*, www.elsevierdirect.com/brochures/hugy/index.html.

Laurier, E. (1998) Geographies of talk: "Max left a message for you." *Area* 31: 36–45.

Laurier, E. (2001) Why people say where they are during mobile phone calls. *Environment and Planning D: Society and Space* 19: 485–504.

Laurier, E. (2004) The spectacular showing: Houdini and the wonder of ethnomethodology. *Human Studies* 27: 377–99.

Laurier, E. (2009) Ethnomethodology/ethnomethodological geographies. In R. Kitchen and N. Thrift (eds. along with co-eds.), *International Encyclopaedia of Human Geography*, www.elsevierdirect.com/brochure/hugy/index.html.

Laurier, E. and Philo, C. (2004) Ethnoarchaeology and undefined investigations. *Environment and Planning A* 36: 421–36.

Laurier, E. and Philo, C. (2006) Possible geographies: a passing encounter in a café. *Area* 38: 353–64.

Laycock, T. (1869) [1860] *Mind and Brain: Or, the Correlations of Consciousness and Organisation*, 2 vols. Edinburgh: Sutherland and Knox.

Laycock, T. (1875) A chapter on some organic laws of personal and ancestral memory. *Journal of Mental Science* 21: 155–87.

Lemert, C.C. and Gillan, G. (1982) *Michel Foucault: Social Theory as Transgression*. New York: Colombia University Press.

Lorimer, H. (2005) Cultural geography: the busyness of being "more-than-representational." *Progress in Human Geography* 29: 83–94.

Lorimer, H. (2009) Caught in the nick of time: archives and fieldwork. In M. Crang, D. Delyser, S. Herbert, and L. McDowell (eds.), *The Handbook of Qualitative Methods in Human Geography*. London: Sage, pp. 150–66.

Morgan, H. (1989) *Burne-Jones, the Pre-Raphaelites and Their Century*, vol. 1: *The Text*. London: Peter Nahum.

Olsson, G. (1980) *Birds in Egg/Eggs in Bird*. London: Pion.

Pavière, S.H. (1968) *A Dictionary of Victorian Landscape Painters*. Leigh-on-Sea: F. Lewis Publishers.

Philo, C. (1989) Thoughts, words and "creative" locational acts. In F.W. Boal, and D.N. Livingstone (eds.), *The Behavioural Environment: Essays in Reflection, Application and Re-evaluation*. London: Routledge, pp. 205–34.

Philo, C. (1992) Foucault's geography. *Environment and Planning D: Society and Space* 10: 137–61.

Philo, C. (2006) Madness, memory, time and space: the eminent psychological physician and the unnamed artist-patient. *Environment and Planning D: Society and Space* 24: 891–917.

Philo, C. (2007) "Bellicose history" and "local discursivities": an archaeological reading of Michel Foucault's *Society Must Be Defended*. In J.W. Crampton and S. Elden (eds.), *Space, Knowledge, Power: Foucault and Geography*. Aldershot: Ashgate, pp. 341–67.

Pocock, D. (ed.) (1981) *Humanistic Geography and Literature*. Beckenham: Croom Helm.

Pred, A. (1989) The locally spoken word and local struggles. *Environment and Planning D: Society and Space* 7: 211–33.

Pred, A. (1990) *Lost Words and Lost Worlds: Modernity and Everyday Language in Late-nineteenth Century Stockholm*. Cambridge: Cambridge University Press.

Said, E. (1978) *Orientalism*. London: Penguin Books.

Thrift, N.J. (1991) Over-wordy worlds? Thoughts and worries. In C. Philo (compiler), *New Words, New Worlds: Reconceptualising Social and Cultural Geography*. Lampeter: St. David's University College, pp. 144–8.

Thrift, N.J. (1997) The still point: resistance, expressive embodiment and dance. In S. Pile and M. Keith (eds.), *Geographies of Resistance*. London: Routledge, pp. 124–51.

Thrift, N.J. (2000a) Afterwords. *Environment and Planning D: Society and Space* 18: 213–55.

Thrift, N.J. (2000b) Introduction: dead or alive? In I. Cook, D. Crouch, S. Naylor, and J.T. Ryan (eds.), *Cultural Turns/Geographical Turns: Perspectives on Cultural Geography*. London: Longman Higher Educational, pp. 1–16.

Thrift, N.J. (2005) From born to made: technology, biology and space. *Transactions of the Institute of British Geographers* 30: 463–76.

Thrift N.J. (2008) *Non-Representational Theory: Space, Politics, Affect*. London: Routledge.

Thrift, N. and Dewsbury, J.D. (2000) Dead geographies: and how to make them live. *Environment and Planning D: Society and Space* 18: 411–32.

Various (1989) *Pre-Raphaelites: Painters and Patrons in the North East*. Newcastle-upon-Tyne: Tyne and Wear Museums Service.

Various (1995) *A Checklist of Painters, c.1200–1994, Representing in the Witt Library Courtauld Institute of Art*, 2nd edn. London: The Courtauld Institute.

Walkerdine, V. and H. Lucey. 1990. *Democracy in the Kitchen: Regulating Mothers and Socialising Daughters*. London: Virago.

Chapter 22

Spiritual Life

Julian Holloway

Introduction

How can we make sense and attend to the spiritual life? One immediate response would be to suggest that the spiritual is a way of ordering and acting towards space and time. Another answer would be to reflect upon how spirituality and religiosity continue to be a presence in societies which, in the west at least, are apparently, although increasingly questionably, dominated by more secular concerns. The pervading focus of this chapter therefore is to outline some of the emergent forms of what could be considered a spiritual life and how these forms seek to intercede in society. I cannot hope to do justice to the multiplicity of the forms that spiritual or religious lives take, yet I hope that through attending to some of the practices and politics of these formations we might begin to build an understanding of belief and faith that holds some resonance across this diversity. To achieve this I begin by exploring some of the performances of ritual which seek to disclose the spiritual and how these are assembled through the intertwining of sociality, sensation, rhythm, and materiality. These patterns of technique are then explored in a more extensive sense through an enquiry into faith as a disposition to space and time. Thus in the third section of this chapter, through an engagement with deconstructive theology and notions of immanence, I seek to elaborate a grammar of faith as a particular sensibility realized in and through different spatialities and temporalities. This allows for, finally, an interrogation into the processes by which spirituality attempts to intervene in secular formations through the affective work of the sermon as a key ritual in the spiritual life.

The Performance and Intensive Relations of the Spiritual Life

Let me begin by focusing attention on the rituals, modes of embodiment, and sensations that presence the spiritual, the holy, the sacred, or the numinous (Bell 1993,

A Companion to Social Geography, First Edition. Edited by
Vincent J. Del Casino Jr., Mary E. Thomas, Paul Cloke, and Ruth Panelli.
© 2011 Blackwell Publishing Ltd. Published 2011 by Blackwell Publishing Ltd.

1998; Asad 1993). In particular I wish to focus on the differentiation and re-presencing that characterizes religious and spiritual ritual (Latour 2001). In order to lay the foundations for a further interrogation of the spiritual life in the following sections, one must acknowledge the centrality of modes of practice and performance in spirituality and religiosity. Put another way, we cannot get away from how the spiritual life is enacted; how it is variously *done*. The spiritual life consists of a continual responding to, calling forth, evoking, or engendering of the sacred, the Ultimate, or the divine, through action. Such practices – worship, prayer, pilgrimage, song, dance, meditation, and all manner of other liturgical ritualizations – appropriate, color, and bind space and time as somehow spiritually ordained or significant: practice and performance open out space–time as resonant with a sense of the divine or sacred. What we might call the different spacings of the spiritual life are the outcomes of practical action and require a continual (and as we shall see, differentiated) practice to gain any degree of consistency and duration.

Take for example the "Church of God with Signs Following" who practice their faith mostly in rural Appalachia and the southern states of the United States and form part of a sect associated with the Pentecostal Holiness Movement. Herein "movement members seek a 'sanctifying moment' in which they accept Jesus Christ as their personal savior; experience an instantaneous, total transformation in which a love of God supplants love of sin; and make a commitment to a godly life. Holiness worship rituals therefore are directed toward a continual renewing of oneness with the sacred" (Bromley 2007: 290; Hood 2008). Principally this commitment to a spiritual life comes through attending and practicing the signs spelt out in Mark 16: 17–18. Whilst these include the laying on of hands for healing, speaking with tongues, and drinking of "deadly things," the practice that has gained most attention is that of serpent handling. Often deemed pathological in both the media and within the social sciences this may seem as a somewhat extreme starting point. However it allows us to think through some of the patterns and relations of ritual that go towards engendering a spiritual life. Bromley (2007: 289) characterizes these rituals as "spiritual edgework," wherein practitioners "regularly and deliberately engaging in ritual practices that apparently risk certain physical or psychological injury or even death" in order to realize a sense of spiritual empowerment.

Through following the strict mandates of the Bible as the only true word of God, this edgework produces in serpent handlers a feeling of oneness with the sacred and a sense of prevailing over death (as Jesus did in resurrection). According to practitioners, the impetus to handle serpents only comes through a sense of being anointed by the power of the Holy Ghost. This anointing involves banishing unbelief and is a mystical achievement partly translated through a sense of strong faith emerging through the sociality of the congregation. As such, a series of events unfold whereby the practitioners approach the serpents with a sense of their danger and through the event of anointing they perform the ritual act and realize their faith and belief.

What is striking to all those who have witnessed serpent handling is the sheer intensity of this ritual practice. The acute sensations and intensities that emerge, and that color and circulate in and through the space of the ritual, we might call affective forces. Religiosity emerges as bodies register these affective forces that swirl and bind this spiritual space together. For example, the moment of anointing is generated through sensations and impressions which practitioners describe as "energy, joy, peacefulness, humbleness, physical numbness, a haze or a bright light, an indescribably pleasant taste, a sense of protectedness" (Bromley 2007: 295). These impressions have the capacity to move practitioners to acts and practices seemingly at or even beyond the edge of normalized behavior. Therefore once anointed, once the sacred as Holy Spirit has affected practitioners and been presenced, the serpents are approached with differentiated sensations and intensities: without the impression of the sacred the bodies of practitioners would turn onto the snakes with an affective and embodied sense of their danger through an impression of, for example, their sharpness (of teeth), toxicity, and writhing movement. Yet the anointing presence of the Holy Spirit allows the snakes to be approached with more joyous, less fearful, impressions. The ritual here becomes an event of relation between bodies human and non-human, and the affective capacities arising through their coming together, touch, and separation. Furthermore, the religious empowerment that is disclosed through anointing and the event of witnessing is further achieved through the rhythm of ritual: a sense of the pacing of the ritual, with its gradual quickening towards the signs following and serpent handling, in turn act as affective forces (on rhythm see Edensor and Holloway 2008). These different elements of the ritual bring about a change toward a sense of the sacred. We can call this relational ordering of the space of ritual, wherein the sacred is brought forth, a *territorialization* or a patterning of the spiritual life.

Another example will allow us to understand these formations further. Rountree (2002, 2006) has studied the pagan Goddess movement and more specifically practices of pilgrimage to Neolithic sites in Malta. The sacralization of these sites emerges through an awareness of the divine in space and through a felt temporal connection to pagan Goddess worshipping societies in the past (which are valorized for their freedom from patriarchal injustice and a more balanced relation to the natural world). Therefore, visits to these ancient sites involve processes of connection and remembering that allow a healing in the present to be brought forth. In similar manner to serpent handlers, these pagan Goddess modes of spiritual life emerge through embodied and affective relations that precipitate a connection to and realization of the sacred. As Rountree (2002: 487) puts it, "the embodied experience of being healed includes physical sensations (for example, an energy opening the heart) and strongly felt emotions (love, joy, longing)." These sensations, felt and territorialized through sacred spaces and times, are central to the skein of ritual and a spiritual life. Indeed, the sensations that realize and are realized through embodied practices become somatic modes of attention to and evocations of the divine or sacred such that "not only does the divine penetrate the site at this node in the landscape, but it (usually thought of as a "she") also becomes one with the pilgrim: human body and landscape both embody the sacred" (Rountree 2006: 101).

What Asad (1993) calls the apt performance of ritual wherein an appropriate sensibility to the sacred is engendered in and through space and time, or what I have called elsewhere a spiritual infralanguage (Holloway 2003a), comes forth in the movement, acts, and affects of pilgrimage. Yet Rountree's account of Goddess pilgrims to Malta adds another layer to the patterning of ritual being developed here. In particular, she cites the account of a pagan pilgrim, Christine Irving, and the transformations she undergoes upon seeing limestone statues of a "fat" Goddess in a Maltese museum: these objects act to heal Irving of the negative associations of having a larger body. This moment of delightful realization and revaluing of her body are further consolidated when she visits Hagar Qim, the Neolithic temple wherein the museum statues were originally housed. Irving enters the temple and:

> by lying down and curving her body into the curved limestone walls of the temple in which the "fat" statues had once stood, the woman maps the Goddess's body onto her own and further embodies her self-recognition as Goddess ... Through encountering the materiality of the site the woman encounters her own materiality in a new way and with renewed intensity. (Rountree 2002: 489)

This account reveals the materiality of ritual. In particular, one must note the shape or even shapedness of the temple's materiality that Irving curves her body to. Here, in a similar move to that between handler and serpent, the ritual is achieved through the body relating to the materiality that is sensed or felt as curvy and rounded (rather than toxic or writhing). Therefore curviness – brought about through statues, temple, and fleshiness – come with its own intensive resonance. Put differently, the texture and form of materiality as curved and curvy generate certain sensations for the ritual and thus the spirituality performed: Irving states the statues "simply delighted me. Their size moved me" (cited in Rountree 2002: 488). As Anderson and Wylie (2009) point out, matter and materiality need to be rethought as active and affective fields for the potential generation of subjects and senses of self which, in this case, involve the spiritual: the objects and space, and their constituent intertwining, become charged as fields of potential through which a sense of the sacred can and is performed. The spiritual life is patterned through affective forces of ritual and different intensities of all that engenders it (see Holloway 2003b). With serpent handlers we see the intensities and event relations of congregations, snakes, and scripture; with pagan Goddess pilgrims we see the different resonances and affective capacities of stone and flesh. In both examples, we see the performative and affective patterning of the spiritual life.

In the example of the pagan Goddess pilgrim performing an embodied connection with and through what the statues and the curviness of the temple offer to her, the body becomes felt and remapped as sacred. Arguably, this sensing of the affective capacities of materiality in the formation of sacred space–times is achieved pre-reflectively. To state this is to recognize that religious belief is always and everywhere imbued by affective forces and intensities. As Wynn (2005: 28) has it, "affectively toned theistic [or pantheistic] experience may constitute a value perception, and may be veridical even if its phenomenal content is purely affective." Thus, the meanings and experiences of the spiritual life are colored and moved by fields of intensity that surface and are assembled in multiple and eventful ways. In light of this we could follow Connolly (2002: 94), drawing on Deleuze, to argue that "affectively imbued thinking is always already under way by the time consciousness intervenes to pull it in this or that direction," where the pull of consciousness here forms religious or spiritual belief. This is not to deny the importance of cognition, texts, and representations upon the way in which bodies and selves are organized in the rituals we have discussed here – for example, it is doubtful if serpent handlers would be as such without the biblical words of Mark 16: 17–18 and Irving's post-ritual account no doubt leaves traces in further performative explorations of the sacred.

However, we need to understand how texts have their own interventional affective force such that they become part of the patterning of a spiritual life enacted in ritual: texts become less about something else, less representational and symbolic, and more concerned with presentation and achieving all sorts of affectively toned motivations, sensations, and thoughts (Asad 1993; Hollywood 2002, 2004). In other words, they become part of the relations and techniques of ritual that bring about spirituality.

Furthermore, focusing on the techniques of the spiritual life allows one to suggest a degree of specificity to religious and spiritual performance. For as Latour (2001) has shown, religion, in its rituals and performances, does not involve a simple repetition, or re-enactment, of a genealogy of action and practice, words, or gestures. Instead each ritual, each event of prayer, liturgy, or praise becomes a re-presencing, a rendering again of sacred presence such that religion and spirituality "does not hesitate to modify the message in order to repeat the same thing – no transport without retranslation" (Latour 2001, 223). Variance is therefore encoded into ritual such that "one must invent in order to remain faithful to what remains always present" (Latour 2001: 223): religious ritual involves repetition and rupture, similarity, and difference. Ritual as a technique for realizing belief needs to be repeated again to be affective and effective whilst what is re-presenced is consistent – in the examples used here, the signs of a suffering and resurrected Christ or the just, healing and loving presence of the Goddess. Thus each time a practitioner approaches the serpent the same presence is felt, the same presencing is performed, yet the tone, patterning, or rhythm of the ritual may vary (and indeed may not always and everywhere be successful). Each time a connection with the divine Goddess is successfully felt, each time the Goddess is iterated in action and performance, the intermediaries and the space may be different (a different temple, a different materiality) but the same presence is felt, the same presencing is enacted. The performative and affective skeins of ritual in the spiritual life cite and re-cite, work on the sensibilities that condition belief, with the "balance between what is kept and what is modified being called a tradition" (Latour 2001: 224).

The Divine Milieus of Faith in the Spiritual Life

I would like to retain this stress on technique in what follows. Yet here I would like to broaden out the focus and scope of this argument beyond the relatively specific times and places of ritual (even if these are multiple and vary across different belief formations) to more extensive spaces and times. Therefore, my aim in this section is to begin an understanding of the spiritual life as realized through a *disposition* to and with divine forces that surface in and color space–times beyond those set-off for ritual. This disposition is, in short, one of *faith*, as that which allows space and time to show up and resonate in particular ways for different believers. I do not wish to claim that faith as disposition is the same for different formations of the spiritual life. However, I think we can tentatively and modestly trace how sensibilities of faith are shaped across many of these formations. With this in mind I wish to look to a form of Christian theology and make it coincide with notions of immanence, drawn from Deleuze, in order to begin sketching this understanding. This encounter may furnish a sense of the dispositions that color, organize, and give space its spiritual resonance. Engaging a particular strand of Christian theology with Deleuze to generate a conceptual grammar of faith is therefore a specific and partial encounter, but hopefully a productive one.

The particular strand of Christian theology that I wish to engage here is itself the ongoing product of a theoretical encounter pursued and developed at the border zone between the traditional confessional mode of theology and a religious studies detached from its "object" of study (Hyman 2004) or, in other words, that which moves and orchestrates belief – what Latour (2001: 230–1) deems those entities with "extraordinarily different ontological status ... Virgin, saints, miracles and gods." Variously known as deconstructive, postmodern, poststructuralist, elimina-tive, "weak" or a/theology, this set of literature engages with poststructuralism and deconstruction in order to refigure the western theological tradition and its central themes and concepts (Griffin et al. 1989; Hart 1989; Robbins 2004; Tilley 1995; Ward 2001; Vanhoozer 2003). As Hyman (2004: 212) describes it:

> Thinking, albeit transgressively, within the resources of the Christian theological tradi-tion, it repudiates the neutral, descriptive, and social scientific stance of religious studies ... On the other hand, however, postmodern a/theology also wishes to distinguish itself from theology, and in doing so draws nearer to religious studies.

Drawing out the tendencies and intersections from deconstructive theology will hopefully produce useful lines of departure for an apprehension and of faith as disposition. I have chosen to focus upon the work of Mark C. Taylor (1982, 1993, 1999) and specifically his work *Erring: A postmodern A/theology* (1984) as this highlights some of the key ways in which deconstructive theologians have sought to work productively with poststructuralism and yet retain, albeit in a radically different form, a theological sensibility. Indeed to think through faith as disposition via deconstructive theology requires us to delimit Taylor's project in some detail.

In a classic deconstructive move, Taylor's major impetus is to invert and subvert the fundamental dyads of western theology, such as God/World, Sacred/Profane, or Presence/Absence, through retaining a sense of continual movement without settling on one side of these divides. This results in the formulation of a/theology, with the "/" indicating that the holy or sacred "occupies that quasi-logical space between theology and atheology" (Nuyen, 1991: 69):

> The / of a/theology ... forms a border where fixed boundaries disintegrate. Along this boundless boundary the traditional polarities between which western theology has been suspended are inverted and subverted ... The a/theologian asks errant questions and suggests responses that often seem erratic or even erroneous. Since his reflection wanders, roams, and strays from the "proper" course, it tends to deviate from well-established ways ... The words of a/theology fall in between; they are always in the middle. (Taylor 1984: 12–13, original emphasis)

Subsequently, this critical nomadism is troublesome. Hence Taylor describes his a/theology as inevitably *erring* – in the sense of being nomadic and roaming and, through various etymological moves, deviant or "straying from the proper course or place" (e.g., erroneous, errant). This wandering, boundless, and dissident depar-ture begins with Taylor's radical deconstruction of theology's central concerns of

God, Self, History, and the Book. In particular, Taylor's subversive project works through revealing the absence that is constitutive of the presence of each of these concerns. For example, God is conceived in most theology as the transcendent and absolute that is totally self-present to Himself and can only be known indirectly through the mediation that is Christ the Logos. Revelation therefore is "necessarily incomplete and forever partial" and thus haunted by the absence of the "mysterious God [who] always manages to escape one's grasp" (Taylor 1984: 36). Furthermore, since the human subject is created in the image of God, the self's "full realization of the *imago dei* necessarily entails the *imitatio christi*," and thus "the self is actually an image of an image, an imitation of an imitation, a representation of a representation, and a sign of a sign" (Taylor 1984: 40, original emphasis). It follows that the self is left in a state of endless anxiety in attempting to achieve full presence that was or is always already haunted by an absence of the self-present God.

Moreover, traditional theology's narrative of History makes plenitude and fullness original and as such the "fallen" subject becomes "wounded" by the need to reaffirm this primal presence. This wounded subject, who seeks fulfilment and original plenitude, is merely a temporary (albeit sinful) wanderer who looks forward to the reaffirmation of presence in the end-times. Yet in Taylor's reading of this story, the displaced need to reaffirm original presence can only lead to the self's "unhappy or lacerated consciousness":

> Although not always obviously distraught, the unhappy person is perpetually discontent. With eyes forever cast beyond, the victim of unhappy consciousness lives in memory and hope. He nostalgically recollects the satisfaction he believes once was and expectantly anticipates the fulfilment he hopes will be. Satisfaction, however, proves to be elusive; it is never present or is *infinitely* delayed ... From this point of view, history amounts to an unending search for a presence that saves. (Taylor 1984: 151, original emphasis)

To this unhappy person "[t]he persistent opposition between "reality" and "ideality" lends a saddened, nay, melancholy, tone to time. To unhappy eyes, even the fresh leaves of spring are tinged with brown" (Taylor 1984: 72).

Against this teleological and eschatological discontent, Taylor offers his deconstructive a/theology which embraces the end of endgames and the death of the (transcendent) God and self through opening up the "possibility of affirming what previously had seemed inadequate and insufficient" (Taylor 1984: 156). In particular, Taylor is keen to embrace the "purposeless process" of what he deems the *divine milieu*. As a "nontotalizable totality," this divine milieu is the acentric "nonoriginal 'origin' of all that is and is not" (Taylor 1984: 118, 168). In other words, the divine milieu is the "generative/degenerative matrix, [wherein] nothing is (merely) itself, for no thing can be itself by itself. Everything is fabricated by the crossing of forces" (Taylor 1984: 111–12). In the divine milieu there can be no origin, self-presence, or transcendence. There can be no point outside the becoming and endless flow of life that we can call *immanence*. Indeed any categorical statements, any distinctions, any "either/or" claims, or any closures are dissolved "into the eternal play of differences" that are composed and decomposed in this milieu (Taylor 1984: 136). Thus for Taylor, divinity is immanent to and presenced through a continual unfolding of the milieu.

The divine milieu is a (de)generative medium where there are no longer self-present or contained selves, but rather *traces*. These traces are never fixed and are always erring. They are always composed in the milieu and crossed (fully aware of the theological sentiment of this term) by divine incarnational flow of forces. Divinity is thus understood by Taylor as immanent to the unending flow of and becoming of life. Hence, tracing in and through the divine milieu is a source of delight and enjoyment, a carefree wandering. What he calls a *"mazing grace"* becomes attendant to this divinity as a practice of faith (Taylor 1984: 168, original emphasis). Taylor's notion of the divine milieu enacts an outpouring of all fixed and self-presence and gives way to a joyful immanence – a sensibility of faith that, in a perpetual state of becoming, is realized in different ways. The tracing of the divine, the performance of faith as "mazing grace," colors and tones experience and, we could argue, how space shows up. Faith is thus formed in and informs the divine milieu such that it becomes a percipient energy, a tracing or a disposition which gives space its spiritual resonance. Put differently, the "mazing grace" in the divine milieu is a way of thinking how space and time is grasped through faith and how the multiplicities of the world are made to resound with (and act back upon) this disposition.

As a grammar of faith Taylor's tracing in the divine milieu affords us an understanding of how faith informs and precipitates the spiritual life: faith is a disposition to and with space and time, a grasping of the geographies and temporalities of life, whereby divinity and the sacred surface and are made present only for that emergence to descend, disappear, and is subsequently in need of being made present again. Faith charges space and time as a divine milieu, it is a disposition which enacts the world for it to reverberate with the mazing, joyful sacred, the spiritual or the divine through a process of re-presencing and re-translation again and again, always with difference. Those with a faithful disposition formed through and with the divine milieu "all faithfully repeat something similar even if they transform it, *because* they transform it" (Latour 2001: 224; original emphasis). In short, the spiritual life resounds with an ongoing and differentiated faithful disposition born out of and with a divine milieu:

> It calls
> Calls daily
> Calls nightly
> Calls (from) without
> Beginning or end
> A whisper so feint
> A murmur so weak
> When to respond
> Where to respond
> How to respond
> To a call that approaches (from) beyond
> Without ever arriving.
>
> (Taylor 1999: 29)

Arguably, the joyful immanence of the divine milieu through which faith emerges is echoed in the work of Deleuze. Such a claim may seem contentious stemming primarily from the "perversity" of employing Deleuzian thought in any project to do with religion (Bryden 2001: 1). Thus any forms of foundationalism or transcendentalism, whether of God or identity, are continually disrupted in Deleuzian thought (Albert 2001). In particular, Deleuze rallies against the *"judgement of God"* and the "order of God" which authorizes and invests presence (Deleuze 1998: 129, original emphasis; 1990: 292; Deleuze and Guattari 1987). Therefore, where Deleuze does engage theology and religion, his is a decidedly anti-transcendental position; resonating with Taylor, nothing can stand outside or have fixed attributes within a radical world of becoming or what Deleuze and Guattari (1994) call a *plane of immanence* (1994). Yet this plane of immanence is different to immanence as thought and developed in "traditional" theology. Immanence in this "Christian philosophy":

> is tolerated only in very small doses; it is strictly controlled and enframed by the demands of an emanative and, above all, creative transcendence ... all philosophers must prove that the dose of immanence they inject into the world and mind does not compromise the transcendence of a God to which immanence must be attributed only secondarily ... Religious authority wants immanence to be tolerated only locally or at an intermediary level, a little like a terraced fountain where water can briefly immanate on each level but on condition that it comes from a higher source and falls lower down. (Deleuze and Guattari 1994: 45)

Instead of immanence as secondary to a primary transcendence, Deleuze (and Guattari here) posit immanence as primary and thus disallow any primal or prior position inside or outside the plane. For example, instead of taking the self as primary – from which everything else flows – we are formed through this immanence, this field of becomings, in multiple ways. As Doel (2000: 126) describes, "one gets swept up by the laying out of a plane of immanence that can be folded, unfolded and refolded in many ways." Thus we as selves and subjects arrive out of or are *actualized* through and in the plane of immanence (Ansell Pearson 1999; Seigworth 2000).

Subsequently, in Deleuzian thought any religious event, any spiritual moment, even faith, belief, and the spaces that they are formed through, become foldings of the immanent becoming of the world and life. It would seem therefore that Deleuze's relation to theology is totally antithetical, particularly as he wishes to develop "the system of the Antichrist" in thought and practice (Deleuze 1990: 298). Yet whereas Deleuze leaves no room for a theology posited through determinate presences and primal organization, there is still potentially room for something like spirituality and even divinity in his thought. This can be exemplified if we entwine Deleuze with Taylor. Thus, the immanent plane arguably resounds with the milieu that is crossed by the divine in Taylor's a/theology. For Deleuze actualization and for Taylor the crossing/tracing of the divine milieu are never finalized, nor can they bring about full self-presence or stasis. Furthermore, these processes are always becoming otherwise: they are always in *potentia*. The divine milieu as the infinite play of generative and degenerative forces echoes the plane of immanence which is folded or actualized in potentially infinite ways. Neither author can abide finitude, self-presence, or origin, and thus in both we have a kenosis that is akin to a spiritual practice attendant to the immanent/divine which "provides an opening to the processes of life itself, to a spiritual dimension wholly immanent to life in which processes of creation and differentiation ... and actualisation are continually *taking place*" (Goddard 2001: 63, emphasis added). In Taylor then we have a taking of place which resounds and reverberates with faith as a sensibility of the divine forever becoming within it: Taylor thus illuminates a practice towards the immanent sacred as it unfolds in time and place. In Deleuze we have the immanent frame through and out of which events unfold and are actualized. Drawing Taylor and Deleuze together here we might think of faith as a disposition towards and emergent from the immanent frame that those whom practice the spiritual life hold as a divine milieu. Faith becomes a framing of divine immanence, a disposition[s] emergent through and actualizing the immanent divine milieu, which presences and re-presences and thus *takes place* spiritually. There is then a continual and unending taking of place as a renewal with the sacred in the divine milieu – a way of coming to and turning on to space which allows it to become spiritually resonant. Faithful dispositions open out the immanent divine milieu in space and time; they bring forth practiced bodies and selves and allow them to affect and be affected by space and time as spiritually meaningful, sacrosanct, or in need of faith.

It is important here to recognize the link between the expansive sense of the spiritual life discussed here and the specific space–times of ritual explored in the previous section. We might want to think, therefore, of the rituals as intensified events of relation where faith is practiced, faithful sensibilities of the divine milieu are affectively enforced, and the immanence of the world is actualized as spiritually significant (see Thrift 2000). In so doing we might see ritualizations as exercises or techniques that attend to or cultivate particular sensibilities which allow participants to move through a wider world with an ongoing grasp or readiness for faith-full action. These actions and practices of the spiritual life will be multiple, have different orderings, yet they are charged and taken hold of through faith. Indeed, these faithful dispositions may be propositional and more conscious ways of ordering space and time (Barnett 2008). So it might be that this faithful disposition involves the proclamation, the wording or representation of faith through texts or speech such that the divine milieu is made reflectively and cognitively apparent. Yet equally these faithful dispositions might remain pre-reflective backgrounds to action that shade practice in divinely attuned ways. Therefore, faithful dispositions, born out of and attendant to the divine milieu are realized, as Taylor argues, through traces and tracing. These traces and the practices of tracing are, as Connolly (2002: 120) notes:

> enough like a thought to affect linguistically sophisticated thoughts and judgements, and not enough like a thought to be susceptible to direct inspection ... [The trace and tracing] becomes, however, marked by affective intensity once triggered by an appropriate event.

Whether these traces subsist as affective sensibilities at a pre-reflective level where faith "grip[s] the senses" and "allow forces and intensities to be focused and channelled" (Thrift 2000: 49, 44), or they are a conscious faith-full grasping or ordering space and time, they amount to an ongoing disposition to the spiritual life realized

in and through multiple geographies and temporalities infused by the sacred immanence of the divine milieu.

The Interventional Orderings of Spiritual Life

A faithful disposition has been characterized here as a reflective and pre-reflective
framing, approaching, and revealing of space–time as immanently infused with
divinity. These sensibilities of faith allow and usher forth a continual presencing
and responding to the divinity of space and time. I have argued how rituals seek to
organize and shape these dispositions, how they engineer the sensibilities of a
spiritual life, such that the faithful color and tone their geographies with and
through a divine frame. Rituals, amongst other practices of spiritual life, can be
seen as interventions into the sensing and opening out of space and time as spiritually resonant. As interventions, then, rituals seek to organize sensibilities and
dispositions. Simultaneously, therefore, rituals and other forms of spiritual interventions must be seen as ongoing and as having varying degrees of political success.
How we might think about success in light of the argument developed here is
the focus of this section. In other words I wish to dwell upon how spiritual and
religious interventions can be seen as political orderings with different degrees of
powerful efficacy.

One of the key ways in which those who practice the Christian spiritual life seek
to order space and time is through the sermon. These proclamations allow us to
draw together some of the different ideas that have been set out in this chapter.
First, sermons are performed across a range of Christian spaces with a continuous
regularity: sermons are thus a key formation in the continual practice of faith and
belief. Second, sermons are events of relation that produce and emerge through the
intertwining of congregation, proselytizer, and the space of its enactment.
Subsequently the sermon is differentiated by the time and space of its performance:
it gains a particular consistency and duration through its specific actualization in
time and space. Indeed, the consistency, the qualitative intensity of the performance,
leads to the third way we might think about sermonizing: how the practice of the
sermon generates particular impressions or how it generates certain affective registers. For it is through the timbre of the voice, the pacing of the delivery, the evoca-
tion of impressions and thought, that the sermon does its work: shot through with
affectively imbued resonance listening to a sermon becomes feeling and sensing, as
well as a practice of re-presencing the divine. Sermons are thus affective interventions that permeate feeling and thinking with a sense of the sacred and a sense of
the spiritual life. They stimulate a sense of divinity and actualize an affectively
textured sense of the spiritual, the sacred. Fourth, therefore, sermonizing is a technique for re-territorializing faith as a sensibility in and towards the world. By this
I mean the sermon seeks to fortify, shade, and tone faithful dispositions both during
the moment of its performance but also beyond as the sense of divinity is carried
out of the event space and re-emerges, is realized again, in the stance the faithful
take to other times and spaces. Fifth, the sermon involves a re-presencing, a telling
again of the same story such that faith is restored and the divine milieu is made
present. Sermons, in light of this, attend to the faithful through a realization that
"the tone is everything and the information content of what is uttered is nothing

... This is after all the very definition of a sermon: the recasting of what was already said through different means of expression" (Latour 2001: 234).

Let us take, briefly, two relatively recent and high profile sermons to illuminate this argument. The first is the Archbishop of Canterbury Rowan Williams' Easter address delivered on April 12, 2009 (Williams 2009). In it Williams, as the Primate of the Church of England and spiritual leader of the Anglican Communion, discusses the knowledge of God. He addresses how this knowledge is not a matter of either objective proof or personal opinion: to reduce faith to opinion is a mistake as it "shrinks the scale of what you're trying to talk about to the dimensions of your own mind and preferences," with the idea of proof similarly erroneous as "it keeps you at arm's length from the whole business by making it impersonal: here are the proofs and it doesn't much matter what I or anyone may be doing about it" (Williams 2009). Instead of these mistakes Williams argues through Christ's resurrection, Christians become aware of Jesus as the son of God and an assured knowledge of their place in heaven. The same message would have been repeated in other churches throughout the United Kingdom and across the world on Easter Sunday. Indeed the same message will be repeated in churches at other times, now and in the future. Repeating one of the central tenets of Christianity re-presences the divinity of life and the world. This presencing reiterates belief whilst simultaneously acting to re-inscribe faithful dispositions. Furthermore, the act of re-presencing binds a community of the faithful. The sermon is a speech act which endorses a "new assemblage of persons in presence" (Latour 2001: 216); the same thing said again to enact a (new) ordering of faith and belief, the faithful and believers. Here the ability of the sermon to affect (and in turn be affected by) dispositional intensities extends beyond content to amplify faith at the level of sensibility and beyond the space of its performance to more extensive geographies.

On the same day Cardinal Cormac Murphy-O'Connor, the Archbishop of Westminster and spokesperson for the Catholic Church in England and Wales, delivered his Easter sermon. His theme is the resurrection of Jesus and the continuing presence of Christ through his Holy Spirit and rituals of Holy Eucharist and Communion: "*Jesus is Lord. Christ is Risen. Alleluia.* It is in the affirmation of that extraordinary fact that we find our meaning and our hope and the purpose to our lives" (Murphy-O'Connor 2009, original emphasis): the same message again and, despite doctrinal difference, the same as Rowan Williams. No doubt the same message will be repeated in 2010 when Murphy-O'Connor's successor Vincent Nichols delivers it. No doubt Nichols will begin his sermon with the address "My dear Brothers and Sisters in Christ" or something similar (Murphy-O'Connor 2009): an opening gambit that seeks to draw the faithful and reaffirm their stance to the world. The performance of this sermon will territorialize faith and affirm belief through an affective event of relation that reiterates, but with a different explanation, a key precept of Christianity. Moreover, Murphy-O'Connor seeks to shape how faithful dispositions should be enacted and frame spaces-times beyond Westminster Cathedral wherein the sermon was delivered. Thus he finishes his sermon through arguing that Christianity's:

> contribution to society is not to impose our Christian faith or values which is something Jesus never did, but rather to reveal the Christian values by the way in which we live.

There are things that we do to create a better world that somehow defy all the sad and shocking and painful things that obscure the love of God. What we offer as caring women and men is the Spirit of the Risen Christ and so a life-style that is redemptive and full of hope. (Murphy-O'Connor 2009)

Correspondingly, Rowan Williams concludes his Easter sermon on how God is known by stating:

We need to hear what is so often the question that's really being asked when people say, "How do you know?" And perhaps the only response that is fully adequate, fully in tune with the biblical witness to the resurrection is to say simply, "Are you hungry? Here is food." (Williams 2009)

Both sermons act as techniques for the affirmation of dispositions and sensibilities, yet both sermons, as they are brought to a (temporary) close, seek to relay the prospective efficacy of these dispositions and sensibilities. Not only do these sermons attempt to order and proselytize through direct enrolling, they seek to affirm faithful dispositions as potentially efficacious interventions. Thus, through witnessing (answering questions of faith with faith) and maintaining the spiritual stance to a world enacted through Christian lives, these dispositions can potentially change the world: just as the sermon itself seeks to resonate with and beyond the congregation and the space of the Cathedral, the sermon seeks to order the spiritual life such that the dispositions carried forth might intercede in the world and make a difference. Sermons are thus doubly politically efficacious: they seek to reaffirm the faithful sensibilities of the congregation *and* extend these apprehensions of the world such that others might be drawn in to presence the immanent sacred.

Sermons tell and reaffirm the faithful whilst simultaneously informing the unfaithful of the possibilities of realizing the divine milieu through faith. Sermons have presentational power to reiterate dispositions and produce these dispositions as interventions; they enact political power to effect change as both reiteration and potential extension. These performances make present and attempt to coalesce a community of witnesses whom through their faith-full actions and deeds, their sensibilities and dispositions, might effect and affect change by bringing others into the practices and performances of faithful sensibilities. Sermons become political attempts at ordering through acting on an orientation to the world, whereby the believers continue to actualize the divine milieu through faith, and by framing those actualizations as interventions in themselves. Therefore, both the sermons and the faithful orientations reiterated through them become part of a series of "disseminative creative relays which may or may not resonate, which may or may not find a hospitable destination" (Dewsbury et al. 2002: 439). The presencing-performative speech act, the mode of witnessing and the exemplary spiritual lives framed in these sermons need to affect alliances of intensity and sensibility to have political efficacy (Connolly 2005). And once again here we need to attend to how these modes of proselytizing seek to imbue and energize thought and action through modes of affect.

For a sermon to be effective therefore it needs to work through affective relays such that sensibilities become orientated towards the divine and the sacred. An example of the attempts of these sermons to do this affective work is made apparent when the Archbishop and Cardinal turn to the current global economic recession and downturn:

> It could hardly be a more propitious time for [witnessing]. The present financial crisis has dealt a heavy blow to the idea that human fulfillment can be thought about just in terms of material growth and possession. (Williams 2009)

> Sometimes we carry resentments, grudges, prejudices, hurts and angers – like security blankets in our lives. This may be particularly pertinent at the present time, when many people are feeling stress and anxiety due to the difficulties resulting from the economic situation locally and globally. But negative feelings and negative attitudes are bundles of death that stand in the way of life. Jesus said love was the greatest commandment, so it is because of Jesus' love that we are able to forgive and pass over things of the past and die to pride and selfishness and begin again a choice for life. (Murphy-O'Connor 2009)

As such these sermons seek to work on and through affective registers that may, if they resonate with efficacy, instill or engineer a faithful disposition of hope or a realization of the divine as a spiritual life. These sermons seek to challenge and re-order through appealing and amplifying feelings of despair with secular concerns and dispositions. And once again this is not only a direct attempt to enroll those with despairing secular dispositions: the work of witnessing through the sensibility of faith offers the alternative disposition of the spiritual life. For this to register, for these dispositions to take hold and to re-order secular anxiety and disquiet to favor the actualization of the immanent sacred, thought and belief needs to be affectively imbued with a sense of the sacred. Such a result would mean these attempts at proselytizing are successful and secular framings of the world are re-amplified as the performance and orderings of the spiritual life.

Conclusions

In this chapter I have sought to think through some of the characteristics of a spiritual life. The focus has been mainly upon practice and performance so that we might begin thinking of the spiritual life as a constellation of spaces and times imbued with a sense of the sacred. Thus, we might want to focus our attention on the set-off spaces of ritual which act as space–times of religious and spiritual techniques. Here rituals are modes of spiritual work which seek to re-presence the sacred in the lives of believer through affective and presentational techniques involving others in the congregation and the materiality and affordance of a whole host of texts, objects, and bodies. However, to limit our understanding of the spatialities of religion to these albeit crucial yet relatively distinct space–times would mean a failure of engagement with the spiritual life as a broader concern that gives a choreography to believers' life worlds. Therefore, we need to think about how faith works and affects the practices of the spiritual life taken in its broader aspects. By working through the deconstructive theology of Taylor in intersection with Deleuze's philosophy of immanence, I have attempted to think of faith as a realization of and disposition towards a world saturated with divinity and the sacred. Through multiple actions in and towards space and time, the faithful become as such through their actualization of the immanent divine: the spiritual life therefore is a realization and a mode of realizing space–time as infused with the divine that the faithful enact and continually re-presence. This faithful disposition is reiterated and assured again through ritual enactment. Sermons, as a constant of at least the Christian faith, attempt this assurance and bolster these dispositions in the hope that they may in turn act as practices of witnessing that draw others into the congregation of the faithful. If the arguments presented here have been successful–if they move the reader to take faith seriously and to take seriously how faith takes place–then they engender a whole series of questions that require further inspection: we need to interrogate rituals, faithful dispositions, and how religious formations seek to effect change in the secular by attuning this-worldy sensibilities to those of the immanent divine. And we need to work through how faithful dispositions are challenged by secular sensibilities and concerns. Therefore, and also, we require new ways of thinking through how the sacred and the secular, and processes of secularization and sanctification, are performed, ordered, and achieved. If we add to the mix the multiplicity of spaces, times, and faiths through which these processes work and are engendered, it is clear there is still much work to be done on the spiritual life.

References

Albert, E. (2001) Deleuze's impersonal, hylozoic cosmology: the expulsion of theology. In M. Bryden (ed.), *Deleuze and Religion*. London: Routledge, pp. 184–5.

Anderson, B. and Wylie, K. (2009) On geography and materiality. *Environment and Planning A* 41: 318–35.

Ansell Pearson, K. (1999) *Germinal Life: The Difference and Repetition of Deleuze*. London: Routledge.

Asad, T. (1993) *Genealogies of Religion: Discipline and Reasons of Power in Christianity and Islam*. Baltimore: Johns Hopkins University Press.

Barnett, C. (2008) Political affects in public space: normative blind-spots in non-representational ontologies. *Transactions of the Institute of British Geographers* 33: 186–200.

Bell, C. (1993) *Ritual Theory, Ritual Practice*. Oxford: Oxford University Press.

Bell, C. (1998) *Ritual: Perspectives and Dimensions*. Oxford: Oxford University Press.

Bromley, D.G. (2007) On spiritual edgework: the logic of extreme ritual performance. *Journal for the Scientific Study of Religion* 46: 287–303.

Bryden, M. (ed.) (2001) *Deleuze and Religion*. London: Routledge.

Connolly, W.E. (2002) *Neuropolitics: Thinking, Culture, Speed*. Minneapolis: University of Minnesota Press.

Connolly, W.E. (2005) The evangelical capitalist-resonance machine. *Political Theory* 33: 869–86.

Deleuze, G. (1990) *The Logic of Sense*. New York: Columbia University Press.

Deleuze, G. (1998) *Gilles Deleuze: Essays Critical and Clinical*. London: Verso.

Deleuze, G. and Guattari, F. (1987) *A Thousand Plateaus*. Minneapolis: University of Minnesota Press.

Deleuze, G. and Guattari, F. (1994) *What Is Philosophy?* London: Verso.

Dewsbury, J.D., Harrison, P., Rose, M. and Wylie, J. (2002) Enacting geographies. *Geoforum* 33: 437–40.

Doel, M. (2000) Un-glunking geography: Spatial science after Dr. Seuss and Gilles Deleuze. In M. Crang and N. Thrift (eds.), *Thinking Space*. London: Routledge, pp. 117–36.

Edensor, T. and Holloway, J. (2008). Rhythmanalysing the coach tour: the Ring of Kerry, Ireland. *Transactions of the Institute of British Geographers* 33: 483–501.

Goddard, M. (2001) The scattering of time crystals: Deleuze, mysticism and cinema. In M. Bryden (ed.), *Deleuze and Religion*. London: Routledge, pp. 53–65.

Griffin, D. R., Beardslee, W.A., and Holland, J. (1989) *Varieties of Postmodern Theology*. Albany, New York: State University of New York Press.

Hart, K. (1989) *The Trespass of the Sign: Deconstruction, Theology and Philosophy*. Cambridge: Cambridge University Press.

Holloway, J. (2003a) Spiritual embodiment and sacred rural landscapes. In P. Cloke (ed.), *Country Visions*. Harlow: Pearson Education, pp. 158–75.

Holloway, J. (2003b) Make believe: spiritual practice, embodiment and sacred space. *Environment and Planning A* 35: 1961–74.

Hollywood, A. (2002) Performativity, citationality, ritualization. *History of Religions* 42: 93–115.

Hollywood, A. (2004) Practice, belief, and feminist philosophy of religion. In K. Schilbrack (ed.), *Thinking through Rituals: Philosophical Perspectives*. London: Routledge, pp. 52–71.

Hood, R.W. (2008) *Them That Believe: The Power and Meaning of the Christian Serpent-handling Tradition*. Berkeley: University of California Press.

Hyman, G. (2004) The study of religion and the return of theology. *Journal for the American Academy of Religion* 72: 195–219.

Latour, B. (2001) "Thou shalt not take the Lord's name in vain" – being a short sermon on the hesitations of religious speech. *Res: Anthropology and Aesthetics* 39: 215–34.

Murphy-O'Connor, C. (2009) 2009 Easter Homily. Diocese of Westminster, www.rcdow.org.uk/cardinal/default.asp?library_ref=1&content_ref=2243. Accessed June 5.

Nuyen, A.T. (1991) Postmodern theology and postmodern philosophy. *International Journal for Philosophy of Religion* 30: 65–76.

Poxon, J. (2001) Embodied anti-theology: the body without organs and the judgement of God. In M. Bryden (ed.), *Deleuze and Religion*. London: Routledge, pp. 42–51.

Robbins, J.W. (2004) Weak theology. *Journal for Cultural and Religious Theory* 5: 1–4.

Rountree, K. (2002) Goddess pilgrims as tourists: inscribing the body through sacred travel. *Sociology of Religion* 63: 475–96.

Rountree, K. (2006) Performing the divine: neo-pagan pilgrimages and embodiment at sacred sites. *Body and Society* 12: 95–115.

Seigworth, G.J. (2000) Banality for cultural studies. *Cultural Studies* 14: 227–68.

Taylor, M.C. (1982) *Deconstructing Theology*. Missoula: Scholars Press.

Taylor, M.C. (1984) *Erring: A Postmodern A/theology*. Chicago: University of Chicago Press.

Taylor, M.C. (1993) *Nots*. Chicago: University of Chicago Press.

Taylor, M.C. (1999) *About Religion: Economies of Faith in Virtual Culture*. Chicago: University of Chicago Press.

Tilley, T.W. (1995) *Postmodern Theologies: The Challenge of Religious Diversity*. Maryknoll: Orbis.

Thrift, N. (2000) Still life in nearly present time: the object of nature. *Body and Society* 6: 34–57.

Vanhoozer, K. J. (2003) *The Cambridge Companion to Postmodern Theology*. Cambridge: Cambridge University Press.

Ward, G. (ed.) (2001) *The Blackwell Companion to Postmodern Theology*. Oxford: Blackwell.

Williams, R. (2009) The Archbishop's Easter Sermon. The Archbishop of Canterbury, www.archbishopofcanterbury.org/2377. Accessed June 5.

Wynn, M.R. (2005) *Emotional Experience and Religious Understanding*. Cambridge: Cambridge University Press.

Chapter 23

Virtual Life

Mike Crang

The Social Geometries of Digital Connectivity

At first glance a social geography of virtual connections seems an oxymoron. For many years, one of the claims behind information and communication technologies (hereafter ICTs) and especially (new) media was to render geography less important. These technologies have long promised, and enabled, proximity without propinquity; they mean people in Birmingham and Bangalore can interact, exchange ideas and information as easily as those in the same city. At least in principle they might. It is clear that in practice there are a range of geographies – thus the scope of communication between Birmingham and Bangalore, even with video conferencing, may still be more restricted than face to face contacts within the cities. There is a consequent question of whether the nature of communication changes when mediated, or whether some kinds of social relations are more easily mediated than others. Secondly, Bangalore just happens to be the high-tech capital of India and is globally well connected. A different empirical situation might prevail if our rhetorical example was Brazzaville in the Democratic Republic of the Congo. The infrastructures, capacities (both hard, like fiber cables, and soft, like skills) and thus the scope of possibilities are (still) geographically uneven – at all scales from global, to national, to regional, to urban, by age, by class, by ethnicity. Third, this uneven geography produces uneven social effects that may compound existing inequalities. This then is a set of geographies around the so called digital divide(s). Fourth, the use of new media may transform and refract existing spatial behaviors, creating new hybrid spaces. Finally, these new media may afford new arenas of social interaction that have their own internal geographies – where "online worlds" of varying types use spatial frameworks to operate. Throughout the chapter it will become clear that much public rhetoric has focused upon technologies causing social changes. At worst this takes the form of a technological determinism, but even when less stark very often depicts society and technology as independent and opposing

A Companion to Social Geography, First Edition. Edited by
Vincent J. Del Casino Jr., Mary E. Thomas, Paul Cloke, and Ruth Panelli.
© 2011 Blackwell Publishing Ltd. Published 2011 by Blackwell Publishing Ltd.

ontological realms that seem to clash together. Instead this chapter works through a theoretical vocabulary that sees technologies and societies co-constituting each other. Instead of seeing a logic of substitution, where ICTs replace "real world" things, it will focus on logics of remediation where we see a layering of socio-technical forms of life – where new media adds to older media, where social action domesticates new technologies, makes them useful and develops them. The approach will ask whether we can have social geographies that are not technosocial geographies. In other words, a great deal of our social world is now enabled by, mediated by, invested in and bound to various communication technologies. It will conclude by noting that developments in technology mean we may have to decenter the human within an internet of things that produce a technological form of life.

Research on the so-called "digital divide" is well established, and thus far has tended to be dominated by aggregate maps depicting differential levels of access to the internet between different income, gender, geographical, and ethnicity groups (Baum et al. 2004; Bromley 2004; Holloway 2005). Such research has successfully traced how the leads and lags of on-line access move, over time, between different social groups and geographical areas, whether regionally through economies, or as rural and urban contrasts, as access to ICTs – most especially internet connected devices – diffuse unevenly through society. Such divides can be charted by various demographic variables, and different spatial categories and scales. To illustrate this one can look to the variations by age as illustrated in a report on Internet Generations for the Pew Internet and American Life Project, which contrasts rates of usage in different generations with around 90 percent of under 25s using the internet, around 60 percent of those in their 50s and 45 percent of those 70 and over (p. 5), which figures are echoed in broadband penetration by age, where of those under 35 around 70 percent had a broadband connection to the internet, while 30 percent or less of those over 70 did so (Jones and Fox 2009: 11). Perhaps this is unsurprising as virtual life has been associated with technological sophistication and novelty – both of which tend to appeal to younger groups and repel older groups equally. As those who grew up with new media age, there will be a cohort effect. That said, the fastest rate of growth in uptake the report shows is amongst the 70–75 year olds.

The social geography of this demographic division also shows regional and spatial divides. Some of these are technically explicable, as in, for instance, the strong urban-rural divide in broadband which is related to the costs of providing fiber optic cable to rural areas in a market based system (though for telephones many countries had a public service obligation to provide connection as a "right" to those who wished it, that entitlement is not yet there for digital media). But there is also a pattern that reflects and compounds patterns of more and less favored regions. For the United Kingdom (see Table 23.1), we can see the absolute variation of regional home access to the internet rates for 2009 of 18 percentage points, a gap that has grown since 1998 when on a low base it was 11 percentage points, though relatively that means that a household in the most connected region was more than three times as likely to be online and is now (only) a third more likely to be so. Those numbers reflect absolute household access, not intensity of use nor inequalities of access within households.

If there is then variation by region, partly reflecting levels of aggregate economic well being, then within a region we also see differences reflecting three sets of digital

Table 23.1 Households with home access to the internet by government office region (UK)

	1998–99	1999–2000	2000–1	2001–2	2002–3	2005	2006	2007	2008	2009
London	16	25	40	49	51	53	63	69	73	80
Eastern	11	22	34	45	52	54	64	67	70	77
South East	13	24	38	48	52	62	66	65	74	75
South West	9	19	37	35	44	55	59	69	67	72
England	11	20	34	41	47	55	59	61	66	71
UK	10	19	32	40	46	55	57	61	65	70
Wales	7	15	22	32	37	54	52	57	67	68
East Midlands	9	19	31	41	49	59	55	59	61	67
North West	9	18	32	39	43	52	54	56	56	67
West Midlands	8	20	33	34	41	56	53	56	61	67
North East	7	14	25	32	41	44	54	52	54	66
Yorkshire and the Humber	8	15	29	34	42	50	52	52	62	64
Scotland	8	14	24	37	42	53	48	60	61	62
Northern Ireland	5	11	20	31	35		50	52	56	na

Source: Family Expenditure Survey (April 1998 to March 2001); Expenditure and Food Survey (April 2001 onwards).

divides: differential access, first, across cities, suburbs, and country and, second, within each city, and, third, consequent divisions in what people do with that access. The reason for emphasizing urban spaces is that connectivity is often concentrated in urban areas, so even in the generally well connected United States only 38 percent of rural American households have broadband access compared to 57 percent and 60 percent for city and suburb dwellers, respectively (Pew Internet & American Life Project 2008).

Second, there is differential access within cities, in terms of provision and infra-structure (for instance the concentration of wireless or broadband coverage in particular districts). These map onto existing zonings and infrastructural provision with points of hyper connectivity and nodes of various flows around economic activities. They also reflect the income, social deprivation, and wellbeing of districts and neighborhoods (Longley and Singleton 2009). In the United Kingdom, the PriceWaterHouseCoopers report for the Digital Inclusion Task Force (2009) suggested that there are at least 10 million adults in the United Kingdom who have never been online. They suggest that every year, £900m of government customer service costs would be saved if all of those 10 million were to interact with the state just once a month via computer, rather than by phone or face to face. Once you get past the rather instrumental view here of saving transaction costs for the state, the report does highlight real issues of how a lack of internet access affects people. Overall, the United Kingdom has 661 out of every 1,000 people connected. Many of the people who are not connected are not only offline, but also poor and have many other social problems with which to contend. The latter group not only costs more to the state, but find themselves facing extra costs because they are not online.

What is apparent is not only a demographic stratification of access, but a strati-fication via the ensuing consequences that impacts upon people in terms of both their life chances and the resources they can utilize (in both social and financial terms) and thus, in the third set of divides, the use made of online access. Thus a study for the UK Post Office found that "the direct financial benefits of broadband in the home are in the order of £70 per month for the average UK household" (SQW Consulting 2008: 1). This figure is derived by calculating the savings on household expenditure achievable by using online comparison services and suppli-ers. That saving therefore was itself unequal ranging from £23 per month for the 10 percent of households with the lowest income to £148 for the 10 percent with the highest incomes as they consume more. In social terms:

> the internet … is a portal which has the potential to liberalise access to a whole host of resources and opportunities, and to increase social connections … It has the capacity to impart knowledge to groups which have tended to be excluded from traditional information sources, to provide new channels of communication, and to open up access to goods and services previously denied or impeded by older technologies or methods of exchange. (Bromley 2004: 73)

Given the scale of divisions and the possible effects, it is no wonder that analysts often see digital media exacerbating existing urban divisions, with critics such as Paul Virilio seeing society split by speed where "one part lives in an electrical world of relative speed – transportation –, the second with absolute speed of transmission of information in real time" (Virilio 1998: 185). Virilio is notably ambivalent whether the absolute speed of electronic interaction is pleasant for those subject to the demands of immediacy in accelerated lifestyles, suggesting the only thing worse is not to be subject to them.

In a similar vein Castells (1996) critiqued the ensuing "dual city" and Boyer (1999) the "min-max" scenario, where the city is sharply divided between prosperous "knowledge workers" and those incapable of finding a place in the "new economy." Processes of social polarization are exacerbated by the unbundling of "public" services through electronic service provision, permitting differential terms of access (Graham 2002). One social geography widely anticipated from this is "a society of cocoons ... where people hide away at home, linked into communication networks" that allow, and increasingly compel, a frenetic globally connected lifestyle, but where people increasingly opt out of the rest of the city through a "spatial closure" (Burrows 1997: 38). Castells sees a space of flows produced through placeless global connections and online networks that touch down in the city, in sealed off spaces, such as corporate offices, hotels and airports, which become real-virtual spaces cocooning the elite, making a "dual city" that is simultaneously "globally connected and locally disconnected" (Castells 1996: 404). This vision thus sees digital media as forming a virtual citadel that excludes participation from many in the so-called connected city.

However, this scenario is too simplistic both in its view of elites, but also its analyses of the level and processes of exclusion. Time-use studies have repeatedly shown that people are not retreating online, but that time spent online has a positive correlation with going out and social activities (except for reducing the time spent watching TV). Cross-tabulating current aggregate time-use figures for the United Kingdom suggests that "each extra minute on the internet is associated with about one-third of a minute reduction in personal care time, one fifth of a minute less visiting, half a minute less watching television ... [and], nearly one-fifth of a minute of *extra* time devoted to going out – eating or drinking in a public place, going to the theater or cinema" – and if one tracks longitudinal changes in time-use amongst people adopting net technologies, that suggests an even stronger positive relationship of going online with increased socializing and specifically with going out to public places (Gershuny 2003: 158, 164). This positive relationship is not surprising if we think of how social media function as social utilities which are intimately bound up in both enacting and organizing many people's daily social lives. That said, online communities and socializing online are also differentiated. Recent work has pointed to racial and class segregation between different social networking sties – with all social networking sites in the United States attracting first a strongly white demographic, but with MySpace in the United States attracting first a lower class user base than Facebook, and recently a rise in African-American usage being concentrated on MySpace, with racially coded language enacting a virtual "whiteflight" to Facebook use (boyd forthcoming). These networks therefore might be seen as enacting social closure rather than opening social groups out to new experiences. They exhibit at the least homophily (that is, birds of a feather flocking together) and reinforce existing social bonds rather than creating diversity. Thus a recent US survey that found one-third of online adults say that the internet has improved their connections with existing friends "a lot" but, by contrast, only 12 percent of internet users feel that the internet has greatly improved their ability to meet new people and nearly two thirds say that it had not enhanced them meeting new people at all (Wellman et al. 2008).

For those connected, online access may also afford new possible patterns of work with more working from home and locational and temporal flexibility enabling new patterns of care work, new hours of work, but also longer hours of work (Perrons 2003; Perrons et al. 2006). Time-use figures suggest that people with web access tend to work longer hours and sleep for fewer, though there is no simple causal inference here. This also points to connections with the "excluded" who are now in increasing demand to provide child care, house cleaning, and similar for the "knowledge workers" to enable them to pursue their frenetically busy lives. The online world of global, instant connectivity is thus often subtended by global flows of poor migrant labor. This vision of the dual city of digital elite and excluded others certainly has resonance, but it too readily equates global connectivity with power.

We could equally look at the way globalized telecommunications are altering the rhythms of urban life to produce new patterns of disjuncture and differentiation (Crang 2007). Thus, global connectivity may indeed bring new times of work not because of increased choice and flexibility, but because people are compelled to follow the rhythms of other places. Here the connected might be the non-elite, the so called "cyber-coolies," for whom global connectivity means exploitation and local disconnection as much as inclusion. For female employees in call centers in Bangalore – answering to the temporality of US markets – the night-shift is a time-trap emphasizing their marginal status in a night time environment that is dominated by men, with women stigmatized and moved around the city by chartered van (Patel 2006) as part of a new global pink collar labor force. This digital economy (with 336 call centers employing 348,000 people in India in 2005) oddly echoes early electronics manufacturing that produced a new urban proletariat in countries like Malaysia, with Malay women brought into urban centers, with fears and desires for urban lifestyles and shift work (Ng and Mitter 2005). These workers are part of both the city where they live and work, and also those, usually western ones, whose populations they service. A new source of urban difference and hybridity is introduced by globe spanning digital linkages that make people share, if unequally, electronic space.

This begins to direct our attention to the greater number of permutations of digitally connected spaces than the "dual city" suggests. Scott Lash (2002: 28–9) contrasts "live" and "dead zones" of cities categorized by "the presence (or relative absence) of the flows" of information within such "zones." He pairs this divide with *responses* to such information, referring to a parallel urban geography of what he calls "tame" and "wild" zones. Thus, Lash produces a quick four-way typology. *Dead/tame* zones are characterized by majoritarian ethnic groups, cut off from digital flows, clinging to traditionalist values in the face of change. *Live/tame* zones are characterized by the "informational bourgeoisie," with affluent connected populations doing comfortably and thus relatively conservative as regards the social order. *Live/wild* zones are characterized by the emergence of new cultural forms driven by the "cultural capital faction of the post-industrial middle class." *Dead/Wild* zones are areas of social decomposition, marked by marginal groups, but connectivity linked to cultural change and fluidity. One might then look at the contrasting geographies of new media workers, who tend to cluster around dynamic and vibrant urban milieu (far from being cocooned) (Pratt 2002) with those of information specialists such as accountants or financiers who opt for "tamer" districts. Clearly there are problems with this fourway division – for instance, Lash's "tame" category is freighted with pejoratives, and then if you unpack what "tame" zones might be like they might appear more frenetic than he implies with speeded up lives, orchestrated by suites of always-on ICTs. Increasingly ICTs here become pervasive and necessary parts of everyday life – not a separate thing. Meanwhile, we need to resist some over-quick assumptions here about how disconnected groups of people actually are. Among those apparently unconnected in their homes, you can find "real life" social networks that enable people to access services online – if only episodically, but enough to benefit from cheaper online deals for consumer goods or travel (Crang et al. 2006). These create multiple geometries of connection – not a singular divide. Indeed geodemographic work has looked to extend the typologies further, building multiple profiles of user types. Instead of four configurations of neighborhood and user, the e-society classification designed by Paul Longley and associates (2008) maps 8 "types" and 23 subtypes of users at a street by street scale. This fine grained typology separates out those without access because they are "unengaged" (do not see the benefit of the technologies) from those who are marginalized (wish to but cannot access them), and those with access according to the intensity of us from "independents" (who calculate what things ICTs might most help with) through to "experts" (who rely upon them massively).

Neotribes and Forms of Digital Life

The differential usage of online media is one thing, but what forms of sociality are thus enabled? In other words, how are virtual spaces and worlds bound into the functioning of socialities? How is the social enacted through the technological? Focusing on the practices of technosociality produces more geographies than typologies of access types do, speaking to neo-tribes (Maffesoli 1996) united by affective bonds, coming together then breaking apart. These neo-tribes are bound together by their use of specific online media and hardware, and by the symbols and information that these enable them to exchange. New media are not just communication channels but also "visible objects" (Morley 2003) where having the right handset, using one social network or another, says a lot, and is meant to say a lot, about the user (e.g., Hjorth and Kim 2005). These neo-tribes form fluid and unstable modes of belonging which relate in complicated ways to the more usual divisions of class, race, and income.

The emergence of less stable social identities through global connections and online socialities has been very worrying for socially conservative groups and governments. For example, the government of Singapore has been eager to foster the image of being a digital hub in an information age. Having on the one hand announced that global flows are inevitable, unstoppable, and an economic necessity, the state has had to face the uncomfortable application of this argument to social realms. The state worked to maintain social control over the new media. The first famous incident came with a scan of all Singaporean email accounts in 1994 ostensibly for offensive imagery followed by outraged messages documenting traces of surveillance posted on alt.soc.singapore (Rodan 1998: 77–8). The ensuing outcry, speedy backtracking, and almost embarrassed scapegoating of bureaucrats, and the global stories confirming Singapore's authoritarian tendencies, make these seem the maladroit steps of a state grappling with a new informational landscape. However, if we look at the effects, then the outcome is not so clear. The state had demonstrated its capacity to search email accounts, and the postings fuelling the rumors of surveillance amplified this message. In other words the state had, despite the technical limits of surveilling huge volumes of data, created the impression that it could do so. The action was followed by the announcement of limits to access to foreign web sites. The state forced all domestic providers to use a state run cache, which had a list of proscribed sites. Ministers were at pains to suggest, to Anglophone audiences and media, that this was a light touch regulation drawing a line in the sand – with but a hundred or so (undisclosed) sites forbidden (Lee, T. 2005).

It is a fairly simple matter for residents to use a Malaysian service provider and bypass the cache. When it was suggested that these limits were unenforceable, Ministers would agree, suggesting it was intended as a signal. Ministers positioned themselves as reasonable people, aware of the new environment, yet unwilling to simply throw in the towel. They pointed to a conservative "heartland" of the island that would not want such a capitulation either.

The promotion of ICTs by the state was enthusiastic but tended to be one directional – suggesting that if only the populace realized what was on offer all would be well, rather than looking at what people wanted or how they used what was provided (Tang and Ang 2002). While Singapore has been a pioneer and exemplar of the electronic delivery of government services to citizens and business, via the E-citizen and GeBiz portals respectively, there has been far less encouragement of political involvement (Sriramesh 2006). In the late 1990s there was a flurry of young educated Singaporeans using the internet – partly self-consciously thinking how it might open a new public sphere, and partly through a wish to stretch Singaporean space to include those doing graduate work in the United States. Initiatives such as Sintercom (Singapore Internet Community) and the Thinkcentre appeared and offered "civil society" discussions, while the pseudonymously and US hosted Singapore Window offered an alternative news portal on happenings in Singapore. They were tolerated, partly due to policies keen to foster, or be seen to foster, a lively society and culture. Simultaneously Singapore, a state that outlaws male homosexuality, became very visible online for the number of gay discussion boards and fora, as inhabitants of the island state used US hosted sites to facilitate a burgeoning, if illicit, gay scene (Ng 1999). Alongside the relaxation of cinema, drama, and entertainment restrictions, Singapore was trying to foster creative thinking and also make the place more vibrant and attractive (as in the "live/wild" rather than its reputation as "live/tame" in Lash's schema) to key informational workers from around the world.

This careful, and successful, taming of digital flows in Singapore contrasts with South Korea where new media have been strongly connected with social transformations. There the new media company Cyworld created the possibility of setting up home pages, rather like blogs, run from mobile phones several years ago (before web enabled phones were common in the west). Such pages became known as "mini-hompys" (as in little homepages). The individualized, often youthful, often female postings stood out from the standard conservative forms of more collective, elder mediated, male dominated self expression:

> Mini-hompy's virtual "homes" and sites for confession and self -presentation are not just an example of an apotheosis of the individual but rather a way in which to negotiate traditional forms of sociality. They are also a space in which gendered types of performance can be both reinforced and subverted. (Hjorth and Kim 2005: 49)

More widely though, it is also a country where digital media have intersected with social and political democratization – for instance OhmyNews, an online portal, has been significant in generating huge crowds for rallies and influencing governmental policy (Kluver and Banerjee 2005: 35). Most notably, after US servicemen killed two teenage girls and were acquitted by a US military court, online postings led to candlelight demonstrations of unprecedented scale and social breadth (Song 2007). Online media have challenged existing conservative media (where three established newspapers have 80 percent of readership) with forms of citizen journalism – from portal/papers like OhmyNews, where three quarters of the content is generated by its 35,000 so called "news guerrillas," to blogs to platforms like Seoprise that host *non-gaek* (polemicist) pages (Chang 2005b, 2005a). This turn to online media fits with a society "democratizing" away from authoritarian rule supported by inter-generational pressure for change. It is also digitally one of the most connected societies in the world where, in 2005, 31 million of 47 million Koreans had access to very high-speed internet (Chang 2005b: 928).

Not surprisingly in Korea new media are seen, by more conservative groups, as threatening to fragment society by undercutting traditional socialities of dense affective ties (*cheong*) (Yoon 2003). However, the media are far from being simply a virtual phenomena and are tied to a rapid growth of youth culture who are connected, symbolically and literally, with digital media. Seen as antagonistic to the state, they are the subject of periodic moral panics about disconnection from Korean norms.

> The use of the mobile phone has caused concern about the increasing dislocation of young people from the Korean norms of harmonious sociality, which derives from the idea that new communication technologies are likely to both defamilialize and individualize human relationships and consequently that the mobile phone may destroy collective and affective relationships. (Yoon 2003: 328)

However, studies of circuits of mobile phone messaging suggest that by offering new means of sharing intimacies (especially among young women), they actually enhance existing patterns of local sociality but without visible public arenas (Yoon 2003: 339–40). Thus mini-hompys adapt the older sense of "cheong," which originally referred to the idea of relationship in terms of degrees of kinship separation, and now fits surprisingly well with the degrees of separation of contacts and friends of friends on social media, so that "pages are not only used to further forge familiarity and connection with friends but also operate as vehicles for meeting new and like-minded people" (Hjorth and Kim 2005: 50). The networks create a form of social space or "cheong space" in which local sociality derived from cultural tradition is embodied in everyday life; thus, it "is a site in which tradition is appropriated as a source of identity formation" (Yoon 2003: 328). On the major mini-hompy provider Cyworld the interface "Administration" button offers statistics, analyzing the frequency of interaction between each "*1-chon*," that is first degree friend directly accepted by the user, and themselves. To show commitment to these virtual bonds, "a new phenomenon has been resulted; *1-chon soonhwe*, or "*1-chon* round (tour)" in English, has become a common practice amongst many Cyworld users, who visit *1-chons* on the list one by one to leave a message in the Guestbook section until they have visited every *1-chon*, mainly as a gesture of courtesy" (Choi 2006: 177).

The digital realm is also transforming specific urban spaces. Sites of connectivity, such as internet cafes, are "technosocial spaces" which bend the physical boundaries of the local space by including actors situated in other villages, cities, and countries. Associated with youth culture, and especially Massively Multi-user Online Worlds/ Games, there were more than 20,000 internet cafés or *PC Bang*, as they are known, across Korea by 2001. This stands in contrast with Japanese youth culture that has rather adapted games to the *keitai* or mobile phone, rather than specific fixed locations (Hjorth 2007). About 50 per cent of all PC Bangs are in the capital region, and almost 25 per cent are located in Seoul (Lee, H. 2005). They foster specific temporalities of behavior and rhythms of activities that become the hallmark not merely of these particular buildings but the wider locale within which they are situated. They might be seen as creating and transecting conventional social orders. Although residential broadband has led to a reduction in the total numbers of PC Bangs, they still thrive precisely as a third place of technosociality – allowing young adults to congregate and share online experiences. Sinchon, a university quarter in Seoul, has developed a new urban consumption landscapes based around these technospaces. Many PC Bangs are clustered there, and many newspaper stories, especially those commenting on "immoral" behavior in PC Bangs and concerned about the social alienation and online "addictions" of youngsters, focus on there. It has become a 24/7 social milieu with all night gaming, but also a social milieu especially for young men that is technologically mediated (Lee, H. 2005). Far from being a virtual cocoon this has become a rebellious, contestatory and "happening" zone in the city. The same area where symbolically student demonstrators fought the police in the 80s under the dictatorships, they now challenge a conservative social order with games. Here there seems something of the liminal – a fragmentation that is allowing the self expression of identity and a transformation of local culture.

Far from a dis-embedding or loss of local belonging, there is a re-embedding and reinscription of the role of older forms of *bang* (as in tea bang, marriage bang) alongside a long established fascination with games (now digital rather than board based). Urban space is being remediated and what is being inscribed here is a third space of neither home nor public for a youth subculture (Chee 2005). But it remains the case that while gamers are often playing people in the next chair or next block, they can also be playing across time zones and the planet. So it is to the online worlds created that this chapter now finally turns.

Virtual Worlds: Going Online

Much of the early writing on online or electronic "spaces" perhaps fetishized the creation of new virtual worlds – seen either as perfections from real life (labeled in distinction RL) – or as degraded, inauthentic copies (Doel and Clarke 1999). A great deal of work has charted the emergence of online realms, that is interactive environments that exist as simulated spaces in shared computer space – especially in terms of identity play. This work has traced how people are able to conceal their identities, work through masquerades, and pass as other genders, races, or sexualities. The last two decades have seen the development of text based interactive multiplayer role playing games from Multi User Dungeons (from online Dungeons and Dragons, the first such game developed from the face to face role playing game at the University of Sussex in the very early eighties) to Multi User Domains (losing the sword and sorcery) to the first graphical interfaces with Multi-User Domains Object Oriented programs (MOOs) to the current Massive Multiply-authored Online Role Playing Games (MMORPGs) – with some leaving out the role playing, such as Second Life, others including it, such as the most popular online world, with up to 10 million players, World of Warcraft, returning to swords and sorcery (Mortensen 2006).

The geographic debate on these virtual worlds picks up a number of earlier themes in this chapter. For instance, they were more precisely predicted to exacerbate processes of socio-spatial polarization, whereby elites would have virtual realms secured from contact with the less privileged, they were also seen as accelerating the disembedding of people from local ("real world") communities and physical neighborhoods, and thus allowing the retreat of people into social cocoons (Robins 1999). On the positive side, many others looked at these as emergent spaces for new identities to emerge – where people could choose their identity, literally for instance by creating an avatar (their online character and persona) allowing plays of identity and belonging (Turkle 1996; Stone 1995). This reflects and enables those who see increasingly fragmented social worlds and indeed fluid, malleable, or fragmented social selves.

For a geographer, one of things that is striking is that these virtual worlds are strangely familiar despite their differences. They are computer simulations that could take almost any form or means of organizing data and interactions, customizing "spaces" to the user, playing with our notions of space, or dispensing with it entirely (Crang 2000). Instead, most are resolutely conventional – they reinvent familiar spatial idioms and rules. So for example, Habbo Hotel uses the hotel metaphor as a meeting space, while Alphaworld has ended up producing something very reminiscent of American suburbia. When faced with effectively a (virtual) isotropic featureless plain, the resultant pattern of building developed a star shape because of a logic over memorable coordinates driving locational development, with people wanting to site at 25E,25N or 0E25N and so on (Ryan 2004). As the online world grew a transport infrastructure for avatars was developed with a monorail system. So we have a virtual public transport system moving virtual figures around a virtual space. This surprising outcome is a result of trying to create realistic spaces, often investing considerable effort (either by game designers or users) in particular features, but also working on specific spatial rules and resources to enable and constrain interaction and game play (such as enabling but limiting the ability to "teleport" between spaces) (McGregor 2006; Anders 1998). Far from being placeless, many games are spatial practices in very strong senses:

[S]pace can be seen as a central trope of any game because the most important activity of the player consists of moving, creating, and sustaining environments. Whether environments are abstract or more recognizable, the player is always involved in an effort to spatially master a game world. Hence, on a very fundamental level an important function of any game is to involve the player in a spatial process and to encourage a strong identification with the spatial dimensions of the game. (Lammes 2008: 260)

Moreover, the mode of representation and spatial experience can vary from different games and online worlds. So some care is needed about making over-general statements. For instance, some online environments allow users to customize the environment, as in Second Life, where you can even hire virtual architects; others have limited formats, as in Disney's Club Penguin for pre-teens, where the decor of the igloos and the dialogue has to use a limited menu. Some game worlds may have a quest structure that echoes colonialist stories of exploration and conquest (Fuller and Jenkins 1995) possibly paralleled with the development of "god" games that enable players to create worlds, often connected to visualization formats on screen that use maps and birds' eye views that invite the player to become a mapmaker, and indeed depersonalize the action (Lammes 2009: 89). Still others may mobilize tropes of cinematic landscapes – as in the Grand Theft Auto series, where socio-spatial clichés of Miami, New York, and Los Angeles are readily mobilized appealing to a "racialized common sense" based on othering different neighborhoods associating race and crime, with women as prostitutes open to abuse (Atkinson and Willis 2007; Longan 2008: 35). Conversely, within online worlds like Second Life the ability to shape the environment has been used by subcultural groups to articulate their identity. For instance, in a study of Gorean role play lifestyle group Bardzell and Odom (2008) note how the ritualized bodily performances and differentiated spaces of city and jungle, market place and tavern were used to enact the social stratifications crucial to that lifestyle.

Rather than just focus upon the space in the games we can also see games as spaces that enable sociality within a "magic circle" that absorbs us and is walled off from everyday life (Lammes 2008). With an average requirement of roughly 2 full months of gaming days to attain Level 60 in World of Warcraft (Ducheneaut et al. 2006: 289), it is clear this can be the major part of someone's social world. This would seem to confirm and amplify the sense of these as isolated realms, but studies show firstly the development of dense socialities within them which are structured through different online world mechanics (such as a proliferation of relatively small "guilds" of likeminded players in World of Warcraft (Williams et al. 2006)) but which increasingly break the magic circle both by para-game functions (with text messaging interfaces of differing kinds on the edge of the game screen (Ducheneaut et al. 2006: 284)), and beyond the confines of the game forming more channels of communication between players through spin-off events, like "Fantasy Fairs," and meetings in real life (Copier 2009).

Indeed, the exchange of online gifts and goods breaks down any easy divide of online or virtual world versus real – with "real" friendships made in online worlds. This link to outside worlds can be extended. There is a flourishing trade in virtual objects and skills whereby people buy virtual objects (magical swords or the like) using real money, and created by real people playing the game. The term for this is so called Real Money Trading and the scale of it is considerable. One company, Internet Gaming Entertainment, at its peak employed 500 doing over a quarter billion dollars of business in trades by acting as a "middleman for Western gamers eager to outsource the boring aspects of play to low-wage third worlders. The people who founded the company realized that scarcity of time and scarcity of virtual resources created a whole new market" (Salo 2008). The result is an entire industry of sweatshop gamers, in low income countries but especially China, playing for hours in cramped conditions aiming not to "play the game" but accumulate specific points or artifacts for sale – what is called "Gold Farming." This online outsourcing offers an uncanny echo of the outsourced geographies and back offices in globalized service sectors.

Conclusions

Electronic media and ICTs are changing the spatialities of sociality and the social life of people. There are real consequences to virtual lives, and virtual spaces can make real differences to people's life chances. The digital divide responds to classic social divisions, but also affects the ways of living and life chances of different people. It produces exclusions, exploitations, and stresses for those included, as well as new social realms and opportunities. The possibilities for self expression are real, but contested. The examples of Korea and Singapore illustrate how the social worlds of virtual lives clash, compete, and integrate with social experience. Online worlds which have often been seen as placeless turn out to be both spatial in themselves, which is perhaps surprising, but also just as "real" as the physical world. Their connection to physical spaces has often been portrayed as negative, but they can form placeworlds, projecting traditional relationships built in physical spaces, into the space of flows (Gordon and Koo 2008).

In this sense new media are becoming the means and mode of a great deal of social life and engagement. They become the mode of experience, social difference, distinction, and disadvantage. Their specific affordances and configurations matter in providing rules and resources for different social actors. They are not an external "driver" of social change. Rather a great deal of our life is now "techno-social" where changes in what we do and the media through which we do it run together. As processing power is located in objects, from predictive lists of favorites learning from our past actions to sensors and programs responding to actions, we are increasingly living in worlds where online and offline blur – both being animated by software algorithms and code (Crang and Graham 2007). In this world we may have to think about how we treat technological and human agency where the environment increasingly records, remembers, recalls, acts, and "thinks."

References

Anders, P. (1998) Envisioning cyberspace: the design of on-line communities. In J. Beckmann (ed.), *The Virtual Dimension: Architecture, Representation, and Crash Culture*. New York: Princeton Architectural Press, pp. 218–33.

Atkinson, R., and Willis, P. (2007) Charting the Ludodrome: the mediation of urban and simulated space and rise of the *flaneur electronique*. *Information, Communication & Society* 10 (6): 818–45.

Bardzell, S., and Odom, W. (2008) The experience of embodied space in virtual worlds: an ethnography of a second life community. *Space and Culture* 11: 239–59.

Baum, S., van Gellecum, Y., and Yigitcanlar, T. (2004) Wired communities in the city: Sydney, Australia. *Australian Geographical Studies* 44 (2): 175–92.

boyd, d. (forthcoming) White flight in networked publics? How race and class shaped American teen engagement with MySpace and Facebook. In L. Nakamura and P. Chow-White (eds.), *Digital Race Anthology*. New York: Routledge.

Boyer, M.C. (1999) Crossing cybercities: urban regions and the cyberspace matrix. In R. Beauregard and S. Body-Gendrot (eds.), *The Urban Moment: Cosmopolitan Essays on the Late-20th-century City*. London: Sage, pp. 51–78.

Bromley, C. (2004) Can Britain close the digital divide? In A. Park, J. Curtice, K. Thomson, C. Bromley, and M. Phillips (eds.), *British Social Attitudes: The 21st Report*. London: Sage, pp. 73–95.

Burrows, R. (1997) Virtual culture, urban social polarisation and social science fiction. In B. Loader (ed.), *The Governance of Cyberspace*. London: Routledge, pp. 38–45.

Castells, M. (1996) *The Rise of the Network Society: Networks and Identity*. Oxford: Blackwell.

Chang, W.-Y. (2005a) The internet, alternative public sphere and political dynamism: Korea's *non-gaek* (polemist) websites. *Pacific Review* 18 (3): 393–415.

Chang, W.-Y. (2005b) Online civic participation, and political empowerment: online media and public opinion formation in Korea. *Media, Culture & Society* 27 (6): 925–35.

Chee, F. (2005) Understanding Korean experiences of online game hype, identity, and the menace of the "Wang-tta." Paper read at Proceedings of DiGRA 2005 Conference, Changing Views: Worlds in Play.

Choi, J. H. (2006) Living in cyworld: contextualising cy-ties in South Korea. In A. Bruns and J. Jacobs (eds.), *Use of Blogs*. New York: Peter Lang, pp. 173–86.

Copier, M. (2009) Challenging the magic circle: how online role-playing games are negotiated by everyday life. In M. van den Boomen, S. Lammes, A.-S. Lehmann, J. Raessens, and M.T. Schäfer, *Tracing New Media in Everyday Life and Technology*. Amsterdam: University of Amsterdam Press, pp. 159–72.

Crang, M. (2000) Urban morphology and the shape of the transmissable city. *City: Analysis of Urban Trends, Culture, Theory, Policy, Action* 4 (3): 303–15.

Crang, M. (2007) Speed = distance/time: chronotopographies of action. In R. Hassan and R. Purser (eds.), *24/7: Time and Temporality in the Network Society*. Stanford: Stanford University Press, pp. 62–88.

Crang, M., and Graham, S. (2007) Sentient cities: ambient intelligence and the politics of urban space. *Information, Communication & Society* 10 (6): 789–817.

Crang, M., Graham, S., and Crosbie, T. (2006) Variable geometries of connection: urban digital divides and the uses of information technology. *Urban Studies* 43 (13): 2551–70.

Doel, M., and Clarke, D. (1999) Virtual worlds: simulation, suppletion, s(ed)uction and simulacra. In M. Crang, P. Crang, and J. May (eds.), *Virtual Geographies: Bodies, Spaces and Relations*. London: Routledge, pp. 261–83.

Ducheneaut, N., Yee, N., Nickell, E., and Moore, R.J. (2006) Building an MMO with mass appeal: a look at gameplay in world of warcraft. *Games and Culture* 1 (4): 281–317.

Fuller, M. and Jenkins, H. (1995) Nintendo and new world travel writing: a dialogue. In S. Jones (ed.), *CyberSociety: Computer-mediated Communication and Community*. London: Sage, pp. 57–72.

Gershuny, J. (2003) Web use and net nerds: a neofunctionalist analysis of the impact of information technology in the home. *Social Forces* 82 (1): 141–68.

Gordon, E., and Koo, G. (2008) Placeworlds: using virtual worlds to foster civic engagement *Space and Culture* 11: 204–21.

Graham, S. (2002) Bridging urban digital divides? Urban polarisation and information and communications technologies (ICTs). *Urban Studies* 39 (1): 33–56.

Hjorth, L. (2007) The game of being mobile: one media history of gaming and mobile technologies in Asia-Pacific. *Convergence* 13 (4): 369–81.

Hjorth, L. and Kim, H. (2005) Being there and being here: gendered customising of mobile 3G practices through a case study in Seoul. *Convergence* 11 (2): 49–55.

Holloway, D. (2005) The digital divide in Sydney: a sociospatial analysis. *Information, Communication & Society* 8 (2): 168–93.

Jones, S. and Fox, S. (2009) Generations online in 2009: Pew Research Center's Internet & American Life Project. Washington: Pew Research Center, www.pewinternet.org/Reports/2009/Generations-Online-in-2009.aspx.

Kluver, R. and Banerjee, I. (2005) Political culture, regulation and democratization: the Internet in nine Asian nations. *Information, Communication & Society* 8 (1): 30–46.

Lammes, S. (2008) Spatial regimes of the digital playground: cultural functions of spatial practices in computer games. *Space and Culture* 11: 260–72.

Lammes, S. (2009) Playing the world: computer games, cartography, spatial stories. *Aether: The Journal of Media Geography* 3: 84–96.

Lash, S. (2002) *Critique of Information*. London: Sage.

Lee, H. (2005) Multimedia and the hybrid city: geographies of technocultural spaces in South Korea. PhD, Durham University.

Lee, T. (2005) Internet control and auto-regulation in Singapore. *Surveillance & Society* 3 (1): 74–95.

Longan, M. (2008) Playing with landscapes: social processes and spatial forms in video games. *Aether: The Journal of Media Geography* 2: 23–40.

Longley, P. A. and Singleton, A.D. (2009) Linking social deprivation and digital exclusion in England. *Urban Studies* 46 (7): 1275–98.

Longley, P. A., Webber, R., and Li, C. (2008) The UK geography of the e-society: a national classification. *Environment and Planning A* 40 (3): 60–82.

Maffesoli, M. (1996) *The Time of the Tribes*. London: Sage.

McGregor, G. L. (2006) Architecture, space and gameplay in World of Warcraft and Battle for Middle Earth 2. Paper read at Proceedings of the 2006 international conference on game research and development at Perth, Australia.

Morley, D. (2003) What's home got to do with it? Contradicatory dynamics in the domestication of technology and the dislocation of domesticity. *European Journal of Cultural Studies* 6 (4): 435–58.

Mortensen, T.E. (2006) WoW is the new MUD: social gaming from text to video. *Games and Culture* 1 (4): 397–413.

Ng, C. and Mitter, S. (2005) Valuing women's voices: call center workers in Malaysia and India. *Gender, Technology and Development* 9 (2): 209–33.

Ng, K.K. (1999) *The Rainbow Connection: the Internet and the Singapore Gay Community*. Singapore: KangCuBine Publishing.

Patel, R. (2006) Working the night shift: gender and the global economy. *ACME: An International E-Journal for Critical Geographies* 5 (1): 9–27.

Perrons, D. (2003) The new economy and the work–life balance: conceptual explorations and a case study of new media. *Gender, Work and Organization* 10 (1): 65–93.

Perrons, D., Fagan, C., McDowell, L, Ray, K., and Ward, K. (eds.) (2006) *Gender Divisions and Working Time in the New Economy : Changing Patterns of Work, Care and Public Policy in Europe and North America*. Cheltenham: Edward Elgar.

Pratt, A. (2002) Hot jobs in cool places: the material cultures of new media product spaces – the case of south of the market, San Francisco. *Information, Communication and Society* 5 (1): 27–50.

PriceWaterhouseCoopers (2009) *Report to the Champion for Digital Inclusion: The Economic Case for Digital Inclusion*. Digital Inclusion Task Force.

Robins, K. (1999). Foreclosing on the city? The bad idea of virtual urbanism. In J. Downey and J. McGuigan (eds.), *Technocities*.London: Sage, pp. 34–60.

Rodan, G. (1998) The internet and political control in Singapore. *Political Science Quarterly* 113 (1): 63–89.

Ryan, B. (2004) AlphaWorld: the urban design of a digital city. *Journal of Urban Design* 9 (3): 287–309.

Salo, D. (2008) How the virtual gold trade works. *Wired* 16 (12).

Song, Y. (2007) Internet news media and issue development: a case study on the roles of independent online news services as agenda builders for anti-US protests in South Korea. *New Media and Society* 9 (1): 71–92.

SQW Consulting (2008) *Broadband in the Home: An Analysis of the Financial Costs and Benefits*, 26. London: Final Report to the Post Office.

Sriramesh, K. (2006) E-government in a corporatist, communitarian society: the case of Singapore. *New Media and Society* 8 (5): 707–30.

Stone, S. (1995) Split subjects, not atoms; or, how I fell in love with my prosthesis. In C. Gray (ed.), *The Cyborg Handbook*. London: Routledge, pp. 393–406.

Tang, P. S. and Ang, P.H. (2002) The diffusion of information technology in Singapore schools: a process framework. *New, Media & Society* 4 (4): 457–78.

Turkle, S. (1996) *Life on the Screen: Identity in the Age of the Internet*. London: Weidenfeld & Nicolson.

Virilio, P. (1998) Critical space. In J. Der Derian (ed.), *The Virilio Reader*. Oxford: Blackwell.

Wellman, B., Smith, A., Wells, A., and Kennedy, T. (2008) *Networked Families*, 55. Washington: Pew Internet & American Life Project.

Williams, D., Ducheneaut, N., Xiong, L., Zhang, Y., Yee, N., and Nickell, E. (2006) From tree house to barracks: the social life of guilds in world of warcraft. *Games and Culture* 1 (4): 338–61.

Yoon, K. (2003) Retraditionalizing the mobile: young people's sociality and mobile phone use in Seoul, South Korea. *European Journal of Cultural Studies* 6 (3): 327–43.

Part IV   Power and Politics

Introduction

Ruth Panelli

The diverse fields of social geography have consistently exposed how power is comprehensively woven through society and space. Foci on geographies of social difference highlight how societies are constituted in unequal ways – how privilige and disadvantage (and shifting combinations of both) are experienced by different groups. Similarly, foci on the social and spatial values of (and struggles over) *matter*

A Companion to Social Geography, First Edition. Edited by
Vincent J. Del Casino Jr., Mary E. Thomas, Paul Cloke, and Ruth Panelli.
© 2011 Blackwell Publishing Ltd. Published 2011 by Blackwell Publishing Ltd.

and meanings emphasize the uneven material and discursive terrains in which everyday lives (and more collective and formal configurations of society) are constructed.

Social geographers can appreciate that representations of a nation, or individual streetscapes and workplaces, are riven with uneven categories of difference (class, gender, ethnicity, sexuality, and so forth), and meanings and materialities of that space (e.g., a national monument, a sidewalk, or a factory floor). Moreover, numerous social geographers are frequently motivated to critique and unsettle existing norms and power relations if a sense of inequality or injustice is identified (e.g., poverty, labor exploitation, gender disadvantage, homophobia, racism, human rights abuses, etc.). Countless books, papers, conferences, journals, networks, and projects have been developed by social geographers to promote their critiques and explore alternative social options and geographic practices that might be pursued. In these instances, the *thinking and doing of social geography* is intimately entwined with questions of *power and politics*. The final part of this Companion illustrates some examples of such endeavors but we cannot be exhaustive in our review of social geographers' critique of power issues. Instead, our collation of chapters here is intentionally broad in order to convey some of the diversity and scope of contrasting interests in the power and politics underpinning social geographic issues. We also present this selection of work as a way to extend our previous emphasis on *thinking and doing social geography* and *matters and meanings;* showing how clearly different trajectories and practices may be developed by geographers.

Across these final six chapters you will find profoundly different examples of how power and/or politics infuse social concerns and the spatialities involved. Equally you will observe contrasting ways in which scholars practice and present their concerns – in research and writing decisions as well as wider professional and political practice. To begin, Nancy Heimstra and Allison Mountz (Chapter 24) provide a review of the classic interface between social and political geography – reviewing the growing interest in *geopolitics* and presenting and illustrating key interventions that have developed from within femininst epistemologies and interests. Canvassing both global and transnational differences in how politics and international relations is played out between nations and across borders, they review the heritage of geopolitics, before moving on to outline the debates and critiques between scholars concerning how geopolitics can (and sometimes should) be refashioned. For instance, the critique of disembodied geopolitical analysis is presented along with illustrations of how geopolitical struggles over migration are profoundly embedded in individual bodies and daily living spaces. In addition, they demonstrate how both theory (e.g., geopolitical and postcolonial) and practice (research activities and writing genres) can be mobilized to engage with power issues within the geography they are producing. Indeed their chapter shows how geographers draw on wide ranging theoretical debate and actively undertake to construct *interventions* that can both enhance academic scholarship and the way wider populations and policies might understand unequal power relations.

In Chapter 25, Dan Trudeau and Chris McMorran review a more generic and broad ranging topic that has been important to many critical social geographies. They unpack the power relations and spatial manifestations and consequences associated with diverse forms of *marginalization*. The social and spatial processes by which groups can identify and marginalize others forms a core concern for many geographers. In this chapter, we see how contrasing approaches have exposed the processes and power relations by which groups may be positioned unequally. The conceptual and material lenses of landscape and exclusion are shown as especially potent tools for showing the power-laden social world that can produce "landscapes of exclusion," "purified space," and "abject" and "enclosed" subjects. Trudeau and McMorran remind us that such forms of inequality and violence do not necessarily go unchallenged. They illustrate responses, noting the way borders may be contested, liminal spaces may be harnessed, or voluntarily maintenance of exclusion is mobilized to navigate and resist mainstream social expectations and controls.

A contrasting engagement with questions of power is presented by David Conradson in his exploration of *care and caring* (Chapter 26). Since many scholars have cared about unequal power relations and material conditions in society, geographies of care and caring can often be seen as an attempt to mobilize understandings about the need for care (e.g., social welfare, care of dependents, and health care). In other cases, some scholars explore how to engage power relations for transformative practices and positive social and spatial outcomes. Conradson notes that many forms of care work (formal and informal) are required for the reproduction of society, but that this results in uneven social and spatial experiences of care labor and receipt of care. He also highlights the concepts of proximity and distance in debates about the transformative ethics and potential for care – raising the issues of how far, and at what scale, care should (or can effectively) be conducted. Cumulatively these geographies of care challenge us to consider "new ways of being together in a ... relatively unequal world."

Turning to a critical consideration of the complex politics of social geography more generally, Chapter 27 provides an example of a set of populations and knowledge systems that have frequently been unacknowledged in social geographies of the West/North. Brad Coombes, Nicole Gombay, Jay Johnson, and Wendy Shaw introduce the challenges surrounding *indigenous geographies*. This chapter illustrates an example of diverse theoretical and methodological practices that can be developed to both acknowledge the worlds of a particular set of frequently excluded populations, and also challenge the power-laden presumptions, closures, and omissions that can often be unreflectively perpetrated in geography. After a discussion of indigenous geographies and indigeneity as a relational construct, two explicit case studies are given of the challenges and learning that are possible when indigenous and geographic traditions engage with each other. These cases give us intimate insights into the practices and reassessment of research that can be experienced in such geographies. In sum, the theoretical and case material in this chapter allow us to understand some of the complexities and rewards of "transcultural engagement." It also highlights the authors' two powerful interventions in (1) unpacking notions of (indigenous) identity and (2) navigating methodological commitments to "storytelling." These constitute important ways that the reproduction of disciplinary power relations may be critiqued and reconstructed if "co-habitation" of indigenous and non-indigenous experiences are to be supported.

Questions of power in universal jurisdiction and international courts provides a broader scale of inquiry in Chapter 28. Here Amy Ross reviews the character of *transnational geographies* and presents the case of human rights. The rise in transnational movements, activities and struggles is illustrated via the politics sur-rounding different human rights concerns. Ross demonstrates how activists mobilize a "spatial fix" using foreign jurisdictions when the politics within a local/national context precludes human rights issues being fully acknowledged. She also illustrates how universal jurisdiction and international courts are engaged and concludes that human rights politics may continue to uphold certain state powers and violence while securing prosecutions elsewhere. Finally, in documenting the intensity of local and national power interests beyond the international arenas, Ross confirms the importance of more nuanced transnational geographies that can expose the "mutual construction of space, power and justice."

In the last chapter of this part, Paul Chatterton and Nik Heynen present an academic review and account of practices surrounding geographies of *resistance and collective social action*. As scholar-activists they illustrate a commitment to both academic geographic review and debate as well as "on the ground" social action. They note the diverse forms of resistance and collective action that can be mobilized against "uneven geographies of … destructive power relations in particular places," but avoid dualistic conceptions of good/heroic/romantic positive resistance against bad/oppressive domination. After a discussion of the contrasting approaches to notions of power, resistance and collective action, Chatterton and Heynen analyse the geographies of resistance and show the importance of nuanced spatial analysis as well as recognition of the resistant "subjects" and "subjectivities" that may be constructed, performed, contested and reassembled. One example of these complexi-ties of resistant spaces and subjectivities is given through the case of climate diso-bedience, and the authors conclude that many new possibilities and repertoires of collective action can be imagined. These include the move away from notion of resistance as reactive and deviating from universal norms, and the adoption of more diverse and creative constructions of resistance and resistant subjects; actions and subjectivities that are able to tangle with the dynamics of power from broad/global through to immediate local/personal concerns.

Cumulatively this part shows how many social geographers "give a damn." The social world is not only there to be mapped, theorized or commentated. Writers in the following six chapters demonstrate how geographic research programmes and scholarly practices can critique and contribute to the way societies shape and contest their worlds. In many ways, this illustrates the power of social geography, for schol-ars and students are frequently drawn not only to an understanding of the diversity and unevenness of society and space, but to the opportunity to ask: "What are we doing about this?" These chapters will give you many stimulating and contrasting examples.

Chapter 24

Geopolitics

Nancy Hiemstra and Alison Mountz

As I planned and conducted a research project on the geopolitical consequences of contemporary US migrant detention and deportation policies, I realized that the project would not fit neatly within any definition of "the political" that did not simultaneously intersect with "the social." I was interested in the spatial, economic, and social reverberations of these deterrence-driven policies far beyond US borders, as well as evaluation of actual policy outcomes alongside policymakers' stated objectives. Epistemologically, I approached this study with the belief that consequences not traditionally classified as "political" can and should be investigated; it is important to look for answers beyond political offices and constituencies, and to ask how the effects of policies extend beyond the state. Some of these answers can and must be found in countries of migrant origin – not just in national government offices, but in the communities, homes and webs of daily life where what are sometimes referred to as "politics with a small p" unfold. Methodologically, I was committed to using primarily qualitative research methods, such as interviews, participant observation, and discursive analysis, to get at the everyday impacts of political decisions. I aimed to talk to people rarely considered when such policies are made. I also wanted to consider the role played by my background and biases in the conduct and interpretation of the research. Fieldwork took place in the southern Andes of Ecuador with migrants deported from the United States as well as with their household members. I paid particular attention to the policies' reverberations in Ecuadorians' everyday life to explore the various far-reaching impacts of political policies. The research analytically blurred domestic and foreign policy, household and public spheres, and the social and the geopolitical. (Nancy Hiemstra)

Within the discipline of geography, the term "geopolitics" has typically been associated with the subdiscipline of political geography. A long history supports this association, as does the very composition of the word. Indeed, some readers of *A*

A Companion to Social Geography, First Edition. Edited by
Vincent J. Del Casino Jr., Mary E. Thomas, Paul Cloke, and Ruth Panelli.
© 2011 Blackwell Publishing Ltd. Published 2011 by Blackwell Publishing Ltd.

Companion to Social Geography may initially wonder why a chapter on geopolitics would be included in this volume. However, as the personal anecdote above shows, contemporary geopolitical questions and research demonstrate a growth in opportunities for intersections between geopolitics and social geography. Indeed, it is becoming clear that geopolitical processes are permeated with and indeed frame sociospatial phenomena.

In order to provide a platform for more conversation between geopolitical inquiry and social geography, this chapter defines the field of geopolitics and then traces its development as various interventions have been made over the years. In the sections that follow, we first offer a brief definition and history of the term and the associated field of "geopolitics." In the second section, we discuss the contributions made by the subdiscipline of critical geopolitics to geopolitical inquiry. The third section reviews critiques of critical geopolitics that essentially stemmed from geopolitics' roots in imperialism. Fourth, we examine a number of feminist interventions into the field. Finally, we illustrate how there is growing room for intersection between geopolitics and social geography with illustrative examples from research by contemporary human geographers.

Definition and Brief History of "Geopolitics"

Outside the discipline of geography, geopolitics is generally understood as the political machinations played out between countries around the world, and more specifically the ways in which geography influences the relationships between states (ÓTuathail 2006; Dahlman 2009). Modern geopolitics came into being as a consequence of the Eurocentric drawing of boundaries around the globe in the nineteenth century, and the partitioning of blank spaces into specific domains of power, resulting in a new order of closed space and associated realms of control (ÓTuathail 1996; Gilmartin and Kofman 2004). Indeed, from its early stages, geopolitics drew on behavioral practices and socio-spatial relationships associated with states. The assumptions held that "natural" aspects of geographical location influenced the roles of states (Dahlman 2009: 89).

One often hears "geopolitical" employed as an adjective loosely synonymous with international relations, and yet the origins of the term have been more precisely traced and defined. Dahlman (2009: 88), for example, notes the two definitions provided by the Oxford English Dictionary as "the influence of geography on the political character of states" and the "pseudo-science developed in Nationalist Socialist [Nazi] Germany." Gilmartin and Kofman (2004: 113) define geopolitics as the "practices and representations of territorial strategies."

The discipline of geography has long been intimately involved in modern geopolitics, from merely mapping new boundaries drawn to furnishing geographical justifications for political hierarchies (Gilmartin and Kofman 2004). Early geopoliticians included German geographer Friedrich Ratzel, who likened states to living organisms in his writing in the late nineteenth century. Similarly, British academic Halford Mackinder was famous for his study of the expansion of Russian power due to its geographic location. Their work related closely to the purview of nation-states and indeed maintained the centrality of the state in conceptual understandings of international or global fields.[1]

Geopolitics – both as a discipline and as a guiding political philosophy – generally functioned as a tool of the state in the nineteenth century and first half of the twentieth century. Geopolitical ideas proved central to the extension of state power and were harnessed to justify imperial aspirations (Gilmartin and Kofman 2004). For example, in his position as the US government's territorial advisor after each of the world wars, geographer Isaiah Bowman was instrumental in the realization of American ambitions for global power (N. Smith 2004).

Geopolitics was also invoked in ways that left the field facing a troubled set of associations within the discipline of geography. In particular, the "pseudo-science" of Nazi Germany, under geopolitician Karl Haushofer, drew on Ratzel's idea of the state as a living organism that must expand its reach of power in order to survive (Gilmartin and Kofman 2004; Dahlman 2009). For decades after World War II and this purported use of *Geopolitik* strategy by the Germans, the term "geopolitics" was tainted by its association with Nazi strategy. Consequently, political geographers carefully avoided the term and sought to portray their work as neutral and objective (Gilmartin and Kofman 2004).[2]

It was not until the 1980s that geographers once again dared to openly engage with concepts of geopolitics. This academic re-engagement occurred subsequent to political developments at the international level that involved a renewed willingness to embrace the term "geopolitics," such as the Nixon administration's (under Secretary of State Kissinger) employment of "Realpolitik," and in tandem with the end of the cold war (F. Smith 2001; Gilmartin and Kofman 2004). Political geographers, partially drawing on the evolution of and debates associated with international relations theory, began to reject the positivist approaches that had dominated human geography since the quantitative revolution. Positivism is a science driven by collection of empirical data which has resulted in social science purporting to be objective and free of social values. Political geographers began to eschew positivism as systematic scientific study of the politics between and within states, instead developing an epistemological approach based in critical theory (Dalby 1991; Dodds 2001). Importantly, geographers started to question the continued centrality of the state in geopolitical analysis (Dalby 1991). These shifts within political geographic thought and practice corresponded with the cultural turn in social geography and set the stage for geography's re-engagement with geopolitics and subsequently for the emergence of critical geopolitics as a subdiscipline in the 1990s (Dodds 2001; Gilmartin and Kofman 2004). In the next section, we delve more deeply into critical geopolitics in order to later make the case for inclusion of particular epistemological and methodological tools in social geographical analysis.

Critical Geopolitics

> Geography is about power. Although often assumed to be innocent, the geography of the world is not a product of nature but a product of histories of struggle between competing authorities over the power to organize, occupy, and administer space. (ÓTuathail 1996: 1)

Simon Dalby (1994: 595) defines critical geopolitics as "the critical and poststructuralist intellectual practices of unraveling and deconstructing geographical and related disguises, dissimulations, and rationalisations of power." The field draws on debates within international relations, critical social theory, political economy, and postmodern theory (Dalby 1991; Dodds 2001). Expanding the work initiated by innovative political geographers in the 1980s, critical geopoliticians soundly reject positivism, taking instead a genealogical and deconstructivist approach that draws heavily on the work of social theorists such as Michel Foucault and Jacques Derrida (Dalby 1991; Gilmartin and Kofman 2004). Drawing inspiration from Foucault, Dalby (1991: 276) explains that "postmodern and poststructuralist approaches ... point to how modes of knowledge are power-related resources, arguing that knowledge of a particular "truth" simultaneously enables and constrains practice."

Understanding the ways in which power works and how powerful actors shape the outcomes and interpretations of events (and geography) is a primary endeavor of critical geopolitics. In his foundational text, *Critical Geopolitics,* Gerard ÓTuathail (1996) illustrates that geography is not natural or automatic, but embedded in broader sets of power relations. ÓTuathail (1996) argues that struggles over geography can be understood as struggles over particular visions of the world and determinations about who belongs and who does not to particular spaces. He redefines "geo-graphing" as an "earth-writing" by ambitious colonizing powers intent on organizing space according to their visions of the world, and he characterizes "geo-politics" as "the politics of writing global space" (p. 18) by powerful intellectuals and institutions. ÓTuathail also uses the work of geographer Halford Mackinder to illustrate how the geopolitical tradition is guilty of erroneously claiming a disembodied, neutral perspective as well as being based on geographical ethnocentrism. The field of *critical* geopolitics, then, exhorts geographers to question the very realities presented by geopolitical processes. Accordingly, ÓTuathail (1996: 68) describes critical geopolitics as "a tactical form of knowledge" through which political discourses can be re-examined under fresh analytical criteria.

The conceptual work of critical geopolitics has often sought to expose the "master narratives" of political relationships as state-centric and rooted in elitism (ÓTuathail 2006), and to open up consideration of political power not necessarily based in the state (Dalby 1991). The "natural" role of geography and its influence on states should no longer be taken for granted, but rather put into place in a broader constellation of relations between power and knowledge.[3] In this global constellation, the writing of space is always an exercise in power (ÓTuathail 1996). Critical geopolitics is centered in this recognition of the power-laden ways in which space is written, and in *how* power moves through particular spatial orders. Critical geopoliticians take special interest in the way the organization of space is discussed and presented. Indeed, they view geopolitics "as a form of political discourse rather than simply a descriptive term intended to cover the study of foreign policy and grand statecraft" (Dodds 2001: 469). Scrutinizing how we talk about politics and the organization of political space is therefore essential to understanding geopolitics. And "talking about" politics brings us to some of the ways in which critical geopolitics has, in turn, itself been critiqued.

Critiques of Critical Geopolitics

Critical geopolitics has made significant contributions to both the discipline of geography and the analysis of geopolitical processes. However, particularly since the publication of ÓTuathail's foundational *Critical Geopolitics*, substantial critiques have been leveled at the field of critical geopolitics itself, many of them made by feminist geographers. There are, as in most innovative fields, a wide range of critiques. In this chapter we focus on those most important to the connections between geopolitics and social geography we wish to illuminate. In this section, we review the subset of critiques that stem primarily from geopolitics' roots in imperialism. The subsequent section concentrates on interventions into critical geopolitics that come largely out of arguments for further engagement between feminism and political geography.

It is important to recognize that there are separate sets of critiques of geopolitics and of critical geopolitics. Indeed, as explained earlier, *critical* geopolitics is grounded in the idea that geopolitics relied on disembodied academic and political epistemology and methodology. However, our review here demonstrates that critiques of geopolitics and critical geopolitics sometimes overlap. Indeed, despite its deconstructionist and analytical intentions, many charge that critical geopolitics has not fully separated from geopolitics' problematic beginnings in imperialism (Gilmartin and Kofman 2004); it remains an elitist, gendered, disembodied mode of inquiry that continues to focus overmuch on "the state" and fails to provide constructive alternatives to present realities. For example, Gilmartin and Kofman (2004: 123) suggest that both traditional and critical geopoliticians "continue to be bound by masculinist modes of analysis and representation that create binary oppositions between elite and popular, between state and local, and between powerful and powerless, and by those who continue to use language that is marked by its apparent objectivity but that masks fundamentally gendered ideas and concepts." This quote illustrates that these critiques are circularly linked, making it difficult to discuss one without coming to others. However, for the sake of clarity here we treat each of these charges briefly.

Critics argue that critical geopolitics tends to focus on the western world (N. Smith 2000; Dodds 2001; Dowler and Sharp 2001). Perhaps in response to Said's (1978) robust assessment of essentializing academic practices that construct "the orient" as region and field of study, many intellectuals have avoided analyses of non-western societies (Dowler and Sharp 2001). Critical geopoliticians have proved no exception; practitioners often fall short in pushing themselves outside the bounds of their western perspective and knowledge, and therefore fail to recognize the Eurocentrism in which their own work is embedded. Dowler and Sharp (2001: 167) charge that though the discourse of critical geopolitics is portrayed as universal, it remains a fundamentally western mode of expression. Accordingly, the subdiscipline has been incapable of evolving beyond the restrictions occasioned by its Eurocentric perspective.

A persistent, troubling elitism is also evident in critical geopolitics (and, of course, traditional geopolitics). Feminists and others charge that despite practitioners' best intentions, the field hypocritically provides its own elitist, masculinist narrative (Sharp 2000; F. Smith 2001; Gilmartin and Kofman 2004), and practitioners tend to pay attention to "the state" and the actions of state actors as traditionally understood – that is, typically privileged, influential males. Consequently, critical geopolitics perpetuates – instead of destabilizes – the inherently gendered nature of geopolitical discourse (Dalby 1994; F. Smith 2001). Furthermore, in her critique of *Critical Geopolitics*, Sharp (2000) claims that ÓTuathail reproduces existing power hierarchies in his focus on the already-powerful; ordinary people are given short shrift while elites are portrayed as having all the power and agency.[4] This oversight prevents critical geopoliticians from considering the influences of popular culture on geopolitical events and actors, falsely separates state actors from non-state (and non-elite) actors (Sharp 2000), and leads them to ignore diverse experiences and resistances (F. Smith 2001). Importantly, it also "silences a whole range of people and groups from the operations of international politics" (Sharp 2000: 363).

Critics further suggest that critical geopoliticians have been unsuccessful in getting away from that which they critique: a disembodied view from nowhere. In *Critical Geopolitics* (1996), ÓTuathail states that practitioners of geopolitics falsely claim personal neutrality in their assessments of political relationships, a claim, he says, rooted in geographical ethnocentrism (ÓTuathail 1996). While ÓTuathail's recognition of some of the faults inherent to the traditional geopolitical gaze is a fair start to bringing marginalized, overlooked people back into analysis (Dowler and Sharp 2001), scholars level the charge of this disembodied gaze right back at the subdiscipline. Neil Smith (2000: 366) argues that *Critical Geopolitics* contains its *own* "symptomatic silences," and ÓTuathail fails to ask what he is *not* seeing and why. Sharp (2000) suggests that ÓTuathail does not acknowledge the unavoidable influence exerted on his scholarship by his social, economic, and political background, or recognize the role that personal biases also inevitably play. This approach – replicated throughout the subdiscipline – falsely endows scholars with the sense that they are somehow immune to cultural and popular influences (Sharp 2000). Thus, critical geopoliticians engage in a "disembodied critical practice" (Hyndman 2004b: 310), typically far removed from the subjects and people they evaluate (Dowler and Sharp 2001). One consequence of these many silences and false neutralities is that critical geopolitics actually contributes to the reproduction of already-existing power arrangements (Dalby 1994).

An additional charge leveled at critical geopolitics suggests that despite its many deconstructions of geopolitics, the subdiscipline does not itself provide recommendations for resistance and change (Sparke 2000), and supplies few suggestions for or examples of alternatives to contemporary state-centered, elitist, gendered political orders (Sparke 2000; Dowler and Sharp 2001). As Dowler and Sharp (2001: 167) state, "Important interventions have been made and there will always be a need to analyse the distortions of powerful geopolitical discourse; however, can there be a more constructive side to critical geopolitics – a more positive politics?" Critical geopoliticians, charge some scholars, are skilled at deconstructing whatever may be their object of study, but they too rarely offer any suggestions for a better reconstruction (Sparke 2000; Dodds 2001; Dowler and Sharp 2001).

A final critical observation holds that traditional geopolitics has silenced, excluded, and "rendered invisible" women (Dalby 1994: 595; Gilmartin and Kofman 2004). That geopolitics remains a patriarchal intellectual endeavor (Dowler and Sharp 2001; Gilmartin and Kofman 2004) is evident in its highly gendered geopolitical discourse (F. Smith 2001),[5] and numerous geographers have charged that the practice of critical geopolitics is steeped in masculine perspectives and methodology (Sharp 2000; Sparke 2000; Hyndman 2004b). For example, Sharp (2000: 363) charges that because ÓTuathail's approach to presenting critical geopolitics centers on a genealogy of "Big Men," it reinforces the masculinism embedded in geopolitics instead of destabilizing it. Consequently, the gendered practice of geopolitics precludes fully understanding how political power works, effectively reproducing existing gender hierarchies (Dalby 1994; F. Smith 2001; Gilmartin and Kofman 2004). Dalby (1994) argues insightfully that critical geopolitics would greatly benefit from a de/reconstruction based on feminist ideas on gender. He notes that there are many gendered assumptions behind the study of international relations (which is foundational to critical geopolitics), and that political spatializations often make women most vulnerable.

In the next section, we take up Dalby's argument by exploring five feminist interventions into critical geopolitics. While it is not our intention to position feminist work on geopolitics as somehow resolving the shortcomings or silences of previous work, we do address feminist contributions in order to explore the fertile connections between geopolitics and social geography as well as some of the exciting directions in which these subfields are moving in tandem.

Feminist Geopolitics and Interventions

[F]eminist and poststructuralist readings of geopolitics both open up the ontological presuppositions of power to critique and in the process reveal how the taken-for-granted specifications of politics are inherently powerful. (Dalby 1994: 603)

Many of the analytical moves made by political geographers and critical geopoliticians follow the "cultural turn" (Dowler and Sharp 2001), and in so doing their work corresponds with that of social geographers. That is, the subjects of analysis have begun to include the more routine ways in which power operates on a daily basis. However, as evident in the review of critiques above, the field of critical geopolitics still has significant room for improvement. In this section, we explore some of the epistemological standpoints and methodological tools that feminist scholars bring to the field of critical geopolitics. We suggest that collectively, feminist critiques and political perspectives offer openings for stronger links between critical geopolitics and social geography. We begin by briefly discussing proposals for a "feminist geopolitics," and then explore five feminist interventions that, we believe, broaden and deepen the tools available for geopolitical analysis. In the final section of this chapter, we illustrate these interventions in practice by looking at contemporary geopolitical research projects in geography, paying special attention to the rich ground such projects provide for links with social geography.

Feminist political geographers have advanced the idea of a "feminist geopolitics," which will strengthen critical geopolitics' powerful tools of discursive analysis and deconstruction by applying lessons from feminist geography (Dowler and Sharp 2001; F. Smith 2001; Hyndman 2004a, 2004b). Feminists have highlighted the arbitrariness of the division of social realms into public and private and asserted that "the political is personal" (Enloe 1989). They also challenge the idea of "capital P Political" as politics that operate only in formal spheres (Peake and Kofman 1990), and turn attention to "little p" political engagement emanating from intimate sites of daily life (e.g., Cope 2004). These epistemological breakthroughs have opened up understandings of political participation and arenas. Feminist geographers Kofman and Peake (1990) assert that "the political" can occur at any scale, and others have mapped fluid boundaries between the personal and the political (Secor 2001; England 2003; Cope 2004).

A feminist geopolitics calls for an expanded conceptualization of scale. The focus of political geography is usually on scales such as the state, region, and city; other scales have often been overlooked (England 2003; Hyndman 2004a, 2004b). Jennifer Hyndman (2004b: 315) calls for the redefinition of scales of inquiry to those "finer and coarser than that of the nation-state and global economy," and she also interrogates the notion of "scale as pre-given and discrete from other levels of analysis" (p. 309). Feminist geographers recognize that it is important to look at previously ignored scales such as the body, the household, the locality, and the supranational organization (Marston 2000; Marston and Smith 2001), to link scales to specific places (Brown and Staeheli 2003), and to focus on the local (Dalby 1994; Cope 2002). Furthermore, Kim England (2003: 612), noting the ongoing conflicts over domestic issues (such as reproductive rights and gendered wage differences) at the borders between public and private, points out that the recent "scale debates" in geography may provide spaces for increased exchange between feminist and political geographies. We contend that these debates also make room for increased exchange between social geography and geopolitics by drawing on the socially constructed nature of scale (Marston 2000).[6]

Feminist geopolitics also reminds scholars to consider the heterogeneity of individual experiences (Moss 2002), and focus on individual, gendered subjects in their daily lives (F. Smith 2001). Ana Secor (2001) proposes an alternative view of the spatialization of politics to that typically explored in critical geopolitics. She focuses on achieving an "understanding of how political life plays out through a multiplicity of alternative, gendered political spaces" (2001: 192), by looking at both formal and informal political practices in Turkey. Her work shows that scholars who privilege the national and international scales in geopolitical analyses risk failure to grasp connections between and across scales. Secor (2001: 193) asserts that "feminist approaches show how the (imminently political) categories of public and private, global and local, formal and informal, ultimately blur, overlap and collapse into one another in the making of political life."

We now draw on the lessons advanced by feminist geopolitics to suggest interventions in political – and social – inquiry.[7] We begin with the recognition that *the state is not unitary*. By incorporating a new range and understanding of scales of political analysis, feminist geographers have significantly destabilized traditional understandings of "the state" (Dowler and Sharp 2001; Brown and Staeheli 2003; England 2003; Desbiens et al. 2004; Hyndman 2004b). Instead, the state functions as "a set of practices enacted through relationships between people, places, and institutions" (Desbiens et al. 2004: 242; Painter 1995). In Alison Mountz's (2004) work, for example, bureaucrats managing immigration along borders are embedded in broader networks of social relations that influence their work, including the daily decisions about which immigrants and refugee claimants succeed in entering sovereign territory. Analyses conducted at the scale of the body further amplify geopolitical inquiry beyond the state (Hyndman 2004a, 2004b). Hyndman (2004a: 175) writes, "From a feminist perspective, shifting the focus of security to that of civilian safety and well-being unsettles the state-centric approaches of conventional geopolitics." Such shifts draw attention to everyday lives instead of high-stakes political actors and their maneuvers (Dodds 2001). Tamar Mayer (2004), for instance, shows how soldiers used rape of women as a tool of ethnic genocide in the Bosnian conflict. Such physical violence to bodies starkly illustrates how the political and the social intersect.

Attention to feminist theory and practice in critical geopolitics has also led to an *expanded understanding of "security"* in ways that could be useful for geographers in general, particularly in the current climate in which "security" proves a persistent framing of public and political discourse. Geopolitics has traditionally placed "security" in the masculine realm, despite the very direct and dramatic impacts "national security" disruptions typically have on women (Dalby 1994, drawing on Enloe 1989). Geopolitical scholars drawing on feminism have suggested that the idea of "human security" is a more expansive idea of "security" (F. Smith 2001; Hyndman 2004a). Dalby (1994: 601) notes that, "By so bluntly posing the question of "whose security?" feminist critiques challenge the territorial presumptions of states" (see also F. Smith 2001). In short, attention to gender challenges the idea of the territorial state as container (Dalby 1994). Hyndman's (2007) work, for example, on the displacement caused by political conflict in Sri Lanka illustrates the lived and gendered realities of losing one's home, community, and family members. Rather than examine only the state as an "actor" in political conflict, Hyndman shifts study of security to center the human suffering and resilience that always accompanies states' moves to securitize.

The *inclusion of feminist political methodological tools offers significant potential for critical political analysis* (Dodds 2001; Dowler and Sharp 2001; Sharp 2004). Despite intentions to break through the narrow, elitist perspective endemic to geopolitics, critical geopoliticians have typically failed to employ methods that facilitate such a breakthrough (Dodds 2001; Dowler and Sharp 2001). While deconstructionist scrutiny of the power-knowledge relationships inherent to discourse is important to critical political projects, it does not draw attention beyond traditionally visible state actors. Feminist political geographers frequently employ more localized, qualitative and/or ethnographic methods such as interviews, participatory techniques, and focus groups, often in addition to quantitative methods (Sharp 2004). Qualitative research attempts to "get at the less formal spaces where hidden and marginalized, but no less important, political identities and processes are formed and reformed" (Sharp 2004: 98; Dowler and Sharp 2001). This can be achieved by encouraging focus on everyday lives (Sharp 2000; Staeheli and Kofman 2004), and explicitly endeavoring to include a range of voices and spaces (Sharp 2000). In addition, methodological commitments to understand the richly embedded nature of lived politics has the potential to ground critical geography's "view from nowhere" (Dowler and Sharp 2001; F. Smith 2001; Gilmartin and Kofman 2004; Hyndman 2004b). Because of feminism's insistence on reflexivity and positionality as these methods are carried out, researchers are encouraged to bear in mind that all knowledge is partial and situated (Moss 2002; Sharp 2004).

Feminist attention to the political reveals the need for *more attention to the everyday*. Indeed, political geographers have long urged that political geographic analysis be attentive to everyday politics, or politics with "the little 'p'" (Kofman and Peake 1990; Staeheli and Kofman 2004). Thrift (2000) similarly calls for attentiveness to "the little things," those mundane practices wherein state power is

reproduced, and Dalby (1994) has argued that political power need not be synonymous with state power. Feminist scholars, however, charge that political geography's
"global visions and grand theorizing" (Sharp 2000: 94) often overlook "little p"
politics. By focusing on women and specifically the "practical power arrangements
of everyday life" (Dalby 1994: 599), critical geopoliticians can better understand
how politics work. Dowler and Sharp (2001: 171) state,

> In order to start to think in terms of a feminist (or post-colonial feminist) geopolitics,
> it is necessary to think more clearly of the grounding of geopolitical discourse in prac-
> tice (and in place) – to link international representation to the geographies of everyday
> life; to understand the ways in which the nation and the international are reproduced
> in the mundane practices we take for granted.

In short, geopolitical analysis must include a focus on daily life as well as the "Big
Men" (Sharp 2000); an interrogation of scales ranging from the global to the body;
and a broad conceptualization of what "the political" can be.

Finally, feminist geopolitics aims to add "*a potentially reconstructive political
dimension* to the crucial but at times unsatisfactory deconstructionist political
impulses" of critical geopolitics (Hyndman 2004b: 309). As discussed in the previous section, assessments of critical geopolitics have found that it provides few
alternative visions of political realities. Feminist geography, with its fundamental
interest in empowerment (Staeheli et al. 2004), has altered modes of knowledge
construction in political geography. Feminist political geography has worked to
"democratize" intellectual endeavors by encouraging scholars to consider positionality and perspective (Staeheli and Kofman 2004: 5). A feminist perspective on the
geopolitical includes a commitment to positive social change that goes beyond the
mere "textual interventions" of critical geopolitics to actually envision and develop
alternative strategies (Dowler and Sharp 2001: 167; Gilmartin and Kofman 2004;
Staeheli and Kofman 2004). Pain and Smith (2008: 2), for example, illustrate poign-
antly the ways that fear has permeated and now shapes every aspect of our everyday
lives, yet lies curiously hidden from discussion. They note the "rarely seriously
unpacked engagement between geopolitics and everyday fears" and see potential in
sustained study of the geopolitics of fear in everyday life in ways that destabilize
the hierarchical fixing of "the global" over "the everyday."

These key interventions by feminist political geographers offer lessons on how
to stretch geographical inquiry in general. Just as feminist inquiry has broadened
the range of subjects explored by critical geopolitics, we suggest that inclusion of
these five analytical moves to the study of geopolitics positions the subdiscipline for
more interactions with social geography, also concerned with how sociospatial
relationships unfold in daily practice. In the next section, we extend our discussion
of the intersecting realms of the geopolitical and the social by examining new directions
in the subfield that bring together the geopolitical, the social, and the everyday.

New Directions: The Geopolitical, the Social, and the Everyday

*The contradictions I witnessed while studying immigration and asylum policies and
their sometimes devastating effects in migrants' lives prompted me to turn to the*

state as subject of analysis. I researched the bureaucracy of Citizenship and Immigration Canada where employees worked daily in the business of making decisions about international border-crossings. I was interested in how they made decisions about who to let in and who to send home. Inspired by the work of feminist theorists on embodiment (e.g., Haraway 1991), I wanted to know how the state was embodied (Mountz 2004) and spent several months observing and interviewing bureaucrats in their day-to-day work on migration. The ethnography of the state examined the response of Canadian federal enforcement to human smuggling from China to North America by boat (Mountz 2010). Immigration policies were implemented unevenly across time and space, and state borders and the very edges of bureaucracy proved blurry. Civil servants interacted daily with other workers in order to do their jobs: immigrants, immigration lawyers, media workers, and refugee advocates, to name but a few. I found this daily process of imagining and enacting the state through bureaucratic work to be intimately bound to the social contexts, rituals, histories, and geographies where workers found themselves. After the federal government intercepted boats carrying migrants from China to Canada, many made claims for refugee status. Although claimants were offered access to the refugee claimant system, mid-level bureaucrats were made aware of the imperative of maintaining smooth diplomatic relations. They knew that the acceptance of refugee claims would embarrass China on the international stage by shaming the country for its poor record on human rights while it was fielding visits by western authorities in anticipation of joining the World Trade Organization. Instead, the claims were aggressively refuted by various arms of the government and most of the claimants ultimately repatriated to China. Ethnographic research exposed important geopolitical contexts that shaped the response to smuggling, but would not have been known without interviews and participant observation. The ethnography of the state uncovered geopolitical forces shaping the outcome of individual refugee claims made by Chinese nationals in Canada, but not written anywhere in official domestic or foreign policy or press releases. (Alison Mountz)

Like other fields of geographic inquiry, geopolitics cuts across subdisciplines and need not remain the sole "jurisdiction" of political geographers. Of course, the field of critical geopolitics is not the only arena where scholars are questioning existing power arrangements. For example, human geography moved toward humanism in the 1970s, made the cultural turn in the 1980s, and saw the rapid development of critical geography in the 1990s. The corresponding wealth of scholarship sought to challenge existing power structures in institutions, in social movements, and with innovative and community-engaged methods. Likewise, as the description of Alison's work shows, the utility of a *feminist* geopolitics far exceeds the work done by scholars who study gender. It does much more than just include women in political geographic analysis; it provides a "lens" for making the lived realities of non-elites more apparent (Dowler and Sharp 2001: 169). We suggest that the incorporation of geopolitical analysis into social geography, and vice versa, can lead to more socially conscious, politically engaged scholarship (see Staeheli and Kofman 2004).

In this section we illustrate the intersection between geopolitics and social geography, drawing examples from recent research on human migration and associated state enforcement practices. Scholarship on human migration offers fertile ground from which to understand the intersections of social geographies and geopolitical forces. The state-level political machinations shape many dimensions of the daily lives of international migrants whose legal status, access to livelihoods, health care, and education are influenced by their citizenship status. Relations between states emerge not only in territorial conflicts along borders, but in everyday practices where the power of nation-states and their borders are reproduced through the lives of migrants who cross them. Wastl-Walter and Staeheli (2004: 144) call attention to how boundaries and territories are socially, politically, and culturally constructed, and highlight that "the ways in which they are constructed, performed, and perceived depends on the cultural, religious, social, and economic contexts in which they are located." Indeed, by challenging commonly accepted ideas of territory and boundaries, the research we highlight here further loosens the state from the center of political geographic and geopolitical analysis (Wastl-Walter and Staeheli 2004), and opens up additional possibilities for epistemological and methodological links with social geography.

Eunyoung Choi (forthcoming) conducted qualitative research with North Korean migrants who crossed the border into China and live and work there without authorization. Mostly women, these are among the most vulnerable migrants in the region and globally precisely because of the geopolitical relationship between North Korea, China, and other states. As a communist country, North Korea has been so isolated since the cold war that its residents live in extreme poverty and widespread hunger. The majority of those who migrate into China are women moving because of their chances of finding particular kinds of work as domestic laborers and sex workers in China. Their criminalization through "illegal" status makes them vulnerable as undocumented workers in China and translates into a need to hide in daily life, exacerbating their exploitation as workers and as women by employers and state authorities alike. By studying geopolitical relations from the scale of the migrant woman's body, Choi shows that the political arrangements between states indeed shape the contours of everyday social lives and economic livelihoods.

In field studies that unfold on different and distant continents both Choi and Mountz bring the policies and practices of states and the global forces in which they are embedded into the realm of the daily lives and lived spaces of transnational migrants. As such, these cases both demonstrate the feminist interventions detailed in the earlier section. Both scholars jump scale, moving between global and intimate scales of the body, while not prioritizing one over the other, an important distinction made by Pain and Smith (2008). Following Hyndman (2004a), they de-center the state to expand notions of security, moving discourse from national to human security by understanding the state through the migrant. Simultaneously, their analyses show that "the state" does not function as a unitary body. Indeed, both the Chinese and Canadian states benefit from maintaining ambiguity around the legal status and the information available to the public about undocumented migrants, whether Chinese in Canada or North Koreans in China. Qualitative methods in the form of semi-structured interviews and participant observation in both studies enabled Choi and Mountz to center the everyday as geopolitical in these studies. Both offer "little p" analyses of the journeys of undocumented workers and their encounters with state authorities, building political potential to "flip" the scripts of the state about migration and enhance political engagement through coalition-building and collaboration.

Conclusions

Critical geopolitics pushes scholars to consider ways in which power is embedded in and influences space. *Feminist* geopolitics promulgates an embodied, situated view that pays attention to private as well as public domains, is capable of recognizing "the political" at all scales, and places importance on studying the everyday lives of non-state actors. This chapter suggests that contemporary geopolitics offers tools whose utility extends beyond political geography into social geography, and we have aimed to provide a knowledge base from which more connections can be made. We outlined geopolitical inquiry: its history, evolution, critiques, and recent intellectual moves. In addition, we explored analytical tools developed within the subdiscipline of critical geopolitics – and particularly feminist geopolitics – that can usefully intersect with, shape, and apply to social geographies in significant ways.

We have drawn on examples from research on human migration to illustrate the central point, but the utility of geopolitics for social geography is certainly not limited to migration research. For example, postcolonial theories that attend to the racialized relationships between colonial powers and decolonized populations offer another field of study where the geopolitical and the social prove fertile, interwoven ground. Transnational feminist scholars such as Chandra Mohanty (2003) and Julia Sudbury (2005) offer analyses that center lived experiences such as cross-border solidarity movements and incarceration that – like our discussion here of human migration – disrupt the imperial fixing of the geopolitical.

In closing, we call on critical social and political geographers alike to disrupt, deconstruct and expand the traditional boundaries of subdisciplines. We challenge scholars to evaluate the ways in which they ontologically, epistemologically, and methodologically frame geographic inquiry and analysis. A willingness to develop points of intersection between social geography and geopolitics will enrich geographic understanding of how the social is interwoven with the political, and vice versa.

Notes

1   The work of several "anarchist geographers," such as Reclus and Kropotkin who actively sought to counter the prevailing views on geopolitics and geography, is considered in more detailed historical reviews (see Gilmartin and Kofman 2004).
2   This avoidance of the term "geopolitics," however, did not preclude geographers from continuing to draw upon what had previously been considered geopolitical ideas (Gilmartin and Kofman 2004).
3   Critical geopolitics' attempts to de-center the state in geopolitical analysis should not be taken to indicate that states are no longer powerful (Dodds 2001). Indeed, we are

reminded of states' continued importance through, for example, their capability for violence, as seen in relatively recent events, such as in Rwanda, Sudan, and Yugoslavia.

4   While many of these critiques have been directed at O'Tuathail's 1996 *Critical Geopolitics*, we believe it is fair to extend these critiques to the subdiscipline as a whole (see Hyndman 2004b).

5   Enloe (1989), for example, points out that although women play an important role in the formation of national and international organizations and identity, this role usually goes unrecognized because women are not among the obvious, conventionally recognized, "political" decision-makers.

6   This point is well-illustrated in the scholarship on migration discussed in the next section.

7   We do not claim that feminist geographers are the only scholars to inspire these interventions. Indeed, the interventions incorporate a range of influences. However, it is accurate to say that feminist geopolitics is a common influence for each point.

References

Brown, M., and Staeheli, L. (2003) "Are we there yet?" Feminist political geographies. *Gender, Place and Culture* 10 (3): 247–55.

Choi, E. (forthcoming) Everyday practices of "bordering" and threatened bodies of undocumented North Korean border-crossers. In D. Wastl-Waters (ed.), *Research Companion to Border Studies*. Aldershot: Ashgate.

Cope, M. (2002) Feminist epistemology in geography. In P.J. Moss (ed.), *Feminist Geography in Practice: Research and Methods*. Oxford: Blackwell, pp. 43–56.

Cope, M. (2004) Placing gendered political acts. In L.A. Staeheli, E. Kofman, and L.J. Peake (eds.), *Mapping Women, Making Politics: Feminist Perspectives on Political Geography*. New York: Routledge, pp. 71–86.

Dalby, S. (1991) Critical geopolitics: discourse, difference, and dissent. *Environment and Planning D: Society and Space* 9 (3): 261–83.

Dalby, S. (1994) Gender and critical geopolitics: reading security discourse in the new world disorder. *Environment and Planning D: Society and Space* 12 (5): 595–612.

Dahlman, C.T. (2009) Geopolitics. In C. Gallagher, C.T. Dalhman, M. Gilmartin, A. Mountz, and P. Shirlow (eds.), *Key Concepts in Political Geography*. Thousand Oaks: Sage, pp. 87–98.

Desbiens, C., Mountz, A., and Walton-Roberts, M. (2004) Introduction: reconceptualizing the state from the margins of political geography. *Political Geography* 23 (3): 241–3.

Dodds, K. (2001) Political geography III: critical geopolitics after ten years. *Progress in Human Geography* 25 (3): 469–84.

Dowler, L. and Sharp, J. (2001) A feminist Geopolitics? *Space and Polity* 5 (3): 165–76.

England, K. (2003) Towards a feminist political geography? *Political Geography* 22 (6): 611–16.

Enloe, C. (1989) *Bananas, Beaches, and Bases: Making Feminist Sense of International Relations*. Berkeley: University of California Press.

Gilmartin, M. and Kofman, E. (2004) Critically feminist geopolitics. In L.A. Staeheli, E. Kofman, and L.J. Peake (eds.), *Mapping Women, Making Politics: Feminist Perspectives on Political Geography*. New York: Routledge, pp. 113–25.

Haraway, D. (1991) *Simians, Cyborgs, and Women: The Reinvention of Nature*. New York: Routledge.

Hyndman, J. (2004a). The (geo)politics of mobility. In L.A. Staeheli, E. Kofman, and L.J. Peake (eds.), *Mapping Women, Making Politics: Feminist Perspectives on Political Geography*. New York: Routledge, pp. 169–84.

Hyndman, J. (2004b) Mind the gap: bridging feminist and political geography through geo-politics. *Political Geography* 23 (3): 307–22.

Hyndman, J. (2007) Conflict, citizenship, and human security: geographies of protection. In D. Cowen and E. Gilbert (eds.), *War, Citizenship, Territory*. New York: Routledge, pp. 241–59.

Kofman, E. and Peake, L. (1990) Into the 1990s: a gendered agenda for political geography. *Political Geography Quarterly* 9 (4): 313–36.

Marston, S. (2000) The social construction of scale. *Progress in Human Geography* 24 (2): 219–42.

Marston, S. and Smith, N. (2001) States, scales and households: limits to scale thinking? A response to Brenner. *Progress in Human Geography* 25 (4): 615–19.

Mayer, T. (2004) Embodied nationalisms. In L.A. Staeheli, E. Kofman, and L.J. Peake (eds.), *Mapping Women, Making Politics: Feminist Perspectives on Political Geography*. New York: Routledge, pp. 153–67.

Mohanty, C. (2003) *Feminism without Borders: Decolonizing Theory, Practicing Solidarity*. Durham, NC: Duke University Press.

Moss, P. (2002) Taking on, thinking about, and doing feminist research in geography. In P. Moss (ed.), *Feminist Geography in Practice*. Malden: Blackwell, pp. 1–17.

Mountz, A. (2004) Embodying the nation-state: Canada's response to human smuggling. *Political Geography* 23 (3): 323–45.

Mountz, A. (2010) *Seeking Asylum: Human Smuggling and Bureaucracy at the Border*. Minneapolis: University of Minnesota Press.

ÓTuathail, G. (1996) *Critical Geopolitics: The Politics of Writing Global Space*. Minneapolis: University of Minnesota Press.

ÓTuathail, G. (2006) Thinking critically about geopolitics. In G. ÓTuathail, S. Dalby and P. Routledge (eds.), *The Geopolitics Reader*. New York: Routledge, pp. 1–14.

Pain, R. and Smith, S. (eds.) (2008) *Fear: Critical Geopolitics and Everyday Life*. Hampshire and Burlington: Ashgate.

Painter, J. (1995) *Politics, Geography and "Political Geography": A Critical Perspective*. London and New York: Arnold.

Said, E. (1978) *Orientalism*. London: Routledge & Kegan Paul.

Secor, A. (2001) Toward a feminist counter-geopolitics: gender, space and Islamist politics in Istanbul. *Space and Polity* 5 (3): 191–211.

Sharp, J.P. (2000) Re-masculinising geo-politics? Comments on Gearoid ÓTuathail's *Critical Geopolitics*. *Political Geography* 19 (3): 361–4.

Sharp, J.P. (2004) Doing feminist political geography. In L.A. Staeheli, E. Kofman, and L.J. Peake (eds.), *Mapping Women, Making Politics: Feminist Perspectives on Political Geography*. New York: Routledge, pp. 87–98.

Smith, F. (2001) Refiguring the geopolitical landscape: nation, "transition" and gendered subjects in post-cold war Germany. *Space and Polity* 5 (3): 213–35.

Smith, N. (2000) Is critical geopolitics possible? Foucault, class and the vision thing. *Political Geography* 19 (3): 365–71.

Smith, N. (2004) *American Empire: Roosevelt's Geographer and the Prelude to Globalization*. Los Angeles: University of California Press.

Sparke, M. (2000) Graphing the geo in geo-political: *Critical Geopolitics* and the revisioning of responsibility. *Political Geography* 19 (3): 373–80.

Staeheli, L.A., Kofman, E., and Peake, L.J. (eds.) (2004) *Mapping Women, Making Politics: Feminist Perspectives on Political Geography*. New York: Routledge.

Staeheli, L.A. and Kofman, E. (2004) Mapping gender, making politics: toward feminist politi-cal geographies. In L.A. Staeheli, E. Kofman, and L.J. Peake (eds.), *Mapping Women, Making Politics: Feminist Perspectives on Political Geography*. New York: Routledge, pp. 1–14.

Sudbury, J. (ed.) (2005) *Global Lockdown: Race, Gender, and the Prison-Industrial Complex*. New York: Routledge.

Thrift, N. (2000) It's the little things. In K. Dodds and D. Atkinson (eds.), *Geopolitical Traditions: A Century of Geopolitical Thought*. London and New York: Routledge, pp. 380–7.

Wastl-Walter, D. and Staeheli, L. (2004) Territory, territoriality, and boundaries. In L.A. Staeheli, E. Kofman and L.J. Peake (eds.), *Mapping Women, Making Politics: Feminist Perspectives on Political Geography*. New York: Routledge, pp. 141–51.

Chapter 25

The Geographies of Marginalization

Dan Trudeau and Chris McMorran

Introduction

How is space fashioned to privilege some groups and marginalize others? How does space contribute to the social exclusion of particular groups? These questions have been at the center of much scholarship on the social geography of marginalization over the past four decades. Concern about social exclusion was excited within geography by multiple tears in the social fabric of societies throughout the world, including the end of colonization, the rise of civil rights movements, the arrival of third world migrants in first world locations, widening gaps between rich and poor, and the increasing feminization of labor. Anglophone social geographers initiated the academic journal *Antipode* in 1969, for instance, in order to provide a specific forum for discussion and debate about the role and effect of social relationships and geographic environments in the processes of marginalization. Geographers initially focused on illustrating the patterns and extent of social inequality and exclusion, often producing maps to illustrate such patterns. Studies of experiences, effects, and causal processes of exclusion were soon added to the growing literature on marginalization and subsequently contributed to theories of marginalization. In the case of racialized ghettos, for instance, social geographers have explored how members of racial groups in these areas experience forms of material deprivation that may result in a lack of access to services, food, or shelter, which may in turn affect individual's health. Since the 1980s, cultural geographers have added to our understanding of this instance of marginalization by documenting the discursive and symbolic ways in which the material conditions of ghetto environments influence the social labels and negative stereotypes that reproduce the marginalization of social groups. Extrapolating from this example, the broader geographic literature on marginalization has consequently produced multiple conceptualizations of

A Companion to Social Geography, First Edition. Edited by
Vincent J. Del Casino Jr., Mary E. Thomas, Paul Cloke, and Ruth Panelli.

exclusion and strategies to analyze it. Scholars working in social and cultural geography have together shown, however, that marginalization entails material and discursive relationships between society and space. In this chapter, we focus specifically on geographical scholarship on the production of landscape in order to trace ways in which geographers working within and between social and cultural geography frameworks have defined and studied marginalization. An important part of our tracing exercise is to show how scholars have drawn from other disciplines to theorize the geographies of marginalization and to highlight some unanswered questions that remain and to which scholars working in both social and cultural geography frameworks may potentially contribute answers.

On the Margins

Marginalization – as a process of becoming peripheral – has been a matter of substantial interest in human geography. Interest in understanding the foment that characterized revolutionary, rights, and reterritorialization movements in the decades after World War II brought the topics of marginalization and marginality (of diverse sets of social difference) onto research agendas in the academy. Descriptions of this process follow a center-edge analogy, in which actors at the edge are disempowered in comparison to actors at the center, who are privileged and socially dominant. Scholars have thus used this concept to describe the ways in which individuals and social groups are relegated to positions of low(er) and unequal standing in society. The study of marginalization is by no means unique to human geography. In fact, many disciplines across the social sciences and humanities have contributed to this field of inquiry. However, geographers' specific focus on the relationships between society and space has shown the ways in which labeling places as marginal and the marginalization of space compound and complicate the social inequalities that marginalization produces (Anderson 1991; Craddock 2000; Hanson and Pratt 1995; Tilly et al. 2001).

In the decades following World War II, scholars working in social geography in particular began to search for ways in which the academy and the discipline of geography could create knowledge relevant to understanding and, hopefully, resolving the systems of social inequality to which the movements were responding. Ironically, this enterprise was initially marginal within the discipline. For instance, the journal *Antipode* was created as a forum for discussion of radical theory and praxis in geography, and as a call to action to address processes of marginalization and their attendant injustices. While *Antipode* was decidedly outside the mainstream and arguably on the margins of academic geography when it was founded, the issues considered in its volumes – social inequality, injustice, marginality, and marginalization – are well within the purview of mainstream human geography today. Indeed, geographers working in a wide array of geography sub-disciplines contribute to what is now a vast set of literature that explores processes of marginalization. Moreover, in the past two and a half decades the interests of a resurgent form of cultural geography, informed by critical theory, have also expanded this literature. Such interests focus on cultural meanings and how these animate the production of space, social difference, and the ways people experience them in everyday life (Del Casino and Marston 2006). While there may be important nuances to distinguish social geography from cultural geography, these two sub-fields together are vital to understanding the geographies of marginalization.

The geographical literature on marginalization thus draws on the wider traditions of social inquiry mentioned above and is as diverse as it is rich. This geographic literature explores patterns and processes of marginalization using a variety of theoretical lenses (e.g., Marxism, Feminism, Structuration theory, non-representational theory), and methodologies (e.g., spatial analysis, hermeneutics, ethnography), and it focuses on particular nodes of social difference (e.g., race/ethnicity, gender, sexuality, class, and dis/ability) and their intersections. Some of the most recent productive discussions of marginalization in geography have cohered around particular concepts, which have served as analytical crucibles and provided theoretical insights about the spatiality of marginalization and empirical descriptions of the social experiences of marginality. Discussion around such interdisciplinary concepts as citizenship (Secor 2004), segregation (Johnston et al. 2007), neoliberalism (Leitner et al. 2007), and landscape (Schein 2006), among others, has enriched the geographies of marginalization and informed interdisciplinary study of these social processes. Reviewing the depth and extent of these contributions is beyond the scope of this chapter. Instead, we focus on the ways in which social and cultural geographers have used the landscape concept, once a fixture in apolitical approaches in American cultural geography, to generate a vibrant set of discussions on the relationship between the production of space and social exclusion.

Landscape and Exclusion

While landscape study has deep roots in cultural geography, recent work by both social and cultural geographers engaging with critical social theory has used the landscape concept to produce provocative and productive insights about marginality and marginalization. Since the 1990s, geographers have explored how particular social groups are excluded from landscapes and the ways in which marginalized groups experience exclusion. Indeed, the incorporation of social theories generated outside of the discipline of geography into landscape study has provided innovative ways through which to conceptualize the material and discursive roles landscapes play in producing marginalization.

This section discusses five exemplary works that investigate the relationship between geographic landscapes and social exclusion. These works focus on how landscapes serve political purposes. Our selection does not attempt to exhaustively or comprehensively analyze the ways in which geographers approach marginalization through landscape study. Rather, we deliberately select exemplars that illustrate some of the different ways social and cultural geographers have drawn on social theory in order to understand the socio-spatial processes of exclusion. Furthermore, our selection facilitates discussion of the ways particular conceptualizations shape understanding of relationships between geographical landscapes and exclusion. Toward these ends, the following discussion evaluates selected works for their conceptualizations of exclusion and landscape, as well as analytical strategies to study connections between them.

We begin with Mitchell's *The Lie of the Land*, which provides a theory of landscape that is both innovative and distinct in its approach to understanding exclusion.

In this text, Mitchell offers two interconnected goals. On the one hand, he seeks to synthesize two seemingly disparate approaches to landscape: the Sauerian approach to studying the processes that shape the morphology of landscape (Sauer 1925) and the iconographic approach to studying landscape as a visual ideology or "way of seeing" that naturalizes particular socio-spatial relations (Cosgrove 1985 [1984]). Indeed, Mitchell endeavors to show how these two approaches are integral to understanding the material and discursive practices that produce both the look of the land and ways of looking at it. On the other hand, Mitchell intends to balance material and discursive approaches to landscape on a point of political economy of place scholarship: nothing about places are "natural"; rather, places are produced through ongoing struggle between different social groups to control how a place appears, how it is represented in geographical imaginaries, who has legitimate access to it, and who benefits from it. Elaborating this point, Mitchell (1996: 34–5) offers a "labor theory of landscape" to conceptualize landscape as

> an uneasy truce between the needs and desires of people who live in it, and the desire of powerful social actors to represent the world as they assume it should be. Landscape is always both a material form that results from and structures social interaction, and an ideological representation dripping with power. In both ways landscapes are acts of contested discipline, channeling spatial practices into certain patterns and presenting to the world images of how the world (presumably) works and who it works for.

Mitchell (1996: 28) offers this cultural materialist approach to landscape in order to understand the labor that produces the shape of the landscape and that attempts to "naturalize a system of domination, order, and control that appropriates the labor." Mitchell thus generates a theory that focuses on the role landscapes play in reproducing conditions of alienation under capitalism.

Mitchell's theory progresses landscape studies of social exclusion in three distinct ways. First, it offers a compelling ontology of landscape as a moment of social reproduction that expresses the dialectic relationship between material and discursive processes. Second, exclusion is understood as a status of omission. The struggles, negotiations, and competing interests that produce the material landscape are not immediately knowable from the look of the land. Likewise, while there may be competing "ways of seeing" the landscape, visual ideology of the dominant class obfuscates relations of production and the alienation of workers from their labor. In this way the landscape naturalizes particular representations of the world. Mitchell draws attention to landscapes as necessarily omitting the material and discursive embodiment of these relationships (hence the *lie* of the land). The role landscapes play in reproducing these omissions leads Mitchell to argue that landscapes are neither neutral nor self-evident. This leads to the third contribution to the study of exclusion in the landscape. Mitchell's conceptualization of landscape generates an epistemology to trace the practices and struggles that engender such forms of exclusion. Mitchell thus offers an approach to puzzle out the lie of the landscape not only for the sake of better scholarship, but also to understand and ultimately rectify the ways landscapes marginalize workers.

Geographies of Exclusion, by David Sibley, offers a decidedly different approach to studying exclusion. Sibley (1995) too provides a conceptual framework for connecting discursive and material formations in the production of space. Yet, we high-light Sibley's attention to the processes that produce exclusionary landscapes. Sibley is concerned primarily with what he calls the "purification of space," which he defines as a process of social control through which a dominant social group con-structs socio-spatial boundaries that contribute to the marginalization of groups judged as deviant and outside the mainstream. For Sibley, exclusion plays a part in the reproduction of social identity. Sibley thus focuses narrowly on the socially con-structed boundaries that contribute to the marginalization of minority groups, espe-cially in advanced capitalist societies. Sibley's thesis is similar to Cresswell's (1996) work on transgression in *In Place/Out of Place*, although Sibley's use of object rela-tions theory to examine the social processes that produce boundaries is novel.

Sibley employs object relations theory to explain connections between individual and group behavior, and behavior and the geographical environment, which are integral to processes of exclusion. Scholars working in the fields of psychoanalysis and social anthropology developed this theory to conceptualize how the self is constructed through an individual's relationships with human and non-human objects in its wider environment. Sibley is particularly interested in the boundaries individuals create to construct the *self* as separate and distinct from *other* objects. These boundaries are created through a self-definition process of "abjection," which Kristeva (1982) refers to as an individual's attempt to distance oneself from objects that represent undesirable, non-conforming, and even antithetical characteristics. Sibley theorizes that abjection operates too on a social level to create boundaries around social groups – sameness and community on one side and the deviant and marginal on the other. Moreover, he draws on Mead's (1934) notion of the "gen-eralized other" to conceptualize a connection between abject things, people, and places. This is all to say, Sibley uses object relations theory to explain why dominant social groups attempt to purify space of other marginal social groups and reproduce boundaries that separate the marginal places with which they are associated. As Sibley (1995: 11) explains it, "the geographies of exclusion, the literal mappings of power relations and rejection, are informed by the generalized other."

Sibley is careful to qualify that the purification of space occurs in limited situa-tions and can take multiple forms. He notes that, in the west, there is a continuum of tolerance for difference and thus variation in response to it. Environments that are already highly ordered and purified of other objects, however, facilitate forms of social control that construct differences as out of place, deviant, and potentially abject. Highly organized homogeneous spaces thus can facilitate a variety of exclu-sionary practices to maintain conformity, purify space, and push non-conforming elements to the margins. Less organized and heterogeneous spaces are less support-ive, if not thwarting, to forces of purification. Furthermore, exclusion can be sym-bolic as well as material. Exclusion can take the form of strong spatial divisions meant to separate people and places – take for example the construction of walls and highways to partition ghetto neighborhoods in urban places. Yet individuals and social groups may also employ measures that are more symbolic in nature in order to purify a place of abject characteristics – the case of prohibiting non-English language on commercial signage in the built environment of nativist municipalities in the United States illustrates the point that social boundaries may still be con-structed in ways that do not manifest in stark spatial divisions.

Sibley's explanation of the purification of space makes an important contribution to the study of exclusionary landscapes. Most significant is his framework for examining the connection between group identity formation and the creation and enforcement of territorial boundaries. This certainly provides another conceptual path to imagine the connections between the discursive and material dimensions of landscape. Sibley (1995) does not, however, provide a clear ontology of landscape. In fact, space is only implicitly defined as Sibley gives more attention to critiquing the Cartesian conceptualization than generating a positive definition of space. Thus, while Sibley has used object relations theory to shed light on social processes that contribute to exclusion, his approach begs further question of how the social and spatial boundaries are etched into geographical landscapes, how they matter in social practice, and how they may be challenged by new inscriptions.

Duncan and Duncan's *Landscapes of Privilege* investigates a most interesting problem in landscape studies of exclusion. In the current global era, transnational flows of people, capital, land development, and ideas threaten to unsettle imagined communities and the imagined geographical boundaries that delineate places. Observing this process, Massey (1994) noted that some social groups work fervently to maintain such boundaries, establish coherence, and protect the imagined authenticity of places in an attempt to ensure the integrity of a place called home. As Duncan and Duncan (2001) have elsewhere discussed, this process of re-inventing places amidst and indeed sometimes against such unsettling changes is an important issue for social and cultural geographers to explore. In *Landscapes of Privilege*, Duncan and Duncan (2004) demonstrate that the attempt to preserve the imagined authenticity of places has important implications for geographies of marginalization and exclusion.

Landscapes of Privilege is empirically concerned with efforts of wealthy social elites to preserve the pastoral landscape of Bedford, a small town that is now a part of the commutershed for the New York metropolitan area. They examine the ways in which elites use environmental conservation and historical preservation to shape the look of Bedford's landscape and have it conform to an idealized notion of what the town (may have) looked like in the nineteenth century – a beautiful pastoral New England landscape unmarked by the development of industrial urbanism. Duncan and Duncan note that concern with the aesthetics of the landscape has unseen consequences for the exclusion of lower-income and racialized groups. Indeed, the effort to preserve a pastoral look to the landscape has been bolstered by land use zoning for very low-density settlement and ordinances to preserve undeveloped land to maintain a landscape rich in tree coverage. These strategies have the effect of making Bedford a highly exclusive place as they make it financially impractical for the development of affordable housing. Consequently, the labor that is essential for the maintenance of the pastoral landscape, which is provided primarily by Latino migrants who work as day laborers, is excluded from residing in Bedford. *Landscapes of Privilege* thus examines how the struggle to control the look of the landscape operates as a subtle yet effective mechanism of exclusion.

Beyond the study of Bedford, *Landscapes of Privilege* has two significant theoretical implications for studies of landscape and exclusion. One the one hand, perhaps the most novel contribution is to see landscapes through a lens of performance. Duncan and Duncan (2004) draw on Bourdieu's (1984) work on cultural capital to see landscapes as a positional good. Shaping the look of the landscape according to distinct tastes and styles is thus a way to perform social identities and show participation and membership in particular communities. Duncan and Duncan (2004: 7) follow Austin (1975) and Butler (1990) in understanding performance as a productive and everyday embodied practice and theorize that "identities are performed in and through landscapes." In their perspective, landscapes are aesthetic productions that provide a symbolic resource privileged social groups employ in the pursuit of social distinction. Prestige and material benefits can accrue to people whose residential location and property are associated with an authentic landscape. On the other hand, Duncan and Duncan point out that the aesthetic production of landscapes is made possible through various legal, political, and economic practices that shape material landscapes according to abstract ideas. "These practices tend to be exclusionary although they are not always acknowledged, or even recognized, as such ... In fact, the goal is not always social exclusion in itself but to preserve the "look of the landscape," which is central to the performance of particular social identities that depend on lifestyle, consumption patterns, taste, and aesthetic sensibilities" (Duncan and Duncan 2006: 159). Social identities that are considered abject or antithetical to the cultivation of particular visual aesthetics in the landscape and the material bodies and buildings with which they are associated are erased or expelled from the visual scene. Moreover, because aesthetics are often treated as apolitical, idiosyncratic taste and style preferences, the exclusive consequences of the aesthetic production of landscapes as well as the complicity in the connections that privileged groups have to such consequences are often overlooked or go unseen.

Lastly, *Landscapes of Privilege* raises a problem that is relevant to further study of exclusion and marginality. This problem concerns the ways in which elites are complicit in unequal, oppressive and otherwise exclusive sets of social relations. Duncan and Duncan (2004) focus on the roles of elites in the aesthetic production of Bedford as a pastoral place and note the importance of further inquiry about how locally and globally powerful groups are involved in producing geographies of exclusion in ways that they may not be individually accountable for. Tracing the web of complicity and placing it in the context of a global geometry of power will greatly contribute to understanding geographies of marginalization and enrich theories of landscape generally and its contribution to social exclusion in particular.

In important ways, Price's *Dry Place* incorporates theories from diverse sources to interrogate the political implications of particular landscapes and to advance landscape theory in general. Price (2004) weaves geographic notions of landscape and place with three threads of scholarship in the humanities – work on sacred places from religious studies, studies of the "New West" from history, and most adeptly, the use of narrative in literary criticism – to investigate the landscape near (and including) the US–Mexico border. She combines these ideas to investigate the tenuous and ever-changing relationships between place, nationalism, and globalization as they are powerfully expressed through landscape narratives.

Like other social and cultural geographers, Price interrogates the relationship between the material and discursive practices that shape landscape. However, she argues that Mitchell (above) and others tend to favor the material landscape over the discursive, implying that landscape representations are merely a veneer that must be stripped away to uncover the material landscape that lies beneath. She argues that the stories people relate about particular landscapes are just as powerful, just as "real," as material practices, and she takes issue with the inherent gendering of research on landscape representations as female, and therefore less legitimate. For Price, the power and "real"-ness of landscapes comes from their essence as narrative constructs, the stories that people tell and retell about themselves and their relationships to place. As such, landscapes are "power constructs, always processual, usually contested, and deeply performative." Price insists that these stories can be "powerfully real and really powerful" (Price 2004: 22).

Price utilizes literary theory to examine the political use of landscapes by introducing the notion of the landscape as palimpsest, or a location of erasure and overwriting by successive groups. Instead of a single accepted reading of a landscape, Price reveals the narrative layering that constitutes all landscapes in politically significant ways and constantly alternates between inclusion and marginalization. Thus Price focuses on how place narratives accumulate over time, leading to a theory of landscape as "a layered text of narratives of belonging and exclusion" (p. 7).

Price's major contribution to the study of landscapes of marginalization comes through her examination of how these narratives of belonging and exclusion ebb and flow to shape particular landscapes. To do so she employs Deleuze and Guattari's (1987) notions of smooth space and striated space. Spaces that lack meaning within a particular narrative are considered smooth, while spaces imbued with meaning are considered striated. Importantly, Price notes that spaces often move between smooth and striated within different landscape narratives. The tendency to smooth certain narratives has led jungle, desert, ocean, and polar spaces to all be conceived as empty and meaningless, allowing them to be colonized without acknowledgment of prior claims to those landscapes. Price shows a similar trend when relating how the landscape surrounding and including the US–Mexico border has served as the centerpiece of national identity narratives by various groups since the mid-1800s. "The West" has been a smooth landscape to the United States throughout history, perceived as an empty space into which the country could grow, despite the fact that this narrative relied on the distinct striation that came from the creation of the geopolitical US–Mexico border in 1848. Price (2004: 41) summarizes, "The smoothing of spaces constructed a blankness that was at the heart of dominant landscape attitudes in colonizing societies more generally." In all cases, the landscapes were actively regarded as featureless and smooth, despite their necessary and inherent striations.

Price asserts that every nationalism attempts to write a narrative in the landscape that resists existing narratives, by implying uniform support among its members. Price (p. 97) refers to this as "homogenization-in-resistance," whereby resistance by a marginalized group is argued to be possible only when the group temporarily ignores (smooths) its internal inequalities and differences (striations) in order to present a unified narrative. However, these inequalities and differences of ethnicity, religion, gender, race, and sexuality naturally resist smoothing. The newly smoothed narrative leads to further marginalization, eventually leading to the narrative's implosion at the hands of the differences constructed as marginal. Such was the case with Aztlán, the utopian landscape encompassing the US–Mexico border that was central to Chicano nationalism of the 1960s and 1970s, which Price (2004: 82) concludes was "haunted by its own ghostly voices from the margins." Even contemporary advocates of globalization fall into this trap by claiming that the world is becoming increasingly borderless, a smooth space in which boundaries to trade and cultural differences continually fall away. However, such a smoothing narrative ignores the internal inconsistencies, the striations of more-heavily patrolled national borders, reinforced claims of cultural and ethnic identity, and increased gaps between rich and poor, all of which threaten to undermine the narrative. For Price, marginality that is constantly written out of the landscape always finds its way back in.

Italian philosopher Giorgio Agamben has inspired a final collection of scholarship in the theoretical integration of landscape and exclusion. Cultural and social geographers have utilized Agamben's theories to describe the ways that political violence is enacted through the physical exclusion of marginalized groups from society. Gregory's (2004) *The Colonial Present* is one example. Gregory applies ideas from Agamben's *Homo Sacer* (1998 [1995]), particularly the notion of spaces of exclusion, to explain the imaginative geographies produced alongside the military campaigns of the United States in Afghanistan, Israel against Palestine, and the United States and Britain in Iraq. In each case, individuals and groups inhabiting each place have come to be regarded as outsiders "occupying a space beyond the pale of the modern," whose rights, protections, and dignities have thus been forfeited (Agamben 1998 [1995]: 28). Agamben refers to such people as *homo sacer*, or sacred man, living outside of both divine law and juridical law, and thus able to be killed with impunity. Agamben traces the notion of *homo sacer* to Roman times when sovereign power was exercised precisely through the process of exclusion, not inclusion. Thus, reminiscent of Sibley and object relations scholars, Agamben contends that the process of marginalization is vital to the formation of political communities.

Geographers have utilized Agamben's notion of spaces of exclusion to understand the state's denial of legal protections to marginalized groups, a condition that typically arises under a state of (perceived) emergency. When a state of emergency threatens to become the rule, it leads to "the enclosure of the subject, its transformation into the direct object of violence" (Secor 2007: 39). Most geographers utilizing Agamben's spaces of exclusion have focused on specific sites of social and political injustice, such as Guantanamo Bay, Cuba. For instance, Hyndman and Mountz (2007) have concentrated on would-be refugees caught in nonsovereign spaces, while Kearns (2007) has used Agamben to conceive of nineteenth century Ireland as a "camp," colonized by Britain and object of Britain's ceaseless violence.[1] Geographers interested in marginalization have readily adopted and adapted Agamben's theories in their continued explorations of the landscapes of exclusion.

Experiencing Exclusion

Geographers have also recently sought to understand marginalized groups' everyday experiences of exclusion, as well as the practices they may engage in to challenge in some ways, and reproduce in others, processes of social exclusion. In this section we begin by focusing on the importance of borders, both material and discursive, in creating landscapes of exclusion. We emphasize both geopolitical borders and the unmarked boundaries that separate the self and the other, to show how marginal groups experience exclusion through multiple nodes of difference that compound and complicate their marginality. Second, we show how marginalized groups engage in subtle forms of resistance in order to contest exclusion in ways that offer mixed and often unpredictable results. The key to this second point is that members of marginal groups are often acutely aware of their exclusion and even engage its processes. This sort of engagement is important to understanding the production and reproduction of marginality. Third, we examine how the practices of marginal groups can lead to the creation of new spaces of exclusion that are at once unanticipated and integral to the processes that construct and relate privilege and marginality.

Borders produce landscapes of exclusion by their very existence, by separating groups, creating and pointing out differences. However, borders can also bridge divides and serve as meeting points, providing locations around which alternative inclusions can occur. This dual nature of borders makes them essential to understanding the complex experience of marginality. The violent materiality of borders can be seen in the fences, walls, and police forces along international borders, like that which separates the United States and Mexico. The effects of such a border can be far-reaching. Wright (2006) argues that the US–Mexico border not only encourages the construction of *maquiladora* factories essential to global capitalism, but also helps produce the myth of the third-world woman. This woman, like her labor, is considered "disposable," as evidenced through her recruitment and limited training, as well as her spatial marginalization from the center of factories themselves. More horrifying, though, has been the murder and disposal of hundreds of young women in the desert near factories in Ciudad Juárez since the late-1990s. Here, a discourse of female disposability produced in part by the US–Mexico border and the political and economic relations it signifies has become etched into a landscape that morally justifies the deadly exclusion of women from public space (see also Wright 2004).

On the other hand, one can also see the inclusiveness of the US–Mexico border, where artists, writers, and others have championed a notion of the border as a seam of commonality to parties on both sides. By reimagining the border as a unifying force, they have rewritten the narrative of the borderland landscape as one of inclusion instead of division. Price (2004: 90) calls this erasure of the geopolitical US–Mexico border "a transgressive, contestatory, liberatory gesture" that creates a landscape "where dichotomous constructions of belonging and exclusion are no longer viable" (see also Wright 2006: 95). This perspective raises a critical question for any geographer: how can one move beyond a simple dichotomy to see the messy realities of how inclusion/exclusion is experienced daily?

Marginality can stem from multiple sources of difference that separate the self and the other, including race, ethnicity, sexuality, gender, nationality, citizenship, class, and religion, among others. The ways that people experience exclusion through these factors can vary immensely, as marginality is often complicated due to the intersection of multiple nodes of difference. Geographers have described the everyday lives of marginal groups, both to understand how these nodes intersect, and to complicate notions of marginality by questioning the simple division between inclusion and exclusion.

Pratt's (2004) work with Filipina domestic workers in Canada is a prime example of the inherent tension between inclusion and exclusion found among marginalized groups. Filipina domestic workers are marginalized on a daily basis due to their nationality, race, and political status. Their professional training and educational attainment from the Philippines is devalued, while their political status as non-citizens on specific work visas provides few opportunities for career advancement, causing them to feel deskilled over time. However, their marginality is complicated by the fact that their labor is constructed as central to the career advancement of middle and upper-class Canadian women, as well as the fact that as domestic workers residing in their workplaces, many of their employers perceive them as "family members." These are significant expressions of the importance of Filipina domestic workers to Canadian society through their exploitation within the international division of labor, making these workers feel both included and excluded in concrete and incredibly intimate ways.

A similar convergence and complication of nodes of difference is found in Faier's (2009) investigation of Filipina migrants to Japan. Triply marginalized as racially and economically inferior (due to the assumed poverty of the Philippines), as well as morally suspect via their labor in bars, the women experience multiple exclusions everyday. Yet, those Filipinas who marry Japanese men become included into rural families in intimate and complicated ways. Many are referred to as *ii oyomesan*, or "good brides," because they dutifully care for family needs in ways refused by many contemporary Japanese women who increasingly delay or avoid marriage. Paradoxically, each Filipina's intimate inclusion in Japanese families re-inscribes a material and discursive landscape of a traditional, gendered, patriarchal Japan that is imagined as racially homogenous, which in turn re-excludes the Filipinas from popular ideas of the rural Japanese landscape in which they live. Like Pratt, Faier complicates marginality, showing the multiple ways that differences interact to create spaces of both inclusion and exclusion, as well as spaces that cannot be clearly delimited as one or the other (see also Chapter 12 this volume).

Borders are constantly being produced, negotiated, challenged, and redrawn. One way of conceptualizing this process is through de Certeau's (1984) theories of "strategies" and "tactics." De Certeau calls "strategies" the top-down meanings given by political, economic, and cultural elites, which are aimed to allow little or no room for differing interpretations. Consumers, on the other hand, employ "tactics," or processes that challenge the assumed meanings of things produced by others, such as books and even city streets. De Certeau calls tactics "the ingenious ways in which the weak make use of the strong" in everyday life (xvii; see also Scott 1985). De Certeau (1984: xvii) deals specifically with marginality in his work *The Practice of Everyday Life*, by first stating, "Marginality is becoming universal," then celebrating the multiple ways that people creatively use and manipulate the cultural meanings imposed on them from productive elites.

The experience of exclusion often elicits tactics that contest borders. However, marginalized groups also engage in practices that reproduce or even compound their exclusion. Literally crossing a geopolitical border can be an act of both resistance to and acceptance of a recognized border narrative. Wright notes the powerful role played by the US–Mexico border in marginalizing Mexican women within global firms. Ambitious women who retain markers of their Mexicanness, through clothing, makeup, and language, are accused of not knowing their proper place and are considered disposable. Others literally and figuratively cross the border to make themselves more "American," and thus increase their value to firms. Here, Mexican women accept the landscape of exclusion created by the border, while also resisting the myth of their disposability by exploring possibilities for inclusion.

Discursive borders can also be resisted. Pratt shows that Filipina domestic workers employ a variety of tactics to challenge their marginalization by Canadian society. One example is the way they respond to such questions as, "Where do you work?" Through clever word play and avoidance of the response, "I'm a nanny," the women resist playing into stereotypes held by most Canadians that would justify their exclusion.

In some cases, these tactics may open liminal, or transitional, spaces between inclusion and exclusion. These are spaces of uncertainty and negotiation that are neither inside nor outside. For instance, Faier highlights the fact that many of Filipinas in rural Japan feel "stuck" between the Philippines and an idealized "America" that symbolizes their desires for economic opportunity, glamor, modernity, and mobility. For them, Japan represents a liminal zone on the way to a desired goal, but one in which some semblance of these desires can be fulfilled. Thus, while in some ways their experience excludes them from Japanese society, in other ways it also includes them in a more cosmopolitan and sometimes romantic life than they could have imagined in the Philippines.

Sometimes, groups excluded from particular landscapes wish to remain that way. Sibley's scholarship on the exclusion of Roma people in urban England speaks well to this point. Roma or Gypsy communities are often located on the margins of urban settlements in derelict spaces at the edge of cities. These communities are pushed to the margins in part because of stereotypes of Roma people as uncivilized: immoral, criminals, and vagabonds, who otherwise embody heresy to law-abiding and property-based assumptions of modern capitalist society. At the same time, Sibley (1995: 68) notes that Roma communities also seek out marginal spaces "in order to avoid control agencies and retain some degree of autonomy." The separation of Gypsy communities further reinforces the discursive boundaries that categorically associate Gypsies with defiled elements of society. The important point in this example is that groups may seek marginal locations as part of an effort to create and maintain boundaries separating the group from the larger society. Such an effort is also evident in the practices of some orthodox religious groups who voluntarily exclude themselves from spaces of mainstream society in order to maintain boundaries between the pure and the defiled.

Towards Inclusion

For more than four decades geographers have examined marginalization in order to understand the social processes and geographical representations that contribute to unjust relationships in society. Geographers have shown that marginalization is an inherently socio-spatial process: the term marginalization itself is a spatial metaphor that correctly draws attention to the geographical aspects of exclusion. As part of this endeavor, some of the best work on this topic has drawn productively on theories and concepts generated outside of geography. We have focused in this chapter on geographers' efforts to trace ways in which landscapes contribute to social exclusion as part of a strategy to illustrate on the one hand the contribution of other disciplines to social and cultural geographies of marginalization and to critically examine on the other hand how this area of scholarship may be advanced, improved, and elaborated. We turn to this last point in this final section of the chapter by considering two issues. First, we describe several insights and an unresolved question concerning geographers' analytical approaches to understanding the production of exclusion in geographical landscapes. We present these primarily to point out pathways along which future work might travel to productively contribute to the literature. Second, we discuss a recent contribution to literature on marginalization that considers how geographical landscapes might be produced in ways that contribute to social inclusion and promote social justice. This is a nascent, but promising area of research that is poised to galvanize scholarship of marginalization in social geography.

Our discussion highlights several important theoretical insights about landscapes and exclusion with which future scholarship on the geographies of marginalization should be concerned. One insight has been discussed elsewhere (Mitchell 2003, 2008), but its importance merits repeating here: landscapes are not just local products. Landscapes may be experienced and imagined as a local phenomenon, yet agents working at a variety of scales produce them. For instance, the production of an authentic and pastoral New England landscape in Bedford, New York, is made possible by the labor of international migrants, the cultural and economic capital of elites who draw their wealth from transnational corporations located in New York City, and the activities of large-scale national institutions, such as the Nature Conservancy and the American Civil Liberties Union, all of which connect Bedford to the wider region, nation, and beyond (Duncan and Duncan 2006). Geographies of marginalization must attend to the ways in which relationships that operate at and across different scales inform, shape, and animate the processes of exclusion. Duncan and Duncan (2004) also raise a corollary insight: social exclusion in the landscape is *not always intentional*; it can often result unconsciously from efforts to construct a particular look of the land. However, social exclusion is *not incidental* in efforts to shape the landscape in particular ways. Duncan and Duncan therefore suggest that landscape theory should adopt a more nuanced understanding of complicity to trace the complex and multi-scaled relationships through which social exclusion operates. These are important insights to carry forward in future examination of exclusion in landscapes.

In addition to these two insights, we also point out that social and cultural geographers have begun to make productive use of boundary analysis in order to study the production and experience of exclusion in landscapes. Boundaries seem well suited to such scholarship, in part, because they highlight the spatiality of exclusion: boundaries are both discursive and material constructs; they operate across a variety of scales to mark the geopolitical borders of national states as well as the socially-scripted roles that men and women embody through gender, race, and so forth; and they remind us that processes of exclusion do not remove marginalized groups altogether. Price (2004) and Sibley (1995) have each theorized that boundaries are always in the process of becoming and that efforts to establish them reflect attempts to order the environment and stabilize particular meanings (e.g., purity, authenticity, community) in a changing world. Scholars utilizing Agamben's insights on spaces of exception also highlight the use of boundaries to create both imagined and real spaces in which exclusion is practiced and justified (Gregory 2004; Secor 2007). Analyzing the production, use, and experience of boundaries by individuals and social groups thus offers an innovative approach to the study of exclusion in the landscape.

We suggest that further study of boundaries in the cultural landscape is poised to enrich our understanding of the geographies of marginalization. Tracing the ways in which boundaries make landscapes (meaningful), as well as how landscapes concretize boundaries constitute an important part of future scholarship on exclusion. Our discussion of geographical scholarship on landscape and exclusion has thus far emphasized the way it has drawn on theory and insight from other disciplines in order to understand the spatiality of boundaries. We further encourage that geographers emphasize how the spatiality of discursive and material boundaries that are concretized in landscape contribute to social relations and experiences of exclusion.

Geographies of everyday practice and social performance provide a useful entry point for further exploration of this research frontier. In his critique of landscape scholarship, Rose (2002) argues that geographers have privileged analysis of what landscapes mean at the expense of questions of how landscapes come to be meaningful. Rose (2002: 456) identifies a series of unanswered questions about landscape that also speak to the research agenda on boundaries that we highlight. Following Rose, we ask how are boundaries sustained in the mental and material worlds? How is it that we imagine and comprehend boundaries in the landscape, and how are these concretized in our everyday experience? How are boundaries called forth to affect social exclusion? And how are liminal spaces surrounding borders used to create new social formations? These questions draw attention to the everyday practices, stories, scripts, performances, and improvisations through which "individual agents [call] the landscape into being as they make it relevant for their own lives, strategies, and projects" (Rose 2002: 457) in ways reminiscent of de Certeau's (1984) assertions about everyday practice. Geographies of marginalization may thus find ontologies of landscape that emphasize human performance and practice (e.g., Duncan and Duncan 2004; Price 2004; Schein 1997) useful to trace the ways in which boundaries are understood, enacted, and inflected in everyday life to (re) produce social exclusion.

In light of this reminder, it is also important for geographers to consider how landscapes can be produced in ways that lead to the social *inclusion* of marginalized groups. Schein (2009: 823) explains, "The very 'everydayness' of the cultural landscape gives us the ability to intervene." Building on Price's (2004) idea of landscape as a palimpsest that accumulates narratives, discourses, and other attempts to stabilize meaning and order the environment, Schein (2009: 823) reminds us that landscapes can also serve as a point of intervention in which "we might seize the opportunity to enact a (slightly) different version of the world." Schein's hopeful suggestion is based on empirical observation of social action and change in Lexington, Kentucky, and not mere conjecture. He follows the efforts of a citizen's group in this racially segregated city to use the construction of a public art garden memorializing local African American citizen Isaac Murphy to interrupt patterns of racial injustice and enact a narrative of African Americans' belonging in Lexington society. The group has taken control of the design process in order to create a memorial art garden that represents the legacy of African American participation in thoroughbred horse racing in Kentucky and elsewhere in the United States. The garden's design is meant to interact with young people to both encourage an historical awareness of African American contributions to the city and support young African American's full membership in society. Schein sees the actions of the citizen's group as effecting change in and though the landscape by shaping its material form and symbolic meaning. In effect, the citizen group's intervention in the public spaces of Lexington may (incrementally) shift the boundaries separating margin from center to include African Americans as full members of society.

Social and cultural geographers have often explored exclusion and marginalization in order to understand their causal processes and find practical ways to work against them. We have discussed in this chapter several recent theoretical approaches specific to the study of social exclusion in the landscape through which geographers are pursuing this project. These approaches offer important insights to studies of landscape and exclusion and also represent new frontiers geographers might productively explore to enrich general understanding of causes and experiences of marginalization. More importantly, we have also highlighted the importance of studying the spatiality of boundaries to further geographical research on exclusion and landscape. We have suggested that one productive way to contribute to this agenda is to examine the spatiality of performances and practices that bring exclusionary boundaries into the theater of everyday life. Lastly we also suggest that additional work is needed that explores the potential for and processes through which landscapes can contribute to inclusion of marginal groups. Schein's (2009) work on belonging through landscape is an important contribution. However, this remains an underrepresented, yet incredibly important avenue of research in social and cultural geography. It is important precisely because it stands to offer new insights to landscape theory and outlines practical ways that geographical landscapes might be produced in order to contribute to social inclusion and promote social justice.

Note

1 Some geographers argue that one misinterprets Agamben when one topographically situates this "enclosure of the subject." For instance, Belcher et al. (2008: 501) argue that Agamben's work best reveals *how* exclusion works, not *where*; that for Agamben the exception is "spatializ*ing*, not spatializ*ed*" (see also Coleman 2007). It is noteworthy that following his explicitly spatial language in *Homo Sacer*, Agamben's 2005 book is titled *State* of Exception (not *Space*). These distinctions continue to be negotiated by geographers drawing on Agamben to understand marginality and exclusion.

References

Agamben, G. (1998) [1995]. *Homo Sacer: Sovereign Power and Bare Life*. Tr. D. Heller-Roazen. Stanford: Stanford University Press.

Agamben, G. (2005) *State of Exception*. Tr. K. Attell. Chicago: University of Chicago Press.

Anderson, K. (1991) *Vancouver's Chinatown: Racial Discourse in Canada, 1875–1980*. Montreal: Queen's University Press.

Austin, J.L. (1975) *How to Do Things with Words*. Cambridge, MA: Harvard University Press.

Belcher, O., Martin, L., Secor, A., Simon, S., and Wilson, T. (2008) Everywhere and nowhere: the exception and the topological challenge to geography. *Antipode* 40 (4): 499–503.

Bourdieu. P. (1984) *Distinction: A Social Critique of the Judgement of Taste*. Cambridge, MA: Harvard University Press.

Butler, J. (1990) *Gender Trouble: Feminism and the Subversion of Identity*. New York: Routledge.

de Certeau, M. (1984) *The Practice of Everyday Life*. Tr. S. Rendall. Berkeley: University of California Press.

Coleman, M. (2007) Review: state of exception. *Environment and Planning D: Society and Space* 25 (1): 187–90.

Cosgrove, D. (1985) [1984] *Social Formation and Symbolic Landscape*. Totowa: Barnes & Noble.

Craddock, S. (2000) *City of Plagues: Disease, Poverty, and Deviance in San Francisco*. Minneapolis: University of Minnesota Press.

Cresswell, T. (1996) *In Place/Out of Place: Geography, Ideology and Transgression*. Minneapolis: University of Minnesota Press.

Del Casino, Jr., V. and Marston, S. (2006) Social geography in the United States: everywhere and nowhere. *Social and Cultural Geography* 7 (6): 995–1009.

Deleuze, G. and Guattari, F. (1987) *A Thousand Plateaus: Capitalism and Schizophrenia*, tr. B. Massumi. Minneapolis: University of Minnesota Press.

Duncan, J. and Duncan, N. (2001) The aestheticization of the politics of landscape preservation. *Annals of the Association of American Geographers* 91 (2): 387–409.

Duncan, J. and Duncan, N. (2004) *Landscapes of Privilege: The Politics of the Aesthetic in an American Suburb*. New York: Routledge.

Duncan, J. and Duncan, N. (2006) Aesthetics, abjection, and white privilege in suburban New York. In R. Schein (ed.), *Race and Landscape in America*. New York: Routledge, pp. 157–76.

Faier, L. (2009) *Intimate Encounters: Filipina Women and the Remaking of Rural Japan*. Berkeley: University of California Press.

Gregory, D. (2004) *The Colonial Present: Afghanistan, Palestine, and Iraq*. Malden: Blackwell.

Hanson, S. and Pratt, G. (1995) *Gender, Work, and Space*. London: Routledge.

Hyndman, J. and Mountz, A. (2007) Refuge or refusal: geographies of exclusion. In D. Gregory and A. Pred (eds.), *Violent Geographies: Fear, Terror, and Political Violence*. New York and London: Routledge, pp. 77–92.

Johnston, R., Poulsen, M., and Forrest, J. (2007) The geography of ethnic residential segregation: A comparative study of five countries. *Annals of the Association of American Geographers* 97: 713–38.

Kearns, G. (2007) Bare life, political violence, and the territorial structure of Britain and Ireland. In D. Gregory and A. Pred (eds.), *Violent Geographies: Fear, Terror, and Political Violence*. New York; London: Routledge, pp. 7–35.

Kristeva, J. (1982) *Powers of Horror: An Essay on Abjection*. New York: Columbia University Press.

Leitner, H., Peck, J., and Sheppard, E. (eds.) (2007) *Contesting Neoliberalism*. New York: Guilford Press.

Massey, D. (1994) *Space, Place, and Gender*. Minneapolis: University of Minnesota Press.

Mead, M. (1934) *Mind, Self, and Society*. Chicago: Chicago University Press.

Mitchell, D. (1996) *Lie of the Land: Migrant Workers and the California Landscape*. Minneapolis: University of Minnesota Press.

Mitchell, D. (2003) Just landscape or landscapes of justice? *Progress in Human Geography* 27: 813–22.

Mitchell, D. (2008) New axioms for reading the landscape: paying attention to political economy and social justice. In J. Wescoat, Jr., and D. Johnston (eds.), *Political Economies of Landscape Change*. Dordrecht, the Netherlands: Springer, pp. 20–50.

Pratt, G. (2004) *Working Feminism*. Philadelphia: Temple University Press.

Price, P. (2004) *Dry Place: Landscapes of Belonging and Exclusion*. Minneapolis: University of Minnesota Press.

Rose, M. (2002) Landscape and labyrinths. *Geoforum* 33: 455–67.

Sauer, C. (1925) *The Morphology of Landscape*. Berkeley: University of California Press.

Schein, R. (1997) The place of landscape: a conceptual framework for interpreting an American scene. *Annals of the Association of American Geographers* 87 (4): 660–80.

Schein, R. (2009) Belonging through land/scape. *Environment and Planning A* 41 (4): 811–26.

Schein, R. (ed.) (2006) *Race and Landscape in America*. New York: Routledge.

Scott, J. (1985) *Weapons of the Weak: Everyday Forms of Peasant Resistance*. New Haven; London: Yale University Press.

Secor, A. (2004) "There is an Istanbul that belongs to me": citizenship, space, and identity in the city. *Annals of the Association of American Geographers* 94 (2): 352–68.

Secor, A. (2007) "An unrecognizable condition has arrived": law, violence, and the state of exception in Turkey. In D. Gregory and A. Pred (eds.), *Violent Geographies: Fear, Terror, and Political Violence*. New York; London: Routledge, pp. 37–53.

Sibley, D. (1995) *Geographies of Exclusion: Society and Difference in the West*. London and New York: Routledge.

Tilly, C., Moss, P., Kirschenman, J., and Kenneley, I. (2001) Space as signal: how employers perceive neighborhoods in four metropolitan labor markets. In A. O'Connor, C. Tilly, and L. Bobo (eds.), *Urban Inequality: Evidence from Four Cities*. New York: Russel Sage Foundation, pp. 304–38.

Wright, M. (2006) *Disposable Women and Other Myths of Global Capitalism*. New York: Routledge.

Wright, M. (2004) From protests to politics: sex work, women's worth and Ciudad Juárez modernity. *Annals of the Association of American Geographers* 94 (2): 369–86.

Chapter 26

Care and Caring

David Conradson

Towards New Ways of Being Together

Care has a particular significance for social geographies, for it speaks of a trans-formative ethic and a relational dynamic that has the potential to transcend self-interest. As an ideal, care invites us to recognize the lived experience of others as worthy of our attention. When these others are vulnerable, marginalized or in need, care suggests that we respond in a way that is helpful and which perhaps facilitates positive change. And although expressions of care are most often conceived of as occurring between two individuals, as part of a dyadic relation, care also has significance for intra- and inter-group sociality. As an orientation and embodied practice, care thus has the capacity to alter the character of social geographies across a range of registers and scales. It holds the possibility, in other words, of facilitating new ways of being together.

For all these reasons, we might assume that care has been a longstanding concern within social geography. Yet some of the earliest or at least most visible work on the subject is to be found within medical and health geography. Here care has generally been approached in terms of *health*care. Studies have examined geographic variations in access to health services and professionals (e.g., Joseph and Phillips 1984; Wilson and Rosenberg 2002; Brabyn and Barnett 2004), spaces of medical practice (e.g., Andrews and Evans 2009), and practices of self-care (Parr 2002). Within the related and growing area of geographical gerontology (Andrews and Phillips 2004), scholars have investigated formal and informal care for the elderly and transitions from home to institutional settings (Milligan 2001; Andrews et al. 2007; Wiles et al. 2009). Together these studies have contributed significantly to our understanding of the interactions between care, health, and place.

A Companion to Social Geography, First Edition. Edited by
Vincent J. Del Casino Jr., Mary E. Thomas, Paul Cloke, and Ruth Panelli.
© 2011 Blackwell Publishing Ltd. Published 2011 by Blackwell Publishing Ltd.

Alongside this health focused work, social geographers have also been active in exploring the phenomenon of care. In this chapter I review this work, taking a broad and relatively inclusive view of what constitutes social geography, recognizing that many studies defy easy categorization in sub-disciplinary terms and that sub-disciplinary categories are themselves shifting and porous. A certain blending of ideas from social, cultural, political, health and urban geography is evident in Gleeson and Kearn's (2001) work on remoralizing landscapes of care, for example, in Parr and Philo's (2003) exploration of rural community care dynamics, and in Cloke et al.'s (2005) investigations of homelessness and welfare provision. So whilst mindful of recent framings of social geography as "the study of social relations and the spatial structures that underpin those relations" (Smith et al. 2010: 1) and the longstanding social geographic engagement with inequality, difference and processes of transformation, I consider scholarship that identifies both directly and less directly with these themes.

The chapter begins with a series of conceptual points, where I suggest that care can be understood as both a disposition and a practice which, in any particular setting, has the potential to alter the socio-natural-material relations that character-ize lived space. I then consider three strands of social geographical work on care. The first concerns care-giving for dependent others, such as children and the elderly, in the context of the work of social reproduction. The second examines the care expressed within various welfare and support organizations, such as homeless shel-ters and community drop-in centers, and the receipt of this care by service users. Studies here have documented the material transactions and relational dynamics amongst staff, volunteers and service users, as well as experiences of dwelling within and passing through these organizational settings. A third and more recent strand of work examines the significance of an ethic of care in and for a globalizing world. Some of this research has been animated by concerns for so-called "distant others," in the sense of individuals who are both socially and geographically apart from us. What is our responsibility to such individuals? How far, as Smith (1998) has expressed it, should we care? Another body of work has framed care as a trans-formative ethic more generally, as an alternative to the capitalist emphasis on acquisition and competition, and as a potential touchstone in our encounters with those who differ from us.

Conceptual Foundations: Care as Disposition and Embodied Practice

In considering the phenomenon of care, Fisher and Tronto (1990) distinguish between *caring about* (being or becoming mindful of the needs of another person or of a situation that requires attention), *care giving* (the actual practice of offering care to another person or persons), and *care receiving*. These distinctions are valu-able insofar as they suggest that care involves awareness of a situation or the needs of another, a cognitive and emotional orientation towards addressing that situation or need and, in some instances, a practical expression of this orientation in terms such as offering physical assistance, giving time or money, listening or simply helping more generally. These dimensions of care may of course occur individually – they are not necessarily co-incident – but it is relatively common to find them woven together. The broader point is that care may take the form of a disposition or affective orientation towards another, but that it also finds expression as a material practice.

From a geographical perspective, the sociological and philosophical dimensions of Fisher and Tronto's (1990) conceptualization of care can be extended in a number of ways. In view of the uneven distribution of care within and between human communities, we might ask why care is present in some locations and relatively absent in others. Such a question suggests the possibility of relatively care-full and relatively care-less places, with perhaps positive and negative consequences for human well-being respectively. Somewhat differently, we might move beyond the implicit humanism of Fisher and Tronto's framework and examine the dynamics of care between different species (Haraway 2007). The relation between humans and their pets is perhaps one example of such care; animal welfare initiatives might be another. And of course there are many other inter-species dynamics that characterize the ecologies in which we dwell (Fox 1990; Fox 2006). We can also consider care in terms of emotional and personal investment in particular objects (e.g., homes, vehicles, money) and creative projects (e.g., art, music, writing, building, performance).

When directed beyond the self, the expression of care normally involves the negotiation of social and geographical distance. This situation arises because others are seldom if ever coincident with ourselves. So even when care is expressed towards those closest to us, we find there is distance to span and perhaps overcome, whether in communicative, imaginative or embodied terms. Smith's (1998) question "how far should we care?" builds upon this realization, referencing both geographical formulations of distance as well as the kinds of travel entailed in bridging gender, ethnicity, race, age, sexual orientation, and citizenship differences. How far we extend our care is a moral and ethical issue, for it reflects our recognition of the needs of others and the value we assign to their livelihood. Even in habitual, apparently unthinking behavior it is possible to locate judgements about the relative deservingness of potential recipients of care.

Beyond the social and geographical dimensions of care, an important issue is its actual existence. Why does care come into being at all? Amongst humans, what is it that motivates or induces people to care, to take an interest in the situations and needs of others? Are there particular ethical standpoints or impulses which support such engagements, especially when the potential recipients are not part of one's immediate family or friendship groups? Such questions are important foci within ethical and moral philosophy (for helpful discussions, see Proctor and Smith, 1999; Held 2006; Slote 2007). Space does not permit a detailed engagement with this work, but one point of relevance here is the extent to which humans are ever genuinely concerned about the needs of others. Some commentators believe not, insisting that altruism is either scarce or ultimately illusory. If we push beyond the surface of an apparently altruistic act, they contend that we will soon discover a calculating and benefit-maximizing impulse. For some socio-biologists, even apparently benevolent acts are linked to genetic selfishness, understood in terms of shoring up one's own fortunes or enhancing the circumstances and prospects of kin (Wilson 1975, 2006; Dawkins 1976; Pinker 2002). The extension of care beyond one's immediate family is then interpreted as a form of favoritism which anticipates a reciprocal or loosely symmetrical future response. In short, it is thought that people only engage in helping behaviors when there is something in it for them.

Evidence for human self-centeredness and selfishness is not, of course, difficult to locate. And yet to interpret all human action in these terms, insisting that care is inevitably selfish seems problematic. For there are forms of caring behavior in which it is difficult to identify the return to the giver. Consider those who chose to assist others to whom they have no immediate or likely future connection, for instance, or those who forgive serious injustice and harm, sometimes in the absence of apology or the discernible admission of responsibility, so that a process of social transformation can continue. In both cases, there may be a commitment to a broader good which provides the motivation for localized care. We also need to account for the thousands of volunteers who give of their time and labor, across diverse settings, responding to what they perceive as social need or valuable causes. Of course, motivation is a complex and multi-faceted phenomenon, but these expressions of care seem difficult to explain solely in terms of the benefits which may accrue to the persons involved, to their kinship group or to an affiliative community with which they identify. The socio-biological account of such actions, in which self-interest is declared to be the underlying motivation, seems incomplete in these cases. As the critics of socio-biology contend, such actions instead suggest the possibility of genuine human interest in others and their needs (Allen et al. 1975; Lewontin et al. 1984).

If we recognize that genuine care exists, then from a social geographical perspective we are led to consider how it might shape the nature of places and social relations within them. What kinds of relational dynamics and environments does care bring into being or make possible? In what ways does care make the world go round? How does care alter over distance? And what happens if care is displaced, devalued or difficult to find? Over the last two decades, social geographers have grappled with these kinds of questions across a range of settings. In what follows, I consider a selection of this work.

Care, Social Reproduction, and the Home

A first strand of social geographic work on care concerns the relationships between care, social reproduction and the home. This nexus is also addressed in sociology and social policy (e.g., Ungerson 1990; Twigg 2000), politics (e.g., Hirshmann 2008) and economics (e.g., Folbre 2001), but the distinctive elements of a geographical approach lie in the attention given to the settings in which care takes place. Some geographers have examined the intertwining of public and private worlds associated with the receipt of care in home environments (Milligan 2000; Wiles 2003a, 2003b), for instance, whilst others have charted the challenges of combining care work with paid employment outside the home (Jarvis et al. 2001; Bailey et al. 2004; McDowell et al. 2005).

Across these studies, social reproduction is typically understood as the "material and social practices through which people reproduce themselves on a daily and generational basis" (Katz 2001: 711). In their study of middle-class households in Britain, Gregson and Lowe (1995) identified three broad types of social reproductive task, relating to: (1) the reproduction of adult labor power through work such as food preparation, washing clothes, and cleaning; (2) generational reproduction (particularly childcare activities); and (3) the structural upkeep of a dwelling (e.g., domestic repairs and gardening). Even a cursory consideration of this schema under-lines the extensive and demanding nature of social reproductive work. In addition, we see that care is intrinsic to social reproduction.

As a form of material and emotional labor, care is often costly in both financial and time terms (Daly and Lewis 2000). In contrast to conventional economic analy-ses, where care-giving is often regarded as a solely private activity, feminist scholars argue that social reproductive tasks should be considered as important as work undertaken in the so-called productive sphere. The formal economy, they note, depends closely upon the ongoing supply of healthy and motivated labor generated by social reproduction. Feminist scholars also draw our attention to the unequal power relations inherent in much care work, both in terms of the gendered division of care labor and its chronic undervaluation by employers and society more gener-ally (Fisher and Tronto 1990).

A distinction can be drawn between informal and formal care. For Soldo et al. (1989: 194), "informal care is personal care provided by a relative, friend, or neigh-bor. By extension, an informal care network is a diffuse primary group characterized by its small size, affectivity, and durable commitment to each member's well-being." Formal care, in contrast, is provided by actors beyond one's family and friends and generally involves payment. Formal carers are often employees of structured bureaucracies, such as nursing or personal care agencies. There is some debate over whether the informal/formal care dualism adequately captures the diversity of care arrangements within contemporary households in industrialized nations (Ungerson 1990), but the two terms continue to be widely used. Perhaps the most important point here is that the great majority of social reproductive care work – whether for children, older people, those with impairments or disabilities – is entirely unpaid and informal in nature. Members of households provide it themselves. In Canada, for instance, it has been estimated that over 80 percent of care for the ill, disabled or frail elderly is provided informally (Wiles 2003a).

In recent years, many individuals and families in western countries have found it increasingly difficult to provide the care needed by the children and adults who depend upon them (Daly and Lewis 2000). The resulting "crisis of care" is in part linked to the significant rise in female labor market participation since the 1960s and 1970s. As increasing numbers of women have taken up paid employment beyond the home, so the time and energy they have available for domestic care work has generally fallen. And although many men now participate more fully in household care work, their contribution has not generally been sufficient to off-set increased female labor force participation. The reduction in labor available for care work is also connected to the growth of dual-earner households, where both members of a couple undertake paid employment. In Britain, the proportion of families dependent upon a single wage earner fell from 42 percent in 1975 to 17 percent in 2002 (McDowell et al. 2005), by which time 65 per cent of mothers with dependent children were in paid work of a part- or full-time nature. The dynamics of this situation are again linked to the gendered nature of care work, where women tend to undertake the bulk of this material and emotional labor. That said, structural constraints in terms of the labor market opportunities available to men for parental leave and part-time employment arrangements also have some effect.

A further factor in the crisis of care has been the reduction in affordable, extra-familial care services. In the context of reduced household labor capacity for care-work, many individuals and couples have turned to external organizations for assistance with their care responsibilities. The state, private and voluntary sectors have all been involved in the provision of nurseries, day centers, childcare facilities, and nannies. In many western countries, however, the rise of neoliberalism during the 1980s and 1990s saw a reduction in governmental involvement in childcare and aged care provision. As a consequence, extra-familial care support has increasingly been sought from the private sector (Player and Pollock 2001). In the case of the UK, however, despite a significant increase in both the number of registered child-minders and their use by middle-class households during the 1980s (Gregson and Lowe 1995), demand for care today still often exceeds local supply. In Britain in 2001, for example, "there was still only one place for every 6.6 children aged under 8 years in either a day nursery, with a registered childminder or in an after-school club" (McDowell et al. 2005: 222). This mismatch between demand and supply has generated upward pressure on childcare costs, rendering it less affordable and making it even harder for many households to manage care-taking responsibilities alongside paid employment. For McDowell et al. (2005), this shortage of public provision highlights the inconsistency of British social policy in recent years. For while paid work has been vigorously promoted as an important route to social inclusion and prosperity, the state has not funded childcare provision to the extent necessary to support such levels of labor market participation.

The home has unsurprisingly been a key site of enquiry in geographical work on care and social reproduction, whether with respect to childcare (Gregson and Lowe 1995; England 1996; Boyer 2003) or care for the elderly (Mowl et al. 2000; Milligan 2001; Angus et al. 2005; Dyck et al. 2005). A number of studies have explored the negotiations surrounding privacy and intimacy when receiving care in one's home, particularly when this is provided by formal carers rather than by family members (England 1996; Milligan 2001). Other research has traced the complex movements undertaken between schools, workplaces, nurseries and childminder's homes in order to coordinate care-work and paid employment (Jarvis et al. 2001; Bailey et al. 2004). The gendered nature of these care and coordination efforts continues to be a major theme for exploration (Jarvis et al. 2009).

The care work inherent in social reproduction has become transnationalized in recent years. A significant amount of childcare and aged care in the West is now provided by migrant workers, particularly individuals from less developed countries in southeast Asia, Africa and the Pacific (Kofman and Raghuram 2004; Wills et al. 2007, 2009; Raghuram 2008; Connell 2009). These individuals form part of global care chains, with the migratory trend being to move from less developed nations to the industrialized west (e.g., western Europe, North America and Australasia). There is a now a significant geographical literature on migrant care workers, including studies of Filipinos (Gibson et al. 2001; Cheng 2004; McKay 2007; Pratt 1997a, 1997b), Mexicans in California (Mattingly 2001), domestic assistants in Singapore (Huang and Yeoh 2007), and healthcare professionals in Britain (Ragurham and Kofman 2002). McGregor's (2007) study of Zimbabwean carers in London high-lights the difficulties many migrant care-workers face in terms of low pay, poor working conditions and vulnerability to abuse. Yet because of the labor shortages in Britain's care industry and the wage differential relative to developing countries, many of the jobs at the "unskilled end of the care labor market are [filled by] migrants who have recently arrived in Britain – particularly women" (p. 802). This work is crucial to the reproduction of daily life in many western cities and yet, as McGregor and others demonstrate, it is often poorly rewarded and regarded (Anderson 2000; Ehrenreich and Hochschild 2003; Pyle 2006). The situation of these migrant carers would seem to reflect the broader western cultural tendency to undervalue care work (McKie et al. 2002; Boyer 2003; Jarvis 2007).

Care and Welfare Provision

A second strand of social geographic work on care arises in relation to welfare provision for marginalized citizens. During the twentieth century, the postwar development of the welfare state was a significant achievement in terms of support for individuals and families in need. As is now well documented, however, the ascendency of neoliberal forms of governance across many western states during the 1990s led to the erosion of these collectivist forms of welfare provision. Governments sought to reduce public expenditure by shifting responsibility for social security to individuals and families, along with community and voluntary organizations. This transfer of responsibility was in part enacted through processes of welfare reform, including benefit cuts, narrowing the eligibility criteria for welfare recipients, and privatizing governmental welfare assets such as public housing (Conradson 2009). Unsurprisingly, such processes have tended to impact disproportionately upon people on low incomes, members of ethnic minorities, the under- and unemployed, and those whose socio-economic position might otherwise be described as marginal or precarious (Le Heron and Pawson 1996; Dorling 1995; Glasmeier 2005).

A number of geographers have sought to document, understand and, in some cases, contest the outcomes of these processes of welfare reform and retrenchment. With respect to psychiatric de-institutionalization, several studies have charted the impacts of facility closures for people diagnosed as being "mentally ill" (Laws and Dear 1998; Kearns and Taylor 1989; Philo 1997). Under care in the community arrangements, previously institutionalized individuals have been expected to access support from a range of out-patient and day center services, against a backdrop of assumed community support. The reception of ex-psychiatric patients amongst local communities has been uneven, however, with some individuals experiencing significant difficulty, stigma and disapproval. Parr (2000) has detailed the post-asylum geographies of mental health in Nottingham, England, for example, noting the dynamics of inclusion and exclusion within an inner city drop-in center. Knowles (2000a, 2000b) documents the way that people in Montreal with mental health problems developed individualized paths between various support organizations and city spaces, whilst experiencing fluctuating levels of vulnerability and well-being. Similar institutional cycling has been observed amongst homeless people in Los Angeles, California (DeVerteuil 2003) and amongst individuals with schizophrenia in Winnipeg, Manitoba (DeVerteuil et al. 2007). The intersections between psychiatric de-institutionalization and care in the community have also been examined in Australasia (Law and Gleeson 1998; Kearns and Joseph 2000). In some cases, ex-psychiatric patients have fallen between the categories of "those able to live relatively independently in a community environment" and "those requiring in-patient services in a hospital setting," and have subsequently ended up homeless, a situation which has tended to further compromise their physical and mental health (Dear and Wolch 1987). In a project which complements this emphasis on the lived experience of deinstitutionalization, Moon et al. (2006) have examined the re-use of asylum buildings. They cite facilities in Britain, Canada, and New Zealand that have been converted into educational, accommodation, and even leisure facilities.

With respect to the support that community and voluntary organizations offer, a number of geographers have employed the notion of "spaces of care" to analyze the experiential dimensions of organizational service environments. Johnsen et al. (2005), for example, have written of soup runs for homeless people in English cities as mobile spaces of care, as a phenomenon enacted by volunteers as they come together to provide food and assistance in particular locations across the city. Conradson (2003) explores the dynamics of a community drop-in center in a peripheral estate in Bristol in this regard, whilst similar ideas are evident in Wilton and DeVerteuil's (2006) study of alcohol-treatment facilities in Canada. This scholarship has highlighted the provisional and fragile nature of spaces of care, with their durability influenced by the disposition and practices of volunteers, staff and service users. Where durable spaces of care do emerge, these seem able to facilitate positive change for those who temporarily inhabit or pass through them, whether in a short-term material capacity or a broader, whole of life sense.

Faith-based and post-secular motivations are a significant influence upon the care offered by community and voluntary organizations in the industrialized west (Cloke et al. 2005; Beaumont 2008; Conradson 2008; Ley 2008; Wills and Jamoul 2008). A range of religious and spiritual motivations can be discerned in vision and mission statements, for instance, and also within the service practices of staff and volunteers. In Cloke et al.'s (2005) work on the ethos of care amongst homeless service organizations in Britain, Christian faith was found to be a significant motivational and dispositional influence for many volunteers. In a somewhat more politicized manner, some of the larger social service organizations in New Zealand have drawn upon Christian values to develop public critiques of neoliberal government policies (Conradson 2008). At the same time, we can identify faith-based organizations that endorse what are arguably regressive policies, such as the Christian right's pro-war stance in the United States and its tendency to pathologize people who are poor, homosexual or members of ethnic minority groups. Other faith-based organizations have sought to link eligibility for welfare and care to particular behavioral requirements, such as being sober whilst on site, or, arguably more problematically, to some form of personal religious commitment. Sagar and Stephens' (2001) study of an agency "serving up soup and sermons" in New York city explores the latter terrain, whilst Hackworth (2009) examines the associations between faith-based organizations and some of the more problematic dimensions of neoliberal welfare policy in the United States.

While religious and spiritually motivated "care" is in some cases exploitative and manipulative, it nevertheless seems important to acknowledge the positive expressions of care and support that at times flow from spiritual and post-secular motivations. These motivations may in various ways be connected to faith and spiritual traditions such as Judaism, Christianity, Islam, Buddhism, and Hinduism. But they may also emerge from more individualized and hybridized forms of spirituality that defy easy description in such terms (Heelas and Woodhead 2004; Partridge 2004; King 2008).

A further aspect of care and welfare provision is the growing popular engagement with places intended to facilitate rest and recovery, such as retreat and relaxation centers. While it is possible to view attendance at retreat centers and spa complexes as an indulgent pastime of the affluent middle classes, such visits might also be said to reflect the need in western societies for time and space away from the demands and intensity of everyday life. As the Norwegian sociologist Eriksen puts it, "a fragmented and rushed temporality is typical of a growing majority of the population in the rich countries" (2001: 48). Franklin (2003: 13) adds that "[t]he tyranny of the present is not boredom or the lack of difference and color and excitement in our lives but the opposite: we are over-excited, bombarded by stimulation, information, possibilities, connections and access." When a person deliberately seeks time apart from such circumstances, this might therefore be considered a form of care for the self.

In relation to these pressures on work-life balance, a number of researchers have been considering the place of stillness in western societies (Bissell and Fuller 2010). Lea (2008) has written about yoga in this respect, describing the affective consequences of regular practice as part of a multi-day retreat or holiday. Conradson's (2007) work on Benedictine monasteries in southern England similarly describes the affective environments that emerged through particular rhythms of spiritual practice, shared meals and silence. A related study has then examined the orchestration of stillness during Buddhist and Christian weekend retreats (Conradson 2010). Across these investigations, places of retreat emerge as sites of rest, reflection and self-care. But taking care of oneself need not be considered individualistic; it might also be considered an act of care for others. For if affect and emotion flow through individuals into the broader socio-ecological fabric in which they dwell, then enhancements to personal well-being and resilience will also have the capacity to impact positively upon collective well-being. What at first glance might appear to be a relatively individualized practice – meditation in a retreat center – thus becomes implicated in a potential recalibration of social geographies more broadly.

Care as a Transformative Ethic in a Globalizing World

This point about the connections between individual affective dynamics and collective well-being links to a third strand of scholarship in which care is framed as a transformative ethic. One focus in this work has been care and responsibility for distant others, particularly in terms of the relationship between relatively affluent people in the global North and those who are materially less fortunate in the global South (Silk 1998). The concern here has been to explore the ways in which "[c]are is not solely localised, but extends to distant, different and unknown others by virtue of theoretical and practical cross-cultural connections" (Silk 2004: 230). Drawing on work in moral geographies, Silk (1998) examines the moral motivation to give to charity and the nature of charitable representations of human need. In a subsequent article, Silk (2000) traces the relationships between donors/carers in affluent nations and those who are supported and cared for in less developed countries. More recently,

Raghuram et al. (2009) have argued that notions of care and responsibility offer valuable ways of rethinking the relationships between places in a postcolonial world.

In various ways, these studies all touch upon the socio-biological contention that care beyond one's family and friends is either rare or undertaken in anticipation of future reciprocity. On this point, a small investigation of charitable donors in England offers an interesting window onto the relationship between a person's perceived deservingness for assistance and their social-geographic distance from the donor (Paxman 2002). Immediate family members were typically regarded as highly deserving of assistance, whilst materially needy yet non-familial others in one's own country, such as homeless people, were generally viewed less positively. The latter group was often judged to be less deserving of help and support because its members were assumed to be responsible to a significant extent for their own difficulties. Interestingly, materially needy people in less developed countries, such as those in sub-Saharan Africa, were usually seen as highly deserving of assistance. Whether these judgements translated into different levels of practical support for particular groups was a matter beyond the scope of the study, but the observed variations in perceived deservedness were nonetheless noteworthy. Moreoever, the findings seem to challenge the socio-biological position that expressions of care are limited to family or those from whom future reciprocity seems likely.

Alongside this work on care and responsibility for distant others, a number of geographers have recently suggested that care has the potential to function as a transformative ethic more broadly. In view of the distress generated by contemporary economic and social arrangements, Lawson (2007, 2009) has argued that an ethic of care might act as a useful guide in developing new forms of society, economy and polity. Her starting point is an ontology of connection which recognizes that individual actions and dispositions have implications for the wider socio-ecological fields in which we are located. In practical terms, this framework then entails "foregrounding social relationships of mutuality and trust rather than dependence" and "structuring relationships in ways that enhance mutuality and well-being" (Lawson 2007: 3). Such a perspective is in contrast to the relational dynamics of contemporary capitalism, in which competition and profitability feature prominently:

> Care ethics questions (neo)liberal principles of individualism, egalitarianism, universalism, and of society organised exclusively around principles of efficiency, competition, and a "right" price for everything. (Lawson 2007: 3)

There is a critique here of capitalism's capacity to concentrate wealth and power in the hands of the few whilst generating difficulty for many others. Informed by both feminist and Marxian perspectives, Lawson's analysis makes it clear that foregrounding care is a form of political action, for it calls for us to do things differently at a range of scales.

In a related analysis of western societies, McDowell (2004: 145) writes that:

> an individualistic ethos pervades both the labor market and the welfare state, undermining notions of collective welfare and an ethic of care, within the wider context of a hegemony of a neoliberal ideology in global as well as national politics.

Building on the work of Nancy Fraser, McDowell's (2004: 157) response is to call for "greater gender equality and a wider distribution of the responsibility for the labor of caring," anti-poverty measures, improved income inequality and reduced exploitation of vulnerable individuals. Their schematic form notwithstanding, it is hard to disagree with such proposals. But there is little consideration in McDowell's paper of what might motivate an individual or organization to move towards these more caring or interdependent forms of social interaction. It seems assumed that such dispositions and practices are latent, awaiting expression if only a sufficiently convincing argument is developed or someone paints a sufficiently moving picture of contemporary injustice and inequality. Of course the representations and arguments of others do have the capacity to shape our dispositions towards others (Silk 2000; Slote 2007). But given that there is as least as much apathy regarding the distress of our fellow humans as there is proactive engagement, an important question is just which arguments and what kinds of dispositions are best able to facilitate relations of care.

A number of scholars have considered this issue. Cloke (2002) explores the significance of Christian notions of agape in this respect, whilst drawing on the work of Hannah Arendt and Melissa Orlie and their recognition and theorization of evil. Other studies have considered the work of Levinas and Derrida with respect to acknowledging the claims of others (Barnett 2005), and the work of Jean-Luc Nancy in relation to community (Welch and Panelli 2007; Popke 2009). These discussions can be set alongside more philosophical considerations of care ethics (e.g., Noddings 1986; Koehn 1998; Robinson 1999; Held 2006). And we should not overlook the many small instances of everyday generosity that, while perhaps not easily mappable onto larger philosophical or conceptual frameworks, are nonetheless significant in shaping the unfolding world (Barnett and Land 2007).

There is also work that explores the development of an ethic of care rooted in Buddhist understandings of socio-ecological interconnectedness (e.g., Fox 1990, 2006; Macy and Brown 1998). This perspective resonates with Lawson's ontology of connection, whilst emphasizing practices of self-extension and compassionate identification with others. As Fox (1990: 230) expresses it, "Every living being is connected intimately and from that intimacy follows the capacity of identification and as its natural consequences, the practice of non-violence."

An ethic of care has implications for our analysis of "the structures and institutions that reproduce exclusion, oppression, environmental degradation" (Lawson 2007: 7). This analysis may occur as part of our everyday sense-making as well as being woven more formally into collective projects. An initiative of the latter kind is the international Critical Global Poverty Studies (CGPS) network. Here a group of scholars is seeking to develop a more unified conceptualization of poverty, in the hope of transcending the existing schism between analyses of poverty in less developed countries and poverty research in the global North.[1]

Within the profession of academic geography, Lawson suggests that a care ethics approach should inform our interactions with students and colleagues, prompting particular forms of community action, and increasing our use of participatory research methodologies. This analytical extension resonates with her view that:

Care ethics focuses our attention on the social and how it is constructed through unequal power relationships, but it also moves us beyond critique and toward the construction of new forms of relationships, institutions, and action that enhance mutuality and well-being. (2007: 8)

This call to step beyond critique and into action of various kinds is consistent with social geography's longstanding interest in critiquing power differentials and facilitating positive change (Panelli 2004). A further point is that Lawson's proposals extend beyond the social and economic domain to encompass the natural environment.

There are some indicative steps towards a more-than-social geography of care, achieved in part through analytical recognition of various non-human others. Such a manoeuvre resonates with the deep ecology perspective advanced by Arne Naess and others (Fox 1990), as well as with various formulations of post-humanism in human geography (Castree et al. 2004; Whatmore 2006).

An ethic of care is also potentially transformative with respect to our negotiations of social and cultural difference. As our cities and towns become more culturally diverse, so we increasingly encounter others who differ from us in ethnic religious, political, and citizenship terms. In these circumstances, a caring and empathetic orientation might sensitize each party to their shared humanity, and this affective imaginary might then in turn constitute a safeguard against antipathy and violence. A number of geographers have grappled with these issues, examining the dynamics of encountering and living with difference (Valentine 2008), spaces of cosmopolitan responsibility (Popke 2007) and an ethic of care among strangers (Amin 2002, 2010). As Sandercock and Attili (2009) express it, an ethic of care may help us to view unfamiliar others – in the sense of those who are not part of our immediate families – less as strangers and more as neighbors. If contemporary tensions have demonstrated anything, it is that our capacity to live in harmony with unfamiliar others is important but surprisingly fragile. As an orientation towards the other, care would seem to be particularly valuable in such circumstances.

Conclusion

In this chapter I have considered three strands of social geographical work on care, relating to social reproduction and the home, welfare provision, and care as a transformative ethic. What this scholarship demonstrates is that care is woven into and through the social fabric in a range of important ways. It is critical to the work of social reproduction, such that children are raised and nurtured and the elderly are appropriately cared for. An ethos of care lies at the heart of many community welfare services, particularly in the voluntary sector, where motivated individuals come together to support and help others with identifiable needs and difficulties. And care has a broader significance as a transformative ethic in and for a globalizing world, whether at the level of our everyday encounters with strangers or as an alternative to the competitive individualism that characterizes so many interactions in market-oriented societies.

Some might argue that care is somehow too soft a sentiment for contending with matters of inequality and injustice in a hard world. This is an issue for philosophical and practical debate, and scholars of different persuasions have certainly argued over the relative merits of concepts such as injustice, equality and care (Slote 2007).

These discussions notwithstanding, it will hopefully be evident here that care has the capacity to confirm existing power relations but also to unsettle and disrupt them. In terms of re-inscribing the status quo, it is widely recognized that the burden of reproductive care work, particularly in relation to children, the elderly and unwell family members, continues to fall disproportionately on women. In recent years there have been small changes to this situation, both at level of individual house-holds and within western societies, but by and large we find that gendered divisions of caring labor continue. At the same time, we can identify instances in which care disrupts rather than reproduces the contours of existing social arrangements. When volunteers help those less fortunate than themselves or a new migrant is treated as a fellow human being rather than as an alien or stranger, for example, then we observe the capacity of care to alter social relations in progressive ways. And some-times, when large numbers of people are mobilized into action because of their shared care for a particular end (e.g., international debt and poverty relief, action against child abuse, efforts to halt cruelty to animals, the maintenance of biodiver-sity), then really quite remarkable things can be achieved. Such outcomes are a basis for hope. For although care is no magical elixir, it certainly has the capacity to foster new ways of being together.

Note

1   See www.cgps.uib.no/ for more details. Site accessed in December 2009.

References

Allen, E., Beckwith, J., Beckwith, B., Chorover, S., Culver, D., Duncan., M., Gould, S.J., Hubbard, R., Inouye, H., Leeds, A., Lewontin, R., Madansky, C., Miller, L., Pyeritz, R., Rosenthal, M., and Schreier, H. (1975) Against "socio-biology." *New York Review of Books*, 2218, November 13, www.nybooks.com/articles/9017?sess=305fe41afae729849 e1e7eb4b004bb8. Accessed December 2009.

Amin, A. (2002) Ethnicity and the multicultural city: living with diversity. *Environment and Planning A* 34 (6): 959–80.

Amin, A. (2010) Cities and the ethic of care among strangers. Unpublished presentation, National Centre for Australian Studies, Monash University, February 19.

Anderson, B. (2000). *Doing the Dirty Work? The Global Politics of Domestic Labour.* London: Zed Books.

Andrews, G. and Evans, J. (2009) Understanding the reproduction of health care: towards geographies in health care work. *Progress in Human Geography* 32 (6): 759–80.

Andrews, G., Cutchin, M., McCracken, K., Philips, D., and Wiles, J. (2007) Geographical gerontology: the constitution of a discipline. *Social Science & Medicine* 65 (1): 151–68.

Andrews, G. and Phillips, D.R. (eds.) (2004) *Ageing in Place.* London: Routledge.

Angus, J., Kontos, P., Dyck, I., McKeever, P., and Poland, B. (2005) The personal significance of home: habitus and the experience of receiving long term home care. *Sociology of Health and Illness* 27 (2): 161–87.

Bailey, A., Blake, M., and Cooke. T.J. (2004) Migration, care and the linked lives of dual-earner households. *Environment and Planning A* 36 (9): 1617–32.

Barnett, C. (2005) Ways of relating: hospitality and the acknowledgement of otherness. *Progress in Human Geography* 29 (1): 1–17.

Barnett, C. and Land, D. (2007) Geographies of generosity: beyond the "moral turn." *Geoforum* 38 (6): 1065–75.

Beaumont, J.R. (2008) Faith action on urban social issues. *Urban Studies* 45 (10): 2019–34.

Bissell, D. and Fuller, G. (eds.) (2010) *Stillness in a Mobile World.* London: Routledge.

Boyer, K. (2003) At work, at home? New geographies of work and care-giving under welfare reform in the US. *Space and Polity* 7 (1): 75–86.

Brabyn, L. and Barnett, J.R. (2004) Deprivation and geographic access to general practitioners in rural New Zealand. *New Zealand Medical Journal* 117: 1–13.

Castree, N., Nash, C., Badmington, N., Braun, B., Murdoch, J., and Whatmore, S. (2004) Mapping posthumanism: an exchange. *Environment and Planning A* 36 (8): 1341–63.

Cheng, S.A.-J. (2004) Contextual politics of difference in transnational care: the rhetoric of Filipina domestics' employees in Taiwan. *Feminist Review* 77 (1): 46–64.

Cloke, P. (2002) Deliver us from evil? Prospects for living ethically and acting politically in human geography. *Progress in Human Geography* 26 (5): 587–604.

Cloke, P., Johnsen, S., and May, J. (2005) Exploring ethos? Discourses of "charity" in the provision of emergency services for homeless people. *Environment and Planning A* 37 (3): 385–402.

Connell, J. (2009) *The Global Health Care Chain: From the Pacific to the World.* London: Routledge.

Conradson, D. (2003) Spaces of care in the city: the place of a community drop-in centre. *Social and Cultural Geography* 4 (4): 507–25.

Conradson, D. (2007) Experiential economies of stillness: the place of retreat in contemporary Britain. In A. Williams (ed.), *Therapeutic Landscapes.* Aldershot: Ashgate, pp. 33–48.

Conradson, D. (2008) Expressions of charity and action towards justice: faith-based welfare in urban New Zealand. *Urban Studies* 45 (10): 2117–41.

Conradson, D. (2009) Welfare reform. In R. Kitchin and N. Thrift (eds.), *International Encyclopedia of Human Geography.* Oxford: Elsevier, pp. 230–4.

Conradson, D. (2010) The orchestration of feeling: stillness, spirituality and places of retreat. In D. Bissell and G. Fuller (eds.), *Stillness in a Mobile World.* London: Routledge, pp. 71–86.

Daly, M. and Lewis, J. (2000) The concept of social care and the analysis of contemporary welfare states. *British Journal of Sociology* 51 (2): 281–8.

Dawkins, R. (1976) *The Selfish Gene.* Oxford: Oxford University Press.

Dear, M. and Wolch, J. (1987) *Landscapes of Despair: From Deinstitutionalisation to Homelessness.* Oxford: Polity Press.

DeVerteuil, G. (2003) Homeless mobility, institutional settings, and the new poverty management. *Environment and Planning A* 35 (2): 361–79.

DeVerteuil, G, Hinds, A., Liz, L., Walker, J., Robinson, R., and Roos, L. (2007) Mental health and the city: intra-urban mobility among individuals with schizophrenia. *Health and Place* 13 (2): 310–23.

Dorling, D. (1995) *A New Social Atlas of Britain.* London: Wiley and Sons.

Dyck, I., Kontos, P, Angus, J., and McKeever, P. (2005). The home as a site for long term care: meanings and management of bodies and spaces. *Health and Place* 11 (2): 173–85.

Ehrenreich, B. and Hochschild, A.R. (2003) *Global Woman: Nannies, Maids and Sex Workers in the New Economy.* London: Granta Books.

England, K. (ed.) (1996) *Who Will Mind the Baby? Geographies of Child-care and Working Mothers.* London: Routledge.

Eriksen, T.H. (2001) *Tyranny of the Moment: Fast and Slow Time in the Information Age.* London: Pluto Press.

Fisher, B. and Tronto, J. (1990) Towards a feminist theory of caring. In E. Abel and M. Nelson (eds.), *Circles of Care.* Albany: SUNY Press, pp. 35ff.

Folbre, N. (2001) *The Invisible Heart: Economics and Family Values.* New York: New Press.

Fox, W. (1990) *Towards a Transpersonal Ecology: Developing New Foundations for Environmentalism.* Boston and London: Shambala Books.

Fox, W. (2006) *A Theory of General Ethics: Human Relationships, Nature, and the Built Environment.* Cambridge, MA: MIT Press.

Franklin, A. (2003) *Tourism: An Introduction.* London: Sage.

Gibson, K., Law, L., and McKay, D. (2001) Beyond heroes and victims: Filipina contract migrants, economic activism, and class transformations. *International Feminist Journal of Politics* 3 (3): 365–86.

Glasmeier, A. (2005) *An Atlas of Poverty in America: One Nation, Pulling Apart.* New York: Routledge.

Gleeson, B.J. and Kearns, R. (2001) Re-moralising landscapes of care. *Environment and Planning D: Society and Space* 19 (1): 61–80.

Gregson, N. and Lowe, M. (1995) "Home"-making: on the spatiality of daily social reproduction in contemporary middle-class Britain. *Transactions of the Institute of British Geographers* 20 (2): 224–35.

Hackworth, J. (2009) Neoliberalism, partiality and the politics of faith-based welfare in the United States. *Studies in Political Economy* 84: 155–79.

Haraway, D. (2007) *When Species Meet.* Minneapolis: University of Minnesota Press.

Heelas, P. and Woodhead, L. (2004) *The Spiritual Revolution: Why Religion Is Giving Way to Spirituality.* Oxford: Blackwell.

Held, V. (2006) *The Ethics of Care: Personal, Political, Global.* Oxford: Oxford University Press.

Hirschmann, N.J. (2008) Mill, political economy and women's work. *American Political Science Review* 102 (2): 199–213.

Huang, S. and Yeoh, B.S.A. (2007) Emotional labour and transnational domestic work: the moving geographies of "maid abuse" in Singapore. *Mobilities* 2 (2): 195–217.

Jarvis, H. (2007) Home truths about care-less competitiveness. *International Journal of Urban and Regional Research* 31 (1): 207–14.

Jarvis, H., Pratt, A., and Cheng-Chong, W. (2001) *The Secret Life of Cities: The Social Reproduction of Everyday Life.* Harlow: Prentice Hall.

Jarvis, H., Cloke, J., and Kantor, J. (eds.) (2009) *Cities and Gender.* London and New York: Routledge.

Johnsen, S., Cloke, P., and May, J. (2005) Transitory spaces of care: serving homeless people on the street. *Health and Place* 11: 323–36.

Joseph, A. and Phillips, D. (1984) *Accessibility and Utilization: Geographical Perspectives on Health Care Delivery.* New York: Harper and Row.

Katz, C. (2001) Vagabond capitalism and the necessity of social reproduction. *Antipode* 33 (4): 708–27.

Kearns, R. and Joseph, A. (2000) Contracting opportunities: interpreting post-asylum geographies of mental healthcare in Auckland, New Zealand. *Health and Place* 6: 159–70.

Kearns, R. and Taylor, S.M. (1989) Daily life experience of people with chronic mental disabilities in Hamilton, Ontario. *Canada's Mental Health* 37: 1–4.

King, U. (2008) *The Search for Spirituality: Our Global Quest for a Spiritual Life.* Canterbury: Canterbury Press.

Knowles, C. (2000a) Burger King, Dunkin Donuts and community mental health care, *Health and Place* 6: 213–24.

Knowles, C. (2000b) *Bedlam on the Streets*. London: Routledge.

Koehn, D. (1998) *Rethinking Feminist Ethics: Care, Trust and Empathy*. London: Routledge.

Kofman, E. and Raghuram, P. (eds.) (2004) Global labour migration: an introduction. *Feminist Review* 77 (1): 4–6.

Law, R.M. and Gleeson, B.J. (1998). Another "landscape of despair"? Charting the "service-dependent ghetto" in Dunedin. *New Zealand Geographer* 54 (1): 27–36.

Laws, G. and Dear, M. (1998) Coping in the community: a review of factors influencing the lives of deinstitutionalised expatients. In C.J. Smith and J.A. Giggs (eds.), *Location and Stigma: Contemporary Perspectives on Mental Health and Mental Healthcare*. London: Unwin Hyman, pp. 83–102.

Lawson, V. (2007) Geographies of care and responsibility. *Annals of the Association of American Geographers* 97 (1): 1–1.

Lawson, V. (2009) Instead of radical geography, how about caring geography? *Antipode* 41 (1): 210–13.

Le Heron, R. and Pawson, E. (1996) *Changing Places in New Zealand*. London: Longman.

Lea, J. (2008) Retreating to nature: rethinking therapeutic landscapes. *Area* 40 (1): 90–8.

Lewontin, R.C., Rose, S., and Kamin, L. (1984) *Biology, Ideology and Human Nature: Not in Our Genes*. Harmondsworth: Penguin

Ley, D. (2008) The immigrant church as an urban service hub. *Urban Studies* 45 (10): 2057–74.

Macy, J. and Brown, M.Y. (1998) *Coming Back to Life: Practices to Reconnect our Lives, our World*. Gabriola Island: New Society Publishers.

Mattingley, D. (2001) The home and the world: domestic service and international networks of caring labor. *Annals of the Association of American Geographers* 91 (2): 370–86.

McDowell, L. (2004) Work, workfare, work/life balance and an ethics of care. *Progress in Human Geography* 28 (2): 145–63.

McDowell, L., Ray, K., Perrons, D., Fagan, C., and Ward, K. (2005) Women's paid work and moral economies of care. *Social and Cultural Geography* 6 (2): 219–35.

McGregor, J. (2007) "Joining the BBC (British Bottom Cleaners)": Zimbabwean migrants and the UK care industry. *Journal of Ethnic and Migration Studies* 33 (5): 801–24.

McKay, D. (2007) Sending dollars shows feeling: emotions and economies in Filipino migration. *Mobilities* 2 (2): 175–94.

McKie, L., Gregory, S., and Bowlby, S. (2002) Shadow times: the temporal and spatial frameworks and experiences of caring and working. *Sociology* 36 (4): 897–924.

Milligan, C. (2000) Bearing the burden: towards a restructured geography of caring. *Area* 32 (1): 49–58.

Milligan, C. (2001) *Geographies of Care: Space, Place and the Voluntary Sector*. Aldershot: Ashgate.

Moon, G., Kearns, R., and Joseph, A. (2006) Selling the private asylum: therapeutic landscapes and the (re)valorization of confinement in the era of community care. *Transactions of the Institute of British Geographers* 31 (2): 131–49.

Mowl, G., Pain, R., and Talbot, C. (2000). The ageing body and the homespace. *Area* 32 (2): 189–97.

Noddings, N. (1986) *Caring: A Feminine Approach to Ethics and Moral Education*. Berkeley: University of California Press.

Panelli, R. (2004) *Social Geographies: From Difference to Action*. London: Sage.

Parr, H. (2000) Interpreting the hidden social geographies of mental health: ethnographies of inclusion and exclusion in semi-institutional places. *Health and Place* 6 (3): 225–37.

Parr, H. (2002) New body-geographies: the embodied spaces of health and medical information on the Internet. *Environment and Planning D: Society and Space* 20 (1): 73–95.

Parr, H. and Philo, C. (2003) Rural mental health and social geographies of caring. *Social and Cultural Geography* 4 (4): 471–88.

Partridge, C. (2004) *The Re-enchantment of the West*, vol. 1: *Alternative Spiritualities, Sacralization, Popular Culture and Occulture*. London: Continuum.

Paxman, M. (2002) Charity, deservingness and care. Unpublished BA dissertation, School of Geography, University of Southampton.

Philo, C. (1997) Across the water: reviewing geographical studies of asylums and other mental health facilities. *Health and Place* 3 (2): 73–89.

Pinker, S. (2002) *The Blank Slate: The Modern Denial of Human Nature*. New York: Viking.

Player, S. and Pollock, A. (2001) Longterm care: from public responsibility to private good. *Critical Social Policy* 21 (2): 231–55.

Popke, J. (2007) Geography and ethics: spaces of cosmopolitan responsibility. *Progress in Human Geography* 31 (4): 509–18.

Pratt, G. (1997a) From registered nurse to registered nanny: discursive geographies of Filipina domestic workers in Vancouver, BC. *Economic Geography* 75 (3): 215–36.

Pratt, G. (1997b) Stereotypes and ambivalence: the construction of domestic workers in Vancouver, British Columbia. *Gender, Place and Culture* 4 (2): 159–77.

Proctor, J. and Smith, D. (1999) *Geography and Ethics: Journeys in a Moral Terrain*. London: Routledge.

Pyle, J. (2006) Globalization and the increase in transnational care work: the flip side. *Globalizations* 3 (1): 297–315.

Raghuram, P. (2008) Thinking the UK's medical labour market transnationally. In J. Connell (ed.), *A Global Health System: The International Migration of Health Workers*. London: Routledge.

Ragurham, P. and Kofman, E. (2002) The state, skilled labour markets and immigration: the case of doctors in England. *Environment and Planning A* 34 (11): 2071–81.

Raghuram, P., Madge, C., and Noxolo, P. (2009) Rethinking responsibility and care for a postcolonial world. *Geoforum* 40 (1): 5–13.

Robinson, F. (1999) *Globalising Care: Ethics, Feminist Theory and International Affairs*. Boulder: Westview Press.

Sager, R. and Stephens, L.S. (2005) Serving up sermons: clients' reactions to religious elements at congregation-run feeding establishments. *Nonprofit and Voluntary Sector Quarterly* 34 (3): 297–315.

Sandercock, L. and Attili, G. (2009) *Where Strangers Become Neighbours: The Integration of Immigrants in Vancouver, Canada*. Dordrecht: Springer

Silk, J. (1998) Caring at a distance. *Ethics, Place and Environment* 1: 165–82.

Silk, J. (2000) Caring at a distance: (im)partiality, moral motivation and the ethics of representation. *Ethics, Place and Environment* 3 (3): 303–22.

Silk, J. (2004) Caring at a distance: gift theory, aid chains and social movements. *Social and Cultural Geography* 5 (2): 229–51.

Slote, M. (2007) *The Ethics of Care and Empathy*. London: Routledge.

Smith, D.M. (1998) How far should we care? On the spatial scope of beneficence. *Progress in Human Geography* 22 (1): 15–38.

Smith, S.J., Pain, R., Marston, S.A. and Jones III, J.P. (eds.) (2010) *The Sage Handbook of Social Geographies*. London: Sage.

Soldo, B.J., Agree. E.M. and Wolf, D.A. (1989). The balance between formal and informal care. In M.G. Ory and K. Bond (eds.), *Ageing and Healthcare: Social Science and Policy Perspectives*. London: Routledge, pp. 193–216.

Twigg, J. (2000) *Bathing: The Body and Community Care*. London: Routledge.

Ungerson, C. (ed.) (1990) *Gender and Caring: Work and Welfare in Britain and Scandinavia*. Hemel Hempstead: Harvester Wheatsheaf.

Valentine, G. (2008) Living with difference: reflections on geographies of encounter. *Progress in Human Geography* 32 (3): 323–37.

Welch, R.V. and Panelli, R. (2007) Questioning community as a collective antidote to fear: Jean-Luc Nancy's "singularity" and "being singular plural." *Area* 39 (3): 349–56.

Whatmore, S. (2006) Materialist returns: practising cultural geography in and for a more-than-human world. *Cultural Geographies* 13 (4): 600–9.

Wiles, J. (2003a) Informal caregivers' relationships to formal support services. *Health and Social Care in the Community* 11 (3): 189–207.

Wiles, J. (2003b). Daily geographies of caregivers: mobility, routine, scale. *Social Science and Medicine* 57: 1307–25.

Wiles, J., Allen, R., Palmer, A., Hayman, K., Keeling, S., and Kerse, N. (2009) Older people and their social spaces: a study of well-being and attachment to place in Aotearoa New Zealand. *Social Science and Medicine* 68 (4): 664–71.

Wills, J. and Jamoul, L. (2008) Faith in politics. *Urban Studies* 45 (10): 2035–56.

Wills, J., Datta, K., Evans, Y., Herbert, J., May, J., and McIlwaine, C. (2007) Keeping London working: global cities, the British state, and London's new migrant division of labour. *Transactions of the Institute of British Geographers* 32: 151–67.

Wills, J., Datta, K., Evans, Y., Herbert, J., May, J., and McIlwaine, C. (2009) *Global Cities at Work: New Migrant Divisions of Labour*. London, Pluto.

Wilson, E.O. (1975) *Sociobiology: The New Synthesis*. Cambridge MA: Harvard University Press.

Wilson, E.O. (2006) *Nature Revealed: Selected Writings, 1949–2006*. Baltimore: Johns Hopkins University Press.

Wilson, K. and Rosenberg, M.W. (2002). The geographies of crisis: exploring accessibility to health care services in Canada. *Canadian Geographer* 46 (3): 223–34.

Wilton, R. and DeVerteuil, G. (2006) Spaces of sobriety/sites of power: examining social model alcohol recovery programs as therapeutic landscapes. *Social Science and Medicine* 63 (3): 649–61.

Chapter 27

The Challenges of and from Indigenous Geographies

Brad Coombes, Nicole Gombay,
Jay T. Johnson, and Wendy S. Shaw

Introduction

The authors of this chapter debated whether we should open with an acknowledgment of Indigenous peoples, their histories, landed connections, and struggles, but some were concerned about the global, non-specific nature of such an address. That there should be reason for debate about a common – if not, institutionalized – gesture mirrors the difficulty in generalizing about Indigenous peoples amidst their conspicuous diversity. Indeed, that difficulty will be a central theme in this chapter wherein we confront the tendency to essentialize Indigenous identities and cultures. Given its capacity to fix identities temporally and spatially, essentialsim can be disempowering, particularly for those urban Indigenous peoples who may have lost contact with country. It locates Indigeneity somewhere in prehistory, bound by tradition and therefore unable to operate fully in the present. Likewise, a monolithic application of "Indigeneity" as a universal label to represent disparate peoples silences their inherent diversity. Even by combining the voices of Indigenous and non-Indigenous authors from distinct "settler societies" – the United States, Canada, Australia, and New Zealand – we cannot present an overarching account of Indigeneity, nor its associated geographies.[1] We also recognize that any attempt to promote a sub-discipline in Indigenous geographies, or even to consider the geographies of Indigenous peoples as if they are conceptually discrete, is contestable. Nonetheless, we engage tentatively with those notions because they encourage consideration, not only of how social geography is challenged by Indigenous geographies, but also of how they may add to, and enrich, our discipline.

Recognition of how diversity seemingly undermines "Indigenous Geography" may explain geography's ambivalent engagement with Indigeneity. According to

A Companion to Social Geography, First Edition. Edited by
Vincent J. Del Casino Jr., Mary E. Thomas, Paul Cloke, and Ruth Panelli.
© 2011 Blackwell Publishing Ltd. Published 2011 by Blackwell Publishing Ltd.

some commentators, it was not until the late 1990s that Geography began to engage critically with Indigenous issues (Castree 2004). More accurately, it can be suggested that geographers *have* engaged critically with Indigenous geographies (refer to the influential publications by Brody 1988 and Gale 1972), but that they have generally done so within area or topic specialisms and associated journals – notably, outside the disciplinary spotlight. The hitherto peripheral status of Indigeneity within geography has meant that research on the topic has been marginalized in mainstream disciplinary debate.

Although the validation of Indigenous G/geographies within the discipline is far from complete, since the cultural turn of the 1990s three strands of geographic scholarship have opened the possibility for new engagements with Indigenous peoples. First, increasing attention to geographies of racialization, rather than preoccupations with "race," stimulated interest in the racialization of Indigenous peoples (Anderson 1993; Peters 1998). The critical study of Whiteness prospered in geography (Bonnett 2000; Bonnett 2008; Shaw 2000; Shaw 2006), broadening in its focus to include Indigenous experiences of White empowerment and how historical geographies of imperialism unjustly rendered otherness knowable for academic consumption. Second, postcolonial scholarship scrutinized further the colonizing work of geographic knowledge and highlighted a need to consider the unique counter-resistances of Indigenous peoples to neo/colonial processes (Jacobs 1996). Accordingly, the focus of research shifted from a politics of survival to a more nuanced evaluation of how the Indigenous other complicates and transforms neo/colonial practice. Third, critical development theory highlighted the resilience and ongoing relevance of Indigenous governance and social systems (Corbridge 1991), something which also became a focus of geographic contributions to early political ecology (Peet and Watts 1996).

While these trajectories progressively implicated Indigeneity in geographic scholarship, they also relegated Indigenous peoples to habitual *case studies* of neo/colonial excess or of heroic resistance thereto. Despite the pervasive exemplar status of Indigenous peoples within their discipline, geographers have been reticent to theorize Indigeneity or, of particular relevance for this chapter, Indigenous identities. A new phase of self-critique has challenged geographers to rethink their discipline, its origins and ongoing complicity with neocolonial structures (Shaw et al. 2006). Building on this critique, emergent social geographies have demanded re-engagement with Indigenous concerns through sensitive, inclusive and emancipatory research projects (refer to Panelli 2008 for an overview). Despite their steadfast attention to associated concerns about representing the other, however, geographers have not agonized over the conceptual status of Indigeneity to the same degree as, for example, social anthropology (refer to the reaction to Kuper 2003). We suspect that disciplinary empathy for Indigenous activism, along with recognition that insensitive theorization may disrupt that activism, has dissuaded geographers from analyzing Indigeneity and Indigenous identities with the full range of available conceptual tools. As a result, this and earlier forms of disciplinary ambivalence have inhibited a deeper engagement with Indigenous affairs. Nonetheless, recent advances in the way social geographers have conceptualized identity could address this ambivalence and enliven Indigenous G/geographies far beyond a suite of paradigmatic examples.

Rejecting assumptions that engagement between settler societies and Indigenous peoples results in unidirectional transformation for the latter, many social geographers now regard identity formation as relational, performative and co-constitutive. However, this challenges some conceptions of Indigeneity, especially because the political projects of Indigenous peoples often rely on essentialized differences and their projection to a global audience which yearns for "authentic" cultures. Researchers confront the difficulty of distinguishing these strategic essentialisms and expressions of deeper cultural concern, whilst reconciling them with contemporary theories about multi-layered, situated and incomplete identities. Moreover, increasing acceptance that cultural engagement creates hybrid outcomes might threaten the political gains which Indigenous peoples have made through representing themselves as authentic, unique, or original. Hence, theoretical and methodological angst characterizes geographical research into Indigeneity at the very time the discipline has become more open to resolving its postcolonial status.

Addressing this anxiety, the chapter has two specific purposes. One is to confront the essentialisms that haunt and sometimes deny notions of Indigeneity and to consider the utility of more "relational" approaches to the question of Indigenous identities. We pursue this, in part, to collapse a reductive dilemma which renders the existence of Indigeneity as *necessarily* essentialist. We draw on critiques of the search for fundamental characteristics within Indigeneity to contextualize ontological and epistemological uncertainties in geographical research with Indigenous peoples. We maintain that geographers need to understand better the challenges of Indigeneity to settler societies and norms if they are to transcend the definitional malaise and methodological inertia in Indigenous studies. Persistent Indigenous claims about the injustice of colonial systems reveal the incompleteness of colonial attempts to substitute endogenous with exogenous cultures. Significantly, Indigenous concerns about the ethics and relevance of social research parallel those claims, revealing gaps in geographical approaches to place, community, and culture.

The research process also, therefore, reflects dilemmas about Indigenous identities. To the extent that research may transform both the objects and subjects of research, we argue that the research process often mimics the transcultural outcomes of engagement between Indigenous and non-Indigenous cultures. Constituting a second purpose for this chapter, we account for some of the methodological shifts which may follow adoption of a relational understanding of identity. Open-ended methods, which include scope for informal storytelling and personal learning, are required if our research is to reflect rather than assume away the inevitably transformative, plural and unpredictable outcomes of cross-cultural engagement. We embrace storytelling in recognition of its important role in the geographic imaginaries of many Indigenous peoples; its part in situating identities in country and place. By acknowledging the importance of storytelling, we wish to demonstrate both how the discipline is challenged by engagement with Indigenous geographies and may conversely be enriched by such engagement. Storytelling represents the dyadic challenge of and from Indigenous G/geographies and, accordingly, we have included short autobiographic accounts in this chapter to illustrate some of the potential hazards and rewards in pursuing an openly transcultural approach to research.

Indigeneity as a Relational Construct

Defining the Indigenous has become problematic within the academy, not least because Indigenous and non-Indigenous actors strategically repackage the "essential" characteristics of Indigeneity to suit fluid political contexts. The application of these discourses of alterity simultaneously reproduces and unsettles "persistent ideas concerning what may be taken to characterize Indigeneity" (Merlan 2009: 303). Niezen (2003) observes that the lack of a rigorous definition for Indigeneity precludes analysis and demands further scrutiny of its social meaning. We agree that the debate must continue but, whilst academics deliberate, the communities which are the focus of their debate must sustain their lives (e.g., Gombay 2005) and engage with political institutions to secure rights and recognition (e.g., Coombes 2007; Porter 2006). Equally, international organizations tread boldly towards defining Indigeneity, closing definitional borders around specific nations and communities. Indigeneity emerges as a performance within the identity expectations of global NGOs and other non-local actors who want for primordialized "authenticity" (Sylvain 2005). Working within this context, Indigenous peoples have exhibited considerable skill at creating autoethnographic representations which engage modern discursive constructs. However, the strategic performance of an essentialized identity for political leverage perplexes academics and non-academics alike. Risking the appearance of contradiction, Indigenous communities often claim that they have been altered irrevocably in the contact zone of colonial relations *and* that they practice "authentic" or "traditional" cultures.

Given the paradoxes in strategic essentialism, it is unsurprising that Indigenous peoples are not always able to control the outcomes of their activism. When the Makah tribe killed a whale off the Washington coast in the name of "cultural survival," mainstream media and environmental groups vilified their actions as inauthentic (Marker 2006). This induced "anti-Indian" sentiment and a denial of rights throughout the state which extended beyond the Makah, suggesting that strategic essentialism is a fraught endeavor with unpredictable rewards. Where cultural difference is deployed successfully to attain public sympathy for Indigenous causes, the liberal democratic state typically manages associated dispute within rights discourses which offer state recognition, apologetic dogma and affirmation of certain cultural practices or access to resources. Lobbying to protect these "authentic" practices may embarrass the state to uphold "traditional" rights, but it seldom leads to self-determination or the realization of Indigenous political autonomy (Corntassel 2008; Porter 2006). Irrespective of whether the rights which emerge from strategic essentialism are sufficiently robust to offer meaningful progress, a reversal in the current outlook of the liberal state now seems to threaten the rights discourse in its entirety (Kowal 2008). Self-identification as "authentic" can sometimes miscarry and its benefits are indefinite.

It is, nonetheless, academically reckless to regard the gains of Indigenous activism with casual indifference. Even though strategic essentialism may fix Indigeneity within an unrecoverable past of fetishized tradition, propinquity with nature and immutable connections to land, it may have been the only available and effective means for Indigenous political mobilization (Blackburn 2007; Bryan 2009).[2] These gains emerged from a conjuncture amongst the Occidental search for non-Occidental solutions to modern predicaments like the environmental crisis, the globalizing work of NGOs which has placed Indigenous activism on an international stage, and unprecedented recognition of the impacts of colonial history. This conjuncture has enabled Indigenous peoples "to use socially recognized difference as a strategy to better their individual lives and communities," even during an era of increasing hostility to Indigenous claims settlement and other forms of historical redress (Valdivia 2005: 290).

Recent years have witnessed increasing conflict between Indigenous peoples and settler institutions. Farmer lobbies, environmental NGOs and recreational associations have contested the return of land, newfound Indigenous rights and what they infer for the White Anglo "good life" (Lane and Hibbard 2005). Whereas Indigenous estates were once considered "marginal," global demand for heritage and recreational resources in rural areas now position them as an irksome hindrance to the expansion of national pastimes and developmental aspirations. We recognize the significance of those material challenges, but focus here on the symbolic contests which accompany them. Indigenous protest about land dispossession and claims to self-determination strain the liberal democratic and rights discourses of settler societies, compelling reflexive evaluation of their unjust premises and assimilative ideals (Johnson 2008). Colonialism envisaged a "mode of erasure" which would displace the longer histories of Indigenous settlement to make space for new cultural and economic assemblages. But conquest has been incomplete, leaving Indigenous and settler claims to the same ground unsettled and requiring defense of their respective legitimacy to occupy it (Garbutt 2006). Indigenous peoples frame their activism as counter-hegemonic narratives of replacement, and their conspicuous demonstration of how the colonial past incriminates the neocolonial present has particular significance for the legitimacy of the state. It exposes the false presumptions of teleological progress from the birth of the nation-state to an imminent condition of multicultural harmony (Blackburn 2007). Dirlik (2003: 25) maintains that Indigenous conceptions of community as statehood – nations within nations – sever "the metonymic relationship the nation-state presupposes between land and national territory." The incompleteness of colonial erasure means that Indigenous and non-Indigenous identities are reliant on the transcultural present, in which settler cultures and Indigeneity co-constitute each other.

Acknowledging the realities of the transcultural present, we maintain that conceptualizing Indigeneity as articulated, relational identities should ultimately strengthen its political momentum. Yet, it may also jeopardize Indigenous political gains in the short term. Those who advocate for Indigenous rights are justifiably cautious about the potential of vain re/theorization to undermine political projects which have only recently achieved traction within the adversarial terrains of civil society and national policy formation. Likewise, recognition that collective Indigenous action has been a significant force in resisting neoliberalism should also discourage casual abandonment of Indigeneity as an absolute category (Wilson 2008). Despite the veracity of these concerns, a *relational* perspective on how Indigenous and settler identities affect each other complements and extends the existing scholarship on postcolonial geographies. First, it draws attention to *how* Indigenous identities are constructed, and that may be a crucial step for conceptualizing other analytical quandaries. Second, focusing on the implications of Indigeneity for settler identities may open conceptual space for another type of transformative politics. If Indigenous and settler cultures are – even only to some degree – co-produced, then that infers new strategies for making the colonizing other take notice. Third, a relational perspective may provide academic legitimacy for the strategies Indigenous peoples adopt to finesse the multi-layered processes of identity formation that they confront in their daily and political lives. From a relational perspective on identity, there will likely be greater empathy for Indigenous peoples' self-representations as both "altered" and "authentic."

Relational thinking offers additional means for conceptualizing those constructs which habitually trouble social geographers and are central to Indigenous geographies. Place, for example, is now routinely treated as the product of interaction: an unfinished consequence of the connections between places and across scales – a "constellation of processes rather than a thing" (Massey 2005: 140).[3] The analysis of identity has moved in a similar direction, but not without reflection on its consequences. Just as a relational conception of place cannot diminish the deeply felt cultural relationships of peoples to places, a relational approach to Indigeneity may honor rather than devalue the customary-modern hybrids which Indigenous peoples may, in fact, *choose* to live, as well as revealing insights to the powerful about their own identities (Pickerill 2009). Francesca Merlan accepts a continuing relevance for "criterial" definitions of Indigeneity, but she claims that a relational perspective better captures the lived experience of Indigenous peoples. She associates this perspective with "definitions that emphasize grounding in relations between the 'indigenous' and their 'others' rather than in properties inherent only to those we call 'indigenous'" (Merlan 2009: 305). The dynamism of transcultural identity formation is contrasted with the intransigent ghosts of noble savagery which afflict Indigeneity in its essentialized forms:

> there is a certain hypostatization of Indigeneity, as if it were a free-standing characteristic of a certain "kind" of people. But … despite its powerful historical and emotional content, Indigeneity does not have meaning on the basis of something that is "simply there" or objectively ascertainable about those we call Indigenous people but, like many other social categories, is a contingent, interactive, and historical product. (Merlan 2009: 319)

Emphasizing how group identities are co-produced through the situated intersection of fragmented social and political processes, relational perspectives on cultural formation have steadily replaced those which assume absolute or fixed identities (Dressler and Turner 2008). Although those who work with Indigenous peoples are cautious about that transition, Bryan (2009: 25) is not alone when he maintains that "indigeneity thus describes a relationship rather than an objective fact."

The emergence of subtle inflections in meaning which redefine autochthony and Indigeneity can be used to distinguish the latter in relational terms. Both words descend linguistically from Classical Greece, although autochthony has inherited nuances from its subsequent application in Francophone territories and their administration (Ceuppens and Geschiere 2005). Historically, the two terms emphasized similar traits: originariness, legitimated authority and fidelity of culture with place. Today, however, autochthony is more closely aligned with the coincidence of people and place, whereas Indigeneity extends to Indigenous peoples' engagement with cultural others and social processes beyond their normatively assumed rangelands (Pelican 2009). Likewise, Indigeneity transcends territorial connection to include moral claims against neo/colonial administration. It constitutes both "a counter-weight to the hegemonic strategies of states" (Niezen 2003: 198) and the solidarity which flows from experience of colonial oppression (p. 10). Endogenous-exogenous, past-present, enchanting-rationalizing and global-local collide in uncertain ways to co-produce Indigenous and globalized discourses of identity. Juxtaposed against the more restricted definition of autochthony, therefore, the meaning of Indigeneity is always situated at the interplay of multi-scalar processes and multi-dimensional performances.

Yet, it seems that relational conceptions of identity continue to be explored as if they are consequential *only* for Indigenous peoples and others who confront hege-monic powers. Morin's critique of those studies which, in her view, misrepresent Bhabha's (1994) *hybridity* and Pratt's (2008) *transculturation* remains germane to this analysis. Early postcolonial geographies disregarded "how European's views of themselves were actually produced in the colonies" and, likewise, Indigenous peoples were "seldom credited with influencing" the colonizers' gaze (Morin 1998: 315). Too often, the relationship between local claims to Indigeneity and globalization has been framed solely as an attempt to impose order on the disruptions of the latter (refer, for instance, to Dunn 2009; Kuper 2003). This unidirectional, "distress-driven" understanding of identity formation fails to acknowledge the agency of Indigenous peoples to appropriate globalized discourses and it also denies their capacity to influence non-Indigenous identities (Sylvain 2005). A genuinely transcul-tural approach to identity recognizes the bidirectional outcomes of cultural engage-ment and, therefore, anticipates that the neo/colonial encounter will produce shifts in *settler* identities.[4] Researchers who work with Indigenous communities are often reticent about adopting this logic, but it has some acceptance beyond the academy. The disproportionate attention of settler states towards the UN Declaration on the Rights of Indigenous Peoples framed it as a pervasive threat to White Anglo culture (Pelican 2009). Lay publics in Canada, the USA, Australia, and New Zealand who pleaded successfully with their governments to hold out the longest against signing the declaration seem to understand that a shift in the relative positioning of Indigeneity portends transfiguration of their own culture and status.

Storytelling as Transcultural Research

Earlier, we summarized the mounting self-critique which geographers apply to their relationships with Indigenous peoples. Deconstruction of the discipline should also extend to more positive reflection on the grounds for commonality between geo-graphical and Indigenous practices. Some have argued, for example, that a shared commitment to place is the basis for such rapprochement (Johnson and Murton 2007; Pickerill 2009). Hence, at times we connect relational perspectives on place and relational approaches to Indigeneity in this chapter, and we commend further work which explores their intersection. For the rest of the chapter, however, we focus on the methodological challenges of working with Indigenous peoples and relate them to our earlier comments about relationality.

We have argued that Indigenous cultures and identities are fluid and dynamic, sometimes unnervingly so for the academy. The mutability in Indigenous cultural practice is most evident in storytelling about place, creation narratives and engagement with country. Mazzullo and Ingold (2008) maintain that Indigenous narratives are so highly contextualized and purposeful that they cannot be understood in a linear, rational nor literal sense. Rather, place and historical narratives are adjusted for particular circumstances; Indigenous stories are living and elastic geographies – a pool of cultural memories which can be applied flexibly to different circumstances. If social geography is to engage fully with Indigeneity, it therefore requires flexible methods which adapt to shifting content and interpretation. This explains why those researchers who work with Indigenous communities have experimented beyond the stock qualitative approaches which have achieved normative, if not fatigued, status within social geography. Geography's engagement with feminist scholarship in the 1980s inspired broad epistemological shifts which are orthodox today (Pain 2004). Similarly, Indigenous critique of geographical practice prompts experimentation, innovations and partnerships which may benefit the wider discipline (e.g., Hodge and Lester 2006; Watson and Huntington 2008), and the specific needs of transcultural research may inspire constructive review of our methodological canons.

Nonetheless, both collaborative and qualitative research have thus far struggled to accommodate the non-material interests and cultural dynamism of Indigenous peoples. Cahill (2007) identifies a doing/doer predilection within participatory research which stresses direct confrontation of social structures. She prefers instead to soften the zeal of participatory research through inclusion of post-structural perspectives to pursue "cultivation of new forms of subjectivity" as a basis for conscientization and, thereby, collective action (Cahill 2007: 269). This distinction may be particularly important in the field of Indigenous G/geographies because for many Indigenous cultures *listening* is the path to becoming an accepted participant and not necessarily *doing*. Researchers who collaborate with Indigenous communities are often required to listen to stories of place, people and history as a rite of passage in their induction into local affairs. Hence, the telling, sharing and, where appropriate, co-production of stories are prerequisites for sympathetically and dynamically relating Indigenous past and transcultural present. As we argue below, this emphasis on storytelling should move beyond the more emotive representation of the other to include acceptance of the need for Indigenous peoples to tell their own stories: "If to tell one's story is to know one's story, it is also to take control over one's representation" (Cahill 2007: 283), suggesting that collaborative storytelling could become a central component of Indigenous G/geographies. Watson and Huntington (2008) demonstrate the capacity of narrative forms to navigate the inbetweenness of research subjects and objects. They maintain that collaborative research has too often assumed distinct cultural knowledges of Indigenous and non-Indigenous participants which must be bridged when, in keeping with our earlier arguments, all cultures are hybrids. Rather than seeking "discovery" of cultural differences through research, Watson and Huntington (2008: 276) employ the sharing of narratives as an appropriate means for bringing "our own hybridities into conversation."

These insights into the capacity of narrative to fulfill the "*intersubjective* form of qualitative research" (Watson and Till 2009: 121, our emphasis) and to negotiate the fluidity in transcultural relations will become increasingly relevant as geographers seek to effect change through their research. Davies and Dwyer (2008) maintain that an unanticipated byproduct of debates about "public geographies" and the alleged requirement to make geographic knowledge more "mobile" within policy domains has been greater emphasis on the capacity of certain types of research to "move" people and thereby effect change. Social science expertise is no longer measured in terms of acquired knowledge or accurate representation but rather through the capacity to *be affected* and to *affect* others. Consequently, the emotive content of narrative forms is commended as a means to explore this affective domain and to produce meaningful scholarship (Richardson and St. Pierre 2005). Reflecting our earlier discussion of the need to accept permeability in identities, Ingold (2008) notes a shift in the social sciences from a preoccupation with "being in place" to the investigation of "becoming along a path" between places and amongst myriad identities. His project to understand our path- rather than place-based life-worlds harmonizes with Indigenous conceptions of becoming, and has prompted him to experiment with methods which incorporate the rich and highly contextualized qualities of narration (Ingold 2007a; Ingold 2007b; Mazzullo and Ingold 2008). Likewise, the more social geography moves away from a preoccupation with "facts and truths" and towards the need to articulate "feeling and knowing," the more it may need to embrace the transcultural qualities of storytelling, particularly by and with Indigenous peoples (Watson and Huntington 2008). In the remaining space, we outline some of the challenges which have confronted the authors in their research with Indigenous communities and, through telling some of our own stories, we relate how that experience has led to re-evaluation of our research practices.

Telling stories, rethinking place?

As a discipline geography is, by definition, founded on the presumption that we are engaged in writing the earth: *geo*, earth; *graphien*, to write. Thus, geographers strive to document the world around them, but such documenting comes at a cost. Its very process embodies, codifies, and solidifies particular ways of making sense of the world. It requires that one steps back, observes, imposes intellectual order, and dispenses with all that lies outside that order. If we are engaged, then, in writing the world into being, what happens when we encounter those who speak it into being, people for whom an understanding of the land is not worked out through the abstraction of pen on paper, but is discerned orally, carried in memory, and experienced sensorially through being *in* the world? Inherent in such encounters are profoundly different ways of understanding and existing in the world. So the exhabitant and the inhabitant confront one another (Ingold 2007a).

For those Indigenous peoples whose world is verbally constituted, the encounter with the fixity of the page can be uneasy (King 2003). In the encounter with colonizers, what gets written down has, in many instances, been used to define and control the world according to a set of assumptions that can be at odds with those of Indigenous peoples. The written word often has political and economic authority which has been used to delimit the powers of inhabitants of the spoken world. As many Indigenous peoples have discovered, they are required to write their world into being in order to establish any semblance of legitimacy with those who rule by the written, rather than the spoken, word. Yet, in the move to the page, some values may be lost.

Imagine a young, non-Indigenous researcher with a freshly minted graduate degree. Keen to put her learning into action, she takes a job working on a land claim by Inuit in northern Canada. Armed with large rolls of maps, colored pens, acetate overlays, and plenty of zeal, she heads to an Inuit settlement – one she has never visited before – to record what is commonly called people's "land use and occupancy." Over several weeks, she and a research assistant from the community spend days on end around a table in the municipal office poring over these maps while asking people, again and again, the various questions that she has been tasked to address. With each man and woman she meets she asks such questions as, "Where were you born? Where did your family live in the winter/spring/summer/autumn? Where did you hunt caribou/whales/geese/wolves, etc. and what season was it in?" With each answer they mark the acetate map overlays, which will later be trans-ferred into digital format and used as proof of Inuit land use and occupancy of the area under claim. Working steadily through the list of people and activities to be recorded, the acetate sheets pile up. She feels a sense of relief that these endlessly repetitive days will soon be over, and, despite a nagging sense that this exercise is not totally fulfilling, at least she can congratulate herself that she has collected the requisite data. These maps will be of use for meetings with government bureaucrats in the South. They will be able to look at the maps and see that yes, Inuit do, indeed, occupy this land.

One evening, towards the close of her time there, at the end of a day during which she had mapped the life history of a man in his 50s, her assistant asked whether she had noticed that the man had, in response to the rote question of where he had hunted wolves in his lifetime, answered that he did not hunt them. The assistant told her that this was because the man's family had made a vow never to hunt wolves. Many years before, when the man's family was still living on the land, before their move to the settlement, they had passed through a period of starvation. Faced with death, they had survived only because they came upon the remnants of a kill left behind by wolves. In thanks for this gift, they had sworn never again to hunt wolves.

All that was reflected of that man's experience was a blank on the map; no, he did not hunt wolves, so no carefully coded mark could be recorded. And the researcher, busy with her colored pens and lists of names and questions, knew nothing beyond the flat surface of the paper and acetate before her, the hum of fluo-rescent lighting overhead, and the four white walls of the office in which they were meeting. She had been busy chronicling his life on the maps without actually telling his story. The reason whether or not he hunted wolves was neither here nor there; all that mattered was that there was no line on the map to record his *occupation* of that land. Questions of *habitation* had no place in such a schema. In one fell swoop, the researcher confronted the gulf between two quite different processes of place-making. As a fly-in-fly-out researcher whose task was only to produce the requisite maps for a land claim, knowledge of the deeper relations of people and place were unnecessary. Instead, what she had been busy creating were the clean, abstract lines of an occupier for whom land and resources may be neatly bounded, and, once accomplished, could then potentially be subsumed by the logic of territorial – and ultimately, economic – control (Brody 2000; Larsen 2006; Thom 2009). Moral obligations to the more-than-human world held no sway, and so, what lay beneath the surface of those lines was effectively immaterial. The hidden power of unspoken agendas guided what could and could not be valued, and by extension, what could and could not be said.

There are stories we tell, often without realizing it, and stories we are told. The challenge is to be prepared to listen to them, and to understand how and why certain narratives are silenced. In many instances, stories represent a vital component of Indigenous geographic imaginaries (Bird 2007; Brody 1988; Collignon 2006; Wirf et al. 2008). For outsiders, in their encounter with Indigenous peoples, the presence of those stories may be apparent, but their meanings can be obscure (Morrow 2002). Even less obvious for those outsiders is the recognition that they, too, bring their own stories to these encounters. Of course geographers are involved in reflecting particular narratives about the world, but the issue is how those narratives are perceived as constituents of truth – or not – and whose narrative holds sway. If part of what is involved in engaging with Indigenous geographies leads us to think about narrative, then, as Cronon (1992: 1350) points out, such an enterprise requires that we acknowledge the power of stories to "make the contingent seem determined and the artificial seem natural."

To varying degrees, geographers have underscored the importance of stories as a means of understanding how the discipline selectively writes the world into being, and in the process, defines its epistemological and teleological terms of reference (Huggins, Huggins, and Jacobs 2008; Willems-Braun 1997). However, the ontological implications of geographical stories receive less attention. On what assumptions of being do our stories rest and what states of being do they construct? How do our notions of different ontological states – the animate and the inanimate, for example – determine our geographical imaginaries, and how are these geo-ontological imaginaries reinforced through practice? What are the political, social, and economic repercussions of our ontological categories and what role do stories play in establishing and maintaining those categories? An engagement with Indigenous geographies challenges us to ask such questions and confront their implications. Ultimately, however, we must consider whose stories get told, whose are silenced, and what this means.

Although Indigenous G/geographies legitimate the role that stories play in place-making and identity formation, what western geographers must also be mindful of are the power dynamics at play in the process of writing the world into being. What are the narratives that we as a discipline are creating and promulgating, and what are their implications? Could we not learn from our engagements with Indigenous peoples for whom the world is not necessarily frozen onto the page and reassess our disciplinary practices? As co-residents, we must recognize what it means to inhabit rather than to occupy the places where we live, and remember that "The inhabitant is … one who participates from within in the very process of the world's continual coming into being" (Ingold 2007b: 91). We must be aware, then, that place and identity are on-going processes of construction that are subject to negotiation.

Collaborative story telling: as producers or directors?

The following narrative reflects one of the authors' experience of the settlement process for Maori land claims in New Zealand. All of his projects for the Waitangi Tribunal have included ethical predicaments about who should be considered *tangata whenua* (people of the land) and, therefore, the principal benefactors of the research. The most vexing of these projects was complicated by contestation of the right of a tribal Trust Board to represent its various constituencies during the research, hearing and redress phases of settlement. More than one hundred years of dispute about which level of Indigenous governance should control resources detonated within the Tribunal process. Doubtless, the need to construct authoritative identities in advance of settlement redress incited that conflict. On the day of greatest volatility, the researcher received two facsimiles. The first, from an officer of the Trust Board, invited him to a *hui* (gathering) to discuss research requirements at 4.00 p.m. on the following day. Attendance would be difficult because the location was relatively isolated, about five hours' drive from the researcher's home. It was further complicated when a newly formed collective of disaffected parties learned of the meeting. That group's Secretary sent a second facsimile with a request to attend another hui which had been scheduled deliberately for the same time as the first: "You have been summonsed to a hui as you are OUR researcher. 4pm, tomorrow. Attendance is COMPULSORY."

Those hui were dutifully attended – to one, arriving two hours early; to the other, two hours late. But the fraught identities of researchers who contribute to the claims process, characterized by Ray (2003: 269) as "hired guns, jackals and whores," defy easy resolution. Irrespective of personal commitments to abstract concepts like "social justice," no amount of meeting, talking and patiently waiting for the insults to end could circumvent its painful reality. The initial coping mechanism was to self-inter within prerequisite archival research; safe territory for a while, but even that would eventually require interpretative work. For whom would the historical narrative be retold? Written history in that part of New Zealand has long been imagined as a case of benevolent administration, whereby enlightened officials faithfully document and contemplate the concerns of the Trust Board. Like most official histories, such imaginings are a convenient simplification: perpetuating the myth of a single authorized voice for a tribe removes any need to engage with those others who constitute an inconvenient sense of disorder. The use of official archives produces a distorted history, so the more the researcher attempted to avoid conflict by concealing himself in archival research, the greater the resultant unease.

Other geographers share this fear of contributing to something patently unjust in the name of claims research (see Wainwright and Bryan 2009). Various methodologies have been debated to reduce the anxiety, but at the time the researcher viewed "collaborative research" with caution. Competing voices with unequal means to express their concerns permeate the claims process, so collaborative research may naively silence rather than enable (Weems 2006). It too may construct singular "partners" amidst irredeemable diversity, resulting in elite capture of research outputs and benefits. Despite acknowledging similar concerns, it was the collective of disaffected parties, rather than the researcher, who proposed the collaborative video project which was to follow. They explained that if the research was restricted to "objective" archival work, their story would not be told, and – because of the tendency to singularize Indigenous interests within the official history – they were doubtless correct. An interview program, focus groups and other such conventions of qualitative research were all discussed and rejected. Like the many hui, they offered only the indefinite possibility of being represented, but these claimants wanted to *direct* their own stories. One had commenced a course in film studies and suggested that "as we're M.I.A. [missing in action] in the paper trail and as we can't drag the Tribunal around our *rohe* [territory], we should bring it to them – make a movie." The researcher was aware of the emerging literature on participatory video (Kindon 2003), but was deeply concerned about the efficacy of such evidence within the hard reality of quasi-judicial process.

The resultant project, *Nga takiwa panekeneke* ("the shifting/flickering landscapes"), had a troubled evolution. The Trust Board protested that it too should receive the right to show landscape transformations within its territory to the Tribunal and, after this was ordered, the movie could easily have become a farcical assortment of competing narratives. As one participant commented in a subsequent hui, "the video project seems dangerous because we're not going to have the same ability to contribute as The Board will." In retrospect, his concern was premature: disaffected, younger claimants who have been dispossessed of rights to land may struggle to access legal experts and other tangible resources in support of their case, but they excelled on screen. They were more passionate and articulate; they developed a collective story rather than a montage of individual narratives; they had the better jokes. Mostly, though, they were superb narrators and directors of their own stories.

The researcher never directed nor operated the camera nor developed the plot; he bought the real directors a video camera, gave them base materials from sterile archives and watched them make history from it. One contributor joked that "somewhere on the way we promoted you from energetic go-for to ... producer but, mate, you'll never sit in that director's chair." That witticism indicates how far geographers will need to remove themselves from authorial control if their experiments with storytelling are to complement Indigenous aspirations. It also suggests that Indigenous identities do indeed shift to accommodate research or other collaborative opportunities, compelling additional methodological caution. If researchers relinquish their directorial inclinations, they are much less likely to coerce involuntary shifts in Indigenous identities. In many respects, a participatory video is a trite response to the complex representational politics which confront Indigenous peoples. In this instance, the exigencies of a formalized process for settling Indigenous land claims forced participants to simplify their complex historical and cultural relations with place, thereby risking self-essentialisation. Yet, because they controlled the directorial aspects of movie production, participants themselves decided upon the optimal balance between strategic simplification and revelation of their complex identities. As is the case with their own identities, the movie was a hybrid of tradition and modernity and, as long as they were in control of the hybridization, it could perform as a culturally appropriate medium.

It is difficult to determine what influence the video will have on the Tribunal, but its best work is already done. Through the project, participants decided what image of themselves they should present to the world, and they related viscerally a history which was otherwise indiscernible within the written record. It may be too wishful to proffer the conclusion, but it seemed like a positive identity-forming moment. The reconstruction of identity was not, however, limited to non-academic partici-pants. Since completion of the project, and despite his own *whakapapa* (genealogy), the researcher has been ever more cautious about investing too much meaning in collective identity phrases like "Indigenous," "Maori" or "iwi" (tribe). Thereafter, he has been less suspicious about collaborative research and more willing to act as *mere* producer. Engaging with Indigenous communities forces us to consider our multiple identities as researchers – story-tellers, activists, producers and directors – but that will surely improve our practice.

Conclusion: Implications for Openly Transcultural Research

We set out to consider how as a discipline, Geography contributes to, and is trans-formed by, engagement with Indigenous geographies. We noted the recognition of some social geographers that, if they are to be *effective*, they ought also to be *affec-tive*. This demands openness to shifts in the positionality of researchers, especially because a/effective research, as with identities, is embedded in relationships. Echoing conclusions in the literature on relational identities, Nicholls (2009: 118) contem-plates the dilemmas of collaborative research with Indigenous peoples, along with the associated "idea of research subjectivity being fluid rather than fixed, although located in a binary of colonial relations." The research stories of this chapter dem-onstrate this fluidity and how it influences those on both sides of the research lens. To engage fully in collaborative research is to open the self to ridicule, enchantment, bewilderment and many other life-changing processes which shape recursively the identity of researchers and their relationships. With similar conclusions, Carter et al. (2008) reveal the inevitably transformative power of critical reflection on the self through involvement in collaborative research. Just like any form of transcultural engagement, research implies flux in identities, both for its objects *and* subjects.

Research with Indigenous peoples is performed within a particular context. It is significant that the personal learning which has been conveyed in the second half of this chapter evolved within the quasi-legal processes which are intended to resolve Indigenous land claims. They are some of the many fora in which Indigenous peoples are forced to plead for their interests and submit to particular identities. To treat Indigeneity as if it is a fixed and absolute construction is to obscure the lived practice of Indigenous peoples and to deny the painful representational decisions they make to secure basic human needs. Identities are and not merely inherited but constructed in ongoing relational processes and struggles. Consequently, researchers who contribute to the claims process and, more generally, those who wish to develop "public geographies" with Indigenous communities must be aware of the implica-tions of their theorizing. A legitimate concern with the adoption of relational per-spectives is that their implications may be used against those newly resourced communities who have adopted a "traditional" identity to perform certain rituals on a stage which is seldom of their making. This should not invalidate the theoriz-ing, but it suggests the need to be particularly cautious when there is a possibility that research will affect Indigenous identities and aspirations.

The twin interventions of this chapter – specifically, the malleability of identity and the methodological significance of storytelling – may have broader relevance to social geographers and their coalescence around place. Massey (2005: 130) has argued that we must see places as "collections of ... stories" and spatial relationships as "a simultaneity of stories-so-far." Place-making is always contingent: a process of becoming which can only be comprehended as a reflection of the changing stories we bring to it. Spatial co-habitation by Indigenous and settler populations entails confrontation of divergent notions of place construction, along with other disorderly ontological categories which underpin epistemological and teleological classifications (Larsen 2006; Sharp 2009). In the face of such inconsistencies, we must strive to find genuine expression of, and reflection amongst, those who inhabit shared spaces. We should learn to recognize our own stories and the ways in which they sustain power relations, and to question their assumptions about the world in which we live. This will reveal our disciplinary presuppositions and give room for speaking back by those whose stories have been silenced. For each of us, engagement with Indigenous G/geographies has caused us to struggle creatively with our existential and disciplinary presuppositions. We conclude, therefore, by offering the expectation that a more mindful consideration of the interplay between story- and place-making may have potential to reconcile Indigenous and non-Indigenous experiences of co-habitation. Yet, we also acknowledge the need to question whose stories are used to construct what sorts of places, and how this is achieved.

Notes

1   We are aware that the term "settler society" is problematic precisely because Indigenous peoples are often part of those societies. The concept tends to hide, and thereby naturalize, the aftermaths of European imperialism. Brody (2000) maintains that such globally expansionist inclinations reflect the "agricultural societies" of so-called settlers and their tendency to uproot themselves from their origin, while the supposedly "nomadic peoples" they encountered in their global expansion do not exhibit such expansionist tendencies.
2   We recognize that institutional sympathies for Indigeneity have enabled some groups to dispossess others of their cultural rights and resources (Dunn 2009; Kuper 2003). We also accept that the conception of Indigeneity which we develop here does not resolve loose and easily manipulated definitions of "indigenous" as a "portmanteau category that gathers together all manner of different peoples," interpolating them to establish "commonality among manifest differences" (Castree 2004: 153). Nevertheless, the alternative of adopting fixed and absolute definitions carries greater risk of injustice.
3   The shift to a relational perspective on place has not been universal within social geography. We accept Jones' (2009) claim that even where relational processes dominate they are seldom the only type of engagement which will be influential. Likewise, to advocate for a relational approach to Indigenous identities is not to disregard such other conceptions of Indigeneity as "blood ties" but, rather, to suggest that within the transcultural present those forms of Indigeneity are subject to new forms of negotiation.
4   We lack space to interrogate the diverse conceptions of transculture and hybridity. We acknowledge that some of those conceptions naively overemphasize freedom of choice in cultural association (Epstein 2009), so disengage with the reality that "hybridization always occurs within a field of power" (Harris 2008: 19). We note, however, that most

scholars who use these concepts reject the view that identities are either wholly chosen or imposed; rather, they emerge as the inchoate products of relational interplay.

References

Anderson, K. (1993) Constructing geographies: "race," place and the making of Sydney's Aboriginal Redfern. In P. Jackson and J. Penrose (eds.), *Constructions of Race, Place and Nation*. London: UCL Press, pp. 81–99.

Bhabha, H.K. (1994) *The Location of Culture*. London: Routledge.

Bird, L. (2007) *The Spirit Lives in the Mind*. Montreal: McGill-Queen's University Press.

Blackburn, C. (2007) Producing legitimacy: reconciliation and the negotiation of aboriginal rights in Canada. *Journal of the Royal Anthropological Institute* 13 (4): 621–38.

Bonnett A. (2000) *White Identities: Historical and International Perspectives*. Harlow: Prentice Hall.

Bonnett, A. (2008) White studies revisited. *Journal of Ethnic and Racial Studies* 31 (1): 185–96.

Brody, H. (1988) *Maps and Dreams*. Vancouver: Douglas and McIntyre Ltd.

Brody, H. (2000) *The Other Side of Eden: Hunters, Farmers and the Shaping of the World*. Vancouver: Douglas & McIntyre.

Bryan, J. (2009) Where would we be without them? Knowledge, space and power in indigenous politics. *Futures* 41 (1): 24–32.

Cahill, C. (2007) The personal is political: developing new subjectivities through participatory action research. *Gender, Place and Culture* 14 (3): 267–92.

Carter, S., Jordens, C., McGrath, C., and Little, M. (2008) You have to make something of all that rubbish, do you? An empirical investigation of the social process of qualitative research. *Qualitative Health Research* 18 (9): 1264–76.

Castree, N. (2004) Differential geographies: place, indigenous rights and "local" resources. *Political Geography* 23 (2): 133–67.

Ceuppens, B. and Geschiere, P. (2005) Autochthony: local or global? New modes in the struggle over citizenship and belonging in Africa and Europe. *Annual Review of Anthropology* 34: 385–407.

Collignon, B. (2006) *Knowing Places: The Innuinnait, Landscapes, and the Environment*. Canada: Canadian Circumpolar Institute Press.

Coombes, B. (2007) Defending community? Indigeneity, self-determination and institutional ambivalence in the restoration of Lake Whakaki. *Geoforum* 38 (1): 60–72.

Corbridge, S. (1991) Third world development. *Progress in Human Geography* 15 (3): 311–21.

Corntassel, J. (2008) Toward sustainable self-determination: rethinking the contemporary indigenous-rights discourse. *Alternatives* 33 (1): 105–32.

Cronon, W. (1992) A place for stories: nature, history, and narrative. *Journal of American History* 78 (4): 1347–76.

Davies, G. and Dwyer, C. (2008) Qualitative methods II: minding the gap. *Progress in Human Geography* 32 (3): 399–406.

Dirlik, A. (2003) Globalization, indigenism, and the politics of place. *Ariel* 34 (1): 15–29.

Dressler, W. and Turner, S. (2008) The persistence of social differentiation in the Philippine uplands. *Journal of Development Studies* 44 (10): 1450–73.

Dunn, K.C. (2009) "Sons of the soil" and contemporary state making: autochthony, uncertainty and political violence in Africa. *Third World Quarterly* 30 (1): 113–27.

Epstein, M. (2009) Transculture: a broad way between globalism and multiculturalism. *American Journal of Economics and Sociology* 68 (1): 327–51.

Gale, F. (1972) *Urban Aborigines*. Canberra: Australian National University Press.

Garbutt, R. (2006) White "autochthony." *Journal of the Australian Critical Race and Whiteness Association* 2: 1–16.

Gombay, N. (2005) Shifting identities in a shifting world: food, place, community, and the politics of scale in an Inuit settlement. *Environment and Planning D: Society and Space* 23: 415–33.

Harris, R. (2008) Development and hybridity made concrete in the colonies. *Environment and Planning A* 40 (1): 15–36.

Hodge, P. and Lester, J. (2006) Indigenous research: whose priority? Journeys and possibilities of cross-cultural research in geography. *Geographical Research* 44 (1): 41–51.

Huggins, J., Huggins, R., and Jacobs, J. (2008) Kooramindanjie: place and the postcolonial. In N. Johnson (ed.), *Culture and Society: Critical Essays in Human Geography*. Aldershot: Ashgate, pp. 283–99.

Ingold, T. (2007a) Earth, sky, wind and weather. *Journal of the Royal Anthropological Institute* 13: S19–S38.

Ingold, T. (2007b) *Lines: A Brief History*. London: Routledge.

Ingold, T. (2008) Bindings against boundaries: entanglements of life in an open world. *Environment and Planning A* 40 (8): 1796–1810.

Jacobs, J.M. (1996) *Edge of Empire: Postcolonialism and the City*. London: Routledge.

Johnson, J.T. (2008) Indigeneity's challenges to the white settler-state: creating a thirdspace for dynamic citizenship. *Alternatives* 33 (1): 29–52.

Johnson, J.T. and Murton, B. (2007) Re/placing native science: indigenous voices in contemporary constructions of nature. *Geographical Research* 45 (2): 121–9.

Jones, M. (2009) Phase space: geography, relational thinking, and beyond. *Progress in Human Geography* 33 (4): 487–506.

Kindon, S. (2003) Participatory video in geographic research: a feminist practice of looking? *Area* 35 (2): 142–53.

King, T. (2003) *The Truth about Stories: A Native Narrative*. Toronto: House of Anansi Press.

Kowal, E. (2008) The politics of the gap: indigenous Australians, liberal multiculturalism, and the end of the self-determination era. *American Anthropologist* 110 (3): 338–48.

Kuper, A. (2003) The return of the native. *Current Anthropology* 44 (3): 389–96.

Lane, M.B. and Hibbard, M. (2005) Doing it for themselves: transformative planning by indigenous peoples. *Journal of Planning Education and Research* 25 (2): 172–84.

Larsen, S. (2006) The future's past: politics of time and territory among Dakleh First Nations in British Columbia. *Geografiska Annaler B* 88 (3): 311–21.

Marker, M. (2006) After the Makah whale hunt: indigenous knowledge and limits to multicultural discourse. *Urban Education* 41 (5): 482–505.

Massey, D.B. (2005) *For Space*. London: Sage.

Mazzullo, N. and Ingold, T. (2008) Being along: place, time and movement among Sami people. In J. O. Barenholdt and B. Granas (eds.), *Mobility and Place: Enacting Northern European Peripheries*. Aldershot: Ashgate, pp. 27–38.

Merlan, F. (2009) Indigeneity: global and local. *Current Anthropology* 50 (3): 303–33.

Morin, K.M. (1998) British women travellers and constructions of racial difference across the nineteenth-century American West. *Transactions of the Institute of British Geographers* 23 (3): 311–30.

Morrow, P. (2002) "With stories we make sense of who we are:" narratives and northern communities. *Topics in Arctic Social Sciences* 4 (1): 17–31.

Nicholls, R. (2009) Research and indigenous participation: critical reflexive methods. *International Journal of Social Research Methodology* 12 (2): 117–26.

Niezen, R. (2003) *Origins of Indigenism: Human Rights and Politics of Identity*. California: University of California Press.

Pain, R. (2004) Social geography: participatory research. *Progress in Human Geography* 28 (5): 652–63.

Panelli, R. (2008) Social geographies: encounters with Indigenous and more-than-white/Anglo geographies. *Progress in Human Geography* 32 (6): 801–11.

Peet, R. and Watts, M. (1996) *Liberation Ecologies: Environment, Development, Social Movements.* London: Routledge.

Pelican, M. (2009) Complexities of indigeneity and autochthony: an African example. *American Ethnologist* 36 (1): 52–65.

Peters, E. (1998) Subversive spaces: first nations women and the city. *Environment and Planning D* 16 (6): 665–85.

Pickerill, J. (2009) Finding common ground? Spaces of dialogue and the negotiation of Indigenous interests in environmental campaigns in Australia. *Geoforum* 40 (1): 66–79.

Porter, L. (2006) Rights or containment? The politics of Aboriginal cultural heritage in Victoria. *Australian Geographer* 37 (3): 355–73.

Pratt, M.L. (2008) *Imperial eyes: travel writing and transculturation,* 2nd edn. New York: Routledge.

Ray, A.J. (2003) Native history on trial: confessions of an expert witness. *Canadian Historical Review* 84 (2): 253–73.

Richardson, L. and St. Pierre, E. (2005) Writing: a method of inquiry. In N. Denzin and Y. Lincoln (eds.), *Handbook of Qualitative Research.* Thousand Oaks: Sage, pp. 959–78.

Sharp, J.P. (2009) *Geographies of Postcolonialism.* London: Sage.

Shaw, W.S. (2000) Ways of whiteness: Harlemising Sydney's Aboriginal Redfern. *Australian Geographical Studies* 38 (3): 291–30.

Shaw, W.S. (2006) De-colonizing geographies of whiteness. *Antipode* 38 (4): 851–69.

Shaw, W.S., Herman, R.D.K., and Dobbs, G.R. (2006) Encountering indigeneity: re-imagining and decolonizing geography. *Geografiska Annaler Series B* 88B (3): 267–76.

Sylvain, R. (2005) Disorderly development: globalization and the idea of "culture" in the Kalahari. *American Ethnologist* 32 (3): 354–70.

Thom, B. (2009) The paradox of boundaries in Coast Salish territories. *Cultural Geographies* 16 (2): 179–205.

Valdivia, G. (2005) On indigeneity, change, and representation in the northeastern Ecuadorian Amazon. *Environment and Planning A* 37 (2): 285–303.

Wainwright, J. and Bryan, J. (2009) Cartography, territory, property: postcolonial reflections on indigenous counter-mapping in Nicaragua and Belize. *Cultural Geographies* 16 (2): 153–78.

Watson, A. and Huntington, O.H. (2008) They're here. I can feel them: the epistemic spaces of indigenous and western knowledges. *Social and Cultural Geography* 9 (3): 257–81.

Watson, A. and Till, K.E. (2009) Ethnography and participant observation. In M. Crang, D. DeLyser, and L. McDowell (eds.), *Sage Handbook of Qualitative Research.* London: Sage, pp. 121–37.

Weems, L. (2006) Unsettling politics, locating ethics: representations of reciprocity in post-positivist inquiry. *Qualitative Inquiry* 12 (5): 994–1011.

Willems-Braun, B. (1997) Buried epistemologies: the politics of nature in (post)colonial British Columbia. *Annals of the Association of American Geographers* 87 (1): 3–31.

Wilson, P.C. (2008) Neoliberalism, indigeneity and social engineering in Ecuador's Amazon. *Critique of Anthropology* 28 (2): 127–44.

Wirf, L., Campbell, A., and Rea, N. (2008) Implications of gendered environmental knowledge in water allocation processes in central Australia. *Gender, Place and Culture* 15 (5): 505–18.

Chapter 28

Transnational Geographies and Human Rights

Amy Ross

Introduction

This chapter discusses contemporary issues in human rights as an example of transnational geographies. Human rights[1] are both the product and producer of transnationalism. This dialect is examined through a discussion of the transnational geographies associated with prosecuting perpetrators of human rights violations. Mapping and analyzing this aspect of human rights serves to illustrate the dynamics of transnational geographies.

Transnationalism has, according to Katharyne Mitchell (1997), gained considerable attention due to recent dramatic increases of borders crossed by human bodies, capital, information, and ideologies. Mitchell argues that transnationalism is a "sexy topic" (p. 101) due to these accelerations, as well as the potential of "epistemological innovations" through research on transnational activities. While recognizing that "studies and theories of the literal transnational movements of capital, commodities and culture have been crucial for furthering our understandings of how globalization works and of the increasing power and effects of systems that operate across borders" (p. 107), Mitchell notes the relative limitations of the field. Specifically, the main emphasis of transnationalism has been the movement of capital, and on immigration (p. 106). While such scholarship is necessary and important, other areas have been neglected. Mitchell calls for "bringing geography back in" so as to "harness the ... potential of transnationalism, and to realize it progressive elements" (p. 108). To do so, she argues for "grounded empirical work" (p. 109) and proposes "transnational spatial ethnographies" (p. 110) as a means of bringing geography back in to transnational discourse.

A Companion to Social Geography, First Edition. Edited by
Vincent J. Del Casino Jr., Mary E. Thomas, Paul Cloke, and Ruth Panelli.

This chapter takes up that challenge through an analysis of the transnational geographies of the prosecution of human rights cases. The chapter proceeds in the following manner. First, the chapter provides a brief overview of why human rights cases are prosecuted transnationally, or in a nation other than where the violence occurred. I argue that impunity at home has led human rights activists to seek out a "spatial fix" of foreign jurisdictions. Secondly, the chapter details prominent instances of the exercise of universal jurisdiction and international courts. The primary examples are the Spanish National Court (in Madrid, Spain) and its treatment of Latin American cases, and the International Criminal Court (in The Hague, The Netherlands) and its involvement in Africa. The transnational biographies of Augusto Pinochet (Chilean general and president), Rigoberta Menchú Tum (1992 Nobel Peace Prize winner), and Luis Moreno-Ocampo (the first chief prosecutor of the International Criminal Court) are used in order to enrich the discussion of politics and institutions with embodied political subjectivities, the transnational lives they have led, and the transformations on social, legal and political landscapes in which they are protagonists. As precedent, peace prize and prosecutor, Pinochet, Menchú and Moreno-Ocampo navigate and create the transnational geographies of human rights. The conclusion links the particular issues arising from the mapping of the cartography of human rights with the field of transnational geographies.

Prosecuting Human Rights Violations "Elsewhere"

Increasingly, certain powerful individuals thought to be responsible for human rights abuses are being investigated for these crimes and served indictments. Some quite prominent figures have even been brought before judicial bodies; and a few prosecuted. Even high-ranking state officials, once thought to be beyond the reach of the law, have faced indictments and arrest on charges of human rights abuses. At the time of this writing, there are dozens of cases before national jurisdictions and international juridical bodies involving charges against high-level officials accused of gross violations of human rights. The extent of the apparent increase in cases in the last 15 years has led some scholars to refer to a "justice cascade" (Lutz and Sikkink 2001).

In many of the most prominent cases (the so-called "famous cases," Sriram 2001) those accused are being indicted, and/or brought to trial, *somewhere other* than where the crimes were committed (see Box 28.1). With the establishment of international courts and the increased willingness of certain national courts to invoke universal jurisdiction, scholars have observed the acceleration of activity concerning lawsuits brought in foreign jurisdictions. This phenomena has been described as the "externalization of justice," "justice without borders," or "doing justice elsewhere" (see, Sriram 2001; Ross 2001).

The emergence of this activity has received considerable attention among political scientists, sociologists, legal scholars and human rights professions (Falk 1999; Keck and Sikkink 1998, 2000; Guidry et al. 2000; Roht-Arriaza 1999).The apparent increase in human rights cases conducted in a nation other than where the violence occurred has provoked debates concerning the growth of global networks, the development, (in numbers and relative power), of Non Government Organizations

---

Box 28.1   The famous cases

October 1998: While, traveling for medical treatment, Chile's President Augusto Pinochet is arrested in London on a warrant issued by a Spanish judge. Pinochet becomes the most significant precedent for the exercise of universal jurisdiction.

June 2001: Serbia's former president, Slobodan Milosevic, is arrested in Belgrade and extradited to The Hague, The Netherlands, where he faces 66 charges before the International Criminal Tribunal for the former Yugoslavia. Milosevic dies while in custody in 2006.

September 2005: A Belgian judge issues an arrest warrant for Chad's deposed dictator Hissene Habre, who resides in exile in Senegal. Senegal refuses to comply with the international order, but under pressure from the African Union, arrests Habre in November 2005 and agrees to prosecute him.

March 2006: Liberia's former president Charles Taylor, is apprehended while fleeing his (once comfortable) exile in Nigeria, as he tried to escape to Cameroon. Taylor was flown briefly to Liberia, then to Sierra Leone and the site of a United Nations' war crimes tribunal (the "Special Court for Sierra Leone, SCSL") designed to prosecute perpetrators such as Taylor. Taylor is soon moved to The Hague, The Netherlands, where the ICC "sublets" space to the SCSL specifically for holding Taylor's trials, which is deemed to dangerous to hold in Sierra Leone.

(NGOs), and tensions between the established order of international relations and an emerging global civil society.

The prosecution of offenders has been welcomed as a series of victories in the battle against impunity. But there have also been concerns. A central issue is the manner in which pursuing a trial in a jurisdiction "far" from where the atrocities have occurred may influence the ability of those closest to the violence to benefit from the prosecutions. Further debate interrogates what "benefiting" actually means. Others have observed that prosecutions in foreign jurisdictions may be potentially disruptive for local efforts to pursue peace and security. Finally, the patchwork of the application of international law has, to date, left an uneven landscape where certain alleged perpetrators enjoy a kind of global impunity, leaving "zones of impunity" (Sriram and Ross 2007) alongside the "successful" cases brought forward.

In contrast to the attention this phenomenon has received in other disciplines, there is a relative lack of geographical scholarship concerned with this transnational activity. Geographers may contribute to the analysis of the phenomenon by exploring its spatiality. Specifically, what counts as "elsewhere"? Why are certain jurisdictions deemed appropriate, and others considered "far" away? Which protagonists hold the power to determine where these trials are held, and why? Which (suspected) perpetrators remain beyond the reach of international justice? How are these activities in the pursuit of justice transforming geographies, and how are geographies transforming justice? What impact does this dialectic have on political subjectivities? These issues arise from the specifics of human rights cases, but also speak directly to broader questions in critical geography, particularly concerning the operations of scale, location, justice and power.

This chapter explores the dialectics of space and justice, and the associated development of transnational geographies. I do so by analyzing the spatiality of the pursuit of justice in order to create a critical cartography[2] of transnational human rights, specifically the establishment of international courts and of national courts invoking universal jurisdiction. This cartography illustrates that transnationalism facilitates a certain kind of human rights movement, with its focus on judicial criminal accountability. In turn, this human rights movement (in its material and discursive incarnations) encourages transnationalism(s) and associated geographies.

David Harvey's notion of "the spatial fix" is a useful lens with which to analysis the role of space in these struggles concerning justice and power. Harvey first invoked the term "spatial fix" in *Limits to Capital* (1982), but it is a consistent theme in his work. Harvey uses the "spatial fix" to describe capitalism's "perpetual turning" to new territories to accommodate its internal crisis. Harvey is interested in more than the merely territorial aspect of capitalism's conquests; in addition to the "seeking out" of new territories, capitalism also transforms space, recreating new environments in the process. Whereas Harvey was expressly interested in flows of capital, the crisis of accumulation, and their relationship to transforming geographies, the "spatial fix" metaphor is salient in understanding the socio-political process of the pursuit of justice. In this usage, the "crisis" is the impunity that accompanies and promotes the mass violation of human rights. The "fix" is the seeking out of new territories/jurisdictions in which to hold the accused accountable.

Why are suspected perpetrators of human violations prosecuted "somewhere else" – i.e., in a jurisdiction or nation other than where the violence has occurred? The observation that "A person stands a greater chance of being tried and convicted for the murder of one than the murder of one-hundred thousand"[3] highlights the dilemmas of accountability in the wake of mass atrocities. Although it is the powerful that are in a position to commit genocide, crimes against humanity and war crimes, the power associated with holding a high-level national office can often protect against prosecution. This condition, known as impunity, has frustrated human rights activists and contributed to the concerns that the violence of the past will be repeated in the future.

Amnesty and the lack of acknowledgement can be more common than prosecutions and accountability. National peace negotiations often result in amnesties for atrocities rather than accountability. In the nation in which the crimes occurred, often the high-level officials most directly associated with the violence may be viewed as war-heros rather than war criminals. Holding power can mean that a person has the ability to manipulate information, control public opinion and contribute to the conditions of impunity that facilitated the violence in the first place. Many state leaders can disguise violence as legitimate warfare – as acts of war rather than crimes of war – a framing process that can have drastic affects on who is held responsible for violence. Impunity is often formalized through the granting of amnesties, which allows for certain offenders to remain in power despite the human rights offenses committed. This is particularly apparent during negotiated peace resolutions, in which protagonists to a national conflict may offer each other amnesty as part of a deal struck in negotiations (such as in South Africa). The point here is that there appears to be a relationship between the scale of the nation and the granting of amnesties.

The international community – as imagined and articulated in international law and the jurisprudence of the international courts and instruments – rejects amnesties. National amnesties therefore create a conflict for international institutions, which are bound to prosecute grave crimes. The response to this conflict between national practices and international law has been the creative use of geography in the "seeking out" of new jurisdictions in order to defeat the impunity of the powerful. As with Harvey, this usage (the spatial fix of justice) invokes both the geographical expansion to new territories: in this case, jurisdictions, other than the territory where the crimes occurred, willing to entertain prosecutions, as well as the spatial re-organization of existing territories.

Key questions arise from considering the relationship between judicial accountability and scale. Are national jurisdictions more likely to settle for amnesties, and if so, why? How are national spaces formed *through* the negotiation process and associated amnesties? How does space provide a fix to this crisis of power, impunity and violence? If transnationalism facilitates this "fix," to the crisis of impunity, why and how? Two aspects of this spatial fix of justice are explored below: (1) universal jurisdiction and (2) international courts. In the following sections, the examples discussed are those brought in European courts concerning crimes committed in Latin America; and the operations of the International Criminal Court (ICC) in Africa.

Universal Jurisdiction: National Courts Address International Crimes

Since the 1980s, human rights activists have increasingly sought out foreign jurisdictions in order to pursue cases that have been stymied by impunity in national arenas. Certain national courts, especially in Europe, have investigated and occasionally prosecuted foreign nationals for crimes committed in other nations. The legal principle that facilitates this activity is universal jurisdiction, which refers to the "… right of a state to prosecute and punish those who commit certain crimes outside its territory and not involving its nationals" (Neier 1998). A criteria for the application of universal jurisdiction, developed by the Princeton University Project on Universal Jurisdiction (2001), states that "universal jurisdiction is criminal jurisdiction based solely on the nature of the crime, without regard to where the crime was committed, the nationality of the alleged or convicted perpetrator, the nationality of the victim, or any other connection to the state exercising such jurisdiction."[4] Underpinning the legal principle of universal jurisdiction is the notion that certain crimes are so serious and massive as to be of "international concern." This category of crime, theoretically, can be prosecuted in any court. Grave breaches of the laws and customs of war, crimes against humanity, genocide and torture are examples of such crimes.

As with other elements of customary international law, its meaning has been contested, its powers developed in practice. Despite its earlier origins,[5] universal jurisdiction received little attention until it ensnared Chile's General Augusto Pinochet in October 1998. While he was in London receiving medical treatment, British officials, acting on a warrant issued by a Spanish judge, placed Mr. Pinochet under house arrest.

A major struggle ensued. Powerful players weighed in on all sides. Pinochet could count on support from heavy hitters such as his old ally Margaret Thatcher, who praised the general as a hero in the struggle against communism. A well-organized community of Chilean exiles in Europe (see, especially, Roht-Arriaza 2005), united with increasingly sophisticated transnational activists, fought just as hard for his prosecution. Having failed to hold Pinochet criminally accountable in Chile, these activists sought recourse in Spain.

Pinochet argued that as a head of state, he enjoyed diplomatic immunity. The British High Court disagreed. In the ruling supporting the extradition of Pinochet to Spain, a member of the High Court, Lord Nicolls, wrote: "International law has made plain that certain types of conduct, including torture and hostage taking, are not acceptable on the part of anyone. This applies as much to heads of state, or even more so, as it does to anyone else. The contrary conclusion would make a mockery out of international law."[6]

At the eleventh hour, an appeals committee in 2000 found Pinochet "unfit" to stand trial on medical grounds. He was flown from England within hours on a Chilean government jet.[7]But Pinochet returned home to a newly invigorated human rights community that had filed suits in his absence. He would spend the last six years of his life[8]fighting efforts to hold him criminally accountable. Although Pinochet avoided the warrant and the transnational political-juridical system it intended to trigger (extradition and trial in Spain), it was clear that he had been released on a technicality – health concerns – while the British High Court's decision had affirmed the merits of the principle of universal jurisdiction as a valid basis to approve the extradition request.

Following the Pinochet affair, other national courts in Europe demonstrated increased interest in invoking universal jurisdiction. For example, the Belgium National Court in Brussels prosecuted and convicted four Rwandans of genocide for their roles in the violence in Rwanda in 1994. Following on the conviction of the four Rwandans, the Belgium National Court opened a judicial process (July 2001) against Israeli Prime Minister Ariel Sharon.[9] Shortly thereafter, the government of Israel put out a travel-advisory to its high-ranking officials, alerting them to foreign countries that might be keen to entertain universal jurisdiction. The list includes many of the western social-democracies.

Given the attention paid to the activities of the Spanish court, the Belgium Court's deployment of universal jurisdiction provoked concerns that the court was transforming Belgium and specifically Brussels into the "crying capital"[10] of the world. Even human rights advocates expressed concern that Belgium's relatively expansive reception of cases on the basis of universal jurisdiction would subject the court in Brussels to a flood of suits.[11] The Belgium legislature, worried about the nation's ability to host international institutions (and the associated diplomats-cum-defendants) attempted to restrict the court's activism. Donald Rumsfeld, as US

Secretary of Defense, threatened to move the North Atlantic Treaty Organization (NATO)'s headquarters from Brussels to Poland. With Brussels as the seat of the capital of the European Union, the Belgium government was particular sensitive to potential violations of diplomatic immunity.[12]The International Court of Justice (ICJ) intervened in the debate, ruling in 2002 that sitting state officials enjoyed diplomatic immunity from foreign courts.[13] The ICJ's decision protects serving state officials, but affirms the fundamental tenets of universal jurisdiction. The point here is that the application of universal jurisdiction has exhibited potential as well as setbacks.

The advancement and the resistance to these cases (and the principle of universal jurisdiction itself) is apparent in the Spanish National Court in Madrid and its treatment of complaints against Guatemalans for genocide. Faced with impunity and inadequate judicial remedies at home,[14] and in collaboration with increasingly sophisticated activists, Guatemalans (like others across the globe) have found certain national courts receptive. In 1999, Mayan activist and Nobel Laureate Rigoberta Menchú Tum and others filed a complaint in the Spanish National Court in Madrid.[15] The complaint charged eight prominent Guatemalans, including former heads of state Efrain Rios Montt, Óscar Humberto Mejía Victores, and Romero Lucas Garcia, with genocide, torture, terrorism, summary execution and unlawful detention. Dozens of plaintiffs joined the case, which addressed atrocities in the communities of Plan de Sanchez and Rio Negro, the murder of members of Ms. Menchú's family, and violations against Spanish nationals living in Guatemala.

Rigoberta Menchú Tum is, herself, a transnational protagonist: A person whose life experience has been shaped by *Crossing Borders* (also the title of her 1998 autobiography: Menchu Tum 1998) and, in turn, a person whose life experience has contributed to further border crossings and specifically the transnational networks organized around indigenous and human rights.[16] Rigoberta was a youth activist, motivated by the conditions of violence, poverty and injustice in Guatemala and inspired by her parents.[17] Rigoberta went into exile in 1981. She managed to link up with Catholic organizations in Mexico who provided sanctuary to Guatemalan refugees. Rigoberta was soon active in giving speeches (*testimonios*) to audiences in Europe and North America in solidarity with the victims of Central America's conflicts. At first the crowds were small.[18] But Rigoberta's superior communication skills and ability to convey the broader context of Guatemalan reality through the narrative of her family's experience drew larger and larger audiences. Her international reputation was firmly established by 1984 with the publication of Elizabeth Debray's biography (Burgos-Debray 1984), "I, RigobertaMenchú" (*"Me llamoRigoberta-Menchú asi nacio me conciencia."*) The best-selling book, which was translated into numerous languages and made its way onto college course syllabi throughout the world, contributed to Rigoberta's growing stature as a spokesperson for indigenous peoples and an advocate for human rights.[19]

Rigoberta Menchú quickly learned how to navigate the institutions of international organizations, particularly the United Nations, such that her personal, "local" experience intersected with global politics. As she observes in her autobiography (*Crossing Borders*) she learned how to maneuver the hallways in Geneva and New York. Rigoberta's political activity is a prime example of how a political/social activist can overcome national obstacles, link with other movements, and have salience in an international context. As such, she has been able to contribute to the struct for justice for crimes in Guatemala by working effectively in international arenas.

The genocide case moved forward as different levels of the Spanish judiciary debated the merits of the jurisdictional tie to Spain, at one point restricting the complaint to victims that held Spanish nationality (Roht-Arriaza, 2006). A further challenge was the attempt to require that the plaintiff proved that the Guatemalan state was unable or unwilling to prosecute the case in its own national courts. In 2005, however, the Spanish Constitutional Court[20] ruled that under the principle of universal jurisdiction, the Spanish National Court had appropriate jurisdiction over the Guatemalan case, as with others concerning genocide, torture and other crimes against humanity, regardless of where the crimes took place. At the time of this writing, the case is proceeding: the Spanish National Court has heard testimony from victims and expert witnesses (February 2008, May 2008).

The fate of the Guatemalan genocide case in Madrid is tied to the fate of the principle of universal jurisdiction, and advances have encountered resistance. For an international order that relies upon guarantees of diplomatic immunity between nations, universal jurisdiction represents a troublesome trend–a disruption of the established order between nations. In dynamics classic to this tension, the Guatemalan case has advanced and retreated before various Spanish courts. The central tension is between the universal imperative to prosecute (and therefore resist and deter) heinous crimes, and the desire to preserve state sovereignty and the international order.

Institutions of International Justice: The International Criminal Court

A further element in the transnational geographies of human rights and justice is the creation of international judicial bodies mandated to prosecute crimes of international concern: genocide, crimes against humanity and war crimes. The establishment and operation of such tribunals has been uneven. To date, international tribunals have been established to address the violence in the former Yugoslavia, Rwanda, Sierra Leone, Cambodia, and a Special Tribunal for Lebanon.[21] The United Nations has invoked the protection of international peace and security as the rational for responding to these particular violent situations with an international instrument. Whereas these bodies operate with specific spatial jurisdictions (and other limitations), the International Criminal Court (the ICC) is the world's first permanent body established to prosecute genocide, crimes against humanity, war crimes, and aggression.[22]

Interest in a permanent international body devoted to prosecuting war crimes has been expressed at various points in history (Bass 2000). During the political polarization of the Cold War period, the super-powers blocked the establishment of an independent international institution devoted to prosecuting such crimes. As the cold war moved towards its closure (1989–91), interest returned in the form of a proposal to prosecute drug-trafficking in the Caribbean as an international crime. Although that particular proposal drifted, the idea of an international court remained in the air. Violence during the breakup of the former Yugoslavia spurred the establishment of a war crimes tribunal in 1993 (the International Criminal Tribunal for the former Yugoslavia – ICTY, located in The Hague, The Netherlands), and a second tribunal was established following the genocide in Rwanda in 1994 (the International Criminal Tribunal for Rwanda–ICTR, located in Arusha, Tanzania).

While there is a relative consensus concerning the establishment of the ad-hoc tribunals (see Bass 2000; Falk 1999; Roht-Arriaza 1999), certain powerful sectors remained averse to a permanent international court that would have jurisdiction over every potential perpetrator on the planet. Among those objecting were the likely subjects – the notoriously violent dictators and despots who had every reason to resist judicial accountability. Objections also came from the US government, which expressed fears that submission to the jurisdiction of an international court would undermine US sovereignty and its potential for global military activity. While the US had supported "international justice" for the peoples of Rwanda, the former Yugoslavia, Sierra Leone and Cambodia, the US government was far less willing to consider such courts for its own nationals.

While the statute for the ICC was being negotiated in the 1990s, certain accommodations and compromises were made in an attempt to bring the US on board. The principle of "complementarity" was among the most significant. Complementarity assures that national jurisdictions have the first obligation to prosecute. It is only when a nation-state proves "unable or unwilling" (terms left undefined) to prosecute its suspected criminals that the ICC can assume jurisdiction.

A notable aspect of the formation of the ICC was the role of international non-governmental organizations (NGOs). A group of more than 800 NGOs established the Coalition for the International Criminal Court, and participated at unprecedented levels in the drafting of the Rome Statute, the lobbying of State parties, and the hard, grass-roots work of bringing diverse national constitutions in line with international aspirations regarding human rights norms. Additionally, victims are to have a greater role in both the prosecution and in terms of the possibility of receiving reparations; provisions supported by the NGOs.

By 2002 the Rome Statute had garnered the requisite 60 signatories and came into affect (see Box 28.2).[23] Although US President Bill Clinton signed the Rome Treaty on December 31, 2000 (on his way out of office), the administrations of George W. Bush adopted a more antagonistic position toward the ICC. In May 2002 (just after the 60th signatures were achieved), Bush "unsigned" the ICC treaty – a move legally dubious – and promoted the American Service-member Protection Act, (ASPA), which authorized the invasion of The Netherlands in the event that a US citizen is held by the ICC.

The biography of the ICC's first chief prosecutor, Luis Moreno-Ocampo, provides another example of the transnationalization of human rights, and how the human rights movement creates transnational spaces and subjectivities. Moreno-Ocampo's career in human rights began as a prosecutor in Argentina, when he bravely (and effectively) participated in the 1985 trials in Argentina against the military officials considered responsible for violence of the "dirty war" (1976–83). Although Moreno-Ocampo managed to try and convict the generals, an amnesty law (passed in the name of national security and stability) revoked the convictions. Moreno-Ocampo developed an extreme aversion to amnesty.[24] Over the next two decades he worked for his private practice, as a law professor at Stanford and Harvard, and participated in Transparency International. When he was elected as the ICC's first chief prosecutor in 2003, Moreno-Ocampo hired many Latin Americans for top positions; the

Box 28.2   Chronology of the International Criminal Court

1947: The United Nations Convention on the Prevention and Punishment of the Crime of Genocide was adopted. Article I of the Convention stated that genocide is "a crime under international law," and article VI indicated that persons charged with the offence of genocide "shall be tried by a competent tribunal of the State in the territory of which the act was committed or by such international penal tribunal as may have jurisdiction." In the same resolution, the General Assembly invited the International Law Commission "to study the desirability and possibility of establishing an international judicial organ for the trial of persons charged with genocide."

1949: The International Law Commission prepared several draft statutes for an ICC but differences of opinions forestalled further developments. The political dynamics of the Cold War block a consensus from the super-powers on an ICC for half a century.

1989: In response to a request by Trinidad and Tobago concerning drug-trafficking, the UN General Assembly directs the International Law Commission to resume work on an international criminal court. Preparatory conferences are held through the 1990s.

July 1998: Adoption of the Statute of the International Criminal Court at the United Nations Conference of Plenipotentiaries in Rome, July 17, with the participation of representatives of 160 States, 33 Intergovernmental Organizations and a Coalition of 236 Non Governmental Organizations. 120 countries voted in favor, 7 against and 21 abstained.

April 2002: The Rome Statute achieves the necessary sixty ratifications (10 nations simultaneously deposit their instruments on April 11).

March 2003: Inauguration of the ICC and swearing-in of the judges.

April 2003: Mr. Luis Moreno-Ocampo (Argentina) elected by consensus by the Assembly of State Parties as first Prosecutor of the International Criminal Court.

January 2004: Ugandan President Museveni and ICC Prosecutor Luis Moreno-Ocampo announce that Uganda has invited the ICC to address the situation of the Lord's Resistance Army in Northern Uganda.

March 2005: The United Nations Security Council refers the situation in Sudan to the ICC.

26 January 2009: Opening of the first trial at the International Criminal Court: *The Prosecutor v. Thomas Lubanga Dyilo*, a Congolese national charged with war crimes, including the recruitment and use of child soldiers.

As of August 2010, 113 nations have joined the ICC.

Source: Adapted from the ICC official website: http://www.icc-cpi.int/Menus/ ICC/About+the+Court/ICC+at+a+glance/Chronology+of+the+ICC.htm.

court in The Hague gave the *abogados* a platform to exercise accountability. Their first attentions were directed at Africa.

When the Office of the Prosecutor opened in 2003, its staff was extremely conscious of the vulnerability of the institution. At his first major press conference in July 2003, Moreno-Ocampo sought to explain how the fledgling court was going to proceed among a universe of possible crimes. Extreme violence was everywhere ... but what to count as legitimate warfare, and what to declare a crime against humanity? Moreno-Ocampo, in his initial statements, stuck to careful ground, narrowly interpreting the ICC's jurisdiction and stressing the limits of its resources. Despite the fact that nearly 25% of the "communications" the ICC received in its first year concerned violence associated with the US invasion of Iraq, Moreno-Ocampo avoided that issue. Instead, he announced that the ICC would focus on Ituri (in the eastern Democratic Republic of Congo) as a site of grave concern.

For Moreno-Ocampo, Ituri was a pragmatic choice as the first activity for his office. The violence associated with the war in the Eastern Congo was indisputably atrocious. An estimated 3.3 million have been killed in the violence since the conflict broke out in 1998, with a significant portions of the crimes committed within the ICC temporal jurisdiction.[25] Human Rights Watch (2003) called Ituri "covered in blood" and documented horrific acts of violence including rape, murder and mutilation. At the July 2003 press conference, Chief Prosecutor Moreno-Ocampo read from a letter his office had received from a thirteen year old female in Ituri, describing her ordeal and begging for international intervention. Investigating the Eastern DRC and its associated warlords also allowed Moreno-Ocampo to act – and he needed to act–without directly confronting any major power, such as the members of the United Nations Security Council.[26]

The interest of the government of Uganda in the ICC was a welcome development for Moreno-Ocampo. In January 2004 the ICC announced that the government of Uganda had referred the situation of the Lord's Resistance Army (the LRA) in Northern Uganda to the International Criminal Court. For Moreno-Ocampo, President Museveni's invitation, a so-called "state-party referral" was seen as the best means for the Office of the Prosecutor (OTP) to take action, much preferred to the "trigger" mechanism by which the prosecutor would move on his own (*proprio motu*). For the newly established ICC, conscious of its critics (including the Bush administration), acting at the invitation of a government was far preferable to being seen as intervening in the internal affairs of a sovereign state. The fact that the ICC was invited by the Ugandan government presented certain important opportunities. The difficulties that the International Criminal Tribunal for the former Yugoslavia (ICTY) had in apprehending fugitives was in part the result of the local governments' failure to cooperate with the ICTY's mission. Operating with Museveni's permission could contribute to its ultimate success in apprehending the indictees, as well as allow the ICC staff to operate in (theoretically) more secure conditions.

Yet, if the ICC seems too "close" to a particular government or faction in a conflict, it risks losing it appearance of impartiality and therefore jeopardizing its ability to appear fair. In the context of the conflict in Northern Uganda, being on ANY side is considered to contribute to the escalation of the violence. Although the ICC's intervention was greeted positively among international human rights enthusiasts, internally its entry into the conflict in Northern Uganda was met with apprehension, confusion, and outright objections by civil society. By accepting Musevini's invitation to investigate the LRA (announced in London at a press conference where Moreno-Ocampo and President Museveni stood side by side), Moreno-Ocampo appeared to the people in Northern Uganda as having taken Museveni's side. As such, the ICC became viewed locally as a protagonist in the conflict rather than a promoter of peace.[27]

The ICC has failed to adequately counter the impression that it is working closely with the Museveni government and the Ugandan army (the UPDF). The ICC's first press release concerning Uganda echoed President Museveni's stance and announced that it would investigation the "Lord's Resistance Army in Northern Uganda." Chief Prosecutor Luis Moreno-Ocampo corrected the terms in a letter addressed to the ICC President Kirsh dated June 17, 2004, explaining that "The letter of referral made reference to the situation concerning the Lord's Resistance Army. My Office has informed the Ugandan authorities that we must interpret the scope of the referral consistently with the principles of the Rome Statute, and hence are analyzing crimes within the situation of Northern Uganda by whomever committed." But this clarification, posted on a rather obscure corner of the ICC's website, fails to adequately rebuff the impression that Moreno-Ocampo and the OTP are operating in tandem with Museveni. It is especially difficult given that the president and the (largely) state-controlled Ugandan press promote the impression that Museveni is directing the ICC's actions.

In March 2005 the United Nations' Security Council referred the case of Sudan to the ICC. This was a victory in that the United States could have vetoed this referral, so the US's refrain was the first sign of détente between the Bush administration and the ICC. The Darfur investigation has also been difficult and contentious, as the OTP lacks the ability to conduct interviews within Sudan. Indicting the sitting Sudanese president also invoked criticisms that the ICC was interfering with the possibility for the African Union to persuade Bashir into some sort of peace settlement.[28]

Moreno-Ocampo has been plagued with problems; some inherent to the job[29] and others more specific to his personality. Moreno-Ocampo has been accused of incompetence in his handling of the first case (against Congolese Thomas Lubanga), arrogance and over-reach in Sudan, and personal misconduct.

Conclusions and Future Directions in the Transnational Geographies of Human Rights

The transnational activities of the human rights social movement has transformed geographies. New institutions – international tribunals – have been formed, and national courts have been transformed into sites that address crimes of international concern. In turn, the creation and transformation of these spaces have impacted the development of the human rights social movement. Transnational geographies are reshaping human rights, just as human rights re-shape geographies. While the mapping of human rights demonstrates this dialectic, it is debatable whether the mutually constituted relationship between transnationalism and human rights has resulted in more progressive politics and/or the advancement of peace and human security.

International justice (in the form of universal jurisdiction and the establishment of international tribunals) has provoked both enthusiasm and anxiety. For some, the apparent proliferation of investigations, indictments and trials is evidence of an advancement of human rights in general and the success of these transnational networks in particular. Critics of universal jurisdiction and international courts argue that such trials impinge on national sovereignty, specifically the right of nation-states to determine how they want to handle their own political, social, economic, and cultural conditions of conflict resolution, justice, and reconciliation. For example, in an essay in *Foreign Affairs* (July 2001), titled "The pitfalls of universal jurisdiction," Kissinger objects to the "Pinochet Precedent" on the grounds of national sovereignty, insisting that Chileans have the right to determine their internal affairs.[30]

Despite the measured increase in prosecutions in the international arena, it is quite apparent that the net of international justice is full of holes. Which violence is prosecuted and which violence is ignored (or deemed legitimate) indicates that the transnational movement for international justice has yet to disrupt international power-relations.

Many of the most famous cases involve third world figures being prosecuted in the courts of former colonial powers. In the exercise of universal jurisdiction, Spain has conducted cases from Argentina, Chile and Guatemala. The overall trend suggests neocolonial relationships. The establishment of international courts have involved an "international community" prosecuting the dictators of less powerful nations – Slobodan Milosevic of Serbia, Charles Taylor of Liberia, aged Khmer Rogue and Rwandans. In its first five years of operation, the four cases taken by the ICC have all been in Africa: Uganda, the Democratic Republic of the Congo (DRC), Sudan and the Central African Republic. There are concerns that the ICC is yet another version of "victor's justice" in which the powerful (in this case the wealthy "western" nations which dominate the institutions of the international community, including the permanent members of the UN Security Council) use institutions such as the police and judiciary to criminalize the actions of the weak, while simultaneously legitimizing their own violence. Of its omissions, the ICC controversially rejected requests to investigate alleged crimes in Iraq.[31]

The International Criminal Court has been lauded as the culmination of millennial aspirations for justice and the future protection of human rights. Yet, to date, the "international" court has indicted only African men. Granted, these suspects are accused of heinous violence: rape, child abduction, forcing family members to kill each other – documented in detail by reputable human rights organizations. Yet their violence and alleged crimes exist alongside those of the State, which happens to be the same authorities that cooperate with the ICC. For example, the ICC has issued indictments against the Ugandan rebels, but has, to date, failed to indict members of the Ugandan government for alleged crimes in Northern Uganda and during the Ugandan military's occupation of Eastern Congo.

The United States has been in favor of international justice – for Rwanda, and Sierra Leone – but far less supportive of an international justice that has a potentially universal reach. Examples of this can be seen in the way that the US has been supportive of the ICTY and the SCSL, but has actively tried to defeat the ICC. Although universal jurisdiction and international tribunals have closed the impunity gap in certain respects, they have yet to challenge leaders from the more powerful nations, such as the five permanent members of the United Nations Security Council.

The nascent international justice movement has yet to demonstrate its ability to prosecute alleged offenders from powerful nations. The transnational human rights network has (largely) failed to challenge state power, especially the strongest states, in the form of prosecuting sitting state figures from powerful countries. Two interesting exceptions concern the ICC's indictment of Sudan's president Omar al Bashir, and investigations into the administration of George W. Bush by national courts invoking universal jurisdiction. Now that he is out of office and has lost the shield of diplomatic impunity, former President George W. Bush might be wise to stay put in Texas.[32] Citizen Bush might very well forego visits to Germany, where suits have been filed on behalf of four Iraqi victims of torture, and Spain, where investigations have been opened into the torture of detainees at Guantanamo.

The ability to pursue justice for human rights violations in a court other than the national jurisdiction where the crimes are committed may be increasing, but it is still to be determined what the implications are for human security. Despite the energies and ambitions of the (relatively new) transnational network, the development of international justice appears to be supporting entrenched power relations. While organized against impunity, the prosecutions to date appear to reaffirm rather than resist a state-centric system. To date the human rights regime has done more to uphold state-power – and the power of certain states – and therefore these states continue to be able to commit violence, and get away with it, precisely through the infrastructure that prosecutes others.

There also remains the epistemological problem of measuring the "success" of human rights in the increasing number of trials. Does having more genocide trials mean we have more justice, or more genocide? After all, a truly effective human rights regime would, theoretically, result in less genocide, crimes against humanity and war crimes – and therefore fewer trials for the same.

Future issues associated with prosecuting and preventing atrocities will likely exhibit complex transnational geographies. International institutions are proliferating, but local and national politics and social movements remain intense; rather than "overshadowed" by the international, local and regional practices seem fully engaged. The pursuit and practices of international justice illustrate complicated transnational geographies and subjectivities. Attention to these transnational geographies widens the focus beyond migrants and money, and therefore contributes to theoretical advances concerning the mutual construction of space, power and justice.

Notes

1 Elsewhere, I have detailed and defined "human rights" as a social movement with its associated geographies (see Ross 2009).
2 I use the term "critical" to emphasize a type of cartography that seeks to expose rather than reproduce violent power-relations.
3 José Ayala Lasso, United Nations High Commissioner for Human Rights (1996). As quoted in Morris and Scharf 1997: xv.
4 www1.umn.edu/humanrts/instree/princeton.html.

5   Universal jurisdiction arose with the development of international law in the seventeenth century, primarily in relation to developments in colonialism and capitalism. Most scholars locate the genesis of international criminal law to states' efforts to contain piracy on the high seas during the era of global merchant capitalism (Rosenbaum 1993: 28). The common interests of nations lay in protecting navigation. By common agreement, states established laws which had the effect of establishing the legal principle of universal jurisdiction (Roht-Arriaza 1995: 25).

6   www.parliament.the-stationery-office.co.uk/pa/ld199899/ldjudgmt/jd981125/pino08.htm.

7   Amongst other ailments, Pinochet was reportedly found to suffer from depression.

8   Pinochet died on December 10, 2006 – International Human Rights Day. He died while under house arrest.

9   The suit was filed by Lebanese and Palestinian plaintiffs, who accuse Sharon of responsibility for crimes against humanity in the 1982 massacres in Sabra and Shatila refugee camps by Israeli -backed Lebanese militia.

10 Professor Stephan Parmentier (personal communication).

11 Simons, Marlise. "Human Rights Cases Begin to Flood Into Belgium Courts," *New York Times*, December 27, 2001: www.nytimes.com/2001/12/27/world/human-rights-cases-begin-to-flood-into-belgian-courts.html?scp=8&sq=universal%20jurisdiction%20brussels&st=cse.

12 Smith, Craig S., "Rumsfeld Says Belgium Law Could Prompt NATO to Leave," *New York Times*, June 12, 2003: www.nytimes.com/2003/06/12/international/europe/12CND-RUMS.html?scp=4&sq=universal%20jurisdiction%20brussels&st=cse.   Donald Rumsfeld himself had a close call with universal jurisdiction in October 2007. See Sriram and Ross 2007b.

13 Arrest Warrant of April 11, 2000 (*Democratic Republic of Congo v Belgium*) (hereafter cited as the *Congo v Belgium* case), www.haguejusticeportal.net/eCache/DEF/6/194.html.

14 While numerous attempts have been made to bring genocide cases forward in the Guatemalan courts, in each instance the cases have met with significant obstacles, from legal maneuvers in the form of appeals to dangerous threats against those involved in such cases. One example is the prosecution of a general and two colonels for the assassination of Guatemalan anthropologist Myrna Mack Chang. That paradigmatic case illustrates the difficulties and dangers of the pursuit of justice within Guatemala (Hale et al. 2002). Other senior Guatemalan officials charged are: ex-Minister of Defense Ángel Anibal Guevara Rodríguez, former Minister of Interior Donaldo Álvarez Ruiz, ex-Chief of the Armed Forces General Staff Manuel Benedicto Lucas García, former Director of National Police Germán Chupina Barahona and head of the police unit Comando Seis, Pedro García Arredondo. Chupina Barahona died in February 2008 after eluding a Guatemalan arrest warrant.

15 For documents related to the case in English, see The National Security Archive, Guatemala project, www.gwu.edu/~nsarchiv/guatemala/genocide/index.htm. See also the Center for Justice and Accountability, www.cja.org/cases/guatemala.shtml, The Rigoberta Menchú Foundation, www.frmt.org/es/, and the Center for Legal Action and Human Rights, www.caldh.org/. Last accessed November 15, 2009.

16 Menchú was born in 1959 in Chimel, a village in the department of El Quiché in the highlands of Guatemala. Like many other indigenous Mayan families at that time and place, Rigoberta's family migrated to work on the plantations on Guatemala's fertile southern coast, a life of hard work and suffering.

17 Rigoberta's father, Vincente Menchú, was a founder of the peasant organization CUC (Comite de Unidad Campesina), and died during an attack on the Spanish Embassy (1980) in Guatemala City that burned him and 37 other protestors to death. Due to the

attack on the Spanish Embassy, Spain has a particular interest in the history of political violence in Guatemala. As a result of that incident, Spain broke diplomatic relations with Guatemala for four years.

18   Professor Beatriz Manz (personal communication, October 1992). Professor Manz was involved in the campaign to promote Menchu's candidacy for the Nobel Peace Prize.

19   "I Rigoberta Menchu," also contributed to her problems, as the veracity of her accounts were challenged.

20   For English or Spanish versions of the "Spanish Constitutional Court Decision Accepting Jurisdiction" (SCCD 237/September 26, 2005, Guatemala Genocide Case 331/1999-10) see www.cja.org/cases/guatemaladocs.shtml. Last accessed November 15, 2009.

21   The Special Tribunal for Lebanon, established by the United Nations, is restricted to trying " all those who are alleged responsible for the attack of 14 February 2005 in Beirut that killed the former Lebanese Prime Minister Rafiq Hariri and 22 others." See www.stl-tsl.org/action/home.

22   The ICC has jurisdiction over the so-called "crimes of international concern:" genocide, crimes against humanity and war crimes, all of which are defined in detail in the Rome Statute: www2.icccpi.int/Menus/ICC/Legal+Texts+and+Tools/Official+Journal/Rome+Statute. htm). A fourth category, the crime of aggression, will be defined and adopted at the review conference in Kampala, Uganda, 2010. The debates concerning how to define aggression proved highly contentious during the 1998 negotiations that led to the Rome Statute; the parties agreed to table the debate in 1998 in order to move ahead with the other issues. For more on the debates concerning aggression, see Politi and Nesi 2004.

23   The Dutch government agreed to host the offices of the ICC in The Hague, a city with an established community of international institutions. The Dutch government also provided substantial financing for the ICC's infrastructure, and passed national legislation associated with the presence on its territory of suspected war criminals and the others who would be participating in such trials. In its gestation, the ICC benefited from its location in The Hague in that many of the staff had significant expertise and experience from working in the ICTY.

24   Interview, June 2003.

25   The ICC can only prosecute crimes that have occurred since the entry of the statute: July 1 2002.

26   Although, as the prosecutor insisted, prosecuting the warlords and their crimes would allow for an investigation into the mineral extraction and smuggling that produced the profit from the violence. In doing so, Moreno-Ocampo stated, he could investigate, and therefore influence, powerful financial interests in Europe (personal communication, July 2003).

27   Fieldwork in Uganda in 2005. Of the scores of interviews conducted, the vast majority cited this press conference as a crucial event in forming Ugandans' opinions concerning Moreno-Ocampo's (and therefore the ICC's) lack of impartiality.

28   This is a common problem for international courts. Louis Arbor, the chief prosecutor for the ICTY, was criticized for openly indicting Serbian President Slobodan Milosevic while NATO was bombing Kosovo; the fact that she had acted in the midst of the conflict was seen as disruptive to NATO's ability to strike a deal with Milosevic in order to end the conflict.

29   As Richard Dicker, head of Human Rights Watch's international justice division commented, "The prosecutor has one of the most difficult jobs in the world" (personal communication, June 2005).

30   It is a startling position coming from one who inserted himself so aggressively in that nation's affairs in the 1970s. In 2001 while touring Europe, Kissinger ducked a summons by a French judge to provide information on the Pinochet case; he left Paris in haste.

See "Criminal Justice on a Global Scale," by Bruce Broomhall, *The New York Times*, July 13, 2001.

31 The ICC issued a statement justifying its decision regarding Iraq: International Criminal Court, "Iraq Response" (February 10, 2006), at www.icc-cpi.int/organs/otp/otp_com.html. The ICC also published a rationale as to its failure to investigate alleged crimes in Venezuela: International Criminal Court, "Venezuela Response" (February 10, 2006), at www.icc-cpi.int/organs/otp/otp_com.html.

32 Although George W. Bush likely has little to fear from the ICC, there are certain indications that he might be vulnerable to universal jurisdiction when traveling abroad. In November 2004, during a visit by Bush, the Chilean government announced that the US president would have immunity during his visit. The unusual declaration was made after lawyers filed a criminal complaint against Bush in court, claiming that he and other US officials were guilty of war crimes in Iraq. On a subsequent trip by Bush to Canada, lawsuits were filed against him in Vancouver under Canada's Crimes Against Humanity and War Crimes Act.

References

Bass, G.J. (2000) *Stay the Hand of Vengeanc: The Politics of War Crimes Tribunals*. Princeton: Princeton University Press.

Burgos-Debray, E. (1984) *I, Rigoberta Menchu: An Indian Woman in Guatemala*. London: Verso.

Falk, R. (1999) The pursuit of international justice: present dilemmas and an imagined future. *Journal of International Affairs* 52 (2): 409–41.

Guidry, J., Michael, A., Kennedy, D., and Zald, M.N. (eds.) (2000) *Globalization and Social Movements: Culture, Power and the Transnational Public Sphere*. Ann Arbor: University of Michigan Press.

Hale, C., Ball, P., Oglesby, E., Manz, B., Nash, J., Ross, A., and Smith, C. (2002) Democracy as subterfuge: researchers under siege in Guatemala. *Latin American Studies Association Forum* 33 (3): 6–10.

Harvey, D. (1982) *Limits to Capital*. London: Blackwell.

Human Rights Watch (2003) Ituri: covered in blood: ethnically targeted violence in the Northeastern DRC. July 2003, www.hrw.org/en/reports/2003/07/07/covered-blood.

Keck, M. and Sikkink, K. (1998) *Activists beyond Borders, Transnational Advocacy Networks in International Politics*. Ithaca: Cornell University Press.

Keck, M. and Sikkink, K. (2000) Historical precursors to modern transnational social movements and networks. In J.A. Guidry, M.D. Kennedy, and M.N. Zald (eds.), *Globalization and Social Movements; Culture, Power and the Transnational Public Sphere*. Ann Arbor: University of Michigan Press, pp. 35–53.

Kissinger, H. (2001) The pitfalls of universal jurisdiction. *Foreign Affairs* 4 (80): 86–96.

Lutz, E. and Sikkink, K. (2001) The justice cascade: the evolution and impact of foreign human rights trials in Latin America. 2 *Chicago Journal of International Law* 1: 1–33.

Menchú Tum, R. (1998) *Crossing Borders*. London: Verso.

Mitchell, K. (1997) Transnational discourse: bringing geography back in. *Antipode* 29: 101–14

Morris, V. and Scharf, M. (1997) *The International Criminal Tribunal for Rwanda*. New York: Transnational Publishers.

Neier, A. (1998) *War Crimes: Brutality, Genocide, Terror and the Struggle for Justice*. New York: Times Books.

Politi, M. and Nesi, G. (eds.) (2004) *The International Criminal Court and the Crime of Aggression*. Aldershot: Ashgate.

Princeton Principles on Universal Jurisdiction, www1.umn.edu/humanrts/instree/princeton.html.

Roht-Arriaza, N. (1999) Institutions of international justice. *Journal of International Affairs* 52 (2): 473–91.

Roht-Arriaza, N. (2005) *The Pinochet Effect: Transnational Justice in the Age of Human Rights*. Pennsylvania: University of Pennsylvania Press.

Roht-Arriaza, N. (2006) Guatemala genocide case judgement no. STC 237/2005. *American Journal of International Law* 100 (1): 207–13.

Roht-Arriaza, N. (ed.) (1995) *Impunity and Human Rights in International Law and Practice*. New York: Oxford University Press.

Ross, A. (2001) The geography of justice: international law, national sovereignty and human rights. *Finnish Yearbook of International Law* 12: 10–19.

Ross, A. (2009) A social geography of human rights. In S. Smith, S. Marston, R. Pain, and J.P, Jones III (eds.), *Handbook of Social Geography*. London: Sage, pp. 488–504.

Sriram, C.L. (2001) Externalizing justice through universal jurisdiction: problems and prospects. *Finnish Yearbook of International Law* 12: 53–77.

Sriram, C.L. and Ross, A. (2007a) Geographies of crime and justice: contemporary transitional justice and the creation of zones of impunity. *International Journal of Transitional Justice* 1 (1): 45–65.

Sriram, C.L. and Ross, A. (2007b) Travel advisory: war criminals beware. *Jurist* October.

Correspondence: Department of Geography, University of Georgia, Athens, GA 30602, e-mail: rossamy@uga.edu.

Chapter 29

Resistance(s) and Collective Social Action

Paul Chatterton and Nik Heynen

Introduction

This chapter addresses resistance and collective social action. Given the continued widespread nature of human suffering, violence, and misery, and ecological destruction the world over, this topic is both crucial and a hotly contested area of intellectual debate and practical action. In this chapter we highlight key themes and intellectual trends evident in studies that have explored organizing resistance and collective action in practice. The background, what we will refer to simply as resistance and collective social action in this chapter, is dynamic and far reaching. Resistance includes all forms of organizing, from social movement formation and struggles against all kinds of corrupt and destructive power relations. From within this framing, we are thinking about a wide range of collective social action across Marxist, feminist, antiracist, postcolonial, anticapitalist perspectives; essentially we are interested in all anti-oppression/anti-authoritarian frames. We are also aware that resistance is far from the preserve of politically progressive groups and that those on the far right, for example, have their own repertoires and histories of resistance. Formal legal roots and the national policy arena are also key and relevant arenas for resistance. What we see in studies from geography in particular as we explore below, is resistance and collective social action employed through a vast mix of areas and concerns including ecology, militarism, human rights, privatization, labor market precarity, and economic restructuring, the neoliberalization and enclosure of lifeworlds and resources, welfare services rights and reform, sexual identities, ethics, identity, and indigenous self-determination (see for example: Gidwani 2006; Wise et al. 2003; Kohl and Farthing 2006; Mackenzie and Dalby

A Companion to Social Geography, First Edition. Edited by
Vincent J. Del Casino Jr., Mary E. Thomas, Paul Cloke, and Ruth Panelli.
© 2011 Blackwell Publishing Ltd. Published 2011 by Blackwell Publishing Ltd.

2003; Cummings 1995; Vasudevan et al. 2008; Glassman 2001, 2006; Hart 2006; Wills 2005).

Within the diversity of forms of resistance and collective social action there are subtleties, nuances, and complexities that make any kind of general discussion problematic. These actions and social forms are always situated, contextual, contested, and made real through practice. We think one of the most important starting points is that resistance and collective social action are always relational; if we unpack them we see how they are constituted through relations across space and time; that is how they are connected to each other. Resistance and collective action shapes, and is shaped by, contemporary practices of globalization as well as local events; again, those that are connected across these spatial scales. It is for this reason that most geographers are keen to stress the situatedness of resistance and collective social action and how both are crafted from the particularities of uneven geographies of patriarchy, racism, capitalism, and other destructive power relations in particular places.

From the onset, we feel that it is important to reject dualistic ways of conceiving of resistance and collective social action – that there is one source and just cause from which all resistance of the oppressed stems. History has shown that moments of resistance and collective social action, more times than not, are not romantic at all; they are instead dirty, mundane, and sometimes banal. But realizing the importance of everyday "weapons of the weak" helps to imagine new forms of collective resistance and counter-hegemonic movements to myriad forms of hegemony (see Scott 1985). Power is now widely conceptualized, not as a state to be acquired, but as a relation. Because of this, we need to assess, compare, and critique different microgeographical tactics of resistance (see Pile and Keith 1997) as they play out in our hearts and minds as well as the streets of the city, and the fields of the country by groups struggling for their own goals as opposed to the objectives we would project upon their efforts from our comfortable classrooms and office spaces.

Understanding how power works is central to understanding resistance and collective power. It is now established to see power, not as a form that can be acquired but as a relation between forces, that is to say that every relation is also a power-relation (Foucault 1980). Power is felt as an effect through practice. As Foucault (1980) reminds us power does not emanate from a central point, and it never totalizes. There are always possibilities of resistance to power and those resisting have to understand the subtle ways that power as a relational effect is deployed. For example, the notion of absolute sovereign and territorial power has given way to bio-power. Bio-power, as Foucault (1994) describes it, is exercised on people as both a general category of population (nation, social groups of one sort or another) and individual bodies (people in their roles as individual citizens, wives, farmers, etc.). Life is controlled and administered, and life itself emerges as a new object of power. In such an all encompassing diagram of power, "where there is power, there is resistance" (Foucault 1980).

This chapter stems from both our personal interests and involvements in wide ranging discussions of what resistance and collective social action is/can be/should be. Within the corpus of our collective work, we have talked about resistance and collective social action within the contexts of making autonomous geographies in

Argentina (Chatterton 2005), public scholarship and cultures of Do It Yourself (DIY) forms of self-managed political activism (Chatterton 2006, 2008; Chatterton et al. 2007; Hodkinson and Chatterton 2006), radical anti-hunger organizing (Heynen 2008, 2009a; Mitchell and Heynen 2009; Heynen 2010), racialization of poverty (Heynen 2009b), and the geographies of survival (Heynen 2006). In this chapter, we take the lessons we have learnt from engagements with resistance movements and from the work and experiences of others in the field of geography. We make a series of points which both reflect contemporary conceptualizations of resistance and the concerns of geographers who have reflected on the practices of resistance and the nature of collective social action.

Approaching Resistance and Collective Social Action

The goal of resistance and collective social action is progressive change at some level defined internally by the group of people who have either come together by their free will or have been forced together as some sort of collective due to oppressive power relations. There has been long disagreement on how progressive social action might be achieved, drawing on different tactics, revolutionary subjects, and discourses. There have been numerous historically significant perspectives from which resistance has come including, but not limited to people interested in seeing an end to colonialism, capitalism, heteronormativity, patriarchy, racism, and war. For different groups struggling for different objectives, social change has seen the development of particular perspectives, not always exclusive of each other, that have formed particular ideologies, tactics, and everyday practices. In their book *Dissident Geographies,* Blunt and Wills (2000, xi) suggest: "In the early days, this awakening of radical geography was mainly characterized by geographical interpretations of anarchist and Marxist thought, but subsequently, over the course of the 1980s and 1990s, the radical agenda has widened to include feminist, sexual liberationist, and postcolonial geographies." Taking a quick look around, it should be obvious to anybody paying attention that there is no shortage of destructive and diverse power relations in need of urgent attention, so this widening of approach is productive and helpful.

The most historically established approach (especially embedded within the Anglophone literature) is that presented as the orthodox Marxist route to revolutionary change. We would argue from the start that this approach is problematic primarily because it puts too much emphasis on a small group of revolutionary leaders, referred to as the vanguardist party, to shape decisions for larger collectivities of people. While we would both agree that understanding class relations offered through many Marxist approaches can be very productive for understanding progressive social change, we tend to feel that orthodoxy of any kind is more difficult than helpful. More open Marxist variations (Bonefeld et al. 1992) have rejected the overly simplistic formulation provided through vangaurdist ideologies and, as Holloway (2002) has pointed out, there is no pre-established road to a better world. They have alerted us to the multiplicity of historical subjects involved in resistance and the need to understand the dynamic decomposition and recomposition of class agency and thus emergence of novel forms of resistance. Instead, as the Zapatistas, echoed by Myles Horton (Horton and Freire 1990) say, "we make the road by walking." Resistance, here, is part of an endless process of negotiation without an obvious plan; it is a perpetual antagonism between what a group accepts and what they reject. What we see in both these formulations, however is that revolution still takes on a task of epic proportions. It is a battle for social change that is systemic, historical, and dialectical. It is a refusal against the dominant system.

Anarchism is a further identifiable, if not as strongly felt, approach to understanding resistance and revolutionary change. This tradition broadens the debate away from economic classes and focuses on the state as the source of oppressive power relations. Anarchist thought has developed a rich vein of enquiry into how best to organize social and economic systems which maximize human freedom and liberation through ideas of federalism, voluntarism, mutual aid, and communalism (see Kropotkin 1987; Cook and Pepper 1990; Marshall 1992; Peet 1978). However, it relies on a largely humanist approach which invests faith in an essentialist subjectivity which would flourish if only the authoritarian tendency of the state were removed.

More recent and more post-structural influenced formulations question the dialectical and negative basis of resistance, suggesting also that resistance can be productive (Day 2004, 2005; May 1994; Newman 2007). Indeed, given the multiplicity and complexity of resistance sites, subject, and practices, the idea that we are simply dealing with the old contradictions of hostile class politics is also called into question. Collective action and forms of resistance reach much further than the realm of economic production, abstract labor, or even the factory gates into the wider "social factory" (see Wright 2006). For instance a poststructural view of class politics acknowledges multiple overlapping identities and struggles – queer, feminist, religious, ethnic. What we are seeking to reject is the binary nature of resistance that sees it only in terms of good against evil, not just by the state but also by our own imagination. Here we seek to also avoid looking for resistance in familiar places – the party, the militant, the spectacular moments.

To expand a little, what we actually call resistance usually in fact refers to a multiplicity of actions involving not just emancipation-oriented struggles or an oppositional consciousness but also "reworkings" and remaking life and organizations, as well as building "resilience" (Katz 2004) and survival tactics. Many nonemancipatory and less spectacular forms of resistance are key to social struggles, especially doing the ground work of building an ethics of care and recognizing the emotional work of resistance (see Naples 1998). Resistance, then, also becomes productive.

Beyond the Marxist and anarchist historical roots within geography, feminist geographers have offered some of the more foundational ideas for thinking about resistance and collective social action, especially when thinking about the intersectional, or interlocking, politics of multiple social power relations (see Peake 1993; Kobayashi 1994; Valentine 2007 in geography, as well as Hill Collins 1990; hooks 1984). Within this literature there has been nuanced attention to many issues ranging between social reproduction (Katz 2004; Mitchell et al. 2004) and youth (Panelli et al. 2007) to work (Hanson and Pratt 1988, 1991) and crime and safety (Pain 1997; Pain and Townsend 2001). Typically feminist scholarship has been productive within this larger context for thinking through diverse details that have been ignored for so long. This literature also has been productive in terms of offering methodological insights for thinking about resistance and collective social action, and how to push from theory to action (see Jones 1997 et al; Moss 2002; Panelli 2004). An important insight from these methodological advances can best be summed up as doing all we can to reducing the distance between "thinking" about our research and actively engaging in the lives of the subjects/people we are researching. This model, we argue, is imperative to doing the most meaningful work related to resistance and collective social action.

Geographies of Resistance

Central to much of the progress within the study of resistance and collective social action has been the recognition that the topic is much more than a study of the state. Social movements help us recognize the degree to which the colonization of "the political" by the state has limited more holistic understandings and stunted our imaginaries of what resistance might entail (Routledge 1996). Resistance then, is much more than the formal sphere of the political, and we need to detach resistance from a dependency on the formative power of the state and explore the other multiple spatial relations that are central to understanding resistance and collective social action. Questions of place, scale, and networks become paramount when we consider resistance in the interplay between different groups striving for power, hegemony, and recognition. Similarly, these geographic processes are important when we consider those tendencies that reject a politics of demand based on the state, and strive not to be defined by their opposition to a hegemonic centre (see Day 2005). Said another way, resistance is a deeply geographical activity.

Spatial understandings of resistance have to deal with and accept the reality of multiple and ever changing geographies (Staeheli 1994). These understandings are deepened through nuanced, locally sensitive, thickly descriptive, and less place-bounded accounts of how resistance practices are constituted (Gibson-Graham 2006). It is now well established that resistance is multi-scalar – that is, it is constituted at many different scales from the local to the global (see Smith 1992; Swyngedouw 1997; Marston 2000; Heynen 2009). But there is also a need to draw upon multiple spatialities, including scale, place, networks, positionality, and mobility, to understand the diversity of contentious politics and practices of resistance (Leitner et al 2008). Furthermore, relational imaginations of resistance are useful for seeing resistance as always extending beyond neat nested scales such as the locale or the nation-state. Particular struggles and moments of resistance are always relationally linked across time and space and rejecting understandings of resistance as merely place-bounded reveals connected, diverse, and surprising sets of practices. Multiple relations of contestation, "maps of grievance" as Featherstone (2008) calls them, stretch beyond the local or politically defined boundaries. This conceptualization is part of reclaiming and exposing much wider, and often invisible, networked practices of resistance than have hitherto been acknowledged. Indeed Featherstone (2008: 2) points to "the significant histories and geographies of networked forms of subaltern political activity that have frequently been omitted from both more totalizing and nation-centred accounts of resistance."

Understandings of resistance and collective social action have recently been enriched by a focus on global justice networks and the transnational solidarities that they engender (see Featherstone 2008; Routledge and Cumbers 2009). The global and transnational nature of resistance spaces and practices weave together complex spatial narratives. However, while these networks are essential to connect places in opposition to neoliberalism and to build effective resistance, their spread and effectiveness are geographically uneven (see Cumbers et al. 2008; Wainwright and Kim 2008). Hence, we are not looking at a simple movement of movements (Mertes 2004), which emerges uniformly and connects struggles across the world, but instead is differently constituted, uneven, and formed from overlapping networks that occasionally come together. We think it is necessary to be particularly aware of the persistent structural inequalities between resistances in the global north and south, especially in terms of issues of representation, organizational capacity, value systems, and access to resources. Sundberg (2007) discusses some of these difficulties and problems developing and maintaining such transnational solidarities. Nevertheless, forging transnational material practices and imaginaries of resistance are essential for building effective movements against the global reach of capital and the extension of patriarchy, racism, and other corrupt power relations.

There is a long standing concern with the possibilities of local resistance and how to scale up its effectiveness to the national or global levels (see Harvey 2001; Featherstone 2005). For David Harvey, a central tension has been how to transform militant particularism "into something more substantial on the global stage" (2001: 175) which can act as a model for the benefit of all society. While resistance, then, is always extra local in some sense, resistance practices, especially amongst the most marginal or in contexts of extreme oppression, often demonstrate a need for creating strong boundaries. Within this context, we think what Escobar (2001: 149) calls "defensive localisation" should not necessarily be seen as regressive, but can be an important act of self defense. While there are certainly conservative tendencies in many forms of local politics, especially when struggles seek to preserve static and exclusionary notions of community, or what Castree (2004) calls a geographical apartheid, we think defensive localization can help call into question broader social and economic issues surrounding equality and progressive politics. The harsh realities that have produced such destructive inequalities within neoliberal forms of capitalism lend some importance for place-bounded, often exclusionary, forms of resistance when other options are more limited. We also need to appreciate the fact that there are many non-progressive movements, such as religious fundamentalists or right wing militias, who create new kinds of exclusionary, autonomous spaces form which to oppose and resist.

In line with Escobar (2001) we think it is important to see many resistance struggles fought over representational and language claims, as much as material claims over rights. Resistance, in this case then, is constituted in and through discursive spaces as much as physical spaces (Routledge 2003). What is also resisted is the power of discourse and of naming things and processes. Much collective action focuses on the ability of people to be recognized, and represent themselves as well as having their identities recognized by certain forms of authority and sources of power. In this context, resistance is a right to define oneself and to determine how one is represented (see McNamara and Gibson 2009). Here we see struggle as also being over what the legitimate objects of resistance are, and the ability of disempowered groups and social movements to discursively construct their own objects of resistance. For instance, Kaup (2008) explores how claims over natural resources, and the motif "the gas is ours," generated resistant identities and collective action, and Peters (1998) shows how power inscribed in landscape and architecture is also a focus for practices of resistance. Many public spaces are masculinized and colonialized creating a sense of exclusion for countless groups, which also in turn become sources of tension and a focus for resistance (see Peters 1998).

We think future work that continues to explore these tensions between particularism and universalism and reveal the complex realities of jumping scale, and its limits and effectiveness on the ground, will continue to inform collective thinking and collective action. For example, Wainwright and Ortiz (2006) urge us to reject simplified narratives of pro and anti-globalization positions which see the state as pro-globalization and those resisting it as anti-state. Moreover, as Massey (2004) reminds us, it is not a simple case of local equals good and global equals bad. The reality is much more complex, especially in the light of anti-imperial, progressive projects of national renewal such as Venezuela. Indeed, what we see is that globalization, seen as a set of technical and social flows, allows oppositional, and not just dominant, narratives to be scaled up. Globalization produces complex and contradictory effects for social movement groups (see Fløysand 2007; BAVO 2008; Gautney et al. 2009). What we see in reality is the framing of resistance within both tendencies towards global civil society and a narrow local parochialism (Marden, 1997). Our conceptualizations of resistance, then, resists seeing it as a bounded entity, or being pushed into false divisions between the local or the global. In particular, we argue for the need to appreciate the problems of narrow and fixed nationalistic formulations of resistances that emerge from heavily localized movements.

In the same vein, we also reject a naïve "pathologisation of place" (Farmer 2003) that suggests places are somehow to blame for the ills they face. Forms of resistance should always strive, we argue, to be formulated on situated and complex analyses of both the structural factors at play in localities as well as the conditions in that locality, and not letting global or structural power relations off the hook simply because they seem too abstract or complicated. Practices of resistance and forms of collective action need to acknowledge the deep roots and structural origins of why resistance is necessary in the first place. One key issue here is the historic and constant movement towards enclosure and dispossession, and the countermove towards the commons (de Angelis 2007; Linebaugh and Rediker 2007). The key point is how we scale up our resistances (Harvey 1996), and whether we simply aim for quantitative growth and massification, or more qualitative growth which pays attention to the nature of relationships (see Alinsky 1972; Moyer 2001). Therefore we call for concrete, grounded, and detailed case analyses of examples of resistances and how they emerge and exist in practice (see: Heynen 2006; Sparke 2008).

Resistant Subjects

Understanding resistance and social action also invites us to ask, who is taking action and doing the resisting? Identifying the resistant subject, then, is an area of some debate. One starting point is to recognize that the Twentieth Century has been preoccupied with the "great man of history," the militant figure who is dedicated to revolutionary change and detached from the mundanity of everyday reality. Such orthodox Marxist-Leninist formulations of resistance entail visible acts which fight an objective oppressor. Sparke (2008: 423) points to the "romance of resistance" where autonomous actions are animated by idealism "which imagines agency in the existential and ageographical terms of some seminal and heroically universalized human spirit." Thoborn (2008: 98) states that the militant is a figure that "persistently returns as the marker ... of radical subjectivity across the spectrum of extra parliamentary politics." This revolutionary agent of history, the god-man (Deleuze 1983), seeks truth and revenge against oppression. For this kind of militant subject, resistance always comes after oppression. The oppressive centre to react against is a fact of life. There is always oppression to fight and a state of grace, once possessed and now lost, to be regained. What drives this subject is the possibility of political completion, a utopian end point.

What are the shortcomings of succumbing to this exceptional subject of resistance? The first is simply the distance from its other – the non-militant, the ordinary citizen. This kind of militant figure estranges us from the job of making here and now into an opportunity for change. This reinforces the a way resistance is invested in the social change specialist who knows best and has the requisite tools and skills to help us achieve redemption (see Chatterton 2006). Second, reliance on binaries such as the militant who resists and the powerful who oppresses can reproduce what Deleuze (1983) called the master-slave relationship. To resist reinforces the slave against the master, in the hope of one day overthrowing the master and replacing him/her. This not only reinforces a way of thinking as slaves amongst the oppressed, but also reproduces our desires to, one day, be masters, and thus perpetuates patterns of social domination.

More recent anarchist, post-structural, feminist, and queer approaches to resistance reject the tragic and the romantic figure of resistance, the purified figures struggling against an impure landscape of oppression, and instead embrace the "invisible (wo)man" of resistance (see Gibson-Graham 1995; Doel 1999; Day 2004, 2005; May 1994; Newman 2007). What this kind of work points to is resistance in all its unromanticized, transnational, messy, impurities. Hardt and Negri (2004) present such a figure in the multitude, but it is still largely undefined, romantic, and totalizing. Thoborn goes furter to propose an "a-militant diagram" constituted through material and immaterial assemblages which are formed in a "complex, intensive and open plane of composition" (2008: 114). Resistance here is not the preserve of the autonomous militant or the *avant garde* on the fringe of society. Indeed inhabiting the political fringe strengthens the idea of the centre against which resistance must act.

Rose (2002) in fact urges us to reject notions of resistance which rely on structuralist notions of power and of agents of resistance who are responding to a dominant system. Resistance should not rely on conceptualization of a pre established system; there is no one point of sovereignty against which there is one point of resistance. Instead, resistance practices need to question the very existence of a dominant system itself. Domination needs to be understood as it is practiced through everyday relations of power, and this is what resistance practices are built from. Finding ways to perform resistance without reference to a dominant or stable centre or without falling back into the false security of the militant subject is an important avenue for progressive political formations. Resistance then is a process of becoming, or resubjectification in the everyday (Gibson-Graham 2006; Grosz 1999). And it is everyday practices of resistance that are vital building blocks for re-imagining alternatives to neoliberal globalization (see Fenton 2004). At its strongest it is a collective process of becoming, where solidarities and affinities are forged between different groups and individuals. These kinds of tendencies can be traced in the case study we outline below.

Climate Disobedience: A Case Study

On June 23, 2009, James Hansen one of the world's most prominent climate change scientists, and director of NASA's Goodard Institute for Space Studies, was arrested along with thirty other activists in West Virginia for impeding the flow of traffic and hindering police officers' effort to disband protesters who converged on Massey Energy Company, the fourth largest coal extractor in the United States. Upon his arrest, Hansen exclaimed, "Civil resistance is not an easy path, but given abdication of responsibility by the government, it is an essential path." Almost one year prior to this event a group of 29 protesters (co-author Chatterton among them) stopped a coal train heading toward the Drax power station in North Yorkshire in England. After flying a banner that read, "Leave it in the ground" from the side of the blocked train, they/we proceeded to shovel more than twenty tonnes of coal off the train and onto the railroad track in an attempt to halt or slow, even for a day, the emissions of carbon from the powerstation.

What do these events, representing a growing range of collective social action and resistance dubbed "climate disobedience," highlight? The first is that they are linked to a longer history of episodes of civil disobedience that have shaped transformative politics for over a hundred and fifty years. In this case, everyday people in the face of extreme injustice and life threatening situations have decided to risk arrest and imprisonment in order to work towards a better world (see Cortright 2006; Sharp 2005; Zunes et al. 1999). For a century and a half, non-violent civil disobedient direct action has been central to efforts to resist oppressive power relations, linking Mohandas Gandhi's anti-colonial efforts for India, Martin Luther King Jr.'s efforts for US civil rights, and Archbishop Desmond Tutu and Steve Biko's anti-apartheid struggles.

Second, both express the need to challenge laws which lock society into life threatening situations. Hansen's claim upon arrest in West Virginia echoes something Howard Zinn wrote during the turbulent 1960s. Zinn said (1968: 4): "[e]xactly when we have begun to suspect that law is congealed injustice, that the existing order hides an everyday violence against body and spirit, that our political structure is fossilized, and that the noise of change – however scary – may be necessary, a cry rises for 'law and order.' Such a moment becomes a crucial test of whether the society will sink back to a spurious safety or leap forward to its own freshening."

A third reason that we find climate disobedience to be such a productive case for this chapter is because we see it emblematic of newer, more visionary, and holistic forms of anti-capitalist, anti-oppressive/anti-authoritarian resistance. In the wake of the Seattle protests in 1999, David Graeber (2002: 61) observed, "It's hard to think of another time when there has been such a gulf between intellectuals and activists; between theorists of revolution and its practitioners. Writers who for years have been publishing essays that sound like position papers for vast social movements that do not in fact exist seem seized with confusion or worse, dismissive contempt, now that real ones are everywhere emerging." By focusing in on the foundations of the climate disobedience movement we suggest the roots of theoretically informed and tactically rich praxis emerges in ways that other movements do not quite demonstrate.

Fourth, many aspects of the climate disobedience movement point to novel organizing principles that begin with clear anti-oppressive organizing models based on participatory and egalitarian principles. Take for instance *Rising Tide North America*'s anti-oppression statement:

> We in Rising Tide have always said that "anti-oppression" is at the core of our organizing, and in the forever ongoing process of self-reflection and growth we are working on figuring out what exactly that means for us. Clearly all oppressions are connected, the killing of the planet is directly tied to the oppression of most people, and all of us with privilege are part of the problem and need to deal with that. But too many folks seem to talk the right talk and use that language to get away with the same ol' shit as always. We're not going to draft up some anti-oppression "platform" or "policy," and we know it's not about wallowing in guilt or making tokenizing and racist attempts at "diversifying." We're asking ourselves the tough questions, and seek to be accountable and always in the process of unlearning the bullshit and decolonizing our minds and hearts.

In a similar way, the final *People's Declaration* which emerged from the coalition of NGOs and civil society organizations at the *Kilmaforum* during the UN COP 15 Climate summit in Copenhagen in 2009 was entitled "system change not climate change" and stated:

> There are solutions to the climate crisis. What people and the planet need is a just and sustainable transition of our societies to a form that will ensure the rights of life and dignity of all people and deliver a more fertile planet and more fulfilling lives to present and future generations. This transition must be based on principles of solidarity – especially on behalf of the most vulnerable – non-discrimination, gender equality, equity, and sustainability, acknowledging that we are part of nature, which we love and respect.[1]

This kind of organizing is in line with the everyday prefigurative political processes that can serve as the foundation for all kinds of substantial resistance and collective social action. What we see in these examples is a recognition of the holistic connection between human life and other forms of life on the planet, upon their organizing to focus on more holistic and egalitarian trajectories as opposed to the more narrow foci we have seen historically through "just" anti-capitalist organizing, or just anti-patriarchal organizing, or just antiracist organizing. Here the recognition that the same power relations behind capitalism that drive the extraction and burning of coal are connected to the same oppressive power relations heterosexual white men have attempted, and mostly succeeded, to maintain over all other populations across the planet over modern history.

Fifth, the movement to stop climate change has led to the emergence of a diverse and challenging set of spatial practices and actions ranging from mass actions near sources of fossil fuel extraction and emission, occupations of multinational head-quarters, powerful image events such as occupying airport runways and climbing powerstations and education/visioning/action spaces at temporary camps. This diversity reflects a growing sensibility that a relational and multi-scalar approach is needed to tackle a problem as complex as climate change. Some of these tactics feel more reformist and lobbyist, others more radical based on direct action, but the point is that climate disobedience has created fertile new ground for many of these kinds of collective action to interact and coexist. The temporary camp has been used a key spatial form recently to galvanize climate disobedience. Camps for Climate Action which began in 2006 in the UK, for example, had become an annual event in fourteen localities across the globe by 2009. The camps have four principle aims of movement building, education, direct action, and sustainable living. The week long camps aim to create a convergence space through which these four aims can be realized. The camps are open, plural spaces where many different kinds of resistances collide. While the camps do not have a set ideological line, they share many principles such as participation, self organization, grassroots democracy, a rejection of ceaseless economic growth and market based solutions. The camps are open and complex spaces, generating collective action which weave together rela-tional and multi-scalar forms of resistance, held together by key transnational activ-ists. However, many problems continue to be debated in such open spaces such as diversity, discrimination, informal hierarchies, the relationship to the state, division of labor, inclusivity, and choice of tactics and messages (Shift Magazine 2010; Turbulence 2009).

Finally, climate disobedience points to subjectivities and practices that are fluid, experimental, and require constant negotiation and work. Resistance is generated through practice, and participants don't enter these kinds of activities with fully formed ideas. The collective actions of climate disobedience form an incomplete terrain where daily struggles are made and remade, both symbolically and materi-ally, and where people live by their beliefs and face contradictions from living between worlds – the actually existing and the hoped for (see Pickerill and Chatterton 2006). In this sense, the overly detached specialist subject of resistance is rejected. Moreover, they combine different practices of contestation against, for example, industrialism, capitalism, patriarchy, racism, heteronormativity, and other power relations that mar human potential. In sum, climate disobedience represents resist-ance and collective social action based on complex collective political identities formed through multiple narratives of oppression and solidarity and geographies of resistance that are intensely local, but also multi-scalar, transnational, and relational.

Conclusions

What we hoped to encourage by writing this chapter are understandings of resist-ance and collective action that illuminate, embrace, and respond to the uncertainty and complexity of our contemporary world, and which can respond meaningfully to material inequalities without becoming fixed, overcoded, or recuperated. This chapter has shown how practices of resistance are necessarily diverse as they emerge out of the uneven conditions of contemporary social power relations and everyday life struggles. We have pointed to a politics of demand and rights around issues such as housing in the sphere of consumption, the organizational work of social movements in their quest for hegemonic power, as well as the less easily definable anarchist inspired movements who refuse to accept a hegemonic centre as a focus for their opposition. Fertile ground for resistance and collective action include both the most horrific conditions of human oppression and suffering and banal everyday moments. What makes resistance powerful is to make surprising yet empowering connections between these very different conditions. Moreover, we urge the need to more fully engage the new possibilities of scaling up resistance, especially through transnational and computer mediated networks, but also pay attention to the process of how we scale up. Finally, we need to see resistance as a richly dynamic and constantly emergent process, rather than an immutable state of being. Cumulatively these features of resistance and collective action inspire us to make several closing reflections.

First, resistance and various forms of collection social action are intensely emotional and spiritual rather than detached practices. They connect to our deep and untamed passions about the world (see Routledge and Simons 1995). Routledge (2003) points to convergence spaces where new solidarities are built. Ideas of local "community" can be used to articulate and embed political resistance, and its construction is used to serve particular political interests of subaltern groups. They are contingent resources which are deployed selectively to resist dispossession. This involves valuing what is authentic and materially important to us and also using symbolic repertoires of resistance used to resist an external other (see Mackenzie and Dalby 2003). It is also important to be aware that new forms of collective associations are continually emerging with tendencies both toward the vertical/majoritarian as well as the nomadic and the horizontal/minoritarian (see also Tormey 2004).

Second, in terms of thinking through the possibilities of the kinds of change which resistance can evoke then, our political imaginaries need to see multiple, co-existing possibilities for action with different rhythms, temporalities, escape routes, and collective agencies. Some seem possible, other impossible. But as Swyngedouw (2009) highlights, the political act is not the art of the possible, but the art of the impossible, and creates interventions that cannot be understood in terms of established symbolic framings. There are both the great ruptures, those knowable and spectacular moments of resistance and revolution, and the more mundane cracks which are disruptive and unknowable (see Holloway 2010). The crack has become a useful motif for thinking through resistance practices as creating spaces that open up as we question and oppose. Instead of facing the great wall of power which we can push down, exposing the heart of power to attack, we more often try to multiply the cracks in power wherever we encounter it in our workplaces, daily lives, homes, relationships, and ourselves. This is a revolution of the everyday (Vaneigem 2003) as there are always cracks which grow and shrink organically. We do not know which way growth will take form, where it will take us or how, but awareness of the cracks of possibility leaves open the opportunity for significant and continued change/collective action.

Thirdly, one key question for geographers involves asking "what would it mean to win?" It would undoubtedly be based as much on processual concerns as those of content. Winning is not about becoming the new master (or hegemon), or inverting the master-slave relation as we discussed earlier. "To win" also equates to looking in very different places for signs of victory. What we have stressed in this chapter is the need to also see the potential in minoritarian as much as majoritarian political activity. Micro-tactics of resistance have potential as they are grounded in immanent modes of association that are capable of inventing alternative forms of social interaction (Krause and Rolli 2008; Gibson-Graham 2006). It is part of the Ghandian philosophy of being the change you want to see, a daily process of prefiguration where revolution becomes part of everyday life (Franks 2003). Experimentation and biodegradability are the lifeblood of this kind of organizing which increases its resilience and adaptability. Opportunities for subaltern resistance exist in the most mundane and difficult circumstances and everyday spaces are key sites for the constitution of alternatives, resistance practices and collective action where resistance does not just necessarily arise in reaction to oppression (see Whitson 2007; Pile and Keith 1997).

Overall, we would urge the need to reconceptualize our activism so we do not simply see oppressive powers as dynamic and resistance as reactive and bounded. Resistance is also a dynamic and open process with its own creative logic which also embraces solidarity, love, humor, and friendship as part of the repertoires of collective action. Those resisting are not simply the "other" of a relentless, dominant neoliberalism – they are the engine of change, productive, dynamic, and constantly shifting and creating subjects. It is from this dynamic generative process of resistance that new possibilities constantly to emerge. Resistance is not a deviation from some universally valid norm; it pre-empts the present and demands the right to be the exception to the rule. At the end of the day, our resistances and forms of collective action have to deal with the fact that they are within, against, and post-capitalist. They embody our dreams and hopes and have to deal and cope with the difficult present.

In conclusion, while this chapter has primarily focused on the world of external ideas, literatures, configurations of resistance, and collective social action, we find it necessary to end by looking at our own labor position and our own role in social change. We are scholar-activists within the university who have the privilege to ponder the meaning of resistance free from the banality or terror of resisting the kinds of external power relations so often discussed within this literature (patriarchy, racism, neoliberalization, globalization, heteronormativity, ecological destruction, inequity of all kinds). As a result, there is much we have a responsibility to do within our own workplaces (supporting those on temporary contracts, joining our own unions, building a knowledge commons, resisting privatization, promoting anti-authoritarian governance structures, sharing liberatory educational practices with students). And there is much you can do in your own lives in terms of building collective action for social change. We hope this chapter has given you ideas and inspiration towards that end.

Note

1   Available at www.klimaforum09.org/.../A_People_s_Declaration_from_Klimaforum09_-_ultimate_version.pdf.

References

Alinksy, S. (1972) *Rules for Radicals: A Pragmatic Primer for Realistic Radicals*. New York: Random House.

BAVO (2008) *Urban Politics Now: Re-imagining Democracy in the Neoliberal City*. Rotterdam: NAi Publishers.

Blunt, A. and Wills, J. (2000) *Dissident Geographies: An Introduction to Radical Ideas and Practice*. Harlow: Pearson.

Bonefeld, W., Gunn, R., and Psychopedis, K. (1992) *Open Marxism*. London: Pluto.

Carlson, C. (2008) *Nowtopia: How Pirate Programmers, Outlaw Bicyclists, and Vacant-lot Gardeners Are Inventing the Future Today*. Oakland: AK Press.

Castree, N. (2004) Differential geographies: place, indigenous rights and "local" resources. *Political Geography* 23: 133–67.

Chatterton, P. (2005) Making autonomous geographies: Argentina's popular uprising and the "Movimiento de Trabajadores Desocupados" (Unemployed Workers Movement). *Geoforum* 36 (5): 545–61.

Chatterton, P. (2006) "Give up activism" and change the world in unknown ways. Or, learning to walk with others on uncommon ground. *Antipode: A Radical Journal of Geography* 38 (2): 259–82.

Chatterton, P. (2008) Becoming a public scholar: academia and activism. *Antipode: A Radical Journal of Geography* 40 (3): 421–8.

Chatterton, P., Cutler, A., and Bryan, K. (eds.) (2007) *Do It Yourself: A Handbook for Changing Our World*. London: Pluto.

Cook, I. and Pepper, D. (eds.) (1990) Anarchism and geography. *Contemporary Issues in Geography and Education* 3 (2): 10–22.

Cortright, D. (2006) *Gandhi and Beyond: Nonviolence for an Age of Terrorism*. Boulder: Paradigm Publishers.

Cumbers, A., Routledge, P., and Nativel, C. (2008) The entangled geographies of global justice networks. *Progress in Human Geography* 32 (2): 183–201.

Cummings, B. (1995) Dam the rivers, damn the people: development and resistance in Amazonian Brazil. *GeoJournal* 35 (2): 151–60.

Day, R. (2004) From hegemony to affinity: the political logic of the newest social movements. *Cultural Studies* 18 (5): 716–48.

Day, R. (2005) *Gramsci Is Dead: Anarchist Currents in the Newest Social Movements*. London: Pluto Press.

Doel, M. (1999) *Poststructuralist Geographies: The Diabolical Art of Spatial Science*. London: Rowman & Littlefield.

De Angelis, M. (2007) *The Beginning of History: Value Struggles and Global Capital*. London: Pluto Press.

Deluze, G. (1983) *Nietzsche and Philosophy*. New York: Columbia University Press.

Escobar, A. (2001) Culture sits in places: reflections on globalism and subaltern strategies of localization. *Political Geography* 20 (2): 139–74.

Farmer, P. (2003) *Pathologies of Power: Health, Human Rights, and the New War on the Poor*. Berkeley: University of California Press.

Featherstone, D. (2005) Towards the relational construction of militant particularisms: or why the geographies of past struggles matter for resistance to neoliberal globalisation. *Antipode: A Radical Journal of Geography* 32 (2): 250–71.

Featherstone, D. (2008) *Resistance, Space, and Political Identities: The Making of Counter-global Networks*. Oxford: Wiley-Blackwell.

Fenton, J. (2004) "A world where action is the sister of dream": surrealism and anti-capitalism in contemporary Paris. *Antipode: A Radical Journal of Geography* 36 (5): 942–62.

Fløysand, H. (2007) Globalization and the power of rescaled narratives: a case of opposition to mining in Tambogrande, Peru. *Political Geography* 26 (3): 289–308.

Franks, B. (2003) Direct action ethic. *Anarchist Studies* 11 (1): 13–41.

Foucault, M. (1980) *Power/Knowledge: Selected Interviews and Other Writings, 1972–1977*. Pantheon: New York.

Foucault, M. (1994) The birth of biopolitics. In P. Rabinow (ed.), *Michel Foucault: Ethics, Subjectivity and Truth*. New York: New York Press, pp. 73–9.

Gautney, H., Dahbour, O., Dawson, A., and Smith, N. (2009) *Democracy, States, and the Struggle for Global Justice*. New York and London: Routledge.

Gibson-Graham, J.K. (1995) Beyond patriarchy and capitalism: reflections of political subjectivity. In B. Caine and R. Pringle (eds.), *Transitions: New Australian Feminisms*. Sydney: Allen and Unwin, pp. 172–83.

Gibson-Graham, J.K. (2006) *The End of Capitalism (As We Knew It): A Feminist Critique of Political Economy*. Minneapolis: University of Minnesota Press.

Gidwani, V. (2006) Subaltern cosmopolitanism as politics. *Antipode: A Radical Journal of Geography* 38 (1): 7–21.

Glassman, J. (2001) From Seattle (and Ubon) to Bangkok: the scales of resistance to corporate globalization. *Environment and Planning D: Society and Space* 19 (5): 513–33.

Glassman, J. (2006) Primitive accumulation, accumulation by dispossession, accumulation by "extra-economic" means. *Progress in Human Geography* 30 (5): 608–25.

Graeber, D. (2002) The new anarchists. *New Left Review* 13 (January/February): 61–73.

Grosz, E. (1999) *Becomings: Explorations in Time, Memory and Futures*. Ithaca: Cornell University Press.

Hanson, S. and Pratt, G. (1988) Reconceptualizing the links between home and work in urban geography. *Economic Geography* 64 (4): 299–321.

Hanson, S and Pratt, G. (1991) Job search and the occupational segregation of women. *Annals of the Association of American Geographers* 81 (2): 229–54.

Hardt, M. and Negri, A. (2004) *Multitude: War and Democracy in the Age of Empire*. New York: Penguin Press.

Hart, G. (2006) Denaturalizing dispossession: critical ethnography in the age of resurgent imperialism. *Antipode* 38 (5): 977–1004.

Harvey, D. (1996) *Justice, Nature and the Geography of Difference*. Oxford: Blackwell.

Harvey, D. (2001) *Spaces of Capital*. Edinburgh: Edinburgh University Press.

Heynen, N. (2006) But it's alright, ma, it's life, and life only: radicalism as survival. *Antipode: A Radical Journal of Geography* 38 (5): 916–29.

Heynen, N. (2008) Bringing the body back to life through the radical geography of hunger: the Haymarket affair and its aftermath. *ACME: An International E-Journal for Critical Geographies* 7 (1): 32–44.

Heynen, N. (2009a) Bending the bars of empire from every ghetto to feed the kids: the Black Panther party's radical antihunger politics of social reproduction and scale. *Annals of the Association of American Geographers* 99 (2): 406–22.

Heynen, N. (2009b) Back to revolutionary theory through racialized poverty: the McGee Family's utopian struggle for Milwaukee. *Professional Geographer* 61 (2): 187–99.

Heynen, N. (2010) Cooking up non-violent civil disobedient direct action for the hungry: Food Not Bombs and the resurgence of radical democracy. *Urban Studies* 47 (6): 1225–40.

Hill Collins, P. (1990) *Black Feminist Thought*. New York: Routledge.

Hodkinson, S. and Chatterton, P. (2006) Autonomy in the city? Reflections on the social centres movement in the UK. *City* 10 (3): 305–15.

Holloway, J. (2002) *Change the World without Taking Power: The Meaning of Revolution Today*. London: Pluto Press.

Holloway, J. (2010) *Crack Capitalism*. London: Pluto Press.

Horton, M. and Freire, P. (1990) *We Make the Road by Walking: Conversations on Education and Social Change*. Philadelphia: Temple University Press.

hooks, b. (1984) *Feminist Theory: From Margin to Center*. Boston: South End.

Jones III, J.P., Nast, H.J., and Roberts, S.M. (1997) *Thresholds in Feminist Geography: Difference, Methodology, Representation*. Lanham: Rowman and Littlefield.

Katz, C. (2004) *Growing Up Global: Economic Restructuring and Children's Everyday Lives*. Minneapolis: University of Minnesota Press.

Kaup, B. (2008) Negotiating through nature: the resistant materiality and materiality of resistance in Bolivia's natural gas sector. *Geoforum* 39 (5): 1734–42.

Kobayashi, A. (1994) Unnatural discourse: "race" and gender in geography. *Gender, Place and Culture: Journal of Feminist Geography* 1 (2): 225–43.

Kohl, B and Farthing L. (2006) *Impasse in Bolivia: Neoliberal Hegemony and Popular Resistance*. London: Zed.

Krause, R. and Rolli, M. (2008) Micropolitical associations. In I. Buchanan and N. Thoburn (eds.), *Deleuze and Politics*. Edinburgh: Edinburgh University Press. pp. 17–25.

Kropotkin, P. (1987) *Fields, Factories and Workshops Tomorrow*. London: Freedom.

Leitner, H., Sheppard, E., and Sziarto, K.M. (2008) The spatialities of contentious politics. *Transactions of the Institute of British Geographers* 33 (2): 157–72.

Linebaugh, P. and Rediker, M. (2007) *The Many-headed Hydra Sailors, Slaves, Commoners, and the Hidden History of the Revolutionary Atlantic*. London: Verso.

Marden, P. (1997) Geographies of dissent: globalization, identity and the nation. *Political Geography*. 16 (1): 57–64.

May, T. (1994) *The Political Philosophy of Post-structuralist Anarchism*. Pennsylvania: Penn State Press.

Mackenzie, A. and Dalby, S. (2003) Moving mountains: community and resistance in the Isle of Harris, Scotland, and Cape Breton, Canada. *Antipode: A Radical Journal of Geography* 35 (2): 309–33.

McNamara, K.E. and Gibson, C. (2009) "We do not want to leave our land": Pacific ambassadors at the United Nations resist the category of "climate refugees." *Geoforum* 40 (3): 475–83.

Marshall, P. (1992) *Demanding the Impossible: A History of Anarchism*. London: Harper Collins.

Marston, S. (2000) The social construction of scale. *Progress in Human Geography* 24 (2): 219–42.

Massey, D. (2004) Geographies of responsibility. *Geografiska Annaler* 86 (B): 5–18.

Mertes, T. (2004) *A Movement of Movements: Is Another World Really Possible?* London and New York: Verso.

Mitchell, D. and Heynen, N. (2009) The geography of survival and the right to the city: Speculations on surveillance, legal innovation, and the criminalization of intervention. *Urban Geography* 30 (6): 611–32.

Mitchell, K., Marston, S., and Katz, C. (2004) *Life's Work: Geographies of Social Reproduction*. Oxford: Blackwell.

Moss, P. (2002) *Feminist Geography in Practice: Research and Methods*. Oxford: Blackwell.

Moyer, B. (2001) *Doing Democracy: The Map Model for Organizing Social Movements*. Gabriola Island: New Society.

Naples, N. (1998) *Grassroots Warriors: Activist Mothering, Community Work, and the War on Poverty*. New York: Routledge.

Newman, S. (2007) *From Bakunin to Lacan: Anti-authoritarian and the Dislocation of Power*. Maryland: Lexington Books.

Pain R.H. (1997) Social geographies of women's fear of crime. *Transactions of the Institute of British Geographers* 22: 231–44.

Pain, R. and Townshend, T. (2001) A safer city centre for all? Senses of community safety in Newcastle-upon-Tyne. *Geoforum* 33 (1): 105–19.

Panelli, R. (2004) *Social Geographies: From Difference to Action*. London: Sage.

Panelli, R., Punch, S., and Robson, E. (2007) *Global Perspectives on Rural Childhood and Youth: Young Rural Lives*. London and New York: Routledge.

Peake, L. (1993) "Race" and sexuality: challenging the patriarchal structuring of urban social space. *Environment and Planning D: Society and Space* 11 (4): 415–32.

Peet, R. (1978) The geography of human liberation. *Antipode: A Radical Journal of Geography* 10 (3): 119–33.

Peters, E.J. (1998) Subversive spaces: first nations women and the city. *Environment and Planning D: Society and Space* 16 (6): 665–85.

Pickerill, J. and Chatterton, P. (2006) Notes towards autonomous geographies: creation, resistance and self management as survival tactics. *Progress in Human Geography* 30 (6): 1–17.

Pile, S. and Keith, M. (1997) *Geographies of Resistance*. London and New York: Routledge.

Rose, M. (2002) The seductions of resistance: power, politics, and a performative style of systems. *Environment and Planning D: Society and Space* 20 (4): 383–400.

Routledge, P. (1996) The third space as critical engagement. *Antipode: A Radical Journal of Geography* 28 (4): 399–419.

Routledge, P. (2003) Voices of the dammed: discursive resistance amidst erasure in the Narmada Valley, India. *Political Geography* 22 (3): 243–70.

Routledge, P. and Cumbers, A. (2009) *Global Justice Networks: Geographies of Transnational Solidarity*. Manchester: University of Manchester Press.

Routledge, P. and Simons, J. (1995) Embodying spirits of resistance. *Environment and Planning D: Society and Space* 13 (4): 471–98.

Ruddick, S. (1996) Constructing difference in public spaces: race, class, and gender as interlocking systems. *Urban Geography* 17: 132–51.

Scott, J.C. (1985) *Weapons of the Weak: Everyday Forms of Peasant Resistance*. New Haven: Yale University Press.

Sharp, G. (2005) *Waging nonviolent struggle: 20th century practice and 21st century potential*. Boston: Extending Horizons Books.

Shift Magazine (2010) Theory into practice, http://shiftmag.co.uk/?p=327.

Smith, N. (1992) Contours of a spatialized politics: homeless vehicles and the production of geographical space. *Social Text* 33: 54–81.

Sparke, M. (2008) Political geography: political geographies of globalization III: resistance. *Progress in Human Geography* 32 (3): 423–40.

Staeheli, L. (1994) Empowering political struggle: spaces and scales of resistance. *Political Geography* 13 (5): 387–91.

Sundberg, J. (2007) Reconfiguring north–south solidarity: critical reflections on experiences of transnational resistance. *Antipode: A Journal of Radical Geography* 39 (1): 144–66.

Swyngedouw, E. (1997) Neither global nor local: "glocalization" and the politics of scale. In K. Cox (ed.), *Spaces of Globalization: Reasserting the Power of the Local*. New York and London: Guilford/Longman, pp. 137–66.

Swyngedouw, E. (2009) The antinomies of the postpolitical city: in search of a democratic politics of environmental production. *International Journal of Urban and Regional Research* 33 (3): 601–20.

Thoburn, N. (2008) What is a militant? In I. Buchanan and N. Thoburn (eds.), *Deleuze and Politics*. Edinburgh: Edinburgh University Press, pp. 98–120.

Tormey, S. (2004) *Anti-capitalism: A Beginner's Guide*. Oxford: Oneworld.

Turbulence Magazine (2009) And now for something completely different, http://turbulence. org.uk/turbulence-5/.

Valentine, G. (2007) Theorizing and researching intersectionality: a challenge for feminist geography, *Professional Geographer* 59 (1): 10- 21.

Vaneigem, R. (2003) *The Revolution of Everyday Life*. London: Rebel Press.

Vasudevan, A., McFarlane, C., and Jeffrey, A. (2008) Spaces of enclosure. *Geoforum* 39 (5): 1641–6.

Wainwright, J. and Kim, S. (2008) Battles in Seattle redux: transnational resistance to a neoliberal trade agreement. *Antipode: A Radical Journal of Geography* 40 (4): 513–34.

Whitson R. (2007) Hidden struggles: spaces of power and resistance in informal work in urban Argentina. *Environment and Planning A* 39 (12): 2916–34.

Wainwright J. and Ortiz, R. (2006) The battles in Miami: the fall of the FTAA/ALCA and the promise of transnational movements. *Environment and Planning D: Society and Space* 24 (3): 349–66.

Whatmore, S. (1997) Dissecting the autonomous self: hybrid cartographies for a relational ethics. *Environment and Planning D: Society and Space* 15 (1): 37–53.

Wills, J. (2005) The geography of union organising in low-paid service industries in the UK: lessons from the T&G's campaign to unionise the Dorchester Hotel, London. *Antipode: A Radical Journal of Geography* 37 (1): 139–59.

Wise, T., Salazar, H. and Carlsen, L. (2003) *Confronting Globalization: Economic Integration and Popular Resistance in Mexico*. West Hartford: Kumarian Press.

Wright, M. (2006) *Disposable Women and Other Myths of Global Capitalism*. New York and London: Routledge.

Zinn, H. (1968) *Disobedience and Democracy: Nine Fallacies on Law and Order*. Cambrige, MA: South End Press.

Zunes, S., Kurtz, L.R., and Asher, S.B. (1999) *Nonviolent Social Movements: A Geographical Perspective*. Malden, MA, and Oxford: Blackwell.

Index

Note: "n." after a page reference indicates the number of a note on that page.

A Companion to Social Geography, First Edition. Edited by
Vincent J. Del Casino Jr., Mary E. Thomas, Paul Cloke, and Ruth Panelli.
© 2011 Blackwell Publishing Ltd. Published 2011 by Blackwell Publishing Ltd.

colonialism
  affective life 338
  and difference 18, 19, 22–3, 25
  indigenous geographies 473–8, 480
  origins of geography 147
communication, ineffectual 242–3
communitarianism 99–100
community 14, 91–2, 102–3
  communitarianism 99–100
  decline 100–2
  defining 95–6
  fieldwork 151–2
  human body as a 266
  identity 96–7
  imagined 99
  purified 97–8
  as relationship 93–5
  scales of 92–3
  territoriality 98–9
community organizations 99
companion species 263
complementarity principle, international
    justice 498
Congo, Democratic Republic of the (DRC)
    500, 502
Connolly, W.E. 388, 394
contextual validity, participatory praxis 224
conversation analysis 376
Coppock, J.T. 237
corncrakes 60
cosmopolitanism 266
Cox, K. 94
Cresswell, T. 441
critical development theory 473
critical geopolitics 423–33
Critical Global Poverty Studies (CGPS)
    network 464
Cronon, W. 482
Crossley, N. 151
Cruikshank, Julie 207, 208–9
cultural diversity and belonging 119
cultural ecology and social natures 57
cultural turn 4
  difference 22, 23
  discursive life 365, 366
  economies 74
  geopolitics 423, 427
  identification 40
  immobilizing effect 238–9
  indigenous geographies 473
  marginalization, geographies of 438–9
  molecular life 259, 263, 264, 265
Cupples, J. 154–5

Dahlman, C.T. 422
Dalby, Simon 423–4, 427, 429, 430
data analysis 186–93
data collection and preparation 182–6
  see also fieldwork
Davidson, J. 288, 350
Davies, G. 281, 284, 480
Day, G. 102
Debord, Guy 286, 289n.2
deconstruction 5
  theology 390
DeLanda, M. 66
Del Casino Jr., Vincent J.
  emotional life 314, 320
  fieldwork 152
  knowing/doing 142n.1
Deleuze, Gilles
  affective life 330, 331, 337
  resistance and collective social action
      515
  sexual life 296
  smooth and striated space 444
  spiritual life 388, 389–90, 392–4
delusion and psychic life 288
Delyser, Dydia 156
Demeritt, D. 59
Democratic Republic of the Congo (DRC)
    500, 502
de Pater, B. 163
Derrida, Jacques 20, 424
descriptive data analysis 187–8, 192–3
Desjarlais, R. 287
deviance 20, 98
Dewsbury, J.D. 146, 153, 368
diasporas
  and belonging 112, 115, 116
  embodied life 350
  molecular life 262
difference 13, 17–18, 31
  affective life 337
  belonging 112
  emotional life 317
  how geographers deal with 22–7
  and identification 40, 96
  molecular life 265–6
  psychic life 278–9
  theorizing 18–22
  thinking differently about 27–31
digital connectivity, social geometries of
    401–7
diplomatic immunity 495, 496
disability 355, 356–7
disciplinary power 356

Printed and bound by CPI Group (UK) Ltd, Croydon, CR0 4YY